W0111673

Wnt Signaling

METHODS IN MOLECULAR BIOLOGY™

John M. Walker, Series Editor

METHODS IN MOLECULAR BIOLOGY™

Wnt Signaling

Volume 2
Pathway Models

Edited by

Elizabeth Vincan, PhD

University of Melbourne, Parkville, Victoria, Australia

 Humana Press

Editor
Elizabeth Vincan
University of Melbourne
Parkville, Victoria VIC 3010
Australia

Series Editor
John M. Walker
School of life Sciences
University of Hertfordshire
Hatfield, Hertfordshire
AL10 9NP, UK

ISBN: 978-1-60327-468-5   e-ISBN: 978-1-60327-469-2
ISSN: 1064-3745   e-ISSN: 1940-6029
DOI: 10.1007/978-1-60327-469-2

Library of Congress Control Number: 2008933593

© Humana Press 2008, a part of Springer Science+Business Media, LLC
All rights reserved. This work may not be translated or copied in whole or in part without the written permission of the publisher (Humana Press, c/o Springer Science + Business Media, LLC, 233 Spring Street, New York, NY 10013, USA), except for brief excerpts in connection with reviews or scholarly analysis. Use in connection with any form of information storage and retrieval, electronic adaptation, computer software, or by similar or dissimilar methodology now known or hereafter developed is forbidden.
The use in this publication of trade names, trademarks, service marks, and similar terms, even if they are not identified as such, is not to be taken as an expression of opinion as to whether or not they are subject to proprietary rights.
While the advice and information in this book are believed to be true and accurate at the date of going to press, neither the authors nor the editors nor the publisher can accept any legal responsibility for any errors or omissions that may be made. The publisher makes no warranty, express or implied, with respect to the material contained herein.

Printed on acid-free paper

springer.com

Preface

Since their discovery some 20 years ago, Wnt signaling molecules have been shown to control key events in embryogenesis, maintain tissue homeostasis in the adult and, when aberrantly activated, promote human degenerative diseases and cancer. Elucidation of Wnt signaling mechanisms has relied on both biochemical methodologies and vertebrate and invertebrate model systems. Therefore, I felt that an issue dedicated to Wnt signaling had to include both assays (biochemical readout) and model systems (functional readout) of Wnt signaling. It is not an exhaustive catalog, but rather a point of reference to current molecular protocols and the diverse model systems employed to study this signaling pathway. The issue is divided into two volumes. The first volume includes assays to measure activation of the diverse Wnt signaling pathways as well as models and strategies used to study mammalian Wnt/FZD function (from protein–protein interaction and simple cell line models to organoid cultures and mouse models). The second volume is dedicated to the diverse vertebrate and invertebrate models that have shaped the Wnt signaling field. It provides an entry point for the novice and an overview of the unique properties of each organism with respect to studying Wnt/FZD function (for example asymmetric cell division in *C. elegans*, epithelial morphogenesis in *Dictyostelium* and so on). Given the collective expertise and knowledge of the contributors, I anticipate that this two-volume issue will be an invaluable resource.

The Wnt field advances at an exceptionally rapid rate for several reasons. First, diverse fields of research converge on this pathway. Second, the Wnt community is very generous: reagents, knowledge, and ideas are shared freely. This is facilitated by informative web sites and regular Wnt meetings that are packed back-to-back with cutting-edge research. The "no-frills" approach to these meetings means that the whole community, including students, can participate. Equally important is the elusive nature of the Wnt pathway itself, which continues to intrigue and fascinate both novice and veteran researchers alike. This book is a testament to all these. It was steered by the generosity and enthusiasm of contributors from diverse fields. I thank them all. Special thanks to Randall Moon and Stefan Hoppler; their suggestions for authors and chapters helped shape this issue.

On a personal note, I would also like to take this opportunity to acknowledge Bill Boyle for being an inspirational mentor during my formative years; his infectious enthusiasm for research set me on this exciting and rewarding career path. I am indebted to Bob Thomas and Rob Ramsay or generously supporting my research into FZD7 function in colon cancer when funding in Australia for the Wnt field was scarce in the early years. Most importantly, I thank my very patient and accommodating children for allowing me to indulge myself!

I thank Tony Goodwin, Scott Bowden, and the University of Melbourne; without their assistance this book would not have been possible. John Walker and all at Humana Press, especially David Casey and Amina Ravi, for their generosity and for the opportunity to edit this issue—a truly rewarding experience.

Contents

PART I: INTRODUCTION

PART II: DICTYOSTELIUM

PART III: CNIDARIANS

PART IV: C. ELEGANS

PART V: DROSOPHILA

Contributors

NATALIA ARBUZOVA, PhD • *Samuel Lunenfeld Research Institute, University of Toronto, Ontario, Canada*

JOANNA M. BINCE, BS • *Department of Zoology, The University of Hawaii at Manoa, Honolulu, HI, USA*

HANS BODE, PhD • *Department of Developmental and Cell Biology, Developmental Biology Center, University of California, Irvine, CA, USA*

CAROLYN M. BROWN, BSc • *Department of Biology, McGill University, Montreal, Quebec, Canada*

ELIZABETH L. CHRISTIE, BSc • *Ludwig Institute for Cancer Research, Parkville, Melbourne, Victoria, Australia*

JENIFER C. CROCE, PhD • *Biology Department, Duke University, Durham, NC, USA*

RAMANUJ DASGUPTA, PhD • *New York University School of Medicine/Cancer Institute, Department of Pharmacology, New York, NY, USA*

TINNEKE DENAYER, PhD • *Department for Molecular Biomedical Research, VIB and Molecular Biology, Ghent University, Ghent, Belgium*

RICHARD I. DORSKY, PhD • *University of Utah, Department of Neurobiology and Anatomy, Salt Lake City, UT, USA*

FRANCOIS FAGOTTO, PhD • *Department of Biology, McGill University, Montreal, Quebec, Canada*

BOB GOLDSTEIN, PhD • *Department of Biology, University of North Carolina, Chapel Hill, Chapel Hill, North Carolina, USA*

FOSTER C. GONSALVES, PhD • *New York University School of Medicine/Cancer Institute, Department of Pharmacology, New York, NY, USA*

ADRIAN J. HARWOOD, PhD • *Cardiff School of Biosciences, Cardiff University, Cardiff, Wales, UK*

JANET HEASMAN, PhD • *Division of Developmental Biology, Cincinnati Children's Hospital Research Foundation, Cincinnati, OH, USA*

JOAN K. HEATH, PhD • *Ludwig Institute of Cancer research, Parkville, Victoria, Australia*

MICHAEL A. HERMAN, PhD • *Division of Biology, Kansas State University, Manhattan, KS, USA*

BERT HOBMAYER, PhD • *Zoological Institute and Center for Molecular Biosciences, University of Innsbruck, A-6020 Innsbruck, Austria*

SAMANTHA VAN HOFFELEN, PhD • *Division of Biology, Kansas State University, Manhattan, KS, USA*

BENJAMIN M. HOGAN, PhD • *Hubrecht Institute for Developmental Biology and Stem Cell Research, Uppsalalaan 8, 3584 CT Utrecht, The Netherlands*

THOMAS W. HOLSTEIN, PhD • *Zoological Institute, University of Heidelberg, Heidelberg, Germany*

STEFAN HOPPLER, PhD • *School of Medical Sciences, University of Aberdeen, Aberdeen, Scotland, UK*

MICHAEL KÜHL, PhD • *University of Ulm, Institute for Biochemistry and Molecular Biology, Ulm, Germany*

SHALIKA KUMBUREGAMA, BSc • *Department of Zoology, The University of Hawaii at Manoa, Honolulu, Hawaii*

DANIELLE L. LAVERY, PhD • *School of Medical Sciences, University of Aberdeen, Aberdeen, Scotland, UK*

KEVIN LEGENT, PhD • *Kimmel Center for Biology and Medicine of the Skirball Institute, New York University School of Medicine, Department of Cell Biology, New York, NY, USA*

TOBIAS LENGFELD, PhD • *Zoological Institute, University of Heidelberg, 69120 Heidelberg, Germany*

GRAHAM J. LIESCHKE, MBBS, BMedSc, FRACP, PhD • *The Walter and Eliza Hall Institute of Medical Research, Parkville, Melbourne, Victoria, Australia*

OLIVIA LUU, BSc • *University of Toronto, Department of Cell and Systems Biology, Toronto, ON, Canada*

DANIEL J. MARSTON, PhD • *Department of Biology, University of North Carolina, Chapel Hill, Chapel Hill, NC, USA*

DAVID R. MCCLAY, PhD • *Biology Department, Duke University, Durham, NC, USA*

HELEN MCNEILL, PhD • *Samuel Lunenfeld Research Institute, University of Toronto, Toronto, Ontario M5G 1X5, Canada*

AMANDA J. MIKELS, PhD • *Howard Hughes Medical Institute, Beckman Center, Stanford University Medical Center, Stanford, CA, USA*

ADNAN MIR, PhD • *Division of Developmental Biology, Cincinnati Children's Hospital Research Foundation, Cincinnati, Ohio, USA*

HIROMASA NINOMIYA, PhD • *University of Toronto, Department of Cell & Systems Biology, Toronto, ON, Canada*

ROEL NUSSE, PhD • *Howard Hughes Medical Institute, Beckman Center, Stanford University Medical Center, Stanford, CA, USA*

PETRA PANDUR, PhD • *University of Ulm, Institute for Biochemistry and Molecular Biology, Ulm, Germany*

ADAM C. PARSLOW, BSc • *Ludwig Institute for Cancer Research, Parkville, Melbourne, Victoria, Australia*

CHIEH-FU PENG, MS • *Department of Biology, The University of Miami, Coral Gables, FL, USA*

LARS F. PETERSEN, PhD • *University of Toronto, Department of Cell & Systems Biology, Toronto, ON, Canada*

MINNA ROH, MSc • *Department of Biology, University of North Carolina, Chapel Hill, Chapel Hill, NC, USA*

HERBERT STEINBEISSER, PhD • *Institute of Human Genetics, University Heidelberg, Heidelberg, Germany*

RAJEEB K. SWAIN, PhD • *CRUK Molecular Angiogenesis Group, Division of Immunity and Infection, Institute for Biomedical Research, University of Birmingham Medical School Edgbaston, Birmingham, UK*

HONG THI TRAN, BSc • *Department for Molecular Biomedical Research, VIB and Molecular Biology, Ghent University, Ghent, Belgium*

JESSICA E. TREISMAN, PhD • *Kimmel Center for Biology and Medicine of the Skirball Institute, New York University School of Medicine, Department of Cell Biology, New York, NY, USA*

HEATHER VERKADE, PhD • *School of Biological Sciences, Monash University, Clayton, Victoria, Australia*

KRIS VLEMINCKX, PhD • *Department for Molecular Biomedical Research, VIB and Molecular Biology, Ghent University, Ghent, Belgium*

GRANT N. WHEELER, PhD • *School of Biological Sciences, University of East Anglia, Norwich, England, UK*

NAVEEN WIJESENA, BSc • *Department of Biology, The University of Miami, Coral Gables, FL, USA*

ATHULA H. WIKRAMANAYAKE, PhD • *Department of Biology, The University of Miami, Coral Gables, FL, USA*

RUDOLF WINKLBAUER, PhD • *University of Toronto, Dept of Cell & Systems Biology, Toronto, ON, Canada*

Contents of Volume 1

Part I

Introduction

Chapter 1

Evolution of the Wnt Pathways

Jenifer C. Croce and David R. McClay

Abstract

Wnt proteins mediate the transduction of at least three major signaling pathways that play central roles in many early and late developmental decisions. They control diverse cellular behaviors, such as cell fate decisions, proliferation, and migration, and are involved in many important embryological events, including axis specification, gastrulation, and limb, heart, or neural development. The three major Wnt pathways are activated by ligands, the Wnts, which clearly belong to the same gene family. However, their signal is then mediated by three separate sets of extracellular, cytoplasmic, and nuclear components that are pathway-specific and that distinguish each of them. Homologs of the Wnt genes and of the Wnt pathways components have been discovered in many eukaryotic model systems and functional investigations have been carried out for most of them. This review extracts available data on the Wnt pathways, from the protist *Dictyostelium discoideum* to humans, and provides from an evolutionary prospective the overall molecular and functional conservation of the three Wnt pathways and their activators throughout the eukaryotic superkingdom.

Key words: Wnt pathway, eukaryote, evolution, canonical, PCP, calcium.

1. Introduction

The founders of the Wnt gene family were identified and sequenced in the late 1900s independently in *Drosophila* and mouse. In *Drosophila*, wingless was identified as a segment polarity gene *(1, 2)*, while in mouse int-1 was cloned as a proto-oncogene, responsible for virally induced mammary tumors *(3, 4)*. The name "Wnt" derived from the combination of both names "wingless" and "int-1" after these two genes were shown to encode homologous proteins *(5)*. Because of the undeniable involvement of these genes

Elizabeth Vincan (ed.), *Wnt Signaling, Volume II: Pathway Models, vol. 469*
© 2008 Humana Press, a part of Springer Science + Business Media, New York, NY
Book doi: 10.1007/978-1-60327-469-2

in cancer and embryonic development, interest was naturally prompted and led to an exhaustive search for Wnt homologs throughout the eukaryotic superkingdom (for a eukaryotic phylogenic representation, *see* **Fig. 1.1**). Combining results from degenerative PCR approaches with bioinformatics surveys using available genomic sequences, more than 100 Wnt genes have now been reported, which comprise the large Wnt multi-gene family. Sequence analyses comparing all of these genes allow the examination of the evolution of the Wnt proteins among eukaryotes. Wnt genes encode secreted ligands that bind to a receptor and activate a signaling pathway. In parallel to the search for Wnt homologs, other studies were thereby carried out to identify the Wnt receptors and the extracellular and intracellular molecules acting alongside or in response to this ligand–receptor interaction. Genetic analyses in *Drosophila, Caenorhabditis elegans*, and zebrafish, as well as biochemical experiments in mammalian cell culture, and *Xenopus* embryos, and gene targeting in mouse, authenticated both the receptors and the molecular pathways activated by Wnts. With these studies it became clear that Wnt signals transduce many distinct pathways, three of which have

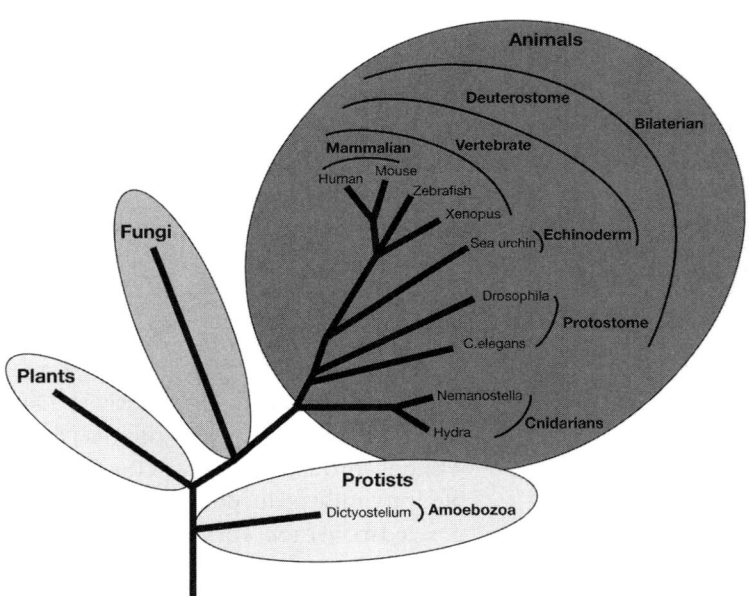

Fig. 1.1. Schematic eukaryotic phylogenetic representation. All organisms discussed in this volume of Methods in Molecular Biology are represented in this tree. Amoebozoa are protists; however, their relative position on the eukaryotic tree relative to plant and fungi remains controversial.

been the most studied these past decades *(6–8)*. Among these three major pathways, the first to be elucidated was the Wnt/β-catenin pathway, often called the "canonical" Wnt pathway. Then, several others called "non-canonical" Wnt pathways were discovered; these involve many of the same components used by the canonical pathway but with molecular relationships between these components that are altered relative to the canonical pathway, or they utilize different transducing molecules. The most studied of these non-canonical Wnt pathways are the planar cell polarity (PCP) pathway and the Wnt-calcium (Wnt/Ca^{2+}) pathway. Like the canonical pathway, these two non-canonical pathways are activated by the same initial events: the interaction of a Wnt ligand with its cognate receptor Frizzled followed by the activation of the cytoplasmic effector Dishevelled (Dsh) *(9, 10)*. However, along with these three molecules, many additional proteins are involved, either to regulate the Wnt/Frizzled interaction or to transduce the signal to the nucleus, and it is the involvement of these various proteins that defines each pathway and provides specificity of function downstream of the individual Wnts. Of these three pathways (canonical, PCP, and calcium), the canonical pathway is the most broadly described throughout living organisms. Equivalent signals have been reported from protists to vertebrates, revealing both a molecular and functional conservation during eukaryotic evolution. In contrast, the PCP and the Wnt-calcium pathways have been less thoroughly studied, reducing the extent of the knowledge on their distribution among eukaryotes; investigations of these pathways have mainly been performed in bilaterian animals. The availability of many genomic sequences for living organisms other than bilaterians permits, however, some speculation on the breadth of the evolutionary distribution of these pathways and the molecular conservation of individual components.

2. The Wnt Family

Wnt genes are generally defined by sequence homology. They are similar in size (350–400 amino acids in length) and provide characteristic structural features that allow their identification by BLAST sequence analysis. Each contains an amino terminal signal sequence followed by a highly conserved pattern of 22 to 24 cysteine residues essential for the function of the Wnts and the specificity of their responses *(13, 14)*. As mentioned previously more than 100 Wnt genes have now been reported and phylogenetic analyses organize them into 13 subfamilies: Wnt-1 to Wnt-11, Wnt-16, and Wnt-A (**Fig. 1.2**). So far, no Wnt genes have been found in organisms other than animals (metazoans). Indeed, they appear to be absent from plants (including *Arabidopsis thaliana*), fungi (such as *Saccharomyces cerevisiae*), and protists (like *Dictyostelium discoideum*), although it should be noted that a transducing pathway related to the canonical Wnt pathway has been reported for *Dictyostelium* (*see* later). By contrast, in cnidarians, one of the most basal of the animal phyla, 14 Wnt genes have been identified that belong to 12 of the 13 Wnt subfamilies *(15, 16)*; the Wnt-9 subfamily is not represented, while the Wnt-7 and Wnt-8 subfamilies have two paralogous genes (**Fig. 1.2**). How these Wnt proteins arose is still an outstanding question. One could propose the existence of a single ancestral animal Wnt protein that gave rise to all cnidarian Wnts through multiple rounds of duplication. However, taking into account the diversity of the cnidarian Wnts, the original duplications must have been ancient, perhaps prior to or linked with the emergence of multicellular animals. More ancient than cnidarians is the porifera phyla that includes animals such as sponges. Although no genome data is yet available from this phylum, EST data from the sponge *Oscarella carmela* revealed that four Wnt genes are expressed in this animal

Genes	Distribution of the Wnt genes in Eukaryotes					
			Protostomes		Deuterostomes	
	Amoebozoans	Cnidarians	Arthropods	Nematodes	Echinoderms	Vertebrates
wnt1	o	■	■	o	■	■
wnt2	o	■	o	o	o	■■
wnt3	o	■	o	o	■	■■
wnt4	o	■	o	■	■	■
wnt5	o	■	■	■	■	■■
wnt6	o	■	■	o	■	■
wnt7	o	■■	■	o	■	■■
wnt8	o	■■	o	o	■	■■
wnt9	o	o	■	o	■	■■
wnt10	o	■	■	■	■	■■
wnt11	o	■	o	o	o	■
wnt16	o	■	o	■	■	■
wntA	o	■	■ (A. gambiae)	o	■	o
wntX	o	o	■	■	o	o

Fig. 1.2. Distribution of Wnt proteins throughout the eukaryotic superkingdom. A square represents each gene found in the corresponding genome by protein subfamily. A 0 designates the absence of homologs of that subfamily in the corresponding genome. A question mark indicates orthology uncertainties. Genomes used are: amoebozoa, *Dictyostelium discoideum*; cnidarians, *Nematostella vectensis*; protostomes, *Drosophila melanogaster* and *Caenorhabditis elegans* (unless stipulated as coming from the mosquito *Anopheles gambiae*); echinoderms, *Strongylocentrotus purpuratus*; vertebrates, *Homo sapiens* (Adapted and Updated from ref. *12*).

(17), supporting the idea that since the emergence of the metazoans multiple Wnts were present. Wnt genes seem therefore to be animal-specific, identified from the most basal phylum to the most complex, and within each phylum highly diverse. How this family has emerged in the first place is still mysterious, and more genome sequencing would therefore be necessary, especially from basal animals, to determine whether all metazoan Wnts derived from a single ancestral gene or multiple ones.

The distribution of the Wnt proteins within the bilaterian groups, compared to the cnidarian Wnts, reveals another interesting level in the complexity of the Wnt repertoire. First, whereas cnidarians possess 14 Wnts, in *Drosophila* and *C. elegans* only 7 and 5 Wnts have been identified, respectively (**Fig. 1.2**; ref. *18)*. Thus, if it is assumed that cnidarians reflect the repertoire inherited by the first bilaterians, then the protostomes have secondarily lost Wnt genes. This putative secondary loss is indeed supported by the identification of only 6 and 7 Wnts in mollusks and platyhelminthes, respectively *(19)*, some of which belong to distinct subfamilies than the subfamilies present in *Drosophila* and *C. elegans* (**Fig. 1.2** and ref. *12)*. However, full genomes of representatives of the lophotrochozoans (such as mollusks and platyhelminthes) have yet to be sequenced, implying that these conclusions might change once full data sets become available. Second, the Wnt gene family is generally represented by 13 subfamilies; however, an additional subfamily exists (noted Wnt-X in **Fig. 1.2**) that is ecdysozoan-specific. This subfamily is represented both in *Drosophila* and *C. elegans* by a unique member that currently does not have an identified homolog in cnidarian, lophotrochozoan, echinoderm, or vertebrate genomes *(18)*. Third, in cnidarians and vertebrates, the Wnt-7 and Wnt-8 subfamilies contain two members, while other phyla, including arthropods, nematodes, mollusks, echinoderms, and cephalochordates, typically contain only one (**Fig. 1.2** and ref. *12)*. This could be the result of two distinct causes: the individual losses of the Wnts in each of the intervening phyla, or, and most likely, duplication events that took place independently in cnidarians and vertebrates. Indeed, supporting the latter is the proposed model of a whole-genome duplication event that would have occurred at about the time of origin of vertebrates *(20)*; although this notion is still debated *(21, 22)*. Finally, in vertebrates, one of the 13 Wnt subfamilies, Wnt-A, is not represented (**Fig. 1.2**), and is actually absent from all chordates investigated including vertebrates, cephalochordates, and urochordates *(12)*. In contrast, a Wnt-A homolog has been described in most of the other metazoan phyla from cnidarians to echinoderms (**Fig. 1.2** and ref. *12)*. Neither the role nor the signaling pathway has yet been identified for any of these Wnt-A genes. However, because this subfamily appears to be non-chordate specific, its study might provide interesting insights into the passage from non-chordate to chordate animals.

Another way to classify Wnt genes could be by looking at their expression pattern and function throughout the animal kingdom. In terms of expression, individual and multi-gene studies have highlighted only few conserved patterns. In addition, in many animals several Wnts are expressed at the same time and sometimes in the same tissue during embryogenesis *(12, 15)*. In contrast, experiments performed in *Xenopus* and mammalian cell culture have established that, based on their ability to induce a secondary body axis and on their oncogenic capacity, Wnt proteins can be grouped into two functional categories: those that induce axis duplication, display transforming activity, and act through the canonical Wnt pathway (Wnt-1, Wnt-3, Wnt-3a, Wnt-7a, Wnt-7b, Wnt-8), and those that do not induce axis duplication, do not have any oncogenic potential, and are activators of the non-canonical Wnt pathways (Wnt-4, Wnt-5a, Wnt-6, Wnt-11) *(13, 23)*. In agreement with this, individual functional analyses performed in other metazoan embryos, such as sea urchin and zebrafish, corroborated these affiliations for some of these Wnts with the canonical or the non-canonical category (e.g., refs. *24* and *25)*. However, it is important to note that the ability of each of the Wnts to activate either the canonical or the non-canonical signals is not absolute. For example, in *Xenopus*, Wnt-11, which mediates non-canonical signaling *(26, 27)*, also activates the canonical pathway during dorsal axis specification *(28)*. Thus, while canonical and non-canonical signaling are distinct pathways with distinct roles, how the individual Wnt ligands use them to control their various signaling functions remains an unresolved issue.

3. The Canonical Wnt Pathway

The canonical Wnt pathway is frequently represented by its six key components that include Wnt and Frizzled (the activators of the pathways), Dishevelled (Dsh) and GSK3 (the cytoplasmic transducer), and β-catenin and TCF/Lef (the nuclear factors). As described in many textbooks and observed in all animals studied, the canonical Wnt pathway is activated by the binding of a Wnt ligand to a Frizzled receptor. That interaction triggers the cytoplasmic effector Dsh that ultimately blocks the antagonistic activity of GSK3, which normally phosphorylates β-catenin and targets it for degradation. Preventing the degradation of β-catenin leads to an increase in its cytoplasmic concentration, and its translocation into the nucleus, where it interacts with a TCF/Lef factor and activates transcription of the Wnt target genes (**Fig. 1.3A**). In the past two decades, searches for homologs of these six main

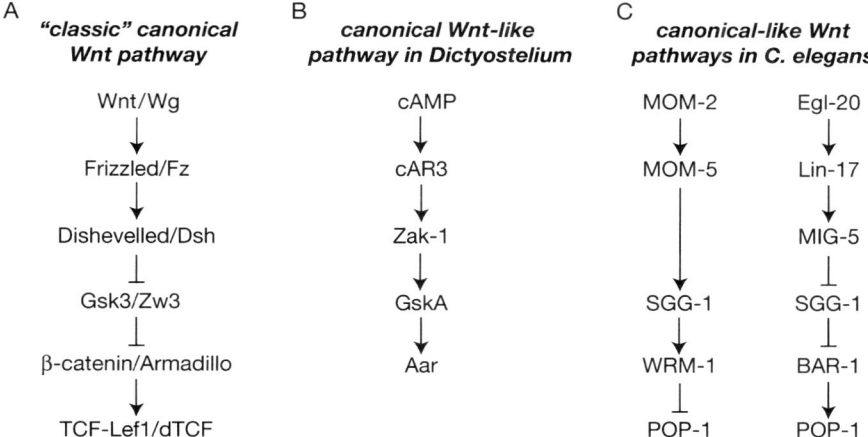

Fig. 1.3. Comparison of canonical Wnt signal transduction models among eukaryotes. (**A**) In most bilaterians, the canonical Wnt pathway is activated by a Wnt/Frizzled interaction, which results in the activation of the cytoplasmic effector Dsh that in turn inhibits the antagonistic GSK3, releasing β-catenin, which now activates TCF/Lef. (**B**) In the protist *Dictyostelium discoideum*, GSK3 and β-catenin homologs, GskA and Aar respectively, are activated by the binding of cAMP to its receptor cAR3. This interaction leads to the activation of GskA, which then phosphorylates and activates Aar. (**C**) In *C. elegans*, two Wnt pathways involving GSK3 and two distinct β-catenin homologs are observed. While the Wnt/BAR-1 pathway evolved as the classic bilaterians canonical Wnt pathway (right), the Wnt/WRM-1 pathway (left) is similar to the GskA/Aar pathway described in *Dictyostelium*. Wnt (MOM-2)/Frizzled (MOM-5) interaction induces the activation of the GSK3 homolog SGG-1, which activates WRM-1 by phosphorylation that in turn inhibits the TCF/Lef homolog POP-1 activity. Arrows within this figure do not necessarily represent direct interaction between the molecules.

Table 1.1
Conservation of the main components of the canonical Wnt pathway within eukaryotes[1]

	Amoebozoa	Cnidarians	Protostomes	Deuterostomes
Dsh	o	x	x	x
GSK3	x	x	x	x
β-catenin	x	x	x	x
TCF	o	x	x	x

[1]x indicates the presence of a vertebrate gene homolog in the corresponding genomes, while o indicates the absence of such a homolog (amoebozoa, *Dictyostelium discoideum*; cnidarians, *Nematostella vectensis*; protostomes, *Drosophila melanogaster* and *Caenorhabditis elegans*; deuterostomes, *Strongylocentrotus purpuratus* and *Homo sapiens*).

components have identified related canonical Wnt pathways in many organisms ranging from *Dictyostelium discoideum* to *Homo sapiens* (**Table 1.1**).

In most plants, fungi, and protists, no Wnts or Frizzled genes have been found (for detail on the evolutionary history of Wnts and Frizzled genes in metazoans *see* ref. *12*), supporting the idea that

Wnt pathways per se are animal-specific. In contrast, homologs of GSK3 and β-catenin have been identified in all eukaryotes, including in the protist *Dictyostelium discoideum* (**Table 1.1**) where they are both acting in a related canonical Wnt-like pathway *(29)*. In this slime mold, GskA and Aar, the GSK3 and β-catenin homologs, respectively, are not activated by a Wnt ligand but by cyclic AMP (cAMP). cAMP binds to its cognate receptor cAR3, which leads to the activation of the tyrosine kinase Zak-1. Zak-1 then phosphorylates and activates GskA/GSK3, which in turn phosphorylates Aar/β-catenin, leading to gene transcription (**Fig. 1.3B**). How Aar/β-catenin induces gene transcription is still unknown: whether it works by itself, by interaction with another factor, or by inhibition of another factor has yet to be elucidated; no TCF/Lef homolog has been identified in the *Dictyostelium* genome (**Table 1.1**). Nevertheless, this pathway involves the same proteins, GSK3 and β-catenin, for a similar outcome, the control of gene transcription. Therefore, although no Wnt or Fizzled homologs have been identified in the *Dictyostelium* genome, key components of the canonical pathway are present and functionally linked in this molecular pathway, indicating that whereas the signal *per se* might have arisen with the emergence of the metazoans, components of the downstream pathway already existed.

Additionally, in contrast to the "classic/textbook" canonical Wnt pathway, activation of this cAMP signal induces activation of GskA, rather than inhibition, and GskA activates Aar/β-catenin, instead of inducing its degradation. These relationships might appear surprising although they are reminiscent of those observed in a "non-canonical" Wnt pathway described in the nematode *C. elegans* that involves the β-catenin homolog WRM-1. In this pathway, signaling is initiated by homologs of Wnt (MOM-2) and Frizzled (MOM-5). Then, analogous to the situation in *Dictyostelium*, receptor ligation directly activates the GSK3 homolog (SGG-1) that, independently of Dsh (which is absent from the pathways in both organisms) activates WRM-1/β-catenin by phosphorylation. In turn, WRM-1/β-catenin inhibits the TCF/Lef homolog POP-1, relieving the transcriptional repression of the Wnt/WRM-1 target genes (**Fig. 1.3C**). Thus, although components of this Wnt pathway are similar to those used by the classic canonical pathway, their relationships to one another are functionally different. Therefore, this raises the interesting possibility that this signal transduction mechanism observed in both *Dictyostelium* and in *C. elegans* might actually represent the ancient form of the textbook canonical Wnt pathway observed in all metazoans from cnidarians to vertebrates. Intriguingly, in *C. elegans* a classic canonical pathway is also present. Unlike most eukaryotes, *C. elegans* possesses three distinct β-catenin homologs, one of which (HMB-2) is dedicated to cell adhesion, while the other two participate in gene regulation. WRM-1 is involved in the pathway described above, while BAR-1 acts in the classic canonical pathway (**Fig. 1.3C**; refs. *30* and *31*). Therefore, in *C. elegans* the potentially ancient GSK3-β-catenin functional relationship coexists alongside the modern, textbook relationship.

In a few model organisms, mostly bilaterians, studies on the composition of the canonical Wnt pathway have been explored further to understand the detailed mechanism of this signaling pathway. The motivation of these studies was primarily to identify potential ways to treat cancer and other human diseases associated with the inappropriate activation of this pathway, and secondarily to assess the level of conservation of the pathway between these model systems. Multiple functional analyses performed in vertebrate embryos have led to the identification of a large number of secondary components of the canonical Wnt pathway acting from the extracellular compartment to the nucleus *(18)*. In *Drosophila*, a high-throughput screen also reported 238 genes involved in regulating the canonical Wnt signal *(32)*. These genes encode for various protein types ranging from membrane-associated proteins to kinases and transcription factors, while a third of them are of unknown function; for many, their involvement has been corroborated by functional analyses *(32)*. In addition, sequence comparisons with human genome databases indicated that 50% of these *Drosophila* genes have human orthologs (18% of which are associated with human disease) *(32)*. Finally in sea urchin, an invertebrate deuterostome, a large-scale analysis carried out on the genome showed that 88% of the molecules examined (a subset of the known vertebrate components) were conserved between sea urchins and humans *(12)*. Thus, at a molecular level, the canonical Wnt pathway appears to be a highly conserved pathway among metazoans, this conservation extending beyond the simple group of its six key components. Not surprisingly, this level of conservation is lower between protostomes and deuterostomes (e.g. Drosophila-humans) than it is among deuterostome organisms (e.g. sea urchin-humans). However, in both cases this assessment simply measures the conservation of the individual components, rather than the conservation of their genetic relationships or functions. Thus, much more remains to be learned to determine the overall conservation of the complete pathway relative to the simplified version discussed previously.

The canonical Wnt pathway is required in many developmental processes, some of which are particularly well-conserved during animal development, such as blastopore determination and endomesoderm specification. However, the canonical pathway also plays a key role during an earlier embryonic event, axis formation, and this role appears to be the most well-conserved and ancient of its various functions. In *Dictyostelium*, the canonical Wnt-like pathway (cAMP/GskA) is required for the regulation of the antero-posterior asymmetry of the multicellular slime mold. Active in the posterior region, the cAMP/GskA pathway promotes posterior cell fates leading to the establishment of the longitudinal body axis *(33)*, suggesting that the axis specification function of the canonical pathway may predate the invention of Wnt ligands. Similarly, in all metazoans, from cnidarians to vertebrates, the Wnt/β-catenin pathway controls the establishment of either the antero-posterior, animal-vegetal, or dorso-ventral axis depending on the model system studied *(34)*. Thus, although the protist canonical Wnt-like pathway is molecularly different from the classic canonical Wnt pathway, they are each essential to promote cell fates and establish body axes. Therefore, since this function is observed from protists to vertebrates, it seems to be the most ancient and perhaps the primordial role of the pathway. Interestingly, in *Dictyostelium* the cAMP/GskA pathway is only required during the multicellular stages of the aggregating slime mold, and not during their unicellular development *(35)*. This exemplifies, therefore, the general assumption that the canonical pathway functions only in the context of multicellularity, suggesting that it may have arisen in particular to facilitate intercellular communication.

4. The Non-canonical Wnt Pathways

The two most studied of the non-canonical Wnt pathways operate through signaling mechanisms that do not involve GSK3 or β-catenin and do not generally regulate cell fate specification in early development, but instead are essential to control morphogenetic cell movements. These pathways are the planar cell polarity (PCP) pathway and the Wnt-calcium (Wnt/Ca^{2+}) pathway. The cell polarity pathway has been named as such because of its function in establishing cell polarity, but it also controls convergent extension movements of cells, which are movements that require polarity guidance for directed displacement. In contrast, the Wnt/Ca^{2+} pathway mediates cytoskeletal dynamics and cell adhesion, through the regulation of intracellular calcium levels. Although the non-canonical Wnt pathways have been characterized in less detail than the canonical pathway, they nevertheless are both defined by a distinct set of proteins that distinguish the PCP and the Wnt-calcium pathways from one another and from the canonical Wnt pathway.

The PCP pathway involves six core genes aside from Wnt, Fz, and Dsh: the transmembrane Frizzled co-receptor Strabismus (or Van Gogh), the cytoplasmic Strabismus partner Prickle, the cytoplasmic small G proteins RhoA and Rac activated by Dsh, and their respective effectors ROCK (Rho-kinase) and JNK

(Jun kinase) (**Fig. 1.4**). Based on published data and examination of genomes databases, all metazoan model systems analyzed from cnidarians to vertebrates possess a homolog for each of these factors (**Table 1.2**). Thus, like the canonical Wnt pathway, the PCP pathway appears well-conserved throughout animal evolution. In support, in the sea urchin genome 93% of the 29 molecules presently known to participate in the regulation and the transduction of the PCP pathway are conserved between humans and sea urchins *(12)*. However, in the non-metazoan *Dictyostelium* genome, Strabismus and Rac were the only PCP components identified (**Table 1.2**). Similar to GskA and Aar/β-catenin, it is possible that these two molecules act in a pathway that is divergent from the PCP pathway but that has a comparative functional outcome; an hypothesis that has yet to be tested.

The PCP pathway was first discovered in *Drosophila* where it controls epithelial planar polarity within the eye, wing, and thorax *(36, 37)*. In the eye, this pathway controls the orientation of the ommatidia, while in the wing and thorax it regulates the pattern of setae. In vertebrates, an equivalent genetic pathway was subsequently described that plays an essential role in patterning the fur and the sensory hairs of the inner ear *(38, 39)*. Further, in vertebrates, the PCP pathway also controls convergent extension

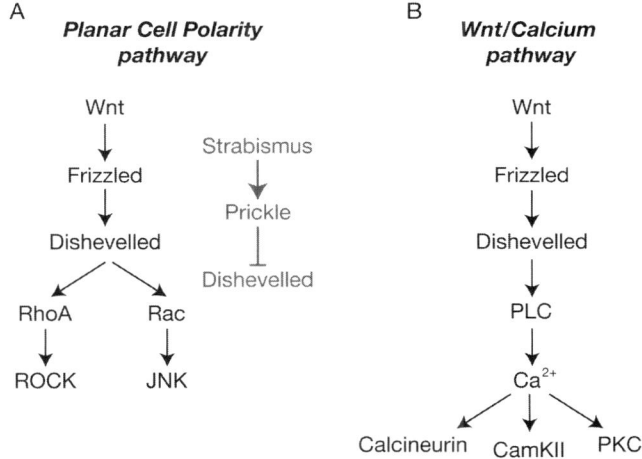

Fig. 1.4. Schematic representation of the two most studied non-canonical pathways. (**A**) The Planar Cell Polarity pathway. Upon interaction of a Wnt ligand with a Frizzled receptor and activation of the cytoplasmic protein Disheveled, the signal is transduced through two distinct small G proteins Rho and Rac, which convey cell polarity via their respective effectors ROCK and JNK. To establish polarity, however, the membrane receptor Strabismus is also activated but in the neighboring cell (in gray), where via Prickle it inhibits the PCP pathway by sequestering Disheveled. (**B**) The Wnt/Ca^{2+} pathway. Activation of the Wnt/calcium pathway leads to the activation of the key component PLC and an increase in intracellular calcium levels, which induces activation of the calcium-sensitive proteins Calcineurin, CamKII, and PKC.

Table 1.2
Conservation of the main components of the non-canonical Wnt pathways within eukaryotes[1]

		Amoebozoa	Cnidarians	Protostomes	Deuterostomes
PCP pathway	Strabismus	x	x	x	x
	Prickle	o	x	x	x
	RhoA	o	x	x	x
	ROCK	o	x	x	x
	Rac	x	x	x	x
	JNK	o	x	x	x
Ca²⁺ pathway	PLC	x	x	x	x
	PKC	o	x	x	x
	CamKII	o	x	x	x
	Calcineurin	x	x	x	x

[1]x indicates the presence of a vertebrate gene homolog in the corresponding genomes, while o indicates the absence of such a homolog (amoebozoa, *Dictyostelium discoideum*; cnidarians, *Nematostella vectensis*; protostomes, *Drosophila melanogaster* and *Caenorhabditis elegans*; deuterostomes, *Strongylocentrotus purpuratus* and *Homo sapiens*).

movements associated with gastrulation and neurulation *(40)*. Convergent extension (CE) movements are polarized rearrangements of cells within a given tissue along an axis (reviewed in ref. *41)*. In CE movements, cells intercalate and elongate, such that a tissue is narrowed in one direction and elongated in the perpendicular direction. Thus, whether associated with ommatidia, hair patterning, or CE movements, the PCP pathway consistently controls cell polarization, underlying a functional conservation of the pathway between protostomes and vertebrates. Reinforcing this idea, analyses performed on some of the PCP pathway components (Fz5/8, RhoA and Prickle) in sea urchins and ascidians demonstrated the role of this pathway in the control of CE movements in these animals *(42–44)*. Further, in *C. elegans* the PCP pathway is involved in regulating the asymmetric division of the B cell (which gives rise to the proctodeum of the larvae), thereby conveying cell polarity at another level *(45)*. Thus, all these studies indicate that the PCP pathway is well-conserved, not only at a molecular level, but also at functional level, at least throughout bilaterians. It will now be interesting, therefore, to determine the role of this pathway in more basal animals such as cnidarians, which have all the main PCP components (**Table 1.2**), although their functions have yet to be investigated.

The Wnt/Ca^{2+} pathway acts through heterotrimeric G-proteins to activate phospholipase C (PLC), which leads to the release of intracellular calcium. This increase in Ca^{2+} activates three calcium sensitive components: PKC (protein kinase C), calcineurin, and CamKII (calcium/calmodulin-dependent kinase II) (**Table 1.2**; for review, *see* refs. *46* and *47*). The Wnt/Ca^{2+} pathway was discovered in *Xenopus* and zebrafish *(48, 49)*, but no equivalent pathways have yet been described in any other model system, raising the possibility that the Wnt-calcium signaling is a vertebrate-specific pathway. Alternatively, protein homologs of all key components of the Wnt-calcium pathway have been reported in a number of metazoan model systems (Cnidarians, sea urchin, *Drosophila*, and *C.elegans*; **Table 1.2**), but whether these molecules act together in a related Wnt-activated pathway has yet to be determined. Indeed PLC, PKC, calcineurin, and CamKII are generic proteins involved in many transduction pathways that play important roles in various physiological processes. Their presence in these more basal metazoans could therefore be due to the conservation of these other functions and not their involvement within a particular Wnt pathway, which then could have arisen as a vertebrate-specific acquisition resulting from the co-option of currently available signaling proteins. Supporting this contention, PLC and calcineurin homologs are present in protists, plants, and fungi, although there is no evidence for any Wnt-like pathway in these groups (other than the similarities alluded to earlier in *Dictyostelium*). In contrast, Cook and his collaborators reported in 1996 the existence of a Wingless–PKC pathway in *Drosophila (50)*, an observation that provides initial support for a potentially broader distribution of the Wnt/Ca^{2+} pathway within metazoans.

5. Conclusions

The three Wnt pathways discussed previously are involved in many aspects of embryonic developmental events, controlling cell fate, proliferation, polarity, migration, and adhesion. As a result, disruption of their functions causes drastic consequences ranging from tumorigenesis to embryonic lethality. These Wnt pathways have been studied in diverse living organisms allowing us to describe them from an evolutionary prospective. As an overview, all three pathways seem to have emerged with the metazoans, although with regard to canonical signaling, a related pathway is present in the protist *Dictyostelium*. Furthermore, based on the identification of individual components of the three Wnt pathways, these pathways appear to be well-conserved among all studied metazoans, from cnidarians to vertebrates, perhaps with the exception of the Wnt-calcium pathway for which the data is still limited. Therefore, because of this important molecular and functional conservation of the Wnt pathways throughout the animal kingdom, their studies across divergent animal phyla add to the richness of our knowledge, although additional investigations will be necessary to determine whether this conservation can be extended to all components, as well as to all of their molecular relationships. Nevertheless, reports currently appearing at high frequency concerning the role and the composition of these pathways in non-vertebrate metazoans (cnidarian, *Drosophila*, *C. elegans*, and sea urchin) are likely to have functional relevant parallels in vertebrates. The synthesis of all these results will lead therefore to a better understanding of the origin of the pathways and the basis of the diseases associated with their malfunction.

Acknowledgements

The authors thank Dr. Cynthia Bradham for critical evaluation of the manuscript. J. C. thanks Athula Wikramanayake for allowing me the opportunity to contribute to this book. This work was supported by grants NIH 61464 and HD 14483.

References

1. Sharma, R. P., Chopra, V. L. (1976) Effect of the Wingless (wg1) mutation on wing and haltere development in Drosophila melanogaster. *Dev Biol* 48, 461–465.

2. Baker, N. E. (1987) Molecular cloning of sequences from wingless, a segment polarity gene in Drosophila: the spatial distribution of a transcript in embryos. *Embo J* 6, 1765–1773.

3. Nusse, R., Varmus, H. E. (1982) Many tumors induced by the mouse mammary tumor virus contain a provirus integrated in the same region of the host genome. *Cell* 31, 99–109.

4. Van Ooyen, A., Nusse, R. (1984) Structure and nucleotide sequence of the putative mammary oncogene int-1; proviral insertions leave the protein-encoding domain intact. *Cell* 39, 233–240.

5. Rijsewijk, F., Schuermann, M., Wagenaar, E., Parren, P., Weigel, D., Nusse, R. (1987) The Drosophila homolog of the mouse mammary oncogene int-1 is identical to the segment polarity gene wingless. *Cell* 50, 649–657.

6. Dale, T. C. (1998) Signal transduction by the Wnt family of ligands. *Biochem J* 329, 209–223.

7. Huelsken, J., Birchmeier, W. (2001) New aspects of Wnt signaling pathways in higher vertebrates. *Curr Opin Genet Dev* 11, 547–553.

8. Korswagen, H. C. (2002) Canonical and non-canonical Wnt signaling pathways in Caenorhabditis elegans: variations on a common signaling theme. *Bioessays* 24, 801–810.

9. Wharton, K. A., Jr. (2003) Runnin' with the Dvl: proteins that associate with Dsh/Dvl and their significance to Wnt signal transduction. *Dev Biol* 253, 1–17.

10. Sheldahl, L. C., Slusarski, D. C., Pandur, P., Miller, J. R., Kuhl, M., Moon, R. T. (2003) Dishevelled activates Ca2+ flux, PKC, and CamKII in vertebrate embryos. *J Cell Biol* 161, 769–777.

11. Huelsken, J., Behrens, J. (2002) The Wnt signaling pathway. *J Cell Sci* 115, 3977–3978.

12. Croce, J. C., Wu, S. Y., Byrum, C., Xu, R., Duloquin, L., Wikramanayake, A. H., Gache, C., McClay, D. R. (2006) A genome-wide survey of the evolutionarily conserved Wnt pathways in the sea urchin Strongylocentrotus purpuratus. *Dev Biol* 300, 121–131.

13. Du, S. J., Purcell, S. M., Christian, J. L., McGrew, L. L., Moon, R. T. (1995) Identification of distinct classes and functional domains of Wnts through expression of wild-type and chimeric proteins in Xenopus embryos. *Mol Cell Biol* 15, 2625–2634.

14. Van Ooyen, A., Kwee, V., Nusse, R. (1985) The nucleotide sequence of the human int-1 mammary oncogene; evolutionary conservation of coding and non-coding sequences. *Embo J* 4, 2905–2909.

15. Kusserow, A., Pang, K., Sturm, C., Hrouda, M., Lentfer, J., Schmidt, H. A., Technau, U., von Haeseler, A., Hobmayer, B., Martindale, M. Q., Holstein, T. W. (2005) Unexpected complexity of the Wnt gene family in a sea anemone. *Nature* 433, 156–160.

16. Lee, P. N., Pang, K., Matus, D. Q., Martindale, M. Q. (2006) A WNT of things to come: Evolution of Wnt signaling and polarity in cnidarians. *Semin Cell Dev Biol* 17, 157–167.

17. Nichols, S. A., Dirks, W., Pearse, J. S., King, N. (2006) Early evolution of animal cell signaling and adhesion genes. *Proc Natl Acad Sci USA* 103, 12451–12456.

18. The Wnt homepage. Available at: www.stanford.edu/~rnusse/wntwindow.html.

19. Prud'homme, B., Lartillot, N., Balavoine, G., Adoutte, A., Vervoort, M. (2002) Phylogenetic analysis of the Wnt gene family. Insights from lophotrochozoan members. *Curr Biol* 12, 1395.

20. Friedman, R., Hughes, A. L. (2003) The temporal distribution of gene duplication events in a set of highly conserved human gene families. *Mol Biol Evol* 20, 154–161.

21. Furlong, R. F., Holland, P. W. (2004) Polyploidy in vertebrate ancestry: Ohno and beyond. *Biological Journal of the Linnean Society* 82, 425–430.

22. Holland, P. W., Garcia-Fernandez, J., Williams, N. A., Sidow, A. (1994) Gene duplications and the origins of vertebrate development. *Dev Suppl* 125–133.

23. Wong, G. T., Gavin, B. J., McMahon, A. P. (1994) Differential transformation of mammary epithelial cells by Wnt genes. *Mol Cell Biol* 14, 6278–6286.

24. Kilian, B., Mansukoski, H., Barbosa, F. C., Ulrich, F., Tada, M., Heisenberg, C. P. (2003) The role of Ppt/Wnt5 in regulating cell shape and movement during zebrafish gastrulation. *Mech Dev* 120, 467–476.

25. Wikramanayake, A. H., Peterson, R., Chen, J., Huang, L., Bince, J. M., McClay, D. R., Klein, W. H. (2004) Nuclear beta-catenin-dependent Wnt8 signaling in vegetal cells of the early sea urchin embryo regulates gastrulation and differentiation of endoderm and mesodermal cell lineages. *Genesis* 39, 194–205.

26. Tada, M., Smith, J. C. (2000) Xwnt11 is a target of Xenopus Brachyury: regulation of gastrulation movements via Dishevelled, but not through the canonical Wnt pathway. *Development* 127, 2227–2238.

27. Heisenberg, C. P., Tada, M., Rauch, G. J., Saude, L., Concha, M. L., Geisler, R., Stemple, D. L., Smith, J. C., Wilson, S. W. (2000) Silberblick/Wnt11 mediates convergent extension movements during zebrafish gastrulation. *Nature* 405, 76–81.

28. Tao, Q., Yokota, C., Puck, H., Kofron, M., Birsoy, B., Yan, D., Asashima, M., Wylie, C. C., Lin, X., Heasman, J. (2005) Maternal wnt11 activates the canonical wnt signaling pathway required for axis formation in Xenopus embryos. *Cell* 120, 857–871.

29. Coates, J. C., Harwood, A. J. (2001) Cell-cell adhesion and signal transduction during Dictyostelium development. *J Cell Sci* 114, 4349–4358.

30. Ruvkun, G., Hobert, O. (1998) The taxonomy of developmental control in Caenorhabditis elegans. *Science* 282, 2033–2041.

31. Korswagen, H. C., Herman, M. A., Clevers, H. C. (2000) Distinct beta-catenins mediate adhesion and signaling functions in C. elegans. *Nature* 406, 527–532.

32. DasGupta, R., Kaykas, A., Moon, R. T., Perrimon, N. (2005) Functional genomic analysis of the Wnt-wingless signaling pathway. *Science* 308, 826–833.

33. Harwood, A. J., Plyte, S. E., Woodgett, J., Strutt, H., Kay, R. R. (1995) Glycogen synthase kinase 3 regulates cell fate in Dictyostelium. *Cell* 80, 139–148.

34. Croce, J. C., McClay, D. R. (2006) The canonical Wnt pathway in embryonic axis polarity. *Semin Cell Dev Biol* 17, 168–174.

35. Plyte, S. E., O'Donovan, E., Woodgett, J. R., Harwood, A. J. (1999) Glycogen synthase kinase-3 (GSK-3) is regulated during Dictyostelium development via the serpentine receptor cAR3. *Development* 126, 325–333.

36. Mlodzik, M. (1999) Planar polarity in the Drosophila eye: a multifaceted view of signaling specificity and cross-talk. *Embo J* 18, 6873–6879.

37. Adler, P. N., Lee, H. (2001) Frizzled signaling and cell-cell interactions in planar polarity. *Curr Opin Cell Biol* 13, 635–640.

38. Guo, N., Hawkins, C., Nathans, J. (2004) Frizzled6 controls hair patterning in mice. *Proc Natl Acad Sci USA* 101, 9277–9281.

39. Dabdoub, A., Kelley, M. W. (2005) Planar cell polarity and a potential role for a Wnt morphogen gradient in stereociliary bundle orientation in the mammalian inner ear. *J Neurobiol* 64, 446–457.

40. Solnica-Krezel, L. (2005) Conserved patterns of cell movements during vertebrate gastrulation. *Curr Biol* 15, R213–228.

41. Mlodzik, M. (2002) Planar cell polarization: do the same mechanisms regulate Drosophila tissue polarity and vertebrate gastrulation? *Trends Genet* 18, 564–571.

42. Croce, J., Duloquin, L., Lhomond, G., McClay, D. R., Gache, C. (2006) Frizzled5/8 is required in secondary mesenchyme cells to initiate archenteron invagination during sea urchin development. *Development* 133, 547–557.

43. Beane, W. S., Gross, J. M., McClay, D. R. (2006) RhoA regulates initiation of invagination, but not convergent extension, during sea urchin gastrulation. *Dev Biol* 292, 213–225.

44. Jiang, D., Munro, E. M., Smith, W. C. (2005) Ascidian prickle regulates both mediolateral and anterior-posterior cell polarity of notochord cells. *Curr Biol* 15, 79–85.

45. Wu, M., Herman, M. A. (2006) A novel noncanonical Wnt pathway is involved in the regulation of the asymmetric B cell division in C. elegans. *Dev Biol* 293, 316–329.

46. Kühl, M., Sheldahl, L. C., Park, M., Miller, J. R., Moon, R. T. (2000) The Wnt/Ca2+ pathway: a new vertebrate Wnt signaling pathway takes shape. *Trends Genet* 16, 279–283.

47. Pandur, P., Maurus, D., Kuhl, M. (2002) Increasingly complex: new players enter the Wnt signaling network. *Bioessays* 24, 881–884.

48. Kühl, M., Sheldahl, L. C., Malbon, C. C., Moon, R. T. (2000) Ca(2+)/calmodulin-dependent protein kinase II is stimulated by Wnt and Frizzled homologs and promotes ventral cell fates in Xenopus. *J Biol Chem* 275, 12701–12711.

49. Sheldahl, L. C., Park, M., Malbon, C. C., Moon, R. T. (1999) Protein kinase C is differentially stimulated by Wnt and Frizzled homologs in a G-protein-dependent manner. *Curr Biol* 9, 695–698.

50. Cook, D., Fry, M. J., Hughes, K., Sumathipala, R., Woodgett, J. R., Dale, T. C. (1996) Wingless inactivates glycogen synthase kinase-3 via an intracellular signaling pathway which involves a protein kinase C. *Embo J* 15, 4526–4536.

Part II

Dictyostelium

Chapter 2

Dictyostelium Development: A Prototypic Wnt Pathway?

Adrian J. Harwood

Abstract

Although Wnt signaling is ubiquitous within the animal phylogenetic group, it is unclear how it evolved. Genes related to the components of Wnt pathway are found in other eukaryotes and one of the most studied of these non-metazoan organisms is the social amoeba *Dictyostelium discoideum*. This organism contains the enzyme GSK-3 and a β-catenin homolog, Aardvark (Aar). Both are required to regulate pattern formation during multi-cellular stages of *Dictyostelium* development. Aar is also required for formation of adherens junctions, as seen in animals. Finally, analysis of the completed *Dictyostelium* genome shows there to be 16 Frizzled (Fz) gene homologs. This chapter discusses *Dictyostelium* development and the role of these proteins.

Key words: *Dictyostelium*, GSK-3, β-catenin, Aar, Frizzleds, pattern formation.

1. Introduction

Wnt signaling is universally found with the metazoan group, but does it exist outside the animals? Results over the last 20 years indicate that elements of the canonical Wnt pathway, involving GSK-3 and β-catenin, exist in a number of major non-metazoan groups. This raises questions about its evolutionary origins and in particular the relationship between the role of β-catenin in cell signaling and the cytoskeleton. One of the best studied non-metazoan Wnt-like pathways is found in the social amoeba, *Dictyostelium discoideum*. This chapter describes the unique biology of the Dictyostelid group and how it can be used to investigate an evolutionarily distinct, perhaps prototypic, signal pathway related to the Wnt pathway of animals.

Elizabeth Vincan (ed.), *Wnt Signaling, Volume II: Pathway Models, vol. 469*
© 2008 Humana Press, a part of Springer Science + Business Media, New York, NY
Book doi: 10.1007/978-1-60327-469-2

2. *Dictyostelium* Biology and Development

Dictyostelium belong to the taxonomic group termed amoebozoa. These are mainly soil amoeba that feed via phagocytosis of particulate matter, bacteria in the case of the Dictyostelids. For many years, based mainly on erroneous molecular phylogenetic data using ssRNA sequences, *Dictyostelium* was considered as a fundamentally different group from other eukaryotes. However, recent evidence produced by Baldauf and Doolittle *(1)* points to a much closer relationship to animals. Through their pioneering work, it is now clear that animals, fungi, and the amoebozoa, belong to a common arm of the eukaryotic evolutionary tree. This much closer relationship to animals is reflected in the genetic composition apparent from the whole genome sequence *(2)*.

The information from the complete genomic sequence is freely accessible through the curated database, known as dictyBase *(3)*. The associated Web pages contain a Blast server for searching both genomic DNA sequences and the complete collection of predicted 12,000 open reading frames (ORFs) and proteins. The database also contains a complete *Dictyostelium* research paper bibliography linked to each gene page. This is the ideal entry point for new researchers entering this field, or those from other fields wishing to make comparative phylogenetics of a protein family of interest.

Utilizing the genetic resources available for *Dictyostelium* is relatively straight forward, as it has a compact, haploid genome with few introns and homologous recombination occurs at high frequency. This means that almost any gene can be isolated by PCR and ablated by targeted gene disruption. The haploid genome makes *Dictyostelium* ideal for insertional mutagenesis, which is most commonly carried out using the plasmid-based REMI method *(4)*. The genetic source material has been further expanded by the generation of a gene bank of more than 6,000 non-redundant sequenced cDNA clones *(5)*. As large numbers of cells can easily be grown from protein extraction, *Dictyostelium* is readily amenable to biochemical analysis. Finally, *Dictyostelium* cells are excellent for molecular cell biological analysis; for example, there are a wide range of GFP and epitope tag vectors for marking proteins and their microscopical analysis. This multidisciplinary approach makes *Dictyostelium* a powerful model system to investigate complex cell signaling processes.

By far the most significant feature of *Dictyostelium* is its unusual survival strategy triggered in response to nutrient starvation. Here the up to 10^5 cells develop into a differentiated and cell patterned multicellular structure, known as a fruiting body *(6)*. This is triggered by a combination of loss of nutrients and high cell density, and can be readily induced in the laboratory by washing cells free of nutrients and then plating them or culturing them in the absence of further nutrients. Cells are either grown in association with bacteria, normally *Klebsiella pneumoniae* or B strains of *Escherichia coli*, or liquid medium. All cell strains can be grown with bacteria; however, only cells that carry a set of three "axenic" mutations, such as the AX2 and AX3 parental strains, can grow in medium *(7)*. Multicellular development can be carried out either on non-nutrient agar; moist nitrocellulose filters or to a limited extent in shaking suspension in buffer. Importantly, growth arrests at the beginning of development offering the opportunity to investigate phenotypic effects without influence of altered cell growth and division.

Dictyostelium development is synchronous and takes 24 hours, except for some variability in the length of the slug stage for cells developed on non-nutrient agar. This basic synchronicity makes it relatively easy to obtain large quantities of cells at the same developmental stage. As *Dictyostelium* cells begin development from unicellular amoebae, the first 8 to 10 hours of development consists of cells coming together through a process of chemotaxis toward pulses of extracellular cAMP. During this aggregation stage, cells are essentially homogeneous. There are however small variations in cell behavior due to differences in nutritional state and position in the cell cycle when starvation begins, which can influence the ultimate cell fate *(8, 9)*. Cells suppress growth phase genes and induce aggregation specific genes during this early phase of development; these latter genes upregulate the cellular machinery to mediate extracellular cAMP signaling and promote cell adhesion *(10, 11)*. As cells can be induced to form terminally differentiated cell types when plated in low density monolayer cultures, these differences in cell history or cell adhesion are not essential for cell development *(12, 13)*, although they could influence morphogenesis during multicellular development.

Multicellular development (**Fig. 2.1**) begins as soon as the cells have formed an aggregate, with cells expressing cell specific genes that mark their ultimate fate in the terminal differentiated structure, the fruiting body. This structure consists of a large spherical head containing 80% of the original cells after they have differentiated into spores. The spore head is supported by a stalk, comprising of dead stalk cells; the remaining 20% of the cell population. Although the most notable feature of the fruiting body is the long, slender stalk of 4 to 5 mm in length, there are other ancillary structures that help support the stalk and spore head. The spore head is cradled between cells that form cup like structures on its top and bottom *(14)*. The stalk is attached to the substratum by the basal disc. The cells within this structure are morphologically the same as those of the stalk itself; however they arise from as different population within the aggregate *(15)*.

Within the newly formed aggregate, cells differentiate into three precursor populations *(16)*. The majority form prespore

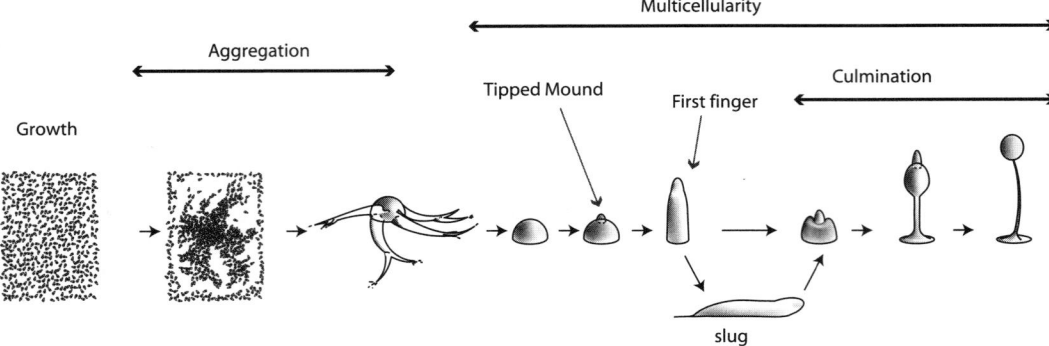

Fig. 2.1. Development of *Dictyostelium*.

cells. A particularly useful marker for looking at GSK-3 control of this process is the gene *pspA*. When linked to a *lacZ* marker gene *(17)*, it can be seen that this gene is expressed in cells as soon as they enter the aggregate. This gene, as with other prespore specific genes, is induced by extracellular cAMP *(18)*. The cells that ultimately form the stalk, express the marker *ecmA* and although this is a secreted matrix protein, its promoter can be used to mark the population of prestalk cells. These are known as pstA cells. The aggregate also contains pstB cells; these express the related gene, *ecmB*, and form the basal disc cells. Stalk cell differentiation and expression of these marker genes requires DIF-1, a dichloronated phenyl hexanone *(19, 20)*. Although this is a potent stalk cell inducer in isolated cells, genetic evidence has revealed that there are other regulatory factors that control the pattern and timing of stalk and basal disc formation in the multicellular structure *(21)*.

Prestalk cells appear to form at random within the aggregate, but quickly sort so that pstA cells fill the tip that forms onto the top of the aggregate (the tipped mound stage), whereas the pstB cells move to the basal region. The tip region has the properties of an embryonic organizer throughout the remaining multicellular development. The tip region of the mound extends so that a long finger structure is formed (first finger stage). This may topple over and form a motile slug structure. This is phototactic and is thought to move the developing structure to locations that will aid spore dispersal.

When migration is finished, developing structures enter the culmination stage. Within the tip region, pstA cells begin to differentiate into stalk cells. They do this in a very defined manner and always in the center of the tip and with newly differentiated cells joining the elongating stalk at the top. As pstA cells join the stalk they begin to express *ecmB*, and hence are known as pstAB cells. The growing stalk penetrates through the prespore region to embed in the basal disc. As the stalk continues to grow, it lifts the prespore cell mass off the substrate. Late during this preculmination stage, as the stalk reaches its full length, the prespore cells differentiate into mature spores *(22)*. Spores can survive for prolonged periods of time without nutrients, but when nutrients are restored, either following dispersal or by plating in the laboratory, they germinate and re-establish the growing population.

Both *gskA* and aar mutants have distinctive effects on the terminally differentiated *Dictyostelium* fruiting body, which makes them ideal for morphologically based genetic screens. As described previously, once obtained, mutants can be easily investigated either within multicellular structures by use of marker genes (*see* **Chapter 3** in **Volume 2**), or by controlling the inductive conditions in isolated cell cultures (*see* **Chapter 4** in **Volume 2**). The phenotypes of *gskA* and *aar* mutants reveal further developmental complexity and are discussed in the following sections.

3. GSK-3

Dictyostelium contains a GSK-3 ortholog, *gskA*, which has very similar properties to the vertebrate GSK-3β enzyme *(23, 24)*. There is no detectable GSK-3 activity in a *gskA* null mutant indicating that there is only one GSK-3 gene in *Dictyostelium*. Analysis of the *Dictyostelium* kinome *(25)* identified a related protein kinase gene, *glkA*, which has an intermediate sequence homology between cdk and GSK-3 kinases. The functions and substrates of this kinase are currently unknown. *glkA* particularly differs from *gskA* in lacking the Axin binding site present in all eukaryotic GSK-3 proteins. Exchange of this binding site for the equivalent region from *glkA* has dramatic effects on GskA function in *Dictyostelium* leading to severe effects on the motility of aggregating cells and altered regulation of gene expression *(26)*.

As seen in animals, GskA is a multifunctional kinase that is involved in cell processes in growing cells and throughout *Dictyostelium* development. GskA is not essential for cell growth, but *gskA* null mutants are compromised in cell growth *(27)*. It is required for regulated gene expression at the transition between growth and development, but with no phenotypic effects on the entry into development *(28)*. However, mutant cells do appear to have a slight chemotaxis effect, and aggregate slightly faster than wild-type cells. The reason for this is unknown. GskA phosphorylates the *Dictyostelium* StatA protein to control the rate of nuclear export *(29)*, a regulatory mechanism similar to that seen for NF-κB. These regulatory functions are very reminiscent of the GSK-3 mediated signaling in animals, but are unlikely to function within a Wnt-related pathway.

The fruiting bodies of *gskA* null have a distinctive phenotype with a small spore head and an enlarged basal disc *(27)*. This phenotype is most pronounced when cells are grown in association with bacterial, and is relatively weak in cells grown in axenic liquid medium. The cause of this difference is unknown. This morphology suggests a defect in pattern formation, and this can be first observed in the aggregate at the mound stage, where the proportion of pstB cells expands at the cost of the prespore cells. Interestingly, by use of the strain differences and the weaker phenotype of cells grown in liquid culture, a second cell patterning defect can be seen in *gskA* null mutants during later developmental stages as *ecmB* is expressed throughout the prestalk zone *(30)*. This suggests that not only does *gskA* control the proportion of prespore to pstB cells, it acts to repress expression of the pstB marker, *ecmB*, in all other prestalk cells.

Extracellular cAMP is a common factor between these events. cAMP is required for prespore and spore induction but represses stalk cell induction and expression of *ecmB* *(31)*. By mediating the effects of cAMP, *gskA* is able to control the induction of all of these cell populations. Consistent with this, GskA activity increases as cells enter the aggregate, and cAMP activates GskA activity in aggregation competent cells *(32)*. Loss of the cAMP receptor cAR3 prevents the cAMP stimulated increase of GskA activity, although it should be noted that it does not decrease the basal level of activity. Furthermore *cAR3* null mutants have a similar cell patterning defect and exhibit a loss of cAMP repression of *ecmB* gene expression and stalk formation to *gskA* null mutants *(32)*.

Biochemically downstream of cAR3 is an unusual protein kinase ZakA that contains both a functional tyrosine kinase and serine kinase domain *(33)*. The tyrosine kinase domain is able to directly phosphorylate GskA on the equivalent of the Tyr-216 amino acid in GSK-3β. Loss of ZakA causes the same phenotype as seen in *gskA* and *cAR3* null mutants. There is a paralog of Zak1 in the *Dictyostelium* genome, known as Dpyk4 *(25)*. Loss of ZakA does not prevent all tyrosine phosphorylation of GskA, only that stimulated by cAMP (unpublished observation). It is possible that this second kinase is responsible for the residual GskA phosphorylation, but this has not yet been tested. Interestingly, loss of another cAMP receptor cAR4, has the opposite phenotype to loss of *gskA*. These mutants have elevated expression of prespore genes, reduced prestalk gene expression and stalk cell formation is hypersensitive to cAMP *(34, 35)*. Biochemically, cAR4 stimulation elevates tyrosine dephosphorylation of GskA, counteracting the effect of ZakA *(36)*. Given that cAR4 has a lower affinity for cAMP than cAR3 does, this may establish a threshold response in which gskA is inactive at low and high levels of cAMP, but active at intermediate concentrations.

4. Aardvark: A β-catenin-related Protein

Dictyostelium contains a β-catenin-like molecule, known as Aardvark (*aar*). Although it does not arrest development, loss of Aar causes a reduction in the expression of the prespore-specific gene *pspA* during multicellular development *(37)*. The effect of loss of *aar* on cells developed in shaking culture is more striking as cAMP is no longer able to induce *pspA* expression. In monolayer, culture spore-cell formation in the *aar* mutant is reduced, although not completely lost. When Aar is overexpressed, the opposite result is seen as cAMP hyper-induces *pspA* expression. This hyper-induction requires GskA activity, linking it to the GskA-mediated events described in **Section 3**. However, loss of *aar* does not lead to increased *ecmB* expression or loss of cAMP expression, restricting the effects of Aar to the prespore induction pathway.

These results show that the GskA-Aar pathway operates differently from the canonical Wnt pathway of animals, where Wnt negatively regulates GSK-3 through protein–protein interactions suppressing the repressive effects of GSK-3 phosphorylation on β-catenin. In *Dictyostelium*, GskA activity is controlled by phosphorylation and acts positively on Aar to induce *pspA* expression (**Fig. 2.2**). The *Dictyostelium* pathway in some respects resembles the non-canonical Wnt pathway that is mediated by mom-2 during early nematode development. Here, genetic evidence indicates that the Wnt protein mom-2 via a Frizzled, mom-5, acts positively on GSK-3 to induce endoderm specification. GSK-3 appears to mediate its effects via a positive interaction with a β-catenin ortholog WRM-1 *(38)*. WRM-1 acheives this by de-repression of the negative effects of a TCF/LEF-related protein, POP-1, via a serine/threonine protein kinase, LIT-1 *(39)*. The mechanism by which Aar regulates transcription is not yet clear.

5. Adherens Junctions

Unexpectedly, given the loose packed nature of the *Dictyostelium* multicellular aggregate, it was discovered that some cells form strong cell–cell junctions that appear very similar to the adherens junctions of animal epithelia *(37)*. These *Dictyostelium* junctions form during the culmination stage and are only present in around 1000 cells situated in a collar-like structure close to the top of the stalk tube. Immuno-EM with antiserum raised against Aar protein localizes to the cell junctions. The junctions were originally missed due to their lability in chemical fixation; however they are persevered after rapid cryopreseveration and freeze substitution and can easily be visualized by EM. They can also be directly seen at

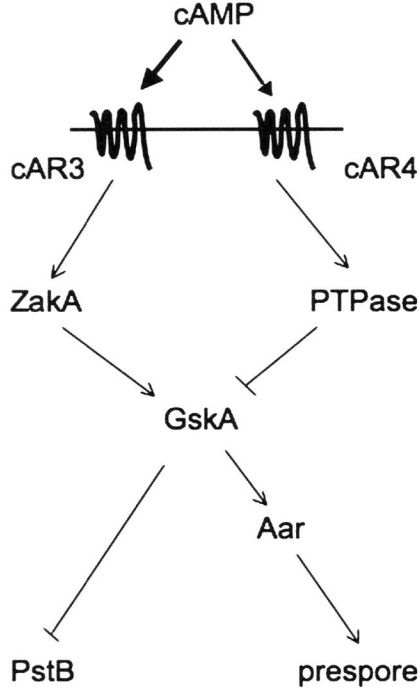

Fig. 2.2. Schematic diagram to summarize GskA-Aar signaling pathway.

a lower resolution in living cells by use of GFP fused F-actin binding proteins *(37)* or GFP-Aar (unpublished results).

Disruption of the *aar* gene causes a complete loss of these cell junctions and Aar overexpression increases their number and size. These observations demonstrate that Aar expression is both necessary and limiting for junction formation. Interestingly loss of *aar* does not block fruiting body formation. Aar mutants however have weak stalks causing the fruiting body to frequently collapse, suggesting that the adherens junctions are required for the mechanical integrity of the stalk tube. The matrix that forms the stalk tube, the extracellular material that surrounds the mature stalk cells, is a complex laminate with individual fibres laid down by both stalk cells and those prestalk cells surrounding the tube. The function of the *Dictyostelium* adherens junctions may be to polarize secretion of matrix components to the cell surface adjacent to the stalk tube (**Fig. 2.3**). To date, however, no direct visualization of vesicle transport to this surface has been possible.

Loss of stalk tube integrity can also explain an unusual phenotype where the *aar*-null mutant forms one or more additional stalks *(40)*. These arise during the culmination stage and branch from the main stalk tube. As they develop, they reorganize

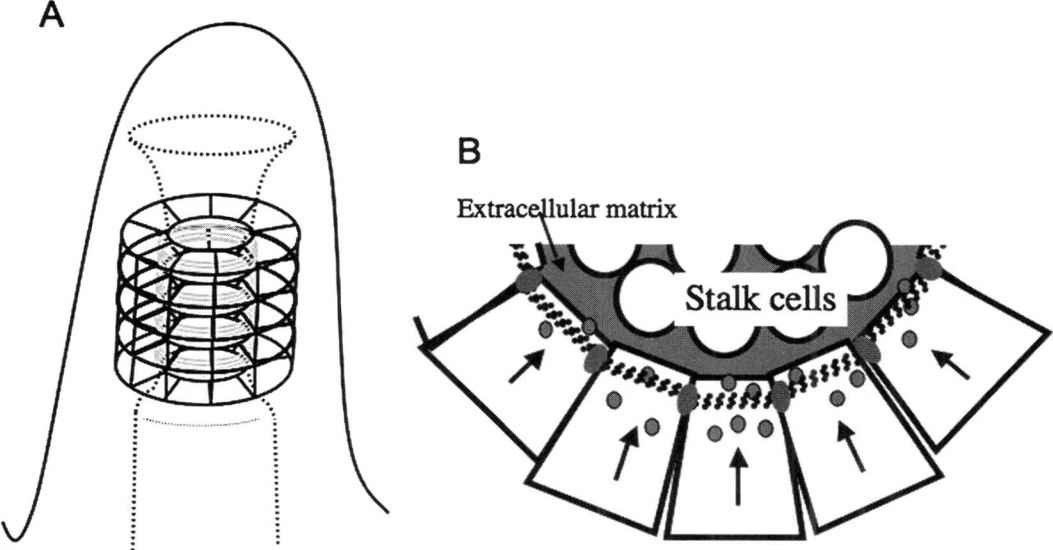

Fig. 2.3. *Dictyostelium* adherens junctions. (**A**) Cells connected by adherens junctions form a collar toward the top of the stalk tube. The junctions of each cell are connected via actin filaments, effectively forming a ring of F-actin that constricts the stalk tube. (**B**) In planar view, the presence of the junctions forms apical/basolateral polarization. This allows vectorial transport of vesicles toward the matrix surrounding the stalk cells.

the cells around them and form additional spore heads. At the cellular level, the new stalks form due to the ecotopic induction of pstAB cells on the outside of the stalk tube. Mixing mutant and wild-type cells demonstrates that super-nummary stalk formation does not arise through axis duplication, but is a cell non-autonomous event due to misplaced signaling. We believe that they arise due to leakage of a stalk inducing signal through damaged sections of the stalk tube.

7. Other Components of Wnt Signaling Pathways

Are there other components of the Wnt signaling pathway in Dictyostelium genome? With one exception, there seems to be no other close protein orthologs specific to the animal Wnt pathways; in fact, the Wnt proteins themselves and downstream components such as Dsh, Axin, and APC proteins have not been found. There are, however, generic components, such as a CK1 protein kinase, and proteins containing Wnt-related domains, such as the DEP domain of *Dictyostelium* ORF DDB0205035. One of the problems with data mining for animal components in *Dictyostelium* is the very high A-T content of the genome. This genome bias is seen at the protein level were there is genetic drift to amino acid sequences encoded by codons that have an A-T bias. Coupled to the large degree of degeneracy of the Wnt signal pathway components seen between animal orthologs, may mean that searching the *Dictyostelium* genome will meet only limited success.

The exception is the Frizzled (Fz) proteins. An initial search of the genome carried out by Prabhu and Eichinger in 2006 suggested that there could be as many as 25 Frizzled-like receptors in the *Dictyostelium* encoded in the genome *(41)*. All genes have homology through their trans-membrane domain, however, only 16 contain putative cysteine-rich domain (CRD) present in the animal Fz protein family. These have been denoted as Frizzled/Smoothened-like (*fsl*) genes. Of these *fsl* genes two, *fslJ* and *fslK*, contain, in addition to the CRD domain, a KTXXXW sequence motif, which is required for activation of the canonical pathway in animals. The functions of these genes are not yet known.

8. Conclusions

Dictyostelium has many of the features associated with animal signaling systems and contains a number of key components of the canonical Wnt pathway. The study of these pathways in *Dictyostelium* is still in its infancy. To date, we have established that there are both similarities and differences between *Dictyostelium* and animal signal pathways. As *Dictyostelium* offers many advantages as a model system to study cell signaling, further study of the function, interactions and cell biology of GskA and Aar is likely to throw up new insights regarding how components of the Wnt signaling pathway operate in a non-metazoan context.

A key observation is that the Aar protein possesses both signaling and cytoskeletal functions, as seen in animals, whereas the signaling function is different from the canonical pathway, its association with adherens junctions appears conserved. This moves the question of which came first to more evolutionary distant groups than the amoebozoa. β-catenin-related proteins exist in other eukaryotic groups, and may be involved in signaling in organisms in which adherens junctions are not present *(42)*. For example, adherens junctions are not found in plants, but in *Arabidopsis* the β-catenin-related proteins Arabidilo-1 and -2 possess a signaling function in the control of lateral root hair development *(43)*.

Chapters 3 and **4** in **Volume 2** provide protocols for two of the key techniques used to investigate the functions of GskA and Aar. There is no doubt that they will prove useful for further study of this signaling system. For more general techniques, *see* the *Dictyostelium discoideum Protocols* volume in the Methods in Molecular Biology series *(44)*.

References

1. Baldauf, S. L., Doolittle, W. F. (1997) Origin and evolution of the slime molds (Mycetozoa) *Proc Natl Acad Sci USA* 94, 12007–12012.

2. Eichinger, L., et al. (2005) The genome of the social amoeba *Dictyostelium discoideum*. *Nature* 435, 43–57.

3. Chisholm, R. L. Gaudet, P., Just, E. M., Pilcher, K. E., Fey, P., Merchant, S. N., Kibbe, W. A. (2006) dictyBase, the model organism database for *Dictyostelium discoideum*. *Nucleic Acids Res* 34, D423–427.

4. Kuspa, A., Loomis, W. F. (1992) Tagging developmental genes in *Dictyostelium* by restriction enzyme-mediated integration of plasmid DNA. *Proc Natl Acad Sci USA* 89, 8803–8807.

5. Morio, T. et al. (1998) The *Dictyostelium* developmental cDNA project: generation and analysis of expressed sequence tags from the first-finger stage of development. *DNA Res* 5, 335–340.

6. Kessin, R. H. (2001) *Dictyostelium: Evolution, Cell Biology, and the Development of Multicellularity*. Cambridge University Press, Cambridge, UK.

7. Watts, D. J., Ashworth, J. M. (1970) Growth of myxameobae of the cellular slime mould *Dictyostelium discoideum* in axenic culture. *Biochem J* 119, 171–174.

8. Zimmerman, W., Weijer, C. J. (1993) Analysis of cell cycle progression during the development of *Dictyostelium* and its relationship to differentiation. *Dev Biol* 160, 178–185.

9. Thompson, C. R., Kay, R. R. (2000) Cell-fate choice in *Dictyostelium*: intrinsic biases modulate sensitivity to DIF signaling. *Dev Biol* 227, 56–64.

10. Iranfar, N., Fuller, D., Loomis, W. F. (2003) Genome-wide expression analyses of gene regulation during early development of *Dictyostelium discoideum*. *Eukaryot Cell* 2, 664–670.

11. Coates, J. C., Harwood, A. J. (2001) Cell-cell adhesion and signal transduction during *Dictyostelium* development. *J Cell Sci* 114, 4349–4358.

12. Town, C. D., Gross, J. D., Kay, R. R. (1976) Cell differentiation without morphogenesis in *Dictyostelium discoideum*. *Nature* 262, 717–719.

13. Kay, R. R. (1982) cAMP and spore differentiation in *Dictyostelium discoideum*. *Proc Natl Acad Sci USA* 79, 3228–3231.

14. Early, A. E., Gaskell, M. J., Traynor, D., Williams, J. G. (1993) Two distinct populations of prestalk cells within the tip of the migratory *Dictyostelium* slug with differing fates at culmination. *Development* 118, 353–362.

15. Jermyn, K., Traynor, D., Williams, J. (1996) The initiation of basal disc formation in *Dictyostelium discoideum* is an early event in culmination. *Development* 122, 753–760.

16. Williams, J. G., Duffy, K. T., Lane, D. P., McRobbie, S. J., Harwood, A. J., Traynor, D., Kay, R. R., Jermyn, K. A. (1989) Origins of the prestalk-prespore pattern in *Dictyostelium* development. *Cell* 59, 1157–1163.

17. Dingermann, T., Reindl, N., Werner, H., Hildebrandt, M., Nellen, W., Harwood, A., Williams, J., Nerke, K. (1989) Optimization and *in situ* detection of *Escherichia coli* β-galactosidase gene expression in *Dictyostelium discoideum*. *Gene* 85, 353–362.

18. Early, A., McRobbie, S. J., Duffy, K. T., Jermyn, K. A., Tilly, R., Ceccarelli, A., Williams, J. G. (1988) Structural and functional characterization of genes encoding *Dictyostelium* prestalk and prespore cell-specific proteins. *Dev Genet* 9, 383–402.

19. Morris, H. R., Taylor, G. W., Masento, M. S., Jermyn, K. A., Kay, R. R. (1987) Chemical structure of the morphogen differentiation inducing factor from *Dictyostelium discoideum*. *Nature* 328, 811–814.

20. Masento, M. S., Morris, H. R., Taylor, G. W., Johnson, S. J., Skapski, A. C., Kay, R. R. (1988) Differentiation-inducing factor from the slime mould *Dictyostelium discoideum* and its analogues. Synthesis, structure and biological activity. *Biochem J* 256, 23–28.

21. Thompson, C. R., Kay, R. R. (2000) The role of DIF-1 signaling in *Dictyostelium* development. *Mol Cell* 6, 1509–1514.

22. Richardson, D. L., Loomis, W. F., Kimmel, A. R. (1994) Progression of an inductive signal activates sporulation in *Dictyostelium discoideum*. *Development* 120, 2891–2900.

23. Ryves, W. J., Fryer, L., Dale, T., Harwood, A. J. (1998) An assay for glycogen synthase kinase 3 (GSK-3) for use in crude cell extracts. *Anal Biochem* 264, 124–127.

24. Ryves, W. J., Harwood, A. J. (2001) Lithium inhibits glycogen synthase kinase-3 by competition for magnesium. *Biochem Biophys Res Commun* 280, 720–725.

25. Goldberg, J. M. Manning, G., Liu, A., Fey, P., Pilcher, K. E., Xu, Y., Smith, J. L. (2006) The *Dictyostelium* kinome—analysis of the protein kinases from a simple model organism. *PLoS Genet* 2, e38.

26. Fraser, E., Young, N., Dajani, R., Franca-Koh, J., Ryves, J., Williams, R. S., Yeo, M., Webster, M. T., Richardson, C., Smalley, M. J., Pearl, L. H, Harwood, A., Dale, T. C. (2002) Identification of the Axin and Frat binding region of glycogen synthase kinase-3. *J Biol Chem* 277, 2176–2185.

27. Harwood, A. J., Plyte, S. E., Woodgett, J., Strutt, H., Kay, R. R. (1995) Glycogen synthase kinase 3 regulates cell fate in *Dictyostelium*. *Cell* 80, 139–148.

28. Strmecki, L., Bloomfield, G., Araki, T., Dalton, E., Skelton, J., Schilde, C., Harwood, A., Williams, J. G., Ivens, A., Pears, C. (2007) Proteomic and microarray analyses of the *Dictyostelium* Zak1-GSK-3 signaling pathway reveal a role in early development. *Eukaryot Cell* 6, 245–252.

29. Ginger, R. S., Dalton, E. C., Ryves, W. J., Fukuzawa, M., Williams, J. G., Harwood, A. J. (2000) Glycogen synthase kinase-3 enhances nuclear export of a *Dictyostelium* STAT protein. *EMBO J* 19, 5483–5491.

30. Schilde, C., Araki, T., Williams, H., Harwood, A., Williams, J. G. (2004) GSK3 is a multifunctional regulator of *Dictyostelium* development. *Development* 131, 4555–4565.

31. Berks, M., Kay, R. R. (1990) Combinatorial control of cell differentiation by cAMP and DIF-1 during development of *Dictyostelium discoideum*. *Development* 110, 977–984.

32. Plyte, S. E., O'Donovan, E., Woodgett, J. R., Harwood, A. J. (1999) Glycogen synthase kinase-3 (GSK-3) is regulated during *Dictyos-telium* development via the serpentine receptor cAR3. *Development* 126, 325–333.

33. Kim, L., Liu, J., Kimmel, A. R. (1999) The novel tyrosine kinase ZAK1 activates GSK3 to direct cell fate specification. *Cell* 99, 399–408.

34. Ginsburg, G. T., Kimmel, A. R. (1997) Autonomous and nonautonomous regulation of axis formation by antagonistic signaling via 7-span cAMP receptors and GSK3 in *Dictyostelium*. *Genes Dev* 11, 2112–2123.

35. Louis, J. M., Ginsburg, G. T., Kimmel, A. R. (1994) The cAMP receptor CAR4 regulates axial patterning and cellular differentiation during late development of *Dictyostelium*. *Genes Dev* 8, 2086–2096.

36. Kim, L., Harwood, A., Kimmel, A. R. (2002) Receptor-dependent and tyrosine phosphatase-mediated inhibition of GSK3 regulates cell fate choice. *Dev Cell* 3, 523–532.

37. Grimson, M. J., Coates, J. C., Reynolds, J. P., Shipman, M., Blanton, R. L., Harwood, A. J. (2000) Adherens junctions and β-catenin-mediated cell signaling in a non-metazoan organism. *Nature* 408, 727–731.

38. Thorpe, C. J., Schlesinger, A., Carter, J. C., Bowerman, B. (1997) Wnt signaling polarizes an early *C. elegans* blastomere to distinguish endoderm from mesoderm. *Cell* 90, 695–705.

39. Lo, M. C., Gay, F., Odom, R., Shi, Y., Lin, R. (2004) Phosphorylation by the beta-catenin/MAPK complex promotes 14-3-3-mediated nuclear export of TCF/POP-1 in signal-responsive cells in *C. elegans*. *Cell* 117, 95–106.

40. Coates, J. C., Grimson, M. J., Williams, R. S., Bergman, W., Blanton, R. L., Harwood, A. J. (2002) Loss of the β-catenin homolog aardvark causes ectopic stalk formation in *Dictyostelium*. *Mech Dev* 116, 117–127.

41. Prabhu, Y., Eichinger, L. (2006) The *Dictyostelium* repertoire of seven transmembrane domain receptors. *Eur J Cell Biol*.

42. Coates, J. C. (2003) Armadillo repeat proteins: beyond the animal kingdom. *Trends Cell Biol* 13, 463–471.

43. Coates, J. C., Laplaze, L., Haseloff, J. (2006) Armadillo-related proteins promote lateral root development in *Arabidopsis*. *Proc Natl Acad Sci USA* 103, 1621–1626.

44. Eichinger, L., Rivero, F. (2006) *Dictyostelium discoideum Protocols*. Humana Press, Totowa, New Jersey.

Chapter 3

Monitoring Patterns of Gene Expression in *Dictyostelium* by β-galacotsidase Staining

Adrian J. Harwood

Abstract

Monitoring the spatial distribution of prespore and pstB cell types is a sensitive method to monitor GSK-3 and Aar activity during *Dictyostelium* development. Cell-specific expression of *lacZ* marker genes can be readily detected using enzymatic cleavage of the substrate X-gal. This protocol describes a simple method for β-galactosidase staining in developed *Dictyostelium* structures.

Key words: *Dictyostelium*, β-galacotsidase staining, multicellular development, ecmB, pspA.

---

1. Introduction

There are many methods for detection of altered gene expression in *Dictyostelium*. Most are quantitative, such as Northern and Western analysis, qRT-PCR and microarray studies, but involve destruction of cells and multicellular structures, followed by extraction of RNA or protein. Although these methods give a good temporal resolution, they loose information on the spatial distribution of gene expression. One of the important aspects of the GSK-3 and Aar mediated signaling in *Dictyostelium* is that it leads to major changes in cell patterning of the prespore gene *pspA* and the prestalk gene *ecmB* (1). Although changes in expression of these genes can be detected by Northern analysis in some mutants, only small overall changes occur in others, such as in mutants lacking the cAMP receptor cAR3 (2). In contrast, changes in the spatial pattern of gene expression provide a sensitive method to detect altered *gskA* and *aar* signaling.

Elizabeth Vincan (ed.), *Wnt Signaling, Volume II: Pathway Models, vol. 469*
© 2008 Humana Press, a part of Springer Science+Business Media, New York, NY
Book doi: 10.1007/978-1-60327-469-2

Spatial gene expression patterns in *Dictyostelium* were first detected by using cell-specific promoters to express a fusion protein containing a fragment of the SV40 Large T antigen to which there are high affinities *(3)*. This marker was quickly superseded by expression of β-galactosidase enzyme from the *Escherichia coli* lacZ gene *(4)*. β-galactosidase staining is robust, easy to perform, and the stained structures can be stored. Furthermore, although the basic substrate, X-gal (5-bromo-y-chloro-3-indolyl galactoside), gives a blue color, other substrates with different chromophores are available. This means that it is possible to combine *lacZ* with a second transgene that expesses β-glucoronidase and carry out double staining using X-Gluc (5-bromo-4-chloro-3-indolyl-glucuronide) and alternatives to X-gal *(5)*. The use of green fluorescent protein (GFP) has to some extent replaced the *lacZ* markers, as they can be used in living cells and give information of the dynamics of morphogenesis and gene expression *(6)*. However, GFP markers require fluorescence microscopy and a sensitive CCD camera. An electronically controlled filter wheel must be fitted if multiple colors are to be used. As samples are usually not fixed, results cannot be stored for later viewing. Consequently, there are still many reasons for using β-galactosidase staining to examine Wnt-like signaling in *Dictyostelium*. The following describes the basic method for staining multicellular *Dictyostelium* structures.

2. Materials

1. *Dictyostelium* cell strains: PspA-lacZ- and ecmB-lacZ-expressing strains are available from the *Dictyostelium* Stock Centre (http://dictybase.org/StockCenter/StockCenter.html). Alternatively, the plasmids are also available and can be used to transform any strain of choice *(7)*.

2. AX medium + G418: Dissolve 14 g of proteose peptone (Oxoid), 7 g of yeast extract (Oxoid), 13.5 g of glucose, 0.5 g of Na_2HPO_4, 0.5 g of KH_2PO_4, 4 µL of folic acid (50 mg/mL), and 4 µL of vitamin B12 (1 mg/mL) to 1 L water. Adjust pH with HCl to pH 6.4 and autoclave to sterilize (the resulting pH should be 6.7). Store at 4°C until use, then warm to room temperature. Add G418 (Invitrogen) to a concentration of 20 µg/mL and streptomycin to 100 µg/mL.

3. KK2: 15.5 mM KH_2PO_4, 3.8 mM K_2HPO_4, pH6.2.

4. KK2 agar: 1.8% (w/v) Bacto-agar (BD Biosciences) in KK2.

5. Z buffer: KK2 supplemented with 1 mM $MgCl_2$.

6. Fix solution: 1% (v/v) gluteraldehyde (electron microscope grade, Sigma) in Z buffer.

7. X-gal stock: 1% X-gal (Sigma) in Dimethyl formamide (DMF).

8. Staining solution: 0.1% X-gal, 2.5 mM $K_4Fe(CN)_6$, 2.5 mM $K_3Fe(CN)_6$ in Z buffer.

3. Methods

1. Grow *Dictyostelum* cells in 50–100 mL of AX medium + G418 in a 500 mL flask to a density of 10^6 cells/mL.

2. Pellet 5×10^7 cells (50 mL of cell culture) by centrifugation at 500 xg in a 50-mL Falcon conical tube (BD Biosciences) for each developmental plate to be set up. Resuspend each pellet in 50 mL of KK2 and re-pellet. Repeat wash and then resuspend cells in 0.5 mL KK2.

3. Plate the cell suspension on a 9-cm KK2 agar plate. Allow to dry until plated cells are just moist. Incubate at 22°C with the plate right-side up.

4. Monitor cell development at 10–20× magnification under a dissecting microscope. At appropriate times (~14 hours for mounds, 16 hours for first fingers and slugs, 20 hours for preculminants), use a scalpel to remove a block of agar containing structures of appropriate developmental stage (*see* **Note 1**).

5. Place the agar block into enough Fix solution (~1 mL) to immerse and fix the structures. These should float off the agar, which can then be removed. Fix for 5–10 minutes (*see* **Notes 2** and **3**).

6. Carefully remove the Fix solution with a micropipette and wash the sample twice in the Z-buffer. Aspirate off all the liquid between washes to remove all traces of gluteraldehyde.

7. Immerse samples completely in staining solution and incubate at 37°C for 1 hour (*see* **Note 4**). Monitor progress of staining by eye or under a dissecting microscope.

8. Staining can be stopped by replacing, or diluting, the staining solution with Z buffer (*see* **Note 5**). Samples can be photographed *in situ* or mounted on a microscope slide for imaging at higher magnification.

This is the most common form of detection of *lacZ* expression. Alternative methods are possible (*see* **Notes 6** and **7**).

4. Notes

1. Samples can be prepared in many different ways. Isolated cells can be plated on glass slides or Petri dishes and fixed *in situ*. Cells can also be developed on 45 µm nitrocellulose filters (Millipore cat no. HAWP047S0) placed on two layers of Whatman 3MM Chr paper and soaked in KK2. Cells and structures on nitrocellulose can be fixed on the filter by placing it sequentially onto each solution, as the solutions can pass through the filter. It is useful to use white filters for staining as it is easier to see the blue stain.

2. Over fixation can produce staining artefacts by reducing penetration of the staining solution.

3. Do not let the samples dry out before end of process. The enzyme will tolerate almost anything, except the sample drying out before staining.

4. For weak staining, samples can be incubated for up to 24 hours.

5. The stained samples are stable even in the absence of staining solution. Samples can be stored at 4°C. If they dry out at this stage, they may be re-hydrated in distilled water.

6. Anti-β-galactosidase antibodies are available from many sources and can be used in conventional immuno-histo-chemical staining. This allows *lacZ* expression to be co-localized with other cell-specific antigens.

7. β-galactosidase enzyme activity can be detected in cell extracts as followed. 10^7 cells are lysed in Lysis buffer (100 mM Na_2PO_4, 8 mM $MgCl_2$, 1 mM EDTA, 1% Triton X-100, pH 7.8) and cleared by centrifugation at 12,000 xg. 100 µL of extract is incubated with 500 µL of o-nitrophenyl-β-D-galactopyranoside (ONPG: 1.6 mg/mL in Z buffer) for 30 minutes or longer at 22°C. Reactions are stopped with 400 µL of Na_2CO_3, cleared by centrifugation at 12000 xg and diluted with 500 µL of Z buffer. The amount of product is read by spectrophotometry at 420 nm.

References

1. Harwood, A. J., Plyte, S. E., Woodgett, J., Strutt, H., Kay, R. R. (1995) Glycogen synthase kinase 3 regulates cell fate in *Dictyostelium. Cell* 80, 139–148.

2. Plyte, S. E., O'Donovan, E., Woodgett, J. R., Harwood, A. J. (1999) Glycogen synthase kinase-3 (GSK-3) is regulated during *Dictyostelium* development via the serpentine receptor cAR3. *Development* 126, 325–333.

3. Jermyn, K. A., Duffy, K. T. & Williams, J. G. (1989) A new anatomy of the prestalk zone in *Dictyostelium. Nature* 340, 144–146.

4. Dingermann, T. *et al.* (1989) Optimization and *in situ* detection of *Escherichia coli*

β-galactosidase gene expression in *Dictyostelium discoideum. Gene* 85, 353–362.

5. Early, A. E., Gaskell, M. J., Traynor, D., Williams, J. G. (1993) Two distinct populations of prestalk cells within the tip of the migratory *Dictyostelium* slug with differ-ing fates at culmination. *Development* 118, 353–362.

6. Prasher, D. C. (1995) Using GFP to see the light. *Trends Genet* 11, 320–323.

7. Eichinger, L., Rivero, F. (2006) *Dictyostelium discoideum Protocols*. Humana Press, Totowa, NJ.

Chapter 4

Use of the *Dictyostelium* Stalk Cell Assay to Monitor GSK-3 Regulation

Adrian J. Harwood

Abstract

GSK-3 activity mediates cAMP repression of stalk induction of cells in low-density monolayer culture. The lower the GSK-3 activity the greater the percentage of stalk cells formed. This protocol describes a robust and quantitative method utilizing an adapted stalk cell monolayer assay to measure GSK-3 activation.

Key words: *Dictyostelium*, stalk cell assay, DIF–1, cAMP repression, GSK-3.

1. Introduction

A feature of *Dictyostelium* development is that, although it normally takes place in the form of multicellular structures of greater than 10^5 cells, terminal differentiation can be induced in isolated cells in low-density monolayer culture. Spore cells can be induced by activation of cAMP-dependent protein kinase A (PKA) either after treatment with the PKA activator 8-bromo-cAMP in mutants that activate the kinase or by incubation with an extracellular spore inducing peptide, SDF-2 *(1–3)*. Initially, induction of stalk cells proved to be trickier, as it requires two processes. First, cells must be induced with extracellular cAMP so that they progress to a stage equivalent of those in the mound. At this point they become sensitive to the stalk cell inducer DIF-1. This is a dichloronated phenyl hexanone, which is a potent stalk cell inducer working in the 1-100 nM range *(4, 5)*. It was discovered however that cAMP represses stalk cell formation, and so must be removed before DIF-1 addition *(6)*. This makes conventional stalk cell induction a 2-day process.

Elizabeth Vincan (ed.), *Wnt Signaling, Volume II: Pathway Models, vol. 469*
© 2008 Humana Press, a part of Springer Science+Business Media, New York, NY
Book doi: 10.1007/978-1-60327-469-2

GSK-3 mediates the cAMP repression pathway and, unlike wild type cells, *gskA* mutant cells will form stalk cells in the presence of cAMP *(7)*. This can form the basis of an assay to assess potential mutants or inhibitors that reduce GSK-3 activity. The assay can also be reversed so that hypersensitivity to cAMP can be used to investigate activating mutations in the GSK-3 regulation pathway *(8)*. The following protocol, describes a variation of the original stalk cell induction assay, which makes use of the loss of cAMP repression in the absence of GskA to compress the basic assay into a single 24-hour induction.

2. Materials

1. *Dictyostelium* cells: An appropriate source of *Dictyostelium* cell strains, either wild type or mutant. Available from the *Dictyostelium* Stock Centre (http//dictybase.org/StockCenter/StockCenter.html; ref. *9*).

2. AX medium: Dissolve 14 g of proteose peptone (Oxoid), 7 g of yeast extract (Oxoid), 13.5 g of glucose, 0.5 g of Na_2HPO_4, 0.5 g of KH_2PO_4, 4 µL of folic acid (50 mg/mL), and 4 µL of vitamin B12 (1 mg/mL) to 1 L water. Adjust pH with HCl to pH 6.4 and autoclave to sterilize (the resulting pH should be 6.7). Store at 4 °C until use, then warm to room temperature. Add streptomycin to a concentration of 100 µg/mL.

3. 100x Stalk salts: 200 mM NaCl, 1 M KCl, 100 mM $CaCl_2$ in distilled water. Store at room temperature.

4. 1 M MES, pH6.3. Store at –20 °C.

5. Stalk medium: For 100 mL, mix 1 mL of 100x Stalk salts, 1 mL of 1 M MES, pH 6.3 with 98 mL of distilled water. Add streptomycin to 100 µg/mL.

6. cAMP stock (× 40): 200 mM cAMP (adenosine 3′:5′-cyclic monophosphate (Sigma) in 1x Stalk medium.

7. DIF-1 stock (× 1000): 100 mM DIF-1 (1-(3,5-Dichloro-2,6-dihydroxy-4-methoxyphenyl) hexan-1-one; BIOMOL International L.P) in 100% ethanol. Store at –20°C.

3. Methods

1. Grow *Dictyostelum* cells in 10 mL or more of AX medium to a density of 10^6 cells/mL (*see* **Note 1**).

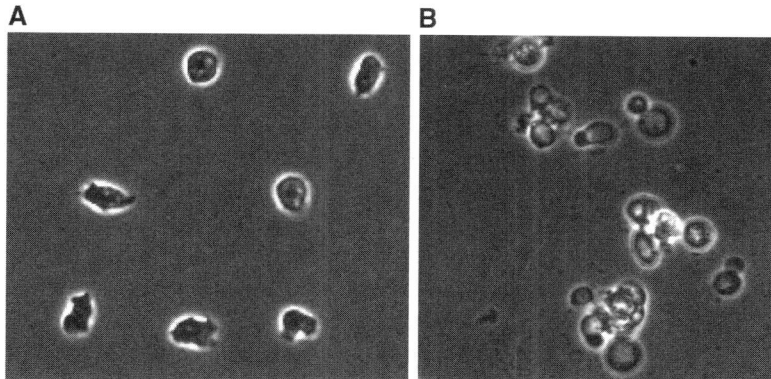

Fig. 4.1. *Dictyostelium* stalk cells formed in low density monolayer culture. (**A**) Undifferentiated cells. (**B**) Stalk cells.

2. Pellet 10^6 cells (1 mL cell culture) by centrifugation at 500 xg in a 1.5-mL centrifuge tube and wash cells twice in 1 mL stalk medium. Resuspend cell pellet in 10 mL of Stalk medium (to give a final cell density of 10^5 cells/mL) in a 50-mL conical Falcon tube (BD Biosciences).

3. Plate 1 mL of cell suspension into 3-cm tissue culture plates (BD Biosciences), and ensure cells dispersed evenly (*see* **Note 2**). A good experimental format is to plate nine monolayers to give three test conditions in triplicate (*see* **Note 3**).

4. Add 1 μL DIF-1 stock (a final concentration of 100 nM) to three monolayer plates; a mixture of 1 μL DIF-1 and 25 μL cAMP stock (a final concentration of 5 mM) to another three plates and leave the remaining cells untreated. Incubate the cells for 24 hours at 22 °C in a sealed and humidified container (*see* **Note 4**).

5. Stalk cells should only form in the presence of DIF and cAMP when GSK-3 activity is inhibited or lost (*see* **Note 5** and **Fig.4.1**). These can be quantified by counting the percentage of cells that have formed into stalk cells; this should be greater than 80% for a *gskA* null mutant strain (*see* **Notes 6** and **7**).

4. Notes

1. Cells can be grown in association with bacteria. To do this, plate a 300-μL saturated culture of *Klebsiella pneumoniae* on 25 mL SM-agar (1% glucose, 1% peptone [Oxoid L34], 0.1% yeast extract [Oxoid L21], 2% Bacto-agar [BD Biosciences], 4 mM $MgSO_4$, 4 mM KH_2PO_4, 6 mM K_2HPO_4) in a bacterial grade 9-cm Petri dish. Spot cells in the center

of the bacterial lawn and grow at 22°C for 4-5 days. This should form a plaque in the center of the plate, comprising of developed cells in the center and a growth zone of cells feeding on bacteria at the edge. Cells can be scraped from the growth zone using a 5-mm loop and dispersed in 1 mL Stalk medium (this is approximately 10^6 cells). Wash the cells twice, or until no visible bacteria, count cells and resuspend at 10^6 cells/mL.

2. This would produce enough for 10 monolayers; the process can easily be scaled up to larger formats.

3. This includes capacity for controls lacking the inducing factors, cAMP and DIF-1. These are important to ensure that developed cells have not entered the assay leading to false-positives; a particular risk for an inexperienced experimenter using cells from a bacterial source.

4. Monolayers can be left for longer than 24 hours, although care must be taken to ensure that the cAMP concentration does not drop, due to phosphodiester activity, to a low enough level that cAMP repression is lost.

5. Stalk cells contain one large vacuole, greater than 50% of the cell, and a distinct cell wall (*see* **Fig.4.1**).

6. It is our experience that the percentage of stalk cell formed is a good reflection of the cellular GSK-3 activity.

7. To test for cAMP hypersensitivity, carry out **steps 3.1–3.4** but only plate cells in 5 mM cAMP. Leave for 24 hours and then aspirate off medium. Wash twice in stalk medium and then add 1 mL Stalk medium containing 100 nM DIF-1. Add decreasing amounts of adenosine-3′, 5′-cyclic mono-phosphorothioate, Rp-isomer (Rp-cAMPS; available from Biolog, Germany) and incubate for a further 24 hours. Quantify stalk cell induction as in **step 3.5**. Rp-cAMPS is a non-hydrolysable analog of cAMP and is used to prevent cAMP degradation.

References

1. Kay, R. R. (1989) Evidence that elevated intra-cellular cyclic AMP triggers spore maturation in *Dictyostelium*. *Development* 105, 753–759.

2. Simon, M. N., Pelegrini, O., Veron, M., Kay, R. R. (1992) Mutation of protein kinase A causes heterochronic development of Dictyostelium. *Nature* 356, 171–172.

3. Anjard, C., Loomis, W. F. (2005) Peptide signaling during terminal differentiation of *Dictyostelium*. *Proc Natl Acad Sci USA* 102, 7607–7611.

4. Morris, H. R., Taylor, G. W., Masento, M. S., Jermyn, K. A., Kay, R. R. (1987) Chemical structure of the morphogen differentiation inducing factor from *Dictyostelium discoideum*. *Nature* 328, 811–814.

5. Masento, M. S., Morris, H. R., Taylor, G. W., Johnson, S. J., Skapski, A. C., Kay, R. R. (1988) Differentiation-inducing factor from the slime mould *Dictyostelium discoideum* and its analogues. Synthesis, structure and biological activity. *Biochem J* 256, 23–28.

6. Berks, M., Kay, R. R. (1988) Cyclic AMP is an inhibitor of stalk cell differentiation in *Dictyostelium discoideum*. *Dev Biol* 126, 108–114.

7. Harwood, A. J., Plyte, S. E., Woodgett, J., Strutt, H., Kay, R. R. (1995) Glycogen synthase kinase 3 regulates cell fate in *Dictyostelium*. *Cell* 80, 139–148.

8. Kim, L., Harwood, A., Kimmel, A. R. (2002) Receptor-dependent and tyrosine phosphatase-mediated inhibition of GSK3 regulates cell fate choice. *Dev Cell* 3, 523–532.

9. Eichinger, L., Rivero, F. (2006) *Dictyostelium discoideum Protocols*. Humana Press, Totowa, NJ.

Part III

Cnidarians

Chapter 5

Wnt Signaling in Cnidarians

Thomas W. Holstein

Abstract

Cnidarians are an ancient group of animals at the base of metazoan evolution. They exhibit a simple body plan with only one well-defined body axis and a small number of cell types. Cnidarians are also well known for their enormous regeneration capacity. Recent work in the freshwater polyp *Hydra* and in the sea anemone *Nematostella* has identified an unexpectedly high level of genetic complexity of *wnt* genes. Canonical Wnt signaling acts in pattern formation and regeneration of *Hydra* and also in gastrulation and early embryogenesis of *Nematostella*. Vertebrate-specific Wnt-antagonists were also identified from cnidarians and exhibit similar conserved functions. The simple cnidarian body plan and the now available genomes from *Hydra* and *Nematostella*, together with new functional approaches, make these animals an attractive model for studying the basic functions of canonical and non-canonical Wnt signaling.

Key words: Wnt signaling, regeneration, axis formation, Hydra, Nematostella, Cnidari.

1. Cnidarians: An Old and Newly Emerging Model System

The Cnidaria are diploblastic, aquatic animals with a simple body plan that is reminiscent to a bilaterian gastrula. The two body layers, an outer ectoderm and inner endoderm are separated by an acellular mesogloea. The fossil records reveal that cnidarians are more than 500 Myr old and molecular phylogenies identified the Cnidaria as a sister-group to the Bilateria *(1, 2)*.

Today there are two major cnidarian genetic models: The first is the freshwater polyp *Hydra*, which is well known for a long time for its high regeneration capacity, and which has been studied in detail on the cellular level *(3)*. *Hydra* has three independent cell lines: ectodermal and endodermal epitheliomuscular cells and interstitial stem cells that give rise to nerve cells (including nematocytes), gland

Elizabeth Vincan (ed.), *Wnt Signaling, Volume II: Pathway Models, vol. 469*
© 2008 Humana Press, a part of Springer Science+Business Media, New York, NY
Book doi: 10.1007/978-1-60327-469-2

cells, and gametes *(3, 4)*. Classical transplantation experiments have revealed gradients of morphogenetic signals that govern the patterning of the oral-aboral body axis *(5–9)*. During regeneration, these morphogenetic signals can be activated or enhanced at the site of wounding so that a new head or foot forms within 36 hours *(10, 11)*. *Hydra* can be also dissociated into a suspension of single cells, and regenerate a new organism after re-aggregation and *de novo* pattern formation *(3, 12, 13)*.

The other model system is the starlet sea anemone *Nematostella vectensis*, which was introduced by the pioneering work of Cadet Hand *(14, 15)*. The great advantage of *Nematostella* is the availability of gametes and embryos that can be harvested in the lab at any time *(16)*. *Nematostella* embryos form a coeloblastula at the stage of 32–64 cells. Afterward, blastulae start to invaginate and form a gastrula, which develops into a planula larva that metamorphosis into a primary polyp *(17–19)*.

The genomes of both cnidarian models have been sequenced. The *Nematostella* genome project revealed that the cnidarian genome is complex, with a gene repertoire, exon-intron structure, and large-scale gene linkage more similar to vertebrates than to flies or nematodes, implying that the genome of the eumetazoan ancestor was similarly complex *(20–35)*. We therefore presume that the genetic repertoire responsible for the formation of specific bilaterian body plan already existed in the common ancestor of cnidarians and bilaterians *(20, 36)*.

In both model systems molecular and genomic tools have been developed to study both organisms on a functional level *(36)*. Besides ISH/IHC protocols and the full set of the genomic tools (BAC, EST, and cDNA libraries), RNAi, morpholinos, and transient and stable transfection are available or under optimization procedure. The transgenic approach is successfully working in all hydra labs and was introduced by the Bosch lab *(37)*. Stable transgenic *Hydra* can be produced by injecting GFP constructs driven by a *Hydra actin* gene promoter (or other promoters) into blastomeres of two- to eight-cell *Hydra* embryos. Thus, due to their morphological simplicity and remarkable regeneration capacity, cnidarians can serve as an important model to understand basic functions of Wnt signaling in gastrulation, axis formation, germlayer specification, regeneration, and cell differentiation *(36)*.

2. *Cnidarian* Wnt Genes

Wnt genes are highly specific for metazoans and, so far, no *wnt* genes have been described from any unicellular eukaryotes, neither from cellular slime molds (*Dictyostelium discoideum*) nor from choanoflagellate Protozoa that are closely related to Metazoa *(36, 38)*. In cnidarians, the canonical Wnt/β-catenin pathway was first identified in *Hydra (39–41)*. Analysis of *wnt* genes in *Nematostella* revealed that all bilaterian *wnt* gene subfamilies, except *wnt 9*, are present in cnidarians *(42–44)*. Recent work from *Hydra* also indicates that most of the *wnt* gene subfamilies found in *Nematostella* can be found in *Hydra* (refs. *36* and *44* and unpublished work from the Bode, Hobmayer, and Holstein labs). A clustering of the Wnt1/6/10/9/3 subfamilies in the phylogenetic analyses that was supported by the syntenic wnt organization between *Drosophila* and vertebrates *(36, 42)* suggested an ancestral cluster of wnts in the evolution of the common ancestor of cnidarians and bilaterians. Recent analysis of the genomic organization revealed that the *wnt6–wnt10* cluster observed in *Drosophila melanogaster* is conserved in the cnidarian lineage *(45)*. Also a novel cluster comprised of *wnt5–wnt7/wnt7b* was observed in *Nematostella (45)*. Despite the intriguing expression patterns of *wnt* genes along the oral aboral axis (*see* **Section 6**); *wnt* genes do not exhibit a Hox-like co-linearity *(45)*. Future work must clarify to what extent wnt genes are expressed under the influence of common enhancers or coordinately regulated by higher-order chromatin organization *(36, 45–47)*.

3. Wnt Receptors

In addition to the putative Wnt receptor *frz7 (41)*, at least six Frizzled (Fzd) receptor genes have been identified in *Nematostella (36, 43)*. Therefore, the general picture is emerging that the radiation of Wnt genes was accompanied by a diversification of wnt receptors. However, it is less clear at present to what extent Wnt co-receptors like LRP5/6 do exist in cnidarians *(36, 44)*.

4. Intracellular Components of Canonical and Non-canonical Wnt-Signaling

Previous cloning work and genomic data from *Nematostella* and *Hydra* databases revealed orthologs of APC, Axin, CamKII, β-catenin, casein kinase 1α, 1δ, 1γ_2, 1γ_3 and 1ε, Dsh, Flamingo, GSK-3β, JNK, PKC, Tcf/Lef, Van Gogh, and Wntless (refs. *21, 39, 40, 43, 48, 49*) and Lengfeld and Holstein, unpublished). These data indicate that, in addition to the canonical wnt pathway, non-canonical wnt pathways, e.g., the Wnt/PCP and the Wnt/Ca^{2+} pathway known from bilaterian animals *(50, 51)* are also present in cnidarians.

5. Wnt Antagonists

Secreted extracellular factors can exhibit a strong antagonizing or modulating effect on Wnts *(52–55)*. Wnt antagonists like the family of secreted Frizzled-related proteins (sFRP), the Wnt inhibitory factor (WIF), Cerberus, and the Dickkopf (Dkk) proteins have been now identified from *Hydra* and *Nematostella (21, 36, 43, 44)*, some of them (Dkk and Cerberus) are absent from the *Drosophila* and *Caenorhabditis elegans* genome databases. In *Nematostella* eight sFRPs exist, and in *Hydra* also several sFRPs have been identified (Hobmayer, unpublished). From *Hydra* and *Nematostella*, two *dickkopf* orthologs have been cloned. One is an ortholog to *dkk-3*, another an ortholog to the *dkk1/2/4* subfamily found in vertebrates *(44, 56)*. In *Hydra*, *dkk-3* is involved in neuronal differentiation *(56)*. *dkk1/2/4* is proposed to be a putative precursors to the *dkk1*, *dkk2*, and *dkk4* subfamilies found in vertebrates *(44, 57)*. Expression analysis in *Hydra* and *Nematostella* reveals that *dkk1/2/4* is expressed opposite to the *wnt* expression domains *(43, 44)* suggesting a Wnt antagonizing function. Overexpression studies of *hydkk1/2/4A* in *Xenopus* embryos indicate that it is a functional ortholog of vertebrate *Dkk1* and *Dkk4* that it is capable of inhibiting Wnt signaling *(36, 44)*. An activation of canonical wnt signaling by alsterpaullone treatment causes the complete, but reversible down-regulation of *hydkk1/2/4*; this shows that *hydkk1/2/4* is negatively regulated by Wnt signaling *(36, 44)*.

6. Function of Wnt Signaling in Cnidarians

The function of cnidarian Wnt signaling has been recently reviewed in detail by Guder et al. and by Lee et al. in two articles emphasizing different expects of wnt function in cnidarians *(36, 43)*. Therefore only a short summary is given here.

One major function of cnidarian Wnt signaling is in axis formation and gastrulation. This is obvious from the striking localization of nearly all *wnt* genes at the site of the blastopore during and after gastrulation of the *Nematostella* embryo. The expression domains for *Nematostella wnt* genes reveal a characteristic pattern in staggered domains that span the oral-aboral axis except for the aboral pole itself. These distinct regional and germ layer-specific expression patterns of the *wnts* suggest that the ancestral role of these genes was in specifying position along the main body axis *(42)*. Although an analysis of the genomic organization of *wnt* genes in Nematostella indicates only a limited syntenic organization of *wnt* genes, it is tempting to speculate that there is a regulative hierarchy of *Nematostella wnt* gene expression during *Nematostella* embryogenesis (refs. *36, 45–47*; *see* **Section 2**).

Functional data from *Nematostella* support a role for canonical Wnt signaling in the gastrulation process. β-Catenin becomes translocated into nuclei in cells at the site of the blastopore during gastrulation *(58)* and Lithium chloride treatment, which blocks GSK-3β-mediated degradation of β-catenin, results in extended gastrulation movements. Lee and co-workers made an important step forward by showing that NvDsh is necessary for the localization of nuclear β-catenin *(19)*. The mechanism of Dsh enrichment at the site of gastrulation is unclear so far, but might be explained by a degradation of Dsh at the opposite pole and/or binding to the cytoskeleton *(19)*. The fact that cnidarians can start gastrulation at the animal pole was surprising because it is different from deuterostomes, which gastrulate at the vegetal pole. These findings stress the importance of understanding the Dsh– β-catenin interaction with the cytoskeleton at the site of gastrulation *(19, 59, 60)*.

The blastoporal signaling center in cnidarian embryos is reminiscent of the other signaling center (i.e., the *Hydra* head organizer), which is located at the apical tip of the polyp's hypostome. Actually, the hypostome of a cnidarian polyp corresponds to the blastopore of a cnidarian gastrula *(18, 19, 60)*. When transplanted to an ectopic site, the head organizer induces a secondary body axis by recruiting host tissue *(5, 61–63)*. *Wnt* genes and β-catenin-dependent signaling play a major role in head induction *(39, 49, 64)*. Treatment with the GSK-3β inhibitor, alsterpaullone, increases the accumulation of β-catenin in nuclei and can stimulate head formation *(63, see also* refs. *65* and *66)*. Alsterpaullone treatment also causes an expansion of the normal expression domains of *hywnt, hytcf,* and *hybra1,* the *Hydra brachyury* ortholog, and induces numerous spots of *hywnt3a* expression seen along the body column Broun *(36, 44, 63)*.

Wnt/β-catenin signaling acts also during Hydra head regeneration *(39)* and during *de novo* pattern formation in re-aggregates *(13)*. The characteristic expression of various *wnt* genes and pharmacological intervention with β-catenin and JNK inhibitors indicate a hierarchy of Wnt signaling in head regeneration (unpublished data).

In summary, one can conclude that the similar activation of overlapping *wnt* expression domains during gastrulation, early embryogenesis, and regeneration indicates various Wnt pathways (i.e., the Wnt/PCP, the Wnt/Ca^{2+}, and the canonical Wnt pathway) act concomitantly during gastrulation, axis formation and regeneration. We are far from having a mechanistic and complete view how these various wnt pathways may interact, and which downstream target genes are activated to induce cell differentiation, (e.g., neuronal differentiation; refs. *67–69*). However, the genomic and molecular tools are on hand to unravel these fundamental processes.

Acknowledgments

This research was supported by funding from the German Research Foundation (Deutsche Forschungsgemeinschaft, DFG).

References

1. Chen, J. Y., Oliveri, P., Gao, F., et al. (2002) Precambrian animal life: probable developmental and adult cnidarian forms from Southwest China. *Dev Biol* 248, 182–196.

2. Conway Morris S. (2000) The Cambrian "explosion": slow-fuse or megatonnage? *Proc Natl Acad Sci USA* 97, 4426–4429.

3. Steele, R. E. (2006) Trembley's polyps go transgenic. *Proc Natl Acad Sci USA* 103, 6415–6416.

4. Steele, R. E. (2002) Developmental signaling in Hydra: what does it take to build a "simple" animal? *Dev Biol* 248, 199–219.

5. MacWilliams, H. K. (1983) Hydra transplantation phenomena and the mechanism of Hydra head regeneration. II. Properties of the head activation. *Dev Biol* 96, 239–257.

6. MacWilliams, H. K. (1983) Hydra transplantation phenomena and the mechanism of hydra head regeneration. I. Properties of the head inhibition. *Dev Biol* 96, 217–238.

7. Bode, M. P., Bode, H. R. (1984) Formation of pattern in regenerating tissue pieces of Hydra attenuata. III. The shaping of the body column. *Dev Biol* 106, 315–325.

8. Bode, P. M., Bode, H. R. (1980) Formation of pattern in regenerating tissue pieces of hydra attenuata. I. Head-body proportion regulation. *Dev Biol* 78, 484–496.

9. Meinhardt H. (2002) The radial-symmetric hydra and the evolution of the bilateral body plan: an old body became a young brain. *Bioessays* 24, 185–191.

10. Holstein, T. W., Hobmayer, E., Technau, U. (2003) Cnidarians: an evolutionarily conserved model system for regeneration? *Dev Dyn* 226, 257–267.

11. Bode, H. R. (2003) Head regeneration in Hydra. *Dev Dyn* 226, 225–236.

12. Gierer, A., Berking, S., Bode, H., et al. (1972) Regeneration of hydra from reaggregated cells. *Nat New Biol* 239, 98–101.

13. Technau, U., Cramer von Laue, C., Rentzsch, F., et al. (2000) Parameters of self-organization in Hydra aggregates. *Proc Natl Acad Sci USA* 97, 12127–12131.

14. Darling, J. A., Reitzel, A. R., Burton, P. M., et al. (2005) Rising starlet: the starlet sea anemone, Nematostella vectensis. *Bioessays* 27, 211–221.

15. Holland, P. (2004) Developmental biology. The ups and downs of a sea anemone. *Science* 304, 1255–1256.

16. Fritzenwanker, J. H., Technau, U. (2002) Induction of gametogenesis in the basal cnidarian Nematostella vectensis (Anthozoa). *Dev Genes Evol* 212, 99–103.

17. Kraus, Y., Technau, U. (2006) Gastrulation in the sea anemone Nematostella vectensis occurs by invagination and immigration: an ultrastructural study. *Dev Genes Evol* 216, 119–132.

18. Fritzenwanker, J. H., Genikhovich, G., Kraus, Y., et al. (2007) Early development and axis specification in the sea anemone Nematostella vectensis. *Dev Biol* 310, 264–279.

19. Lee, P. N., Kumburegama, S., Marlow, H. Q., et al. (2007) Asymmetric developmental potential along the animal-vegetal axis in the anthozoan cnidarian, Nematostella vectensis, is mediated by Dishevelled. *Dev Biol* 310, 169–186.

20. Putnam, N. H., Srivastava, M., Hellsten, U., et al. (2007) Sea anemone genome reveals ancestral eumetazoan gene repertoire and genomic organization. *Science* 317, 86–94.

21. Technau, U., Rudd, S., Maxwell, P., et al. (2005) Maintenance of ancestral complexity

and non-metazoan genes in two basal cnidarians. *Trends Genet* 21, 633–639.

22. Miller, D. J., Ball, E. E., Technau, U. (2005) Cnidarians and ancestral genetic complexity in the animal kingdom. *Trends Genet* 21, 536–539.

23. Martindale, M. Q., Pan, K., Finnerty, J. R. (2004) Investigating the origins of triploblasty: 'mesodermal' gene expression in a diploblastic animal, the sea anemone Nematostella vectensis (phylum, Cnidaria; class, Anthozoa). *Development* 131, 2463–2474.

24. Finnerty, J. R., Pang, K., Burton, P., et al. (2004) Origins of bilateral symmetry: Hox and dpp expression in a sea anemone. *Science* 304, 1335–1337.

25. Scholz, C. B., Technau, U. (2003) The ancestral role of Brachyury: expression of NemBra1 in the basal cnidarian Nematostella vectensis (Anthozoa). *Dev Genes Evol* 212, 563–570.

26. Technau, U., Scholz, C. B. (2003) Origin and evolution of endoderm and mesoderm. *Int J Dev Biol* 47, 531–539.

27. Magie, C. R., Pang, K., Martindale, M. Q. (2005) Genomic inventory and expression of Sox and Fox genes in the cnidarian Nematostella vectensis. *Dev Genes Evol* 215, 618–630.

28. Torras, R., Yanze, N., Schmid, V., et al. (2004) Nanos expression at the embryonic posterior pole and the medusa phase in the hydrozoan Podocoryne carnea. *Evol Dev* 6, 362–371.

29. Extavour, C. G., Pang, K., Matus, D. Q., et al. (2005) Vasa and nanos expression patterns in a sea anemone and the evolution of bilaterian germ cell specification mechanisms. *Evol Dev* 7, 201–215.

30. Matus, D. Q., Thomsen, G. H., Martindale, M. Q. (2006) Dorso/ventral genes are asymmetrically expressed and involved in germ-layer demarcation during cnidarian gastrulation. *Curr Biol* 16, 499–505.

31. Technau, U., Bode, H. R. (1999) HyBra1, a Brachyury homologue, acts during head formation in Hydra. *Development* 126, 999–1010.

32. Spring, J., Yanze, N., Middel, A. M., et al. (2000) The mesoderm specification factor twist in the life cycle of jellyfish. *Dev Biol* 228, 363–375.

33. Spring, J., Yanze, N., Josch, C., et al. (2002) Conservation of Brachyury, Mef2, and Snail in the myogenic lineage of jellyfish: a connection to the mesoderm of bilateria. *Dev Biol* 244, 372–384.

34. Fritzenwanker, J. H., Saina, M., Technau, U. (2004) Analysis of forkhead and snail expression reveals epithelial-mesenchymal transitions during embryonic and larval development of Nematostella vectensis. *Dev Biol* 275, 389–402.

35. Hayward, D. C., Miller, D. J., Ball, E. E. (2004) Snail expression during embryonic development of the coral Acropora: blurring the diploblast/triploblast divide? *Dev Genes Evol* 214, 257–260.

36. Guder, C., Philipp, I., Lengfeld, T., et al. (2006) The Wnt code: cnidarians signal the way. *Oncogene* 25, 7450–7460.

37. Wittlieb, J., Khalturin, K., Lohmann, J. U., et al. (2006) Transgenic Hydra allow in vivo tracking of individual stem cells during morphogenesis. *Proc Natl Acad Sci USA* 103, 6208–6211.

38. King, N., Hittinger, C. T., Carroll, S. B. (2003) Evolution of key cell signaling and adhesion protein families predates animal origins. *Science* 301, 361–363.

39. Hobmayer, B., Rentzsch, F., Kuhn, K., et al. (2000) WNT signaling molecules act in axis formation in the diploblastic metazoan Hydra. *Nature* 407, 186–189.

40. Hobmayer, E., Hatta, M., Fischer, R., et al. (1996) Identification of a Hydra homologue of the beta-catenin/plakoglobin/armadillo gene family. *Gene* 172, 155–159.

41. Minobe, S., Fei, K., Yan, L., et al. (2000) Identification and characterization of the epithelial polarity receptor "Frizzled" in Hydra vulgaris. *Dev Genes Evol* 210, 258–262.

42. Kusserow, A., Pang, K., Sturm, C., et al. (2005) Unexpected complexity of the Wnt gene family in a sea anemone. *Nature* 433, 156–160.

43. Lee, P. N., Pang, K., Matus, D. Q., et al. (2006) A WNT of things to come: Evolution of Wnt signaling and polarity in cnidarians. *Semin Cell Dev Biol* 17, 157–167.

44. Guder, C., Pinho, S., Nacak, T. G., et al. (2006) An ancient Wnt-Dickkopf antagonism in Hydra. *Development* 133, 901–911.

45. Sullivan, J. C., Ryan, J. F., Mullikin, J. C., et al. (2007) Conserved and novel Wnt clusters in the basal eumetazoan Nematostella vectensis. *Dev Genes Evol* 217, 235–239.

46. Spitz, F., Duboule, D. (2005) Developmental biology: reproduction in clusters. *Nature* 434, 715–716.

47. Spitz, F., Gonzalez, F., Duboule, D. (2003) A global control region defines a chromosomal regulatory landscape containing the HoxD cluster. *Cell* 113, 405–417.

48. Philipp, I., Holstein, T. W., Hobmayer, B. (2005) HvJNK, a Hydra member of the c-Jun

NH2-terminal kinase gene family, is expressed during nematocyte differentiation. *Gene Expr Patterns* 5, 397–402.

49. Rentzsch, F., Hobmayer, B., Holstein, T. W. (2005) Glycogen synthase kinase 3 has a proapoptotic function in Hydra gametogenesis. *Dev Biol* 278, 1–12.

50. Strutt, H., Price, M. A., Strutt, D. (2006) Planar polarity is positively regulated by casein kinase Iepsilon in Drosophila. *Curr Biol* 16, 1329–1336.

51. Price, M. A. (2006) CKI, there's more than one: casein kinase I family members in Wnt and Hedgehog signaling. *Genes Dev* 20, 399–410.

52. Kawano, Y., Kypta, R. (2003) Secreted antagonists of the Wnt signaling pathway. *J Cell Sci* 116, 2627–2634.

53. Davidson, G., Mao, B., del Barco Barrantes, I., Niehrs, C (2002) Kremen proteins interact with Dickkopf1 to regulate anteroposterior CNS patterning. *Development* 129, 5587–5596.

54. Mao, B., Niehrs, C. (2003) Kremen2 modulates Dickkopf2 activity during Wnt/LRP6 signaling. *Gene* 302, 179–183.

55. Jones, S. E., Jomary, C. (2002) Secreted Frizzled-related proteins: searching for relationships and patterns. *Bioessays* 24, 811–820.

56. Fedders, H., Augustin, R., Bosch, T. C. (2004) A Dickkopf-3-related gene is expressed in differentiating nematocytes in the basal metazoan Hydra. *Dev Genes Evol* 214, 72–80.

57. Augustin, R., Franke, A., Khalturin, K., et al. (2006) Dickkopf related genes are components of the positional value gradient in Hydra. *Dev Biol* 296, 262–270.

58. Wikramanayake, A. H., Hong, M., Lee, P. N., et al. (2003) An ancient role for nuclear beta-catenin in the evolution of axial polarity and germ layer segregation. *Nature* 426, 446–450.

59. Momose, T., Houliston, E. (2007) Two oppositely localised frizzled RNAs as axis determinants in a cnidarian embryo. *PLoS Biol* 5, e70.

60. Momose, T., Schmid, V. (2006) Animal pole determinants define oral-aboral axis polarity and endodermal cell-fate in hydrozoan jellyfish Podocoryne carnea. *Dev Biol* 292, 371–380.

61. Browne, E. N. (1909) The production of new hydranths in hydra by the insertion of small grafts. *J Exp Zool* 7, 1–37.

62. Broun, M., Bode, H. R. (2002) Characterization of the head organizer in hydra. *Development* 129, 875–884.

63. Broun, M., Gee, L., Reinhardt, B., et al. (2005) Formation of the head organizer in hydra involves the canonical Wnt pathway. *Development* 132, 2907–2916.

64. Cramer von Laue, C. (2003) Analysis of the dual function of beta-Catenin in Wnt signaling and in Cadherin-mediated cel l adhesion of Hydra, Darmstadt University of Technology, Darmstadt.

65. Muller, W., Frank, U., Teo, R., et al. (2007) Wnt signaling in hydroid development: ectopic heads and giant buds induced by GSK-3beta inhibitors. *Int J Dev Biol* 51, 211–220.

66. Plickert, G., Jacoby, V., Frank, U., et al. (2006). Wnt signaling in hydroid development: Formation of the primary body axis in embryogenesis and its subsequent patterning. *Dev Biol* 298, 368–378.

67. Lie, D. C., Colamarino, S. A., Song, H. J., et al. (2005) Wnt signaling regulates adult hippocampal neurogenesis. *Nature* 437, 1370–1375.

68. Onai, T., Sasa, N., Matsui, M., et al. (2004) Xenopus XsalF: anterior neuroectodermal specification by attenuating cellular responsiveness to Wnt signaling. *Dev Cell* 7, 95–106.

69. Teo, R., Mohrlen, F., Plickert, G., et al. (2006). An evolutionary conserved role of Wnt signaling in stem cell fate decision. *Dev Biol* 289, 91–99.

Chapter 6

Detecting Expression Patterns of Wnt Pathway Components in *Nematostella vectensis* Embryos

Shalika Kumburegama, Naveen Wijesena, and Athula H. Wikramanayake

Abstract

The anthozoan cnidarian *Nematostella vectensis* has emerged as a key model system for evolutionary developmental biology studies, and this animal's usefulness will grow with the recent sequencing of its genome. In particular, work done in *Nematostella* is providing insight into the role of the Wnt pathway in the evolution of pattern formation. This chapter describes methods to maintain and spawn these animals, and detailed protocols to detect expression patterns of Wnt pathway components in *Nematostella* eggs and embryos.

Key words: *Nematostella*, cnidarian, Wnt signaling, whole-mount *in situ* hybridization, immunocytochemistry.

1. Introduction

The evolution of gastrulation and acquisition of the primary germ layers were arguably among the key developmental events that led to the morphological diversification of ancestral metazoans *(1)*. The onset of gastrulation movements would have allowed the interaction of different tissue layers leading to new gene expression patterns. These interactions most likely played an important role in the evolution of novel tissues and organs during evolution *(1)*. At present, however, the mechanisms that led to the evolution of gastrulation and the evolution of the germ layers are fundamental questions that are yet to be answered. It is likely that a deeper molecular understanding of embryonic development of key basal metazoan phyla such as the Cnidaria will provide insight into these important evolutionary questions.

Elizabeth Vincan (ed.), *Wnt Signaling, Volume II: Pathway Models, vol. 469*
© 2008 Humana Press, a part of Springer Science+Business Media, New York, NY
Book doi: 10.1007/978-1-60327-469-2

Due to the pioneering work of the late Dr. Cadet Hand, the starlet sea anemone, *Nematostella vectensis*, has been established as an important model system for developmental and evolutionary studies *(2)*. *N. vectensis* belongs to the class Anthozoa, a group believed to be basal within the phylum Cnidaria *(3)*. The availability of the *N. vectensis* genome sequence *(4)*, the ease with which *Nematostella* eggs and embryos can be obtained and manipulated, in addition to its basal position within the phylum makes it an ideal cnidarian model system to help elucidate some of the fundamental questions in evolutionary developmental biology. Recent studies carried out in *Nematostella* have shown that it is an especially useful system for investigating the role of the Wnt signaling pathway in the evolution of pattern formation. These studies have shown that the canonical Wnt signaling pathway is activated asymmetrically in the early *Nematostella* embryo, and serves to regulate the initial segregation of the ectoderm from the endoderm *(5, 6)*. Studies in hydrozoans, a more derived class within the Cnidaria have shown that the Wnt signaling pathway is utilized in a similar manner during embryogenesis in these animals *(7)*, suggesting that this pathway was used in the ancestral cnidarian to segregate the germ layers. Thus, a more complete understanding of how the Wnt pathway is asymmetrically activated during early cnidarian development may provide insight into the evolution of polarity in eggs and embryos. The following describes how to maintain *N. vectensis* in the laboratory and spawn them to obtain gametes, and provides detailed protocols to carry out expression analysis for endogenous mRNA and proteins of the Wnt signaling pathways.

2. Materials

2.1. Spawning Nematostella

1. 1/3 artificial seawater: artificial seawater is made by dissolving commercially available sea salt in reverse osmosis (RO) water according to the manufacturer's specifications. Dilute artificial seawater in an appropriate volume of RO water to make 1/3 seawater. Use unfiltered 1/3 artificial seawater to maintain adult *Nematostella* cultures. Filtered 1/3 artificial seawater is used for all embryo cultures.

2. Food for *Nematostella*: Live brine shrimp, mussels, or clams.

3. 4% cysteine: Dissolve an appropriate amount of cysteine to make a 4% (w/v) cysteine solution in 1/3 filtered artificial seawater. Adjust pH to 7.4–7.6 using NaOH. Store at room temperature.

4. Light box: A light-tight box for spawning *Nematostella*. A 10-gallon glass aquarium lined with aluminum foil will suffice for this purpose.

5. A halogen light source connected to a timer.

2.2. Whole Mount In Situ *Hybridization*

1. Sterile 24-well polystyrene plates.

2. Fixative I: 0.2% (w/v) glutaraldehyde (Electron Microscopy Sciences), 3.7% (w/v) formaldehyde in 1/3 filtered seawater made fresh. Store at 4°C.

3. Fixative II: 3.7% (w/v) formaldehyde in 1/3 filtered seawater made fresh. Store at 4°C.

4. PTw (1L): 1X PBS, 0.1% (v/v) Tween-20 in 0.1% (v/v) diethyl pyrocarbonate (DEPC)-treated water. Autoclave solution prior to adding Tween-20. Allow the autoclaved solution to cool before adding Tween-20. Store at room temperature.

5. Ethanol: 70% (v/v) ethanol in 0.1% (v/v) DEPC-treated water. Store at room temperature.

6. MEGAscript T7 or SP6 High Yield Transcription Kit (Ambion, Austin, TX).

7. Digoxygenin-labeled UTP (Dig-UTP) (Roche), Fluorescein-labeled UTP (Roche).

8. Hybridization Buffer: 50% (v/v) formamide, 5X SSC, pH 4.5, 50 µg/mL heparin, 0.1% (v/v) Tween-20, 1.0% (w/v) SDS, 100 µg/mL single-stranded DNA (Sigma). Store at –20 °C.

9. 20X SSC: 3 M NaCl, 0.3 M sodium citrate. Adjust pH to 7.0 and autoclave. Store at room temperature.

10. PBT: 1X PBS, 0.2% (v/v) Triton X-100, 0.1% (w/v) Bovine Serum Albumin (BSA). Dissolve BSA on a magnetic stirrer and filter sterilize the solution. Store at 4°C.

11. Proteinase K: 0.01 mg/mL Proteinase K in PTw made fresh. Store at 4°C.

12. Glycine: 2 mg/mL glycine in PTw made fresh. Store at room temperature.

13. Triethanolamine: 1% triethanolamine in PTw made fresh.

14. Fixative III: 3.7% (w/v) formaldehyde in PTw made fresh. Store at room temperature.

15. 10X Boehringer-Mannheim blocking stock solution (Roche): Dissolve blocking reagent in maleic acid buffer to a final concentration of 10% (w/v) using low heat while stirring. Autoclave and store at 4°C.

16. 1X Boehringer-Mannheim blocking buffer: Dissolve 10X Boehringer-Mannheim blocking reagent in maleic acid buffer to a final concentration of 1X.

17. Maleic acid buffer: 0.1 M maleic acid and 0.05 M NaCl. Adjust pH to 7.5 with NaOH. Store at room temperature.

18. Anti-Digoxygenin-AP antibody (Roche), Anti-Fluorescein-AP antibody.

19. Alkaline Phosphatase (AP) Buffer: 100 mM NaCl, 50 mM MgCl$_2$, 100 mM Tris-HCl, pH 9.5, and 0.5% (v/v) Tween-20. Prepare fresh and store at room temperature.

20. AP substrate (Roche): 3.3 μL/mL NBT (stock 100 mg/mL) and 3.3 μL/mL BCIP (stock 50 mg/mL) in AP buffer made fresh.

21. Fast-Red buffer: 0.4 M NaCl, 0.1 M Tris-HCl, pH 8.2.

22. FAST RED substrate: Dissolve 1 tablet of Fast Red (Roche) in 1 mL of Fast-Red buffer.

2.3. Immunohisto-chemistry of Nematostella *Eggs and Embryos*

1. 10X salt solution: 1 M MOPS, 20 mM EGTA, 10 mM MgSO$_4$. Prepared fresh and stored at room temperature. Add 6 mL of water to MOPS, EGTA and MgSO$_4$ salt mix. Adjust the pH to 7.4 with 10 N NaOH. Make up to 10 mL with autoclaved water.

2. MEMPfa fixative (2.5 mL): 250 μL 10X salt solution, 100 μL of 16% (w/v) Paraformaldehyde (Electron Microscopy Sciences), and bring up to 2.5 mL with autoclaved water. Prepare fresh and store at 4°C.

3. DENT's: 80% methanol and 20% DMSO. Prepare fresh and store at 4°C.

4. Bleaching solution: 2:1 of DENT's and 30% hydrogen peroxide. Store at 4°C.

5. Blocking buffer: 5% (v/v) normal donkey serum, 0.05% (v/v) Tween-20 in 1X PBS. Store at 4°C.

6. DAB kit (Invitrogen): Solution A (20X buffer concentrate), solution B (DAB solution), and solution C (20X concentrated hydrogen peroxide). To 1 mL of water add one drop of solution A and mix well. Next add one drop each of solution B and C. Mix well and store at room temperature in the dark.

7. Ethanol series (50, 70, 90, and 100%) or Isopropanol for dehydration of samples.

8. Methyl salicilate or 1:2 benzyl alcohol: benzyl benzoate (toxic) as clearing agents.

3. Methods

3.1. Spawning Nematostella

1. The day before spawning feed the adult animals with live brine shrimp or finely chopped mussels or clams. Allow animals to feed for at least 20–30 minutes after adding the food (*see* **Note 1**).

2. Wash bowls with 1/3 seawater. Use a plastic pasteur pipette to gently squirt water to clean the animals. Decant the water and add fresh 1/3 sea water.

3. Place the animals in the light-tight box (at room temperature), with the halogen light source connected to a timer. Set the timer to turn off the light 15–30 minutes after the animals are placed in the box. Set the timer so that there is a 12-hour dark period followed by a 12-hour light period. The animals will spawn the next day, 30 minutes to 1 hour after the light goes off (*see* **Note 2**).

4. Place about 2 mL of the egg masses in a 15-mL conical plastic tube with as little of the 1/3 seawater as possible. Add about 12 mL of 4% cysteine and gently mix by either inverting or by placing the tube on a rotator for 5–8 minutes (*see* **Note 3**).

5. Allow the eggs to settle and decant the cysteine solution by pipetting as much of the cysteine solution as possible taking care not to expose the eggs to air. Wash eggs at least three times with 1/3 filtered artificial seawater to remove all traces of cysteine. The de-jellied eggs can now be used for experiments (*see* **Note 4**).

3.2. Whole-mount In Situ Hybridization for Nematostella Eggs and Embryos

3.2.1. Linearization of Plasmid Vectors for Anti-sense and Sense RNA Probe Synthesis

1. The gene of interest should be cloned into an appropriate expression vector that allows its transcription (*see* **Note 5**).

2. The cDNA has to be linearized prior to setting up the transcription reaction. This can be accomplished by PCR using primers that span the promoter sites or by restriction enzyme digestion (*see* **Note 6**).

3.2.2. Riboprobe Synthesis

1. Thaw, vortex, and spin down 10X transcription buffer and ribonucleotide solutions (from Ambion MegaScript Transcription Kit) and keep on ice.

2. Mix 1.0 μL of 10X buffer, 1.0 μL each of ATP, CTP and GTP, 0.8 μL of UTP, 2.1 μL Dig-UTP, 2.1 μL linear DNA template and 1.0 μL of enzyme mix, centrifuge and incubate at 37°C for 4–6 hours (*see* **Note 7**).

3. Add 0.5 µL of RNAse free DNAse I (from Ambion Mega-Script Transcription kit), mix contents, centrifuge and incubate at 37°C for 15 minutes.

4. Precipitate the RNA by adding 10 µL of RNAse-free water and 10 µL of lithium chloride precipitation solution (from the Ambion MegaScript Transcription kit). Mix the contents thoroughly and chill at –20°C for 30 minutes or overnight (*see* **Note 8**).

5. Centrifuge the sample at maximum speed for 15 minutes at 4°C.

6. Remove the supernatant and wash the RNA pellet with 500 µL of RNAse-free 70% ethanol (*see* **Note 9**).

7. Re-centrifuge the samples at maximum speed for 15 minutes at 4°C. Remove as much ethanol as possible, dry the pellet briefly at 65°C and resuspend in 10 µL of nuclease-free water.

8. For long-term storage, dilute the riboprobe in an appropriate volume of the hybridization buffer to make a 50 ng/µL stock solution. Store at –20°C.

3.2.3. Fixation of Nematostella *Eggs and Embryos for Whole Mount* In Situ *Hybridization*

1. Fix embryos in 500 µL of Fixative I for 4 minutes at room temperature (*see* **Note 10**).

2. Remove Fixative I and fix embryos in 500 µL of Fixative II for 1 hour at room temperature.

3. Remove Fixative II and wash embryos five times in PTw.

4. Wash embryos once in sterile dH$_2$O for 10 minutes at room temperature.

5. Dehydrate embryos by washing two times in methanol and store samples in methanol at –20°C (*see* **Note 11**).

3.2.4. In Situ Hybridization

Unless otherwise stated all solutions need to be made with nuclease-free water (0.1% DEPC treated and autoclaved) and all equipment needs to be sterile and RNAse-free through the hybridization step (**step 14**). All washes, unless otherwise stated, are carried out for 5 minutes at room temperature on a rocker. Whole-mount *in situ* hybridization protocol described later has been adapted from the work carried out in the Martindale Lab *(8, 9)*.

1. Transfer fixed embryos into a sterile, 24-well plate.

2. Bleach embryos in two parts methanol and one part hydrogen peroxide for 30 minutes at 4°C.

3. Rehydrate embryos through the following washes: 60% methanol and 40% Ptw, 30% methanol and 70% PTw, four times in 100% PTw.

4. Incubate embryos in 0.01 mg/mL Proteinase-K for 20 minutes without the rocker.

5. Stop reaction with two glycine washes.

6. Wash two times with 1% Triethanolamine in PTw. During each Triethanolamine wash, add 1.5 μL of acetic anhydride twice at 5-minute intervals.

7. Wash two times in PTw.

8. Refix the embryos in 500 μL of Fixative III for 1 hour at room temperature.

9. Wash five times in PTw.

10. Remove as much liquid as possible and add 500 μL of hybridization buffer. Incubate for 10 minutes at room temperature. Remove hybridization buffer and add 500 μL of fresh hybridization buffer and incubate at hybridization temperature overnight (*see* **Note 12**).

11. Dilute probe to a final concentration of 10–0.05 ng/μL in hybridization buffer (*see* **Note 13**).

12. Denature probe at 80–90 °C for 10 minutes.

13. Remove prehybridization buffer and add the riboprobe diluted in hybridization buffer. Hybridize overnight or for 2 days at hybridization temperature (*see* **Note 14**).

14. Remove the riboprobe and wash one time for 10 minutes and one time for 40 minutes in hybridization buffer at hybridization temperature (*see* **Note 15**).

15. Wash for 30 minutes in 75% Hybridization buffer, 25% 2X SSC at hybridization temperature.

16. Wash for 30 minutes in 50% Hybridization buffer, 50% 2X SSC at hybridization temperature.

17. Wash for 30 minutes in 25% Hybridization buffer, 75% 2X SSC at hybridization temperature.

18. Wash for 30 minutes in 100% 2X SSC at hybridization temperature.

19. Wash three times, each for 20 minutes in 0.05X SSC at hybridization temperature (*see* **Note 16**).

20. Wash for 10 minutes in 75% 0.05X SSC, 25% PTw at room temperature.

21. Wash for 10 minutes in 50% 0.05X SSC, 50% PTw at room temperature.

22. Wash for 10 minutes in 25% 0.05X SSC, 75% PTw at room temperature.

23. Wash for 10 minutes in 100% PTw at room temperature.

24. Wash five times in PBT at room temperature.

25. Make a 1X Boehringer-Mannheim Blocking buffer solution by diluting a 10X solution with maleic acid buffer. Block in 1X Boehringer-Mannheim Blocking buffer for 1 hour at room temperature on the rocker (*see* **Note 17**).

26. Dilute the anti-Digoxygenin-AP antibody in Boehringer-Mannheim blocking buffer to 1:5000. Incubate with Boehringer-Mannheim anti-Dig/AP solution at 4 °C overnight on a rocker (*see* **Note 18**).

27. Wash 10 times for 30 minutes each in PBT at room temperature.

28. Wash 3 times for 10 minutes each in AP buffer (*see* **Note 19**).

29. Develop samples by incubating in AP substrate solution, in the dark at room temperature and by changing the AP substrate solution as often as required. Monitor color development (*see* **Note 20**).

30. Stop reaction by washing five times with PTw.

31. When carrying out double *in situ* hybridization, follow **steps 32–41** to develop the second probe or go directly to **step 41** for single-probe detection.

32. Incubate two times for 15 minutes each in hybridization buffer at Hybridization temperature.

33. Wash for 10 minutes in 75% hybridization buffer, 25% PTw at room temperature.

34. Wash for 10 minutes in 50% hybridization buffer, 50% PTw at room temperature.

35. Wash for 10 minutes in 25% hybridization buffer, 75% PTw at room temperature.

36. Wash seven times each for 10–20 minutes in 100% PTw.

37. Block for 1 hour in 1X Boehringer-Mannheim Blocking buffer at room temperature.

38. Incubate samples as appropriate in either anti-Digoxygenin-AP (1:5000) or anti-Fluorescein-AP (1:3000) overnight at 4 °C.

39. Wash 10 times, 20–30 minutes each, in PBT at room temperature.

40. Wash and develop in AP buffer with NBT and BCIP or in Fast-Red buffer (*see* **Note 21**).

41. Store in 70% Glycerol in PTw (*see* **Note 22**).

3.3. Immunohistochemistry of Nematostella *Eggs and Embryos*

The fixation protocol has been modified from a protocol used in *Xenopus* embryos *(10)*.

1. Place eggs or embryo samples in microcentrifuge tubes. Remove as much of the 1/3 sea water as possible without

exposing the embryos to air. Add 500 μL MEMPfa fixative and fix for 1 hour at room temperature (*see* **Note 23**).

2. Remove fixative and post fix in 500 μL DENT's for 15 minutes at room temperature.

3. (Optional) Bleach fixed samples with 500 μL of two parts DENT's and one part 30% hydrogen peroxide for 30 minutes at 4°C (*see* **Note 24**).

4. Remove DENT's and wash samples three times for 15 minutes each in 1X PBS by placing tubes on a rocker at room temperature.

5. Block for 30 minutes to 1 hour at room temperature in blocking buffer (*see* **Note 25**).

6. Dilute primary antibody in blocking buffer and incubate samples for 1 hour at room temperature or overnight at 4°C.

7. Remove primary antibody and rinse each sample three times for 20 minutes each in blocking buffer on a rocker at room temperature.

8. Dilute secondary antibody in blocking buffer and incubate samples for 1 hour at room temperature on a rocker.

9. If the secondary antibody is a peroxidase-conjugate, follow **steps 10–15**. If the secondary antibody is a fluorescent-conjugate go to **step 16**.

10. Remove the secondary antibody and wash each sample three times for 20 minutes each in blocking buffer on a rotator. Wash once more in 1X PBS for 10 minutes on a rotator.

11. Preparation of DAB solution: add one drop of solution A into a microcentrifuge tube covered with foil containing 1 mL of autoclaved water. Mix well. Then add one drop each of solutions B and C from the DAB kit. Mix well.

12. Incubate samples 5 minutes in 250 μL of DAB solution in the dark.

13. Wash samples three times for 15 minutes each in 1X PBS on a rocker in the dark.

14. Dehydrate samples using an ethanol series (50, 70, 90, and 100%). Incubate samples in each of the ethanol series for 3–5 minutes.

15. Clear in methyl salicylate. Note: If the samples are not completely dehydrated, the addition of methyl salicylate will cause the embryos to become opaque.

16. If the secondary antibody has a fluorescent conjugate, wash the samples three times for 20 minutes each in blocking buffer in the dark on a rocker (*see* **Note 26**).

17. Dehydrate samples in either an ethanol or isopropanol series and clear in 1:2 of Benzyl alcohol and Benzyl Benzoate (BA:BB).

4. Notes

1. Adult *N. vectensis* cultures are kept in glass bowls containing unfiltered 1/3 artificial sea water at 15–18°C at all times except when they are required to spawn at which time they are placed at room temperature (~24°C). Animals fed regularly (three to four times a week) will spawn consistently.

2. *N. vectensis* are dioecious but sex cannot be determined by external morphological features. When induced, males and females will spawn synchronously. We keep most of the animals in bowls containing both males and females. But females can be easily recognized and separated at the time of spawning. Hence unfertilized eggs can be obtained by maintaining females in separate glass bowls. If unfertilized eggs are required, make sure to use a separate dish and pasteur pipettes to collect these eggs from the "females only" bowl.

3. *N. vectensis* eggs are approximately 200–250 μm in diameter and they are spawned in a gelatinous matrix. Cysteine can be used to remove the jelly mass and this is required prior to most experimental manipulations. But care should be taken not to let the eggs sit in the cysteine solution for too long, as this will result in the disintegration of eggs and embryos. The cysteine solution needs to be made fresh just prior to use since old cysteine solutions tend to precipitate.

4. De-jellied eggs readily stick to polystyrene plastic ware. Hence place de-jellied eggs in glass dishes or microcentrifuge tubes prior to fixing.

5. We commonly use the pCS2+ expression vector *(11)* or the pGEM-T easy (Promega) vectors for cloning genes or gene fragments for riboprobe synthesis. Ideally the gene fragment cloned should be at least 750 bp or longer and this may include 5′ or 3′ UTR sequences.

6. To obtain sense or anti-sense mRNA, it is essential to have either of the promoter sites in the linear DNA. Therefore if PCR is utilized select primers that span the promoter sites of the vector. After the PCR, gel extract the linear DNA prior to riboprobe synthesis. If restriction enzyme digestion is utilized, digest the vector 3′ of the promoter site and extract the linear DNA using phenol-chloroform extraction.

7. From 500 ng up to 1 μg of linear template DNA is sufficient for the *in vivo* transcription. But a higher RNA yield can be obtained by maximizing the amount of linear template DNA (2.1 μL) instead of adding nuclease-free water to bring the total volume of the reaction up to 10 μL. Fluorescein-labeled UTP may be used instead of DIG-labeled UTP or

when carrying out double *in situ* hybridization to label a second RNA probe. If fluorescein-labeled UTP is used then remember to use the appropriate antibody and substrate, as described later in the protocol.

8. If you anticipate low yields do not dilute reaction with nuclease-free water prior to precipitation. Instead add half the volume (5 µL) of lithium chloride precipitation solution.

9. It is advisable to remove the supernatant by pipetting because the mRNA forms a clear pellet that may be hardly visible. The mRNA pellet can be stored at –80°C in 70% ethanol overnight or up to 1 week.

10. The polyps can contract quickly. Therefore when fixing polyp stages relax the animals in a solution of 20% $MgCl_2$ in 1/3 seawater for ~5 minutes prior to adding the fixative.

11. Store fixed embryos in either screw cap tubes or in microcentrifuge tubes sealed with parafilm to prevent evaporation.

12. The optimum hybridization temperature may vary for each riboprobe used. But for most riboprobes a hybridization temperature ranging from 55–60°C is sufficient. A lower hybridization temperature may increase background staining.

13. Digoxygenin labeled probe should be stored as a 50 ng/µL stock solution in hybridization buffer at –20°C. Before making the probe dilution, the stock solution should be warmed in a 50°C water bath as SDS in the hybridization buffer precipitates when chilled. We find that a final concentration of 1 ng/µL of the riboprobe is adequate for a robust color change for most riboprobes. The riboprobes stored at –20°C can be reused four to five times. When carrying out double *in situ* hybridization to detect the expression pattern of two mRNAs, the probes for the two individual mRNAs can be added together at the optimum concentration of each probe.

14. Care should be taken while changing solutions throughout the procedure especially after the addition of the hybridization buffer since the embryos become more translucent. A dissecting microscope may be used to remove the hybridization buffer and hence minimize the loss of embryos.

15. The hybridization buffer and all solutions used for washes at the hybridization temperature should be pre-warmed to hybridization temperature before adding to the embryos.

16. The SSC solution can range from 0.2 to 0.05%. When carrying out double *in situ* hybridization, it is recommended that at this step a 0.2% SSC solution be used instead of the 0.05% SSC solution. The percentage of SSC may also depend on

the robustness of the second probe that is detected. If the second probe is robust then a lower SSC percentage (0.05%) is recommended and if the affinity of the second probe to its target mRNA is low then a less stringent or a higher SSC percentage is recommended.

17. If necessary, the blocking step can be carried out overnight at 4°C instead of 1 hour at room temperature.

18. The incubation with Boehringer-Mannheim anti-Dig/AP can be carried out at room temperature for 1–4 hours instead of overnight at 4°C. If the first probe is detected with Fast-Red substrate then add anti-Fluorescein-AP antibody (1:3000) overnight at 4°C or 1–4 hours at room temperature.

19. The eggs/embryos tend to stick together in the presence of $MgCl_2$ in the AP buffer. Therefore it is recommended that the samples are first washed for 10 minutes in AP buffer without $MgCl_2$ and then twice for 10 minutes each in AP buffer, which contains $MgCl_2$.

20. The AP substrate should be changed every 3–4 hours as leaving it in the same substrate solution for extended periods of time will result in an increase in background or nonspecific staining. Color development is considerably slower at 4°C. Therefore if desired the reaction may be carried out at 4°C for a longer period of time. If fluorescein-labeled UTP is used then wash samples three times for 10 minutes each in Fast-Red buffer and develop in Fast Red Substrate at room temperature in the dark and monitor color development.

21. If anti-Digoxygenin-AP antibody is used, first wash the samples once for 10 minutes in AP buffer without $MgCl_2$ and twice for 10 minutes each in AP buffer with $MgCl_2$. Develop the samples in AP buffer containing NBT (3.3 µL/mL) and BCIP (3.3 µL/mL). If anti-fluorescein-AP is used to detect the second probe then wash embryos in Fast-Red buffer three times for 10 minutes each and develop in Fast Red Substrate at room temperature in the dark.

22. Leaving the embryos in 70% glycerol for a longer period of time helps clear embryos. Samples can be stored in 70% glycerol for long time storage at 4°C.

23. In addition to this fixative protocol, embryos may be fixed using the whole-mount *in situ* hybridization protocol mentioned earlier in place of **steps 1–3**. This method may especially be suitable if a peroxidase-conjugate is used instead of a fluorescent conjugate as a secondary antibody since glutaraldehyde results in an increase in auto-fluorescence in eggs and early embryo stages.

24. Bleaching of the embryos is recommended if the secondary antibody utilized is a peroxidase-conjugate.

25. The serum used for the blocking buffer is determined by the secondary antibody. For example, donkey serum is used if the host animal of the secondary antibody is donkey.

26. *Nematostella* eggs are rich in yolk. As a result the eggs and early developmental stages auto-fluoresce. Therefore imaging of these eggs and early embryos under a fluorescence microscope difficult. But confocal imaging can still be performed, which allows flexibility in controlling background color levels.

Acknowledgments

The authors thank Kevin Pang and David Matus from the Martindale Lab for sharing their *in situ* hybridization protocols for *Nematostella*. This work was supported by NSF grant IOS 0720365 to AHW.

References

1. Martindale, M. Q. (2005) The evolution of metazoan axial properties. *Nature Rev Genet* 6, 917–927.

2. Hand, C., Uhlinger, K. R. (1992) The culture, sexual and asexual reproduction, and growth of the sea anemone Nematostella vectensis. *Biol Bull* 182, 169–176.

3. Collins, A. G., Schuchert, P., Marques, A. C., et al. (2006) Medusozoan phylogeny and character evolution clarified by new large and small subunit rRNA data and an assessment of the utility of phylogenetic mixture models. *Syst Biol* 55, 97–115.

4. Putnam, N. H., Srivastava, M., Hellsten, U., et al. (2007) Sea anemone genome reveals ancestral eumetazoan gene repertoire and genomic organization. *Science* 317, 86–94.

5. Wikramanayake, A. H., Hong, M., Lee, P. N., et al. An ancient role for nuclear β-catenin in the evolution of axial polarity and germ layer segregation. *Nature* 426, 446–450.

6. Lee, P. N., Kumburegama, S., Marlow, H. Q., et al. (2007) Asymmetric developmental potential along the animal-vegetal axis in the anthozoan cnidarian, *Nematostella*

vectensis, is mediated by Dishevelled. *Dev Biol* 10.1016/j.ydbio.2007.05.040

7. Momose, T., Houliston, E. (2007) Two oppositely localised frizzled RNAs as axis determinants in a cnidarian embryo. *PLoS Biol* 5, e70.

8. Martindale, M.Q., Pang, K., Finnerty, J.R. (2004) Investigating the origins of triploblasty: 'Mesodermal' gene expression in a diploblastic animal, the sea anemone Nematostella vectensis (phylum, Cnidaria; class, Anthozoa). *Development* 131, 2463–2474.

9. Matus, D. Q., Thomsen, G. H., Martindale, M. Q. (2006) Dorso/ventral genes are asymmetrically expressed and involved in germ-layer demarcation during cnidarian gastrulation. *Curr Biol* 16, 499–505.

10. Klymkowsky, M. W., Hanken, J. (1991) Whole-mount staining of *Xenopus* and other vertebrates. *Methods Cell Biol* 36, 419–441.

11. Rupp, R. A., Snider, L., Weintraub, H. (1994) *Xenopus* embryos regulate the nuclear localization of XmyoD. *Genes Dev* 8, 1311–1323.

Chapter 7

Detection of Expression Patterns in *Hydra* Pattern Formation

Hans Bode, Tobias Lengfeld, Bert Hobmayer, and Thomas W. Holstein

Abstract

Cnidarians are simple metazoans with only two body layers and a primitive nervous system. They are famous for their nearly indefinite regeneration capacity. Recent work has identified most of the Wnt subfamilies and Wnt antagonists known from vertebrates in this basal animal model. Wnt signaling and BMP signaling have been shown to act in *Hydra* pattern formation and regeneration. Because recent genomic work in *Hydra* and *Nematostella* revealed many genes for vertebrate signaling pathways and transcription factors to be present in this more than 500 Myr-year-old phylum, future work will focus on the function and expression of these genes in *Hydra* pattern formation and regeneration. This chapter presents an *in situ* hybridization protocol, which is largely based on a lab protocol of the Bode lab that has proven to be extremely useful in the characterization of many developmental genes from *Hydra*.

Key words: Wnt, Wnt/beta-catenin target genes, *Hydra*, Cnidaria, regeneration.

1. Introduction

In situ hybridization (ISH) was first introduced by Gall and Pardue in 1969 *(1)*. With this method specific nucleic acid sequences within cells can be detected by hybridizing a labeled RNA or cDNA probe to target transcripts *(1, 2)*. Optimized ISH protocols are important for detecting mRNAs, which exist normally in small quantities. Thus, the ISH technique must be sensitive and specific enough to detect a low number of transcripts in a single cell. A particular challenge is the transcripts for Wnt genes, which are normally expressed in a small region of the entire embryo and which are also expressed at low levels. Nonradioactive protocols were initially introduced by Tautz and

Elizabeth Vincan (ed.), *Wnt Signaling, Volume II: Pathway Models, vol. 469*
© 2008 Humana Press, a part of Springer Science+Business Media, New York, NY
Book doi: 10.1007/978-1-60327-469-2

Pfeifle by using digoxygenin-labeled probes and by detection of the bound probe in tissue with an anti-digoxygenin antibody conjugated to alkaline phosphatase (3). Meanwhile, a number of published protocols exist for all model systems including flies, mice, frogs, zebrafish, and others.

An *in situ* protocol using riboprobes was developed for hydra by Hans Bode and Lydia Gee (4, 5). This protocol is based on published protocols from flies and frogs and has been successfully used in all hydra labs. Important insights into the expression patterns of transcription factors and genes encoding for signaling molecules resulted from these studies (4–14). The protocol presented in this chapter is a slightly modified version of Bode's protocol for *in situs* on hydra whole mounts. It can be easily modified for double *in situs* with different probes (15, 16), but also with antibody stainings for specific antigens or in combination with cell cycle studies (17). The hydra ISH protocol was also the basis for an ISH protocol for *Nematostella* embryos, which was developed in our lab and has been used for analyzing *wnt* genes in *Nematostella* embryos (18–20).

2. Materials

2.1. Animals and Hydra Culture

1. Adult polyps of the European *Hydra vulgaris,* or the Japanese *Hydra magnipapillata* strain 105 (21) are cultured in a number of hydra labs worldwide. They can be purchased on request. There exist a number of mutant strains (e.g., *Hydra magnipapillata* sf-1), which eliminates the interstitial stem cell population within 24 hours after a heat shock, or the head regeneration-deficient mutant strain *Hydra magnipapillata* reg-16 (22). Animals are fed daily with brine shrimp (*Artemia salina*) and washed twice each day with hydra medium (1 hour and 8 hours after feeding) to remove dead or undigested brine shrimps.

2. Hydra medium (HM): 1.0 mM $CaCl_2$, 0.1 mM $MgCl_2$, 0.1 mM KCl, 1.0 mM NaH_2CO_3, pH 7.8.

2.2. Synthesis of Anti-sense and Sense Riboprobes

Because riboprobes are extremely sensitive to RNase, minimization of the introduction of RNase is critical for a successful application of this protocol. To do so, gloves, RNase-free tips and tubes should be always used. All aqueous, solutions, except Tris buffer, are treated with DEPC to kill RNase, and then autoclaved to remove DEPC. Non-aqueous solutions are not treated with DEPC.

2.2.1. Preparation of Linearized DNA

1. T3 and T7 RNA polymerase: (Boehringer, Promega); 20 U/µL; stored at –20°C.

2. Phenol:chloroform, phenol equilibrated to pH 7.5; chloroform:isoamyl alcohol in a 24:1 ratio.

3. Chloroform.

4. TE-buffer: 10 mM Tris-HCl, pH 8.0, 1 mM EDTA. Use Tris-base and adjust the pH to 8.0 using HCl.

5. Ethanol.

2.2.2. Transcription Reaction

1. 10x transcription buffer (Roche #1465384; premixed): 400 mM Tris-HCl, pH 8.0, 60 mM $MgCl_2$, 100 mM DTT, 20 mM spermidine. Assume it is RNase-free; store at –20°C.

2. 10x digoxygenin labeling mix (Roche #11277073910, premixed): 10 mM ATP, 10 mM CTP, 10 mM GTP; 6.5 mM UTP; 3.5 mM dig-UTP, pH 7.5 (at 20°C); store at –20°C.

3. DEPC water: To sterilize solutions and minimize exposure to RNase, some of the aqueous solutions are made up and sterilized with DEPC (diethylpyrocarbonate, Sigma #40718) as follows. Sterilize by adding 0.2% (v/v) DEPC and stirring overnight at room temperature in a hood. Thereafter, autoclave to destroy DEPC. Store at 4°C. DEPC-treated solutions are indicated by an asterisk.

4. RNase inhibitor (Promega, RNasin #N2111): 20 units/µL; store at –20°C.

5. DNase (Promega #M610A).

6. Micro Bio-Spin P-30 Tris, RNase-free (Biorad #732-6250).

2.3. Preparation of Tissue

1. Urethane solution (Sigma): 2% solution in HM. Stored at room temperature.

2.3.1. Fixation of Tissue

2. Paraformaldehyde fixative: 8% (w/v) paraformaldehyde in HM. To prepare 8% paraformaldehyde solution, heat about 100 mL water for 1 minute or less at a high setting in a microwave, add 2 g paraformaldehyde to 25 mL heated water using a magnetic stirrer; while stirring add 1 drop 1.0 N NaOH to clear the solution. Let the solution cool to room temperature and adjust the pH to 7.5 with HCl. Store at 4°C. It will stay good for 1 week.

3. Lavdowsky's fixative: ethanol: formalin: acetic acid: water = 50:10:4:40.

4. 3.7% (w/v) Formaldehyde: Dilute concentrated formaldehyde 1:10 with water. Store at room temperature.

5. Ethanol.

6. PBS: 1.5 M NaCl, 0.08 M Na_2PO_4, 0.021 M NaH_2PO_4, pH 7.34.

7. PBT: PBS with 0.1% (v/v) Tween-20.

8. Tween-20.

2.3.2. Treatments After Fixation

1. 100x Proteinase K stock solution: 1 mg/mL Proteinase K (Sigma) in water. To prepare stock solution, dissolve 5 mg Proteinase K in 5 mL of water, place 0.5 mL Eppendorf tubes in dry ice, and aliquot 50 µL into each tube; it will freeze instantly, and store at –70°C. Avoid refreezing aliquots.

2. Glycine stock solution: 40 mg/mL glycine in DEPC-treated water.

3. Glycine working solution: 1:10 dilution of glycine stock solution in PBT. Final concentration of glycine is 4 mg/mL, DEPC-treated.

4. Acetic anhydride (Sigma-Aldrich #242845).

5. Glycine.

6. TEA: triethanolamine 10x, pH 7.8 (Sigma #T9534).

2.4. Hybridization

1. Formamide: Fluka (Sigma-Aldrich #47671) deionized, stored at 4°C.

2. Hybridization solution: 50% formamide, 5x SSC, 200 µg/mL tRNA, 0.1% (v/v) Tween-20, 0.1% CHAPS, 1x Denhardt's, 100 µg/mL heparin, DEPC water. Make fresh for each experiment from the following stock solutions: 100% formamide, 20x SSC, 10 mg/mL tRNA, 100% Tween-20, 1% CHAPS, 50X Denhardt's, 10 mg/mL heparin. Make enough for pre-hyb, hyb, and post-hyb steps. tRNA is very expensive. A less expensive alternative is to use *Torula* RNA (Sigma #R3629; i.e., ribonucleic acid from torula yeast), which needs to be purified with phenol extraction to remove the protein and other contaminants. Use the purified material at the same concentration as tRNA.

3. Heparin: Heparin sodium salt/Grade I-A, ~170 USP units/mg (Sigma #H-3393).

4. 50x Denhardt's: 1% (v/v) PVP (polyvinyl pyrrolidone) (Sigma #PVP360), 1% (w/v) Ficoll (Sigma F 2637), and 1% (w/v) BSA fraction V (Sigma #05488) in water. Filter-sterilize and store at 4°C. Alternatively, 50x Denhardt's Solution is commercially purchased (Invitrogen #750018).

5. 20X SSC: 3 M NaCl, 0.3 M NaCitrate in distilled water. pH is normally around 7.0 without adjusting; store at room temperature.

6. tRNA (yeast, from Sigma or Fisher): 10 mg/mL in water, aliquot in 5 mL samples. Store at –20°C. Store working solution at 4°C.

7. CHAPS: 3-[(3-Cholamidopropyl)dimethylammonio]-1-propanesulfonate (Sigma # C9426), 1 or 10% solution made up in water. Store at room temperature.

2.5. Antibody Binding

1. Antibody: Fab fragments from an anti-digoxigenin antibody from sheep, conjugated with alkaline phosphatase (Roche #11 093 274 910), stored frozen at –20°C.

2. MAB: 100 mM maleic acid (Sigma, M0375), 150 mM NaCL, pH7.5. Adjust pH with NaOH.

3. MAB-B: MAB with 1% (w/v) BSA fraction V (Sigma; stored at 4°C). Make up fresh and store at 4°C for up to a week.

2.6. Staining Reaction, Post-staining, and Mounting

1. Cell culture dishes: 24-well cell culture dishes.

2. MAB: 100 mM maleic acid (Sigma, M0375), 150 mM NaCL, pH 7.5. Adjust pH with NaOH.

3. MAB-B: MAB with 1% (w/v) BSA fraction V (Sigma; stored at 4°C). Make up fresh and store at 4°C for up to a week.

4. Levamisole stock solution: 1 M levamisole (Tetramisole hydrochloride, Sigma #L9756; i.e., 6 mg/25 μL).

5. Blocking solution: 80% (v/v) MAB-B, 20% (v/v) sheep serum (Sigma # S2263). Sheep serum is heat-inactivated at 55°C for 30 minutes before adding to MAB-B. Store at 4°C with 0.01% (w/v) sodium-azide.

6. NBT/BCIP stock solution (Roche #11681451001): 18.75 mg/mL NBT (Nitro blue tetrazolium chloride) and 9.4 mg/mL BCIP (5-Bromo-4-chloro-3-inphate, toluidine salt) in 67% (v/v) DMSO, store at –20°C. Roche stock solution is aliquoted as 0.5 mL samples and stored frozen, BCIP may begin to precipitate with repeated freezing–thawing cycles. Discard at that point.

7. BM Purple stock solution (Roche #11 442074001): 100 mL in dark brown bottle. Store in box at 4°C.

8. NTMT: 100 mM NaCL, 100 mM Tris, pH 9.5, 50 mM MgCl$_2$, 0.1% (w/v) Tween-20 (Sigma #P7949). The sterilized solution without Tween-20 can be stored indefinitely at room temperature.

9. Formaldehyde.

10. Methanol.

11. Ethanol.

12. Euparal: Neutral mounting medium (Carolina Biological Supply #86-1890), stored at room temperature.

13. PBS:glycerol: 1:7 solution of PBS:glycerol.

2.7. Optional RNase Treatment

1. RNase A (Sigma #R6513) 10 mg/mL in water. Boil for 10 minutes to remove DNAase and store at –20°C.

2. 20X SSC: 3 M NaCl, 0.3 M NaCitrate in distilled water; pH is normally around 7.0 without adjusting. Store at room temperature.

3. CHAPS: 3-[(3-Cholamidopropyl)dimethylammonio]-1-propanesulfonate (Sigma # C9426), 1 or 10% solution made up in water. Store at room temperature.

4. PBS: 1.5 M NaCl, 0.08 M Na_2PO_4, 0.021 M NaH_2PO_4, pH 7.34.

5. PBT: PBS with 0.1% (v/v) Tween-20.

3. Methods

3.1. Synthesis of Anti-sense and Sense Riboprobes

3.1.1. Preparation of Linearized DNA

1. Linearize samples of plasmid DNA containing insert. Linearize one sample downstream of the insert for T3 (or SP6) RNA polymerase and the other for T7 polymerase.

2. Verify the digestion by agarose gel electrophoresis (*see* **Note 1**).

3. Carry out a phenol-chloroform extraction followed by a chloroform extraction.

4. Pool aqueous phases and precipitate with ethanol.

5. Resuspend DNA in minimal volume TE or water; use aliquot to determine amount by measuring OD_{260}.

6. Store at –20°C.

3.1.2. Transcription Reaction

1. A single transcription reaction mix (20 µL) contains: DEPC-water to 20 µL, 2 µL 10x transcription buffer, 2 µL 10x digoxygenin labeling mix, 1 µL RNasin, 1 µg linearized DNA, and 2 µL T3 or T7 polymerase. Add components in this order (DEPC-water first and enzyme last; *see* **Note 2**).

2. Prewarm all components, except enzymes to 37°C before assembling reaction mix. Enzymes should be kept on ice (*see* **Note 4**).

3. Mix and incubate at 37°C for 2 to 3 hours.

4. Verify RNA synthesis: run 1 µL of reaction mix on MOPS/formaldehyde gel. By running standards, one can estimate the amount synthesized (*see* **Note 5**).

5. Treat remainder of reaction mix with 2 µL (i.e. 20 units) of RNase-free DNase I solution for 15 minutes at 37°C and stop DNase reaction with 2 µL 0.2 M EDTA.

    a. Optional purification: Purify RNA through RNase-free Micro Bio-Spin P-30 columns.

    b. Optional precipitation (*see* **Note 6**).

6. Precipitate RNA by adding 2.5 µL 4 M LiCl and 75 µL pre-chilled ethanol. Incubate for 30 minutes at –70°C or for 2 hours at –20°C (*see* **Note 6**).

7. Remove ethanol.

8. Resuspend at 10 ng/µL in DEPC water (*see* **Note 7**).

9. Aliquot into small volumes (*see* **Note 8**).

10. Store at –20°C.

3.2. Tissue Preparation

Fix 6–10 animals per sample. Fixation and postfixation treatments (Proteinase K treatment, acetic anhydride treatment, and heat treatment) are carried out in 1-mL volumes in 2-mL well cell culture plates. All incubation steps are carried out at room temperature and on a nutator, unless otherwise indicated. For a given transcript, use separate samples for anti-sense probe, sense probe and no probe (*see* **Notes 9** and **10**).

3.2.1. Fixation of Tissue

1. Relax 24 to 36 hours starved animals in 2% of urethane solution for 1 minutes at room temperature (*see* **Note 11**).

2. Replace urethane solution with 1.0 mL 4% paraformaldehyde at room temperature (*see* **Notes 12–14**).

4. Fix overnight at 4°C (*see* **Note 15**).

5. Replace fixative with 100% ethanol (*see* **Note 16**) and let specimens sit at room temperature for 10 minutes.

6. Rehydrate tissue carefully with the following set of washes (each 5–10 minutes): 75% ethanol, 25% PBT; 50% ethanol, 50% PBT; 25% ethanol, 75% PBT; 3x PBT.

3.2.2. Treatment After Fixation

Partial digestion of fixed tissue aids in penetration of RNA probe and re-fixation is required to stabilize digested tissue. A treatment with acetic anhydride blocks amino groups in proteins and is presumed to reduce background. A heat treatment reduces or eliminates endogenous alkaline phosphatase activity. All steps, except the heat step, are carried out at room temperature.

1. Add 10 µL 100x Proteinase K stock solution (final concentration is 10 µg/mL) to fixed animals in 1 mL PBT.

2. Incubate for 10 minutes at room temperature (*see* **Note 17**).

3. Rinse once in 1 mL glycine working solution (final concentration is 4 mg glycine/mL) to stop enzyme activity (*see* **Note 18**).

4. Replace with 1 mL glycine working solution and incubate for 10 minutes at room temperature.

5. Wash twice for 5 minutes in PET.

6. Wash twice for 5 minutes in 1 mL 0.1 M triethanolamine, pH 7.8 (acetic anhydride treatment; *see* **Note 19**).

7. Replace with 1 mL triethanolamine and 2.5 µL acetic anhydride; shake gently for 5 minutes (*see* **Note 20**).

8. Add another 2.5 µL acetic anhydride; shake gently for 5 minutes (*see* **Note 21**).

9. Wash twice for 5 minutes in 1 mL PBT.

10. Refix in 4% paraformaldehyde (*see* **Note 22**). Replace PBT with 1 mL 4% paraformaldehyde in PBT, and fix for 20 minutes at room temperature.

11. Wash five times for 5 minutes in PBT.

12. Transfer animals into 1.75 mL Eppendorf tubes.

14. Heat animals at 80°C for 30 minutes in PBT (*see* **Notes 23** and **24**).

3.3. Hybridization

Animals are first exposed to hybridization solution without RNA probe to block sites where probe can stick nonspecifically; then to hybridization solution with the RNA probe to hybridize, and finally washed to remove unhybridized probe. 1.75 mL Eppendorf tubes are placed in a plastic tube rack, or Styrofoam float with holes, and then placed in a water bath. All incubations are carried out at 55°C. Divide animals into samples for anti-sense, sense, and no probe.

1. Wash animals in 0.5 mL of a 1:1 mix of PBT:hybridization solution for 10 minutes at room temperature or 55°C.

2. Wash with 0.5 mL hybridization solution for 10 minutes (*see* **Note 25**).

3. Incubate in 0.5 mL hybridization solution for 2 hours (*see* **Note 26**).

4. Replace with 100 µL hybridization solution containing 2 µL anti-sense probe, sense probe, or no probe (*see* **Notes 27** and **28**).

5. Incubate for 3 days (*see* **Notes 29** and **30**).

6. Post-hybridization washes: each is 5 minutes in 0.5 mL solution containing: 100% hybridization solution; 75% hybridization solution and 25% 2x SSC; 50% hybridization solution and 50% 2x SSC; 25% hybridization solution and 75% 2x SSC.

7. Wash twice for 30 minutes in 1 mL 2x SSC + 0.1% CHAPS.

3.4. Antibody Binding

The antibody is an anti-digoxygenin antibody conjugated to alkaline phosphatase (AP), with digoxygenin conjugated nucleotides of the probe. Unless indicated otherwise, carry out incubation in 0.7 mL volumes in 1.75 mL Eppendorf tubes. Shake gently at all steps.

1. Wash samples twice for 10 minutes in maleic acid buffer (MAB) at room temperature.

2. Wash once with MAB-B for 60 min at room temperature (*see* **Note 31**).

3. Expose to 1.0 mL blocking solution for 2 hours at 4°C (*see* **Note 32**).

4. Remove blocking solution from animals; replace with 0.5 mL diluted preabsorbed antibody solution (*see* **Note 33**).

5. Incubate overnight at 4°C (*see* **Notes 35** and **36**).

3.5. Staining Reaction, Post-staining, and Mounting

Cary out in 1 mL volumes in 2 mL-well cell culture plates. The BM-purple procedure is more sensitive than the NBT-BCIP procedure. However, BM-purple precipitates might slightly diffuse, which might be critical if a precise cellular localization of the transcripts is required. Periodic observation with a dissecting microscope is necessary to decide when the signal is strong enough, and when different enough between the anti-sense and sense probes. After removing the stain, the tissue is treated with fixative to hold the precipitation product in place. Light background can be removed with methanol. Permanent samples are prepared by dehydrating the samples and mounting them on slides.

3.5.1. NBT-BCIP Staining

1. Wash eight times for 1 hour at room temperature in MAB.

2. Wash overnight at 4°C in MAB (*see* **Note 36**).

3. Wash twice for 5 minutes in NTMT at room temperature (*see* **Note 37**).

4. Wash once for 5 minutes in 1 mL NTMT + 1 µL levamisole stock solution (final concentration is 1 mM; *see* **Note 38**).

5. Incubate in 0.5 mL staining solution: 4.5 µL of NET + 3.5 µL of BCIP per mL of NTMT + 1 mM levamisole.

6. Incubate at room temperature in the dark: enclose plate with samples in tinfoil. Gently shake for first 20–30 min. Then let sit on bench top.

7. Examine periodically to determine extent of color reaction. Stain until anti-sense sample has a strong signal and is clearly different from sense sample (*see* **Note 39**).

8. Wash 2x with 1 mL NTMT for 10 minutes at room temperature to stop reaction (*see* **Note 40**).

9. Fix in 3.7% formaldehyde for 30 minutes at room temperature.

10. If background is present, rinse out formaldehyde two to three times with methanol.

11. Incubate in methanol for 5–10 minutes until background disappears (*see* **Notes 41** and **42**).

12. Dehydrate in an increasing ethanol series with 2 min at each step: once 70% ethanol; once 95% ethanol; once 100% ethanol (*see* **Note 43**).

13. Mount on glass slides with Euparal: put one drop of Euparal on slide, place animals on top of drop, put cover slip with small drop of Euparal on top of animals, place brass weight on cover slip to flatten samples before Euparal hardens (*see* **Notes 44** and **45**).

14. Store in slide box at room temperature.

3.5.2. BM-purple Staining

1. Wash eight times for 1 hour at room temperature in MAB

2. Wash overnight at 4°C in MAB (*see* **Note 36**).

3. Wash twice for 5 minutes in NTMT at room temperature (*see* **Note 37**).

4. Wash once for 5 minutes in 1 mL NTMT + 1 μL levamisole stock solution (final concentration is 1 mM; *see* **Note 38**).

5. Replace last NTMT + levamisole solution with 0.5 mL BM-purple.

6. Cover samples with tin foil and incubate at 37°C.

7. Examine periodically to determine extent of color reaction. Stain until anti-sense sample has a strong signal and is clearly different from sense sample (*see* **Note 39**).

8. Replace BM-purple with 1 mL 100% ethanol. Gently shake for 10 minutes at room temperature.

9. Replace ethanol with 3.7% formaldehyde in water for 30 minutes at room temperature.

10. Dehydrate in an increasing ethanol series with 2 minutes at each step: once 70% ethanol; once 95% ethanol; once 100% ethanol (*see* **Note 43**).

11. Mount on glass slides with Euparal: put one drop of Euparal on slide, place animals on top of drop, put cover slip with small drop of Euparal on top of animals, place brass weight on cover slip to flatten samples before Euparal hardens (*see* **Notes 44** and **45**).

12. Store in slide box at room temperature.

3.5.3. Optional RNase Treatment

RNase treatment is an optional step to improve signal strength by using RNase after the post-hybridization washes to remove unhybridized or mishybridized RNA probe. Steps carried out in 1 mL volumes in 2 mL-well cell culture plate after the hybridization and before the antibody-binding sections.

1. Incubate samples in last post-hybridization wash at 37°C for 10 min.

2. Treat with 20 µg/mL RNase A in 2x SSC + 0.1% CHAPS for 30 minutes at 37°C.

3. Wash twice for 10 minutes with 2x SSC + 0.1% CHAPS at room temperature.

4. Wash twice 30 minutes with 0.2 x SSC + 0.1% CHAPS at 55°C.

5. Wash twice 10 minutes with PBS at room temperature.

6. Wash twice 5 minutes with PBT at room temperature.

4. Notes

1. Linearization of DNA is essential to avoid background. Remaining circular DNA will result in transcripts containing sequences that hybridize to RNA other than target RNA. Alternatively, purified PCR products can be used (e.g., by using M13 primers flanking the regions coding for T3/SP6 and T7 RNA polymerase, pGEM-T vector).

2. Transcription buffer and digoxygenin labeling mix assumed to be RNase-free. Note that RNasin will be inactivated if put in solution without DTT.

3. Reaction can be scaled up to a maximum of 100 µL if needed. Larger volumes will result in less effective transcription rates.

4. Prewarming components of mix prevents that the polyamine spermidine precipitates DNA at 4°C.

5. Transcription reaction should yield 2.5–10 µg RNA depending on quality and transcriptability of template and total reaction time.

6. In case of small amounts of RNA, incubate overnight at −20°C. To remove remaining free nucleotides present in the *in vitro* transcription-mix, precipitate 10 µL *in vitro* transcription-mix with 48 µL H_2O, 118 µL 7.5 M Ammonium acetate, and 203 µL 100% ethanol, incubate for 50 minutes at room temperature and centrifuge for 20 minutes at 4°C at maximum speed in an Eppendorf centrifuge; wash pellet with ethanol (70%) dry at room temperature and resuspend in 10 µL H_2O.

7. Estimate the amount of RNA synthesized from gel using standards (Digoxygenin absorbs at 260 nm wavelength; therefore one cannot determine mass using OD_{260}).

8. To avoid degradation by RNase, do not freeze and thaw RNA. Aliquot the RNA; the content of a single tube should be used for one experiment.

9. Because of variation among animals, it is best to use 6–10 animals.

10. Nonspecific binding of antibody and endogenous AP activity is analyzed by no probe controls. The specificity of the antisense probe is analyzed by sense controls. Both controls are necessary only until the conditions for the optimal probe concentration and time of staining have been established.

11. Treatment of hydra with urethane for more than 1 to 2 minutes causes damage to the plasma membrane, and 30 seconds to 1 minute is enough. Subsequent fixation with warm or cold fixative causes only minimal contraction.

12. Cross-linking fixatives, (such as glutharaldehyde and paraformaldehyde) preserve tissue morphology and prevent loss of mRNA from cells. Precipitating fixatives (such as ethanol/acetic acid) function by precipitating proteins to trap the RNA inside cells. These types of fixatives provide a good probe penetration. However, tissues fixed by precipitating fixatives have a poor cell morphology and also loose mRNA *(23)*. Cross-linking fixatives are therefore considered better fixatives for *in situs* and they are routinely used to fix tissue samples for ISH studies. Lavdowsky's fixative contains both formaldehyde and ethanol, and can be also used.

13. In protocols for other animals this fixation step is frequently carried out at 4°C to minimize the action of any RNases released during fixation. Fixation of hydra at either 4°C or room temperature yielded similar results and animals can be routinely fixed at room temperature.

14. It is critical that the animals remain stretched out after adding fixative. Remove almost all of the urethane solution, but leave the animals covered in a residual amount of fluid, and add the fixative very rapidly to the animals.

15. Fixation overnight compared to 1 to 2 hours at room temperature provides more resistance to proteinase K treatment. However, fixation for longer times (3 or 4 days) led to a deterioration of signal. Animals can be fixed up to 2 days without loss of signal.

16. The ethanol steps are included to help dissolve membranes so that probes and antibody can move in and out of tissue more easily.

17. The appropriate concentration and time of incubation must be optimized. Too strong a treatment will result in animals loosing tentacles, peduncles, or even ectoderm through the

many steps of the procedure. Optimal treatment may vary among strains and species.

18. Make up working solution for all samples; then distribute.

19. Treatment with acetic anhydride helps to reduce background and increase signal possibly by neutralizing NH_2^+ groups exposed by the proteinase K treatment. Amine groups in proteins are believed to induce nonspecific binding. Acetylation of amine groups by acetic anhydride maybe important to reduce backgrounds for large probes 2.0 kb *(23)*.

20. Because these two solutions do not mix readily, add together and vortex vigorously just before adding to animals.

21. Mix well but without disrupting animals.

22. Unlike the 4% paraformaldehyde used for the initial fixation, this 4% paraformaldehyde solution is made up in PBT instead of hydra medium.

23. Treatment at 95°C for 5–10 minutes will knock out the endogenous AP also. However, the morphology is strongly affected, as the ectoderm is often lost.

24. This will cause animals to shrink by about half in size.

25. Various protocols use temperatures ranging from 40 to 65°C. Some use different temperatures for prehybridization and for hybridization. 55°C for all steps are used for convenience; if stringency must be increased higher temperatures (60 to 65°C) can be chosen.

26. One can store samples at –20°C before hybridization, but this has not been tested for hydra.

27. Optimal probe concentration will depend on target RNA concentration. When using the standard substrates (NBT and BCIP) for alkaline phosphatase (AP), we are using probes in the concentration range of 0.1 to 1.0 ng/μL. 1.0 ng/μL is often too high, and 0.3 ng/μL may be a better starting point. For strongly expressed genes, 0.01 ng/μL is sufficient. When using BM-purple as the substrate, use probe concentrations that are diluted by 50x.

28. Place no more than 10 or 15 animal pieces in a tube. Too many animals cause incomplete hybridization of probe to target and patchy patterns.

29. Incubation time depends on target concentration and must be optimized for each probe. If rare, incubation over the weekend (3 days) is convenient.

30. Probe solution can be stored at –20°C and reused two to three times with no loss of activity. If using the BM-purple reagent, this is probably unnecessary.

31. The -B part of MAB-B is 1% BSA. It is also possible to use the 2% Roche reagent, which is casein treated with RNase and DNase instead of BSA.

32. Blocking solution can be reused. Return solution to working solution. Store working solution at 4°C and reuse up to five times, but not longer than 1 month.

33. The purpose is to remove antibody that will stick non-specifically to fixed tissue. Preabsorb antibody with fixed hydra by adding 20 fixed, washed hydra to 0.5 mL at a 1:400 dilution of anti-dig-AP antibody in blocking solution, gently shake at room temperature for 2 hours, and dilute to a final concentration of 1:2000 with 2.0 mL blocking solution. Store remaining preabsorbed antibody at 4°C with 0.01% (w/v) NaAzide for a maximum of 2 weeks. Preabsorbed antibody can probably be used five times or for up to 2 weeks without loss of activity. Carefully wash after antibody incubation because remaining NaAzide can inhibit phosphatase activity during the substrate reaction. The preabsorption step can be carried out in parallel with the blocking step.

34. The antibody is commonly diluted in the 1:1–5000 x range.

35. Probably necessary for rare mRNAs. For less rare transcripts, 3 to 4 hours at room temperature is enough.

36. Numerous washes are necessary to wash out non-specifically bound antibody. The last overnight wash is for convenience so that the staining reaction can be observed all the next day. Can shorten steps 1–8 to 10 times 30-minute washes, and subsequently run the staining reaction on the same day. As signals for genes examined so far come up in 1–3 hours, the overnight step is most likely unnecessary. The number and length of washing steps may be optimized for each probe.

37. High pH is necessary for the AP reaction.

38. Levamisole inhibits endogenous AP (probably not necessary when heat-treated samples are used).

39. Periodically depends on probe. Initially may be well to check every 5–15 minutes with dissecting microscope. If stain is slow in developing, then check every 30 minutes.

40. Keeps pH high, while washing out unused substrate.

41. Methanol is effective in removing light background (brown to light purple means small precipitations, but is ineffective with large precipitations [purple to black]).

42. If background is not removed or not sufficiently diminished, the methanol treatment may be extended for up to overnight.

43. For temporary (a few days to a couple of weeks) one can mount samples in PBS: glycerol at 1:3 or 1:9.

44. Alternatively, pipette ethanol-washed animals on slide: remove ethanol by draining ethanol onto tissue paper: add a drop of Euparal and gently arrange animals by pushing with forceps.

45. Prevents air bubbles from getting trapped in or on mounted samples.

References

1. Gall, J. G., Pardue, M. (1969) Formation and detection of RNA-DNA hybrid molecules in cytological preparations. *Proc Natl Acad Sci USA* 63, 378–383.

2. John, H. A., Birnstiel, M. L., Jone, K. W. (1969) RNA-DNA hybrids at the cytological level. *Nature* 223, 582–587.

3. Tautz, D., Pfeifle, C. (1989) A non-radioactive in situ hybridization method for the localization of specific RNAs in Drosophila embryos reveals translational control of the segmentation gene hunchback. *Chromosoma* 98, 81–85.

4. Grens, A., Mason, E., Marsh, J. L., et al. (1995) Evolutionary conservation of a cell fate specification gene: the Hydra achaete-scute homolog has proneural activity in Drosophila. *Development* 121, 4027–4035.

5. Grens, A., Gee, L., Fisher, D. A., et al. (1996) CnNK-2, an NK-2 homeobox gene, has a role in patterning the basal end of the axis in hydra. *Dev Biol* 180, 473–488.

6. Grens, A., Shimizu, H., Hoffmeister, S. A., et al. (1999) The novel signal peptides, pedibin and Hym-346, lower positional value thereby enhancing foot formation in hydra. *Development* 126, 517–524.

7. Martinez, D. E., Dirksen, M. L., Bode, P. M., et al. (1997) Budhead, a fork head/HNF-3 homologue, is expressed during axis formation and head specification in hydra. *Dev Biol* 192, 523–536.

8. Gauchat, D., Kreger, S., Holstein, T., et al. (1998) prdl-a, a gene marker for hydra apical differentiation related to triploblastic paired-like head-specific genes. *Development* 125, 1637–1645.

9. Technau, U., Bode, H. R. (1999) HyBra1, a Brachyury homologue, acts during head formation in Hydra. *Development* 126, 999–1010.

10. Smith, K. M., Gee, L., Bode, H. R. (2000) HyAlx, an aristaless-related gene, is involved in tentacle formation in hydra. *Development* 127, 4743–4752.

11. Broun, M., Sokol, S., Bode, H. R. (1999) Cngsc, a homologue of goosecoid, participates in the patterning of the head, and is expressed in the organizer region of Hydra. *Development* 126, 5245–5254.

12. Takahashi, T., Koizumi, O., Ariura, Y., et al. (2000). A novel neuropeptide, Hym-355, positively regulates neuron differentiation in Hydra. *Development.* 127, 997–1005.

13. Technau, U., Cramer von Laue, C., Rentzsch, F., et al. (2000) Parameters of self-organization in Hydra aggregates. *Proc Natl Acad Sci USA* 97, 12127–12131.

14. Hobmayer, B., Rentzsch, F., Kuhn, K., et al. (2000) WNT signalling molecules act in axis formation in the diploblastic metazoan Hydra. *Nature* 407, 186–189.

15. Guder, C., Pinho, S., Nacak, T. G., et al. (2006) An ancient Wnt-Dickkopf antagonism in Hydra. *Development* 133, 901–911.

16. Rentzsch, F., Guder, C., Vocke, D., et al. (2007) An ancient chordin-like gene in organizer formation of Hydra. *Proc Natl Acad Sci USA* 104, 3249–3254.

17. Lindgens, D., Holstein, T. W., Technau, U. (2004) Hyzic, the Hydra homolog of the zic/odd-paired gene, is involved in the early specification of the sensory nematocytes. *Development* 131, 191–201.

18. Scholz, C. B., Technau, U. (2003) The ancestral role of Brachyury: expression of NemBra1 in the basal cnidarian Nematostella vectensis (Anthozoa). *Dev Genes Evol* 212, 563–570.

19. Fritzenwanker, J. H., Saina, M., Technau, U. (2004) Analysis of forkhead and snail expression reveals epithelial-mesenchymal transitions during embryonic and larval development of *Nematostella vectensis*. *Dev Biol* 275, 389–402.

20. Kusserow, A., Pang, K., Sturm, C., et al. (2005) Unexpected complexity of the Wnt gene family in a sea anemone. *Nature.* 2005 433, 156–60.

21. Sugiyama T, Fujisawa T. (1978) Genetic analysis of developmental mechanisms in Hydra. II. Isolation and characterization of an interstitial cell-deficient strain. *J Cell Sci* 29, 35–52.

22. Achermann J, Sugiyama T. (1985) Genetic analysis of developmental mechanisms in hydra. X. Morphogenetic potentials of a regeneration-deficient strain (reg-16). *Dev Biol* 107, 13–27.

23. Lawrence, J. B., Singer, R. H. (1985) Quantitative analysis of in situ hybridization methods for the detection of actin gene expression. *Nuc Acids Res* 15, 1777–1799.

Part IV

C. Elegans

Chapter 8

Analysis of Wnt Signaling During *Caenorhabditis elegans* Postembryonic Development

Samantha Van Hoffelen and Michael A. Herman

Abstract

Wnts play a central role in the development of many cells and tissue types in all species studied to date. Like many other extracellular signaling pathways, secreted Wnt proteins are involved in many different processes; in *C. elegans* these include: cell proliferation, differentiation, cell migration, control of cell polarity, axon outgrowth and control of the stem cell niche. Perturbations in Wnt signaling are also key factors in cancer formation, and therefore of interest to oncobiologists. Wnts are secreted glycoproteins, which bind to Frizzled transmembrane receptors and signal either through, or independently of β-catenin. Both β-catenin-dependant (Wnt/β-catenin) and -independent pathways function during postembryonic development in *C. elegans* and allow Wnt researchers to explore aspects of Wnt signaling both in common with other organisms and unique to the nematode. **Chapter 9** in **Volume 2** discusses various processes controlled by Wnt signaling during *C. elegans* embryonic development; this chapter discusses Wnt-controlled processes that occur during postembryonic development, including an overview of methods used to observe their function.

Key words: *C. elegans*, Wnt signaling, cell polarity, cell fate specification, cell migration, cell lineage.

1. Introduction

1.1. The Wnt Pathway in C. elegans

The nematode *C. elegans* contains many components of the Wnt signaling pathway, most with homologous counterparts conserved in other organisms. *C. elegans* has five genes encoding the secreted Wnt ligand: *lin-44*, *egl-20*, *mom-2*, *cwn-1*, and *cwn-1*. These proteins signal through the Frizzled family of Wnt receptors: LIN-17, MOM-5, MIG-1 and CFZ-2, and Disheveled proteins: MIG-5, DSH-1, and DSH-2. The many Wnt signaling components then function through the four β-catenin proteins

Elizabeth Vincan (ed.), *Wnt Signaling, Volume II: Pathway Models, vol. 469*
© 2008 Humana Press, a part of Springer Science+Business Media, New York, NY
Book doi: 10.1007/978-1-60327-469-2

BAR-1, WRM-1, HMP-1, and SYS-1. In the Wnt/β-catenin pathway, activation of the Wnt pathway causes Disheveled to inactivate GSK3 which frees β-catenin from the "destruction complex," allowing β-catenin to translocate to the nucleus where it complexes with TCF/LEF family members to activate target genes. β-catenin independent Wnt pathways include the Wnt/calcium and Wnt/Jnk or PCP pathways that signal through frizzled type receptors and disheveled, but function independent of β-catenin.

This chapter focuses on Wnt signaling during *C. elegans* postembryonic development and is organized by specific Wnt-controlled processes. These postembryonic processes could be used to assay potential Wnt function during development. The following discusses the known Wnt components of each developmental process and the methods employed to observe and assay it.

1.2. Methods to Detect Wnt Signaling in C. elegans

Wnts have been shown to be involved in the developmental processes of differentiation, polarity, and migration. *C. elegans* has proven to be an excellent genetic model organism because of its rapid generation time, small size, known cell lineage and hermaphroditic mating, making the identification and characterization of genes involved in developmental processes rather easy. In order to design a genetic screen to isolate genetic mutations that affect a specific biological process of interest, namely the Wnt pathway, one must develop an assay that accurately determines whether the process has been disrupted. Knowledge of *C. elegans* lineages, cell positioning, cell migrations, and anatomy during wild-type development is essential in developing such an assay. Many *C. elegans* screens begin with a population of wild-type hermaphrodites exposed to ethyl methane sulphonate (EMS). EMS induces random mutations in the sperm and oocytes of wild type hermaphrodites. Hermophroditic fertilization will generate a heterozygous F1 individual, one quarter of whose progeny will become homozygous for any given mutation, (for the art and design of screens, see ref. *1*). One can also analyze the development of a known Wnt mutant and discover an alternative process controlled by a given Wnt protein.

Introduction of double-stranded RNA (dsRNA) has the ability to interfere with gene function in *C. elegans (2, 3)*. dsRNA with sequence homology to an endogenous gene or transgene induces silencing of the corresponding gene by a process called RNA interference (RNAi). RNAi is an excellent tool for studying gene function and dissecting the role of genes in specific developmental pathways *(4)*, in our case the Wnt pathway. A full *C. elegans* RNAi library was generated by Julie Ahringer's group and is available (www.geneservice.co.uk/products.rnai/Cel-egans.jsp); the library covers an estimated 87% of the genome.

An initial RNAi screen of chromosome I increased the number of genes with phenotypes from 70 to 378 on Chromosone I *(4)*, showing the power and efficiency of such a screen. Since then, many additional RNAi screens have continued to identify new gene functions, including those in the Wnt pathway *(5)*.

2. Migration of the Descendants of the QL Neuroblast

The QL and QR neuroblasts are cells born on the left and right side of *C. elegans* at the same relative anterior/posterior positions. During the first larval stage (L1) QL and QR generate similar progeny, three neurons and two cells that undergo programmed cell death. The QR descendants, referred to collectively as the QR.d, migrate anteriorly and the QL descendants (QL.d) migrate posteriorly. The migration of the QL.d is dependent on the HOX gene *mab-5*; cells that express *mab-5* migrate posteriorly, whereas cells with no *mab-5* expression migrate anteriorly *(6)*. The expression of *mab-5* and therefore the migration of QL.d and QR.d is controlled by Wnt signaling *(7–9)*. Quantification of QL neuroblast migration is an excellent tool for studying the Wnt/β-catenin pathway in potential mutants, as all components of the pathway are involved. **Fig. 8.1** illustrates QL migration and shows the phenotypes of known members of the pathway. Positions of cells in the Q lineage are recorded based on their location relative to the adjacent epidermal cells or V cells. During wild-type development QR migrates a short distance from its birth position between V4 and V5, divides and its descendants, which do not express *mab-5*, continue to migrate anteriorly toward V1. The QL cell migrates a short distance, begins to express *mab-5*, divides and QL.ap migrates to the tail. QL.paa and QL.pap do not migrate.

Loss-of-function mutations in *mig-14/Wls* (10), *egl-20/Wnt*, *lin-17/Fz*, *mig-1/Fz*, *mig-5/Dsh* and *bar-1/β-catenin*, *pop-1/Tcf*, and *mab-5/Hox*, all cause the loss of *mab-5* expression in the QL lineage and resultant anterior migration of both the QL.d and QR.d, overexpression of *C. elegans* TCF, POP-1, also causes the anterior migration of the QL.d. In addition, mutations in the Axin homologs *pry-1* and *axl-1* cause posterior migrations of QR.d whereas overexpression of *pry-1*, *axl-1*, or *sgg-1* an ortholog of GSK-3 causes anterior migration of QL.d, suggesting they function negatively in this pathway *(9, 11, 12)*. Interestingly, it is not the level of Wnt signal that determines the migratory route but the sensitivity to the EGL-20 signal, such that signaling is activated in the QL.d but not in the QR.d *(13)*.

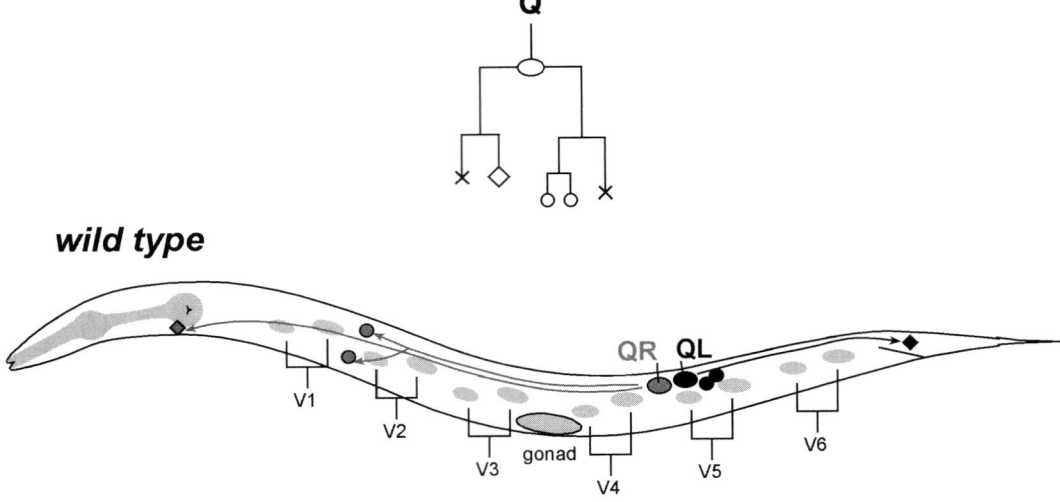

wild type

mig-14, egl-20, lin-17, mig-1, mig-5, bar-1, pop-1, mab-5(lf)
overexpression of sgg-1, pry-1, axl-1

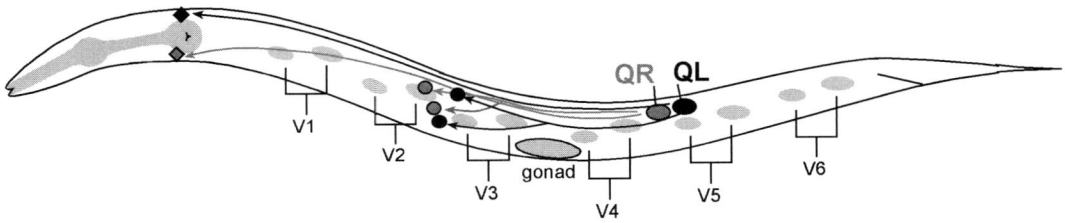

pry-1, pry-1; axl-1, mab-5(lf)

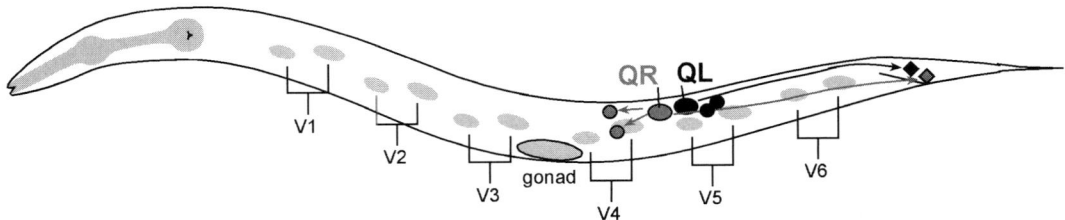

Fig. 8.1. Wnt/β-catenin signaling controls migration of the Q descendents. The Q neuroblast lineage is shown on top. QL (black oval) and QR (dark gray oval) each divide to generate three neurons (diamond and circles) and two cells that undergo programmed cell death (X). Migrations of the QL.d and QR.d in wild-type and mutant animals are shown below. Both are projected onto the views of the left side of a mid-L1 hermaphrodite larva. Positions of the gonad primordium (outlined large light gray oval) and V cell daughters (small light gray ovals) are shown and used for spatial references. The initial positions of QL and QR and the final positions of the QL.d and QR.d are indicated.

Analysis of QL neuroblasts requires Nomarski differential interference contrast (DIC) microscopy and an understanding of *C. elegans* anatomy. In addition, *mec-7::gfp* is expressed in one of the migrating cells in each Q lineage: AVM (QR.paa) and PVM

(QL.paa) and can be used to score or screen for Q migration defects *(14)*.

3. Vulval Fate Specification and Vulval Development

Twelve P cells; six cells on each side of the worm, migrate to the ventral midline, interdigitate and divide. The six central cells P3.p to P8.p express the Hox gene *lin-39*, determining them to be the vulval precursor cells, the VPCs. The six VPC cells are multipotent and can adopt any of three vulval fates: 1°, 2°, or 3° (reviewed in ref. *15*). During the L3 stage, an inductive signal from the anchor cell in the overlying gonad activates receptor tyrosine kinase (RTK) pathway that specifies P6.p to take on the 1° fate. P6.p then signals neighboring P5.p and P7.p through a Notch-related pathway to specify them to take on the 2° fate. Wnt signaling appears to regulate the competence of the VPCs to participate in vulval development as well as the polarity of P7.p (reviewed in refs. *15* and *16*). Mutations in Wnt pathway components that promote signaling such as *bar-1/β-catenin*, *mig-14/Wls* and *pop-1/Tcf* causes fewer VPCs to adopt vulval cell fates, whereas mutations in components that inhibit signaling such as *pry-1Axin* and *apr-1/APC* cause too many VPCs to adopt vulval fates. Redundancy is a key feature of Wnt signaling in vulval development. For example, another Axin, AXL-1, was recently found to function redundantly in the pathway *(12)*. In addition, all five Wnt signaling proteins seem to play a role in VPC specification, with one, CWN-1 functioning antagonistically to two others, LIN-44 and MOM-2 *(17)*. Loss of vulval cell fates results in a Vulvaless (Vul) defect and too many VPCs participating in vulval development results in a Multivulva (Muv) defect in which several protuberances form along the ventral surface of the animal, both of which can be observed in the compound or dissecting microscopes. In addition, VPC fates can be monitored by the expression of several cell-type-specific markers. These include: *egl-17::gfp* for the 1° cell fate, *ceh-2::gfp* and *cdh-3::gfp* for the 2° cell fate *(18)*.

The polarities of divisions of the P5.p and P7.p cells, which take on 2° fate, are mirror symmetric relative to the center of the vulva. Wnt signaling causes P7.p to have a polarity opposite to that of P5.p. This includes the cell division pattern as well as nuclear level of POP-1/Tcf in the P7.p daughters (see **Section 5**; ref. *19*). Mutations in *lin-17* and *lin-18/Ryk* cause P7.p to have the same polarity as P5.p. Three Wnts, LIN-44, MOM-2 and CWN-2, function redundantly through LIN-17 and LIN-18 to control P5.p and P7.p polarities. Specifically, LIN-44 appears to function through LIN-17 and MOM-2 functions through LIN-18 to regulate a common process *(20)*.

4. P12 Fate Specification

P11 and P12 are the most posterior pair of ventral nerve cord precursors and are located laterally in the worm at hatching. In wild-type animals P12 is usually on the right side and P11 on the left. During the L1 stage, P11 and P12 migrate into the ventral nerve cord and divide, each generating a different patterned lineage. Either cell can adopt the fate of P12 before migration *(21)*. Similar to the case for vulval precursor cell fate specification, the Wnt and RTK/Ras pathways interact to regulate the specification of P12 fate. The genes: *lin-3*, *let-23*, and *let-60* of the Ras signaling pathway and *lin-44*, *lin-17*, and *bar-1* of the Wnt pathway as well as the Hox genes *mab-5* and *egl-5* are all involved *(22)*. Observation of cell lineages is the most accurate method to score P12 cell fate. However, P11.p does not divide in hermaphrodites, whereas P12.p divides to generate a cell that dies (P12.pp) and a hypodermal cell (P12.pa) with a unique morphology, which can be used as a marker of P12 fate. In addition, a fragment from the *egl-5* promoter region has been shown to drive *gfp* expression within the P12 lineage but not the P11 lineage *(23)*, and can be used a marker for P12 cell fate.

5. The Control of Cell Polarity and Asymmetric Protein Localization

Wnt signals control the polarities of several cells during *C. elegans* development: The EMS blastomere (see **Chapter 9**), the T and B cells in the tail, V5 in the lateral epidermis, and the Z1 and Z4 cells in the developing gonad (reviewed in refs. *16* and *24)*. One key advantage to using *C. elegans* in studies of cell polarity is the ability to observe division and lineage specification in the living worm. Such studies require no special equipment, protocols, or methods to assay Wnt-controlled processes.

POP-1/Tcf is asymmetrically localized during most asymmetric cell divisions in *C. elegans*, with anterior daughter having higher nuclear levels of POP-1 than posterior daughters *(25)*. This is assayed either by immunolocalization with a POP-1 monoclonal antisera *(25)* or expression of a GFP::POP-1 transgene (**Fig. 8.2**; ref. *26)*. WRM-1/β-catenin, LIT-1/MAPK and POP-1/Tcf, are all asymmetrically distributed to the T-cell daughters

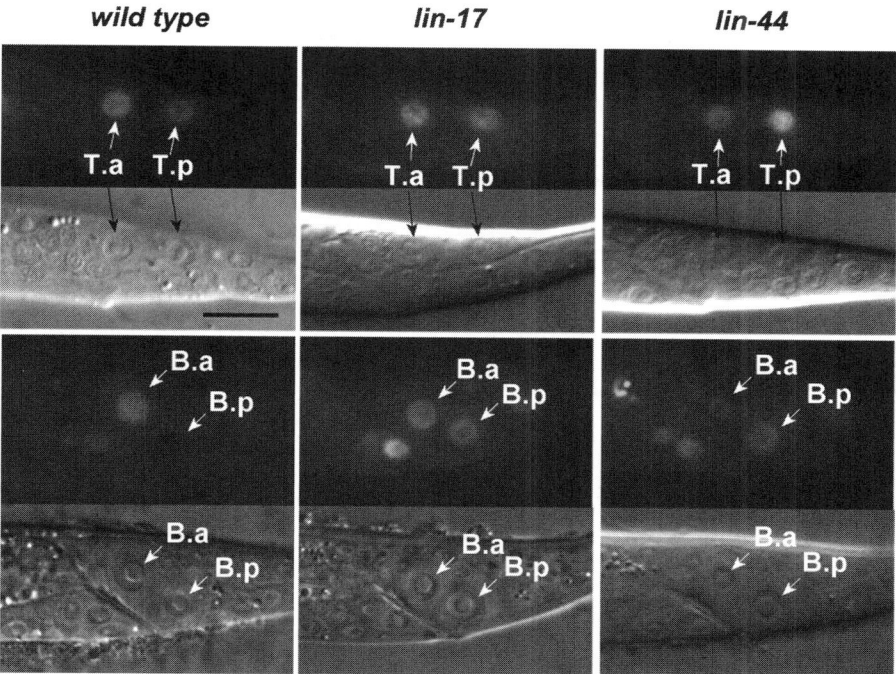

Fig. 8.2. Asymmetric nuclear localization of POP-1/Tcf is controlled by Wnt signaling. Nuclear distribution of GFP::POP-1 is shown for the T-cell (upper) and B-cell (lower) daughters in *wild type*, *lin-44*, and *lin-17* animals carrying *qIs74*. For each pair of cells in each strain, fluorescence and corresponding DIC images are shown. The nuclear levels of GFP::POP-1 are higher in the anterior T- or B-cell daughter than in the posterior daughter in *wild-type* animals, are equal in both daughters in *lin-17* mutants and are higher in the posterior T- or B-cell daughter than in the anterior daughter in *lin-44* mutants. Bar equals 10 µm in all panels.

(8, 27, 28). WRM-1 interacts with and activates LIT-1 kinase, which phosphorylates POP-1 and regulates its subcellular localization *(29, 30)*. The asymmetric distribution of POP-1 to nuclei in the early embryo is controlled by differential nuclear export mediated by the 14-3-3 protein PAR-5 and nuclear exportin homolog IMB-4/CRM-1 *(31, 32)*. Furthermore, LIT-1 modification of POP-1 was shown to be required for its asymmetric nuclear distribution in a process that required WRM-1 *(31)*. Both LIT-1 and WRM-1 were also differentially localized in a reciprocal pattern to that of POP-1, with higher levels in the posterior E cell nucleus *(31, 32)*. WRM-1 was also localized to the anterior cortex of the anterior MS cell in a process that required MOM-5/Fz *(32)* and both WRM-1 and LIT-1 were localized to the anterior cortex before and during postembryonic lateral hypodermal V5 cell division *(28)*. Nakamura et al. proposed that

Wnt and Src signaling leads to the phosphorylation and retention of WRM-1 in the posterior E nucleus, where it phosphorylates POP-1 in a LIT-1-dependent manner *(32)*. Thus asymmetric nuclear retention of WRM-1 appears to drive the control of cell polarity during the EMS and T-cell divisions. Asymmetric cortical localizations of several proteins were recently shown to be involved in generating WRM-1 nuclear asymmetry in the V5 and T-cell divisions. Specifically, APR-1/APC, WRM-1, LIT-1, and PRY-1/Axin were anteriorly localized, whereas LIN-17, DSH-2, and MIG-5/Dsh were posteriorly localized *(28, 33–35)*. Anterior cortical localization of LIT-1 and WRM-1 appear to inhibit the anterior nuclear localization of WRM-1, leading to higher relative WRM-1 levels in the posterior nucleus, which in turn leads to low POP-1 nuclear level *(34)*.

SYS-1 is another β-catenin homolog that functions to control the polarities of the somatic gonad precursor cells by interacting with POP-1 to control cell fates *(36, 37)*. SYS-1 interacts with POP-1 leading to the activation of gene expression and appears to play a positive role in the specification of cell fates *(37)*. SYS-1 is also asymmetrically localized during the EMS, T and somatic gonad precursor (SGP) cell divisions in a pattern opposite to that of POP-1 in a process that requires LIN-17 and Dsh function *(38)*. Thus low POP-1 and high SYS-1 nuclear levels lead to the activation of target genes, such as CEH-22/Hox in the SGPs *(39)*. Interestingly, SYS-1 nuclear asymmetry is not controlled by WRM-1 and possibly not the other anteriorly localized cortical proteins either, suggesting another mechanism may be operating *(38)*. It is not yet clear whether SYS-1 plays a role in WRM-1 localization.

6. T-cell Polarity

In hermaphrodites the TL and TR cells collectively known as T cells lie in the tail on each side of the animal. They divide asymmetrically to give an anterior daughter T.a that generates primarily epidermal cells, and a posterior daughter; T.p which divides to generate neural cells (**Fig. 8.3**). Mutations in *lin-44/Wnt* cause the asymmetric division of T to be reversed *(40)*. Among the cells generated by T.p are the phasmid sockets, which are glial cells that extend out to the surface of the tail and whose presence can be assayed by their ability to allow the phasmid neurons, PHA and PHB, to take up the lipophilic dye DiO. Phasmid dye filling is used as an indicator of normal T-cell lineage, and is an easy assay to screen for mutants in the laboratory. Briefly, worms are

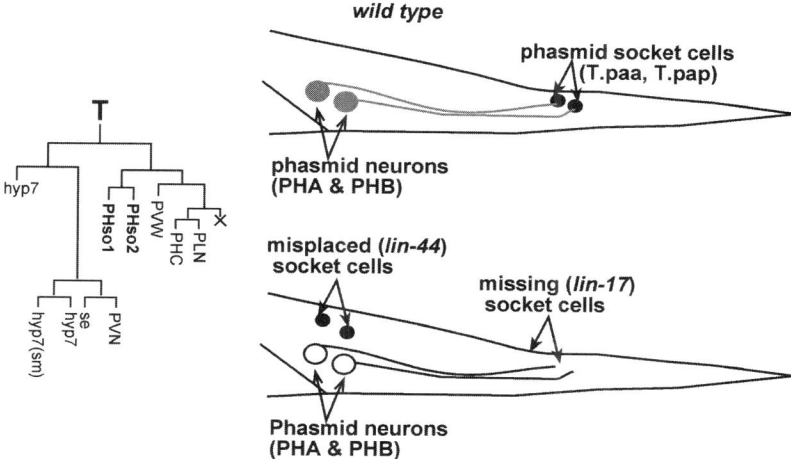

Fig. 8.3. Dye-filling and position of the phasmid socket cells can be used to score -ell polarity. The T-cell lineage is shown on the left and schematics illustrating the positions of the phasmid neurons and the socket cells in *wild type* (above) as well as *lin-44* and *lin-17* (below) on the right. Shading indicates the dye filling status of the phasmid neurons that can be determined by DiO filling. While DiO filling can be done at any life stage, it is easiest to observe in adult hermaphrodites. The phasmid socket cell nuclei can be recognized by their nuclear morphology and position as the most posterior neuronal nuclei on the lateral sides of the animal. They can be seen anytime after their birth in the mid-L1 stage but are easiest to find in L3 animals.

soaked in a 10 µg/mL solution of DiO (in M9 buffer) on a shaker for 2 hours, rinsed multiple times with M9 buffer and plated on *E. coli* OP50 coated NGM plates. After a period of time feeding to clear excess dye from the gut, worms are scored with FITC fluorescence for the presence (WT) or absence of socket cells; such worms are said to be Phasmid dye (Pdy)-filling defective. A more accurate indication of T-cell lineage and polarity defects is the position of the phasmid socket cells, T.paa and T.pap, by DIC optics. This defect is called Psa for phasmid socket cell absent *(41)*.

lin-17/Fz, is also involved in the control of T-cell polarity. Whereas mutations in *lin-44* cause a reversal of polarity, mutations in *lin-17* causes a loss of polarity and both daughters adopt epidermal fates. Mutations in other known Wnt components, *wrm-1*, *lit-1*, *sys-1*, *pop-1* as well as *egl-27*, *tcl-2* and *tlp-1*, also cause a loss of T-cell polarity *(8, 29, 37, 42–44)*. The difference in the phenotype between ligand (LIN-44) and receptor (LIN-17) is of interest and has yet to be explained, one possibility is there is an additional anterior signal working with LIN-17.

7. Gonad Polarity

The gonad of *C. elegans* hermaphrodites is bilaterally symmetric and lies in the center of the developing animal. The somatic gonad develops from two somatic gonad precursor (SGP) cells, Z1 and Z4, on the ends of the gonad primordium that each generates one of the bilaterally symmetric gonad arms. Each SGP divides asymmetrically along the proximal-distal axis with distal cell fates lying at the ends of the primordium and proximal fates lying in the center. The SGPs divide three times in the first larval stage to generate the 12-cell gonad primordium. Ten of these cells have invariant fates, and two become the distal tip cells (DTCs) on each end, which leads the elongation of the gonad arms. The remaining two cells, Z1.ppp and Z4.aaa have variable cell fates, with one becoming the anchor cell (AC) and the other a ventral uterine precursor (VU). However, which cell becomes the AC and which becomes the VU varies from animal to animal *(45)*.

Although a Wnt has not yet been shown to control the asymmetric SGP divisions, Wnt pathway components are involved. Specifically, mutations in *lin-17/Fz, wrm-1/β-catenin, sys-1/β-catenin, lit-1/Nlk* and *pop-1/TCF*, lead to a loss of asymmetry and the generation of two daughters with proximal fates *(36, 46, 47)*. The SGP daughters exhibit POP-1 nuclear asymmetry, with higher nuclear GFP::POP-1 in proximal daughters than in distal ones, which is lost in Wnt pathway mutants *(26)*. SYS-1 is also asymmetrically localized during the EMS, T and somatic gonad precursor (SGP) cell divisions in a pattern opposite to that of POP-1 in a process that requires LIN-17 and Dsh function *(38)*. Thus low POP-1 and high SYS-1 nuclear levels lead to the activation of target genes, such as CEH-22/Hox in the SGPs *(39)*. Interestingly, SYS-1 nuclear asymmetry is not controlled by WRM-1 and possibly not the other anteriorly localized cortical proteins either, suggesting another mechanism may be operating *(38)*. However, it is not yet clear whether SYS-1 plays a role in WRM-1 localization. Asymmetric nuclear localization of GFP::POP-1 and asymmetric expression of CEH-22::GFP can be used to monitor the polarities of the SGP divisions.

8. V5 Polarity

Six epidermal V cells along each side of the postembryonic larvae in *C. elegans* divide asymmetrically. Each anterior daughter Vn.a becomes a syncytial cell and has high POP-1 levels, and each posterior daughter Vn.p becomes a seam cell and continues to divide;

these cells have low nuclear POP-1 levels. The polarity of the V5 cell in the posterior lateral epidermis is controlled by *egl-20/Wnt*. The polarity of the V5 division is reversed in approximately 50% of *egl-20/Wnt* mutants. EGL-20 has been shown to be a permissive rather than an instructive signal for V5 polarity as *egl-20* expressed from a heat shock promoter and a pharynx-specific promoter can rescue V5 polarity. Furthermore, the reversal of polarity seen in *egl-20* mutants seems to require a lateral signal functioning through *lin-17/Fz* and *pry-1/Axin* suggesting an interaction of more than one Wnt pathway *(48)*.

The asymmetric division or polarity of V5 can be visualized by Nomarski differential interference contrast microscopy in L1 stage animals, or by POP-1 localization. Individuals are scored to determine whether Vn.a fuses with the epidermal syncytium and whether Vn.p divides to generate proliferative seam cells. An interaction of Wnt pathways functions to control V cell polarity. Therefore, using the V cells as an assay for Wnt signaling could elucidate a number of molecular signaling components.

9. Postereid and Male Ray Formation

Wnt signaling also plays a role in interactions that occur among the lateral epidermal cells that generate sensilla. In hermaphrodites and males, descendants of the V5 cells, V5.pa, generate the postderid sensilla. Postderid formation by V5.pa is regulated by interactions among the lateral epidermal cells *(49–52)* that regulate *mab-5* expression *(53)*. Specifically, killing the V cells anterior or posterior of V5, causes extopic expression of *mab-5* and V5.pa makes a seam cell instead of a postdereid *(53)*. Ectopic expression of *mab-5* occurs by activation of a Wnt pathway that includes *egl-20/Wnt, lin-17/Fz, bar-1/β-catenin* and *pry-1/Axin (9)*.

In males, different types of sensilla, the sensory rays, are also produced such that each side of the animal generates nine rays: the V5.pp, V6 and T cells generate one, five, and three rays, respectively. Interactions among the lateral epidermal cells regulate the numbers of rays that are generated *(54)*. V4 can generate rays in the absence of V5 and V6 and the number of rays produced by V5 increases in the absence of with V4 or V6, however postderid no postderid in formed *(49, 51, 52)*. *mab-5* is required for generation of the V5- and V6-derived rays. If V6 is killed or if *mab-5* is expressed from a heat-shock promoter, the V5 cell takes on the V6 cell fate, producing extra rays but no postderid *(55)*. However, following the killing of V6, Wnt signaling involving *egl-20Wnt, lin-17/Fz,* and *bar-1/β-catenin* is required for V5 to take on the V6 fate *(53)*. Furthermore, *pry-1* mutations cause ectopic *mab-5* expression (which requires *bar-1*) leading to loss of the postderid and ectopic ray formation *(9)*. This Wnt pathway appears to be inhibited by cell contacts and requires *dpy-22/sop-1/mdt-12* and *sop-3/mdt-1.1*, which encode homologs of the transcriptional Mediator complex components MED12/TRAP230 and MED1/TRAP220, respectively *(56–58)*. The targets of this pathway, either direct or indirect, include *pal-1/caudal* and the Hox genes *mab-5* and *egl-5*.

10. B-cell Polarity

Male tail development involves complex postembryonic cell lineages many involving regulated asymmetric cell divisions. The B, Y, U, and F cells of the male tail divide postembryonically transforming the simple posterior tube of the worm into a complex array or spicules, postcloacal sensillae, proctodeum, and the gubernaculum. Remodeling of this region occurs during late L3 and L4 and results from a series of asymmetric cells divisions and complex cell-cell interactions. The polarity of the B cell and the B-cell lineage is a well-studied tool for the analysis of asymmetric cell divisions. The B cell divides asymmetrically with a large anterior-dorsal daughter B.a and a smaller posterior-ventral daughter B.p (**Fig. 8.2**). B.a divides to produce 40 cells and generates male copulatory spicules, and B.p divides to produce 7 cells *(54)*. Furthermore, B-cell polarity is controlled by Wnt signaling. In *lin-44* mutant males B-cell polarity is reversed *(40, 59)*, while in *lin-17* mutant males B-cell polarity is lost *(46, 60)*.

B-cell polarity is controlled differently than T-cell polarity in that a planar cell polarity (PCP)-like pathway is involved. There are some similarities, such as the involvement of the asymmetrical distribution of POP-1/Tcf. However, there are clear differences. Specifically, MIG-5/Dsh, RHO-1/RhoA, and LET-502/ROCK appeared to play major roles, while other PCP components appeared to play minor roles *(61)*. Furthermore, none of the five *C. elegans* β-catenin homologs *(62)* plays a role in B-cell polarity. While disruption of *wrm-1/β-catenin*, *lit-1/MAPK*, or *sys-1β-catenin* functions caused T-cell polarity defects, including the distribution of GFP::POP-1; little or no B-cell polarity defect was observed and neither *lit-1* nor *wrm-1* affected the asymmetric distribution of GFP::POP-1 to the B.a and B.p cells *(61)*. In addition, the Wnt/β-catenin pathway does not function in the control of B-cell polarity *(61)*. Finally, LIN-17/Fz and MIG-5/Dsh were asymmetrically localized during the B-cell division. Asymmetric localization of LIN-17::GFP was dependent upon LIN-44/Wnt and MIG-5 function whereas asymmetric localization of MIG-5::GFP was dependent upon LIN-44 and LIN-17. This suggests that a Wnt/PCP-like pathway is involved in the regulation of B-cell polarity *(63)*.

11. Conclusions

In *C. elegans*, Wnt pathways control the migration, polarity and fate decisions of many cells. One key advantage of using *C. elegans* in Wnt research is the ability to observe developmental processes as they occur in the living animal. Some of the developmental processes we have discussed share many components and features of the Wnt pathways first described in *Drosophila* and vertebrates, others share some molecular components but signal in a different manner; the method of target gene activation therefore seems to differ among some Wnt pathways. It is clear that *C. elegans* has a conserved Wnt/β-catenin pathway that functions during QL.d migration and vulval development. In addition, in the case of the regulation of cell polarity in *C. elegans*, Wnt pathway components can be asymmetrically localized by at least two different conserved mechanisms. Exactly how asymmetric protein regulation leads to the generation of cell polarity is not yet clear. Finally, the components involved in each developmental process are still being identified. The approaches described here may allow the reader to participate in the search for new Wnt pathway components in *C. elegans*.

References

1. Jorgensen, E. M., Mango, S. E. (2002) The art and design of genetic screens: *Caenorhabditis elegans. Nat Rev Genet* 3, 356–369.

2. Fire, A., Xu, S., Montgomery, M. K., et al. (1998). Potent and specific genetic interference by double-stranded RNA in *Caenorhabditis elegans. Nature* 391, 806–811.

3. Montgomery, M. K., Xu, S., Fire, A. (1998). RNA as a target of double-stranded RNA-mediated genetic interference in Caenorhabditis elegans. *Proc Natl Acad Sci USA* 95, 15502–15507.

4. Fraser, A. G., Kamath, R. S., Zipperlen, P., et al. (2000) Functional genomic analysis of *C. elegans* chromosome I by systematic RNA interference. *Nature* 408, 325–330.

5. Coudreuse, D. Y., Roel, G., Betist, M. C., Destree, O., Korswagen, H. C. (2006) Wnt gradient formation requires retromer function in Wnt-producing cells. *Science* 312, 921–924.

6. Salser, S. J., Kenyon, C. (1992) Activation of a *C. elegans Antennapedia* homologue in migrating cells controls their direction of migration. *Nature* 355, 255–258.

7. Harris, J., Honigberg, L., Robinson, N., Kenyon, C. (1996) Neuronal cell migration in *C. elegans*: regulation of Hox gene expression and cell position. *Development* 122, 3117–3131.

8. Herman, M. (2001) *C. elegans* POP-1/TCF functions in a canonical Wnt pathway that controls cell migration and in a noncanonical Wnt pathway that controls cell polarity. *Development* 128, 581–590.

9. Maloof, J. N., Whangbo, J., Harris, J. M., et al. (1999) A Wnt signaling pathway controls hox gene expression and neuroblast migration in *C. elegans. Development* 126, 37–49.

10. Banziger, C., Soldini, D., Schutt, C., et al. (2006) Wntless, a conserved membrane protein dedicated to the secretion of Wnt proteins from signaling cells. *Cell* 125, 509–522.

11. Korswagen, H. C., Coudreeuse, D. Y. M., Betist, M., et al. (2002) The Axin-like protein PRY-1 is is a negative regulateor of a canonical Wnt pathway in *C. elegans. Genes Dev* 16, 1291–1302.

12. Oosterveen, T., Coudreuse, D. Y., Yang, P. T., et al. (2007) Two functionally distinct Axin-like proteins regulate canonical Wnt signaling in *C. elegans. Dev Biol* 308, 438–448.

13. Whangbo, J., Kenyon, C. (1999) A Wnt signaling system that specifies two patterns of cell migration in C. elegans. *Mol Cell* 4, 851–858.

14. Ch'ng, Q., Williams, L., Lie, Y. S., et al. (2003) Identification of genes that regulate a left-right asymmetric neuronal migration in *Caenorhabditis elegans. Genetics* 164, 1355–1367.

15. Sternberg, P. W. (2005) Vulval development in (Wormbook, ed.) *The C. elegans Research Community*, WormBook, doi/10.1895/wormbook.1.6.1, www.wormbook.org.

16. Eisenmann, D. (2005) Wnt signaling in (Wormbook, ed.) *The C. elegans Research Community*, WormBook, doi/10.1895/wormbook.1.7.1, www.wormbook.org.

17. Gleason, J. E., Szyleyko, E. A., Eisenmann, D. M. (2006) Multiple redundant Wnt signaling components function in two processes during *C. elegans* vulval development. *Dev Biol* 298, 442–457.

18. Inoue, T., et al. (2002) Gene expression markers for *Caenorhabditis elegans* vulval cells. *Mech Dev* 119 Suppl 1, S203–S209.

19. Deshpande, R., Inoue, T., Priess, J. R., et al. (2005) *lin-17/Frizzled* and *lin-18* regulate POP-1/TCF-1 localization and cell type specification during *C. elegans* vulval development. *Dev Biol* 278, 118–129.

20. Inoue, T., Oz, H. S., Wiland, D., et al. (2004) C. elegans LIN-18 is a Ryk ortholog and functions in parallel to LIN-17/Frizzled in Wnt signaling. *Cell* 118, 795–806.

21. Sulston, J. E., Horvitz, H. R. (1977) Post-embryonic cell lineages of the nematode, *Caenorhabditis elegans. Dev Biol* 56, 110–156.

22. Jiang, L. I., Sternberg, P. W. (1998) Interactions of EGF, Wnt and HOM-C genes specify the P12 neuroectoblast fate in *C. elegans. Development* 125, 2337–2347.

23. Teng, Y., Girard, L., Ferreira, H. B., et al. (2004) Dissection of cis-regulatory elements in the *C. elegans* Hox gene *egl-5* promoter. *Dev Biol* 276, 476–492.

24. Herman, M. A. (2003) Wnt signaling in *C. elegans* in (M Kühl, ed.) *Wnt Signalling in Development*, Landes Biosciences, Georgetown, TX, pp. 187–212.

25. Lin, R., Hill, R. J., Priess, J. R. (1998) POP-1 and anterior-posterior fate decisions in C. elegans embryos. *Cell* 92, 229–239.

26. Siegfried, K. R., Kidd, A. R., 3rd, Chesney, M. A., et al. (2004) The *sys-1* and *sys-3* genes cooperate with Wnt signaling to establish the proximal-distal axis of the *Caenorhabditis elegans* gonad. *Genetics* 166, 171–186.

27. Herman, M. A., Wu, M. (2004) Noncanonical Wnt signaling pathways in *C. elegans* converge on POP-1/TCF and control cell polarity. *Front Biosci* 9, 1530–1539.

28. Takeshita, H., Sawa, H. (2005) Asymmetric cortical and nuclear localizations of WRM-1/beta-catenin during asymmetric cell division in *C. elegans. Genes Dev* 19, 1743–1748.

29. Rocheleau, C. E., et al. (1999) WRM-1 activates the LIT-1 protein kinase to transduce anterior/posterior polarity signals in C. elegans. *Cell* 97, 717–726.

30. Maduro, M. F., Lin, R., Rothman, J. H. (2002) Dynamics of a developmental switch: recursive intracellular and intranuclear redistribution of *Caenorhabditis elegans* POP-1 parallels Wnt-inhibited transcriptional repression. *Dev Biol* 248, 128–142.

31. Lo, M. C., Gay, F., Odom, R., et al. (2004) Phosphorylation by the beta-catenin/MAPK complex promotes 14-3-3-mediated nuclear export of TCF/POP-1 in signal-responsive cells in C. elegans. *Cell* 117, 95–106.

32. Nakamura, K., et al. (2005) Wnt signaling drives WRM-1/beta-catenin asymmetries in early *C. elegans* embryos. *Genes Dev* 19, 1749–1754.

33. Goldstein, B., Takeshita, H., Mizumoto, K., Sawa, H. (2006) Wnt signals can function as positional cues in establishing cell polarity. *Dev Cell* 10, 391–396.

34. Mizumoto, K., Sawa, H. (2007) Cortical beta-Catenin and APC Regulate Asymmetric Nuclear beta-Catenin Localization during Asymmetric Cell Division in C. elegans. *Dev Cell* 12, 287–299.

35. Wu, M., Herman, M. A. (2006) Asymmetric localizations of LIN-17/Fz and MIG-5/Dsh are involved in the asymmetric B cell division in C. elegans. *Dev Biol*

36. Miskowski, J., Li, Y., Kimble, J. (2001). The *sys-1* gene and sexual dimorphism during

gonadogenesis in *Caenorhabditis elegans*. *Dev Biol* 230, 61–73.

37. Kidd, A. R., 3rd, Miskowski, J. A., Siegfried, K. R., et al. (2005) A beta-catenin identified by functional rather than sequence criteria and its role in Wnt/MAPK signaling. *Cell* 121, 761–772.

38. Phillips, B. T., Kidd, A. R., 3rd, King, R., et al. (2007). Reciprocal asymmetry of SYS-1/{beta}-catenin and POP-1/TCF controls asymmetric divisions in *Caenorhabditis elegans*. *Proc Natl Acad Sci USA*

39. Lam, N., Chesney, M. A., Kimble, J. (2006) POP-1/TCF and SYS-1/β-catenin control expression of the CEH-22/Nkx2.5 homeodomain transcription factor to specify distal tip cell fate in *C. elegans Curr Biol* 16, 287–295.

40. Herman, M. A., Horvitz, H. R. (1994) The *Caenorhabditis elegans* gene *lin-44* controls the polarity of asymmetric cell divisions. *Development* 120, 1035–1047.

41. Sawa, H., Kouike, H., Okano, H. (2000) Components of the SWI/SNF complex are required for asymmetric cell division in C. elegans. *Mol Cell* 6, 617–624.

42. Herman, M. A., Ch'ng, Q., Hettenbach, S. M., et al. (1999) EGL-27 is similar to a metastasis-associated factor and controls cell polarity and cell migration in *C. elegans*. *Development* 126, 1055–1064.

43. Zhao, X., Yang, Y., Fitch, D. H., et al. (2002). TLP-1 is an asymmetric cell fate determinant that responds to Wnt signals and controls male tail tip morphogenesis in *C. elegans. Development* 129, 1497–1508.

44. Zhao, X., Sawa, H., Herman, M. A. (2003) *tcl-2* encodes a novel protein that acts synergistically with Wnt signaling pathways in C. elegans. *Dev Biol* 256, 276–289.

45. Kimble, J., Hirsh, D. (1979) The postembryonic cell lineages of the hermaphrodite and male gonads in *Caenorhabditis elegans*. *Dev Biol* 70, 396–417.

46. Sternberg, P. W., Horvitz, H. R. (1988) *lin-17* mutations of Caenorhabditis elegans disrupt certain asymmetric cell divisions. *Dev Biol* 130, 67–73.

47. Siegfried, K. R., Kimble, J. (2002) POP-1 controls axis formation during early gonadogenesis in *C. elegans. Development* 129, 443–453.

48. Whangbo, J., Harris, J., Kenyon, C. (2000) Multiple levels of regulation specify the polarity of an asymmetric cell division in *C. elegans. Development* 127, 4587–4598.

49. Austin, J., Kenyon, C. (1994) Cell contact regulates neuroblast formation in the *Caenorhabditis elegans* lateral epidermis. *Development* 120, 313–323.

50. Sulston, J. E., White, J. G. (1980) Regulation and cell autonomy during postembryonic development of *Caenorhabditis elegans*. *Dev Biol* 78, 577–597.

51. Waring, D. A., Kenyon, C. (1991) Regulation of cellular responsiveness to inductive signals in the developing *C. elegans* nervous system. *Nature* 350, 712–715.

52. Waring, D. A., Kenyon, C. (1990) Selective silencing of cell communication influences anteroposterior pattern formation in C. elegans. *Cell* 60, 123–131.

53. Hunter, C. P., Harris, J. M., Maloof, J. N., et al. (1999) Hox gene expression in a single Caenorhabditis elegans cell is regulated by a caudal homolog and intercellular signals that inhibit wnt signaling. *Development* 126, 805–814.

54. Sulston, J. E., Albertson, D. G., Thomson, J. N. (1980) The *Caenorhabditis elegans* male: postembryonic development of nongonadal structures. *Dev Biol* 78, 542–576.

55. Salser, S. J., Kenyon, C. (1996) A *C. elegans* Hox gene switches on, off, on and off again to regulate proliferation, differentiation and morphogenesis. *Development* 122, 1651–1661.

56. Moghal, N., Sternberg, P. W. (2003) A component of the transcriptional mediator complex inhibits RAS-dependent vulval fate specification in *C. elegans. Development* 130, 57–69.

57. Zhang, H., Emmons, S. W. (2000) A *C. elegans* mediator protein confers regulatory selectivity on lineage-specific expression of a transcription factor gene. *Genes Dev* 14, 2161–2172.

58. Zhang, H., Emmons, S. W. (2001) The novel *C. elegans* gene *sop-3* modulates Wnt signaling to regulate Hox gene expression. *Development* 128, 767–777.

59. Herman, M. A., Vassilieva, L. L., Horvitz, H. R., et al. (1995) The C. elegans gene *lin-44*, which controls the polarity of certain asymmetric cell divisions, encodes a Wnt protein and acts cell nonautonomously. *Cell* 83, 101–110.

60. Sawa, H., Lobel, L., Horvitz, H. R. (1996) The *Caenorhabditis elegans* gene *lin-17*, which is required for certain asymmetric cell divisions, encodes a putative seven-transmembrane protein similar to the *Drosophila frizzled* protein. *Genes Dev* 10, 2189–2197.

61. Wu, M., Herman, M. A. (2006) A novel noncanonical Wnt pathway is involved in

the regulation of the asymmetric B cell division in *C. elegans*. *Dev Biol* 293, 316–329.

62. Natarajan, L., Witwer, N. E., Eisenmann, D. M. (2001) The divergent *Caenorhabditis elegans* beta-catenin proteins BAR-1, WRM-1 and HMP-2 make distinct protein interactions but retain functional redundancy *in vivo*. *Genetics* 159, 159–172.

63. Wu, M., Herman, M. A. (2007) Asymmetric localizations of LIN-17/Fz and MIG-5/Dsh are involved in the asymmetric B cell division in *C. elegans*. *Dev Biol* 303, 650–662.

Chapter 9

Wnt Signaling During *Caenorhabditis elegans* Embryonic Development

Daniel J. Marston, Minna Roh, Amanda J. Mikels, Roel Nusse, and Bob Goldstein

Abstract

Wnt signaling has been demonstrated to regulate diverse cell processes throughout the development of the *Caenorhabditis elegans* embryo. This chapter describes methods that have been used to investigate some of these Wnt-dependent processes: endoderm specification, mitotic spindle orientation, and cell migration.

Key words: *C. elegans*, Wnt signaling, embryonic development.

1. Introduction

Wnt signaling has been demonstrated to regulate diverse cell processes throughout the development of the *Caenorhabditis elegans* embryo. A Wnt signal was identified as the signal produced by germ line precursors required to induce endoderm from mesendoderm precursors at the four-cell stage *(1, 2)*. Wnt signaling has been shown to regulate cell fate specification at many other points during *C. elegans* embryonic development *(3)*. More recently, Wnt signals have been shown to regulate other cell biological processes such as orientating mitotic spindles and thereby cell division axes, and regulating cell migrations *(4–8)*. The following

Elizabeth Vincan (ed.), *Wnt Signaling, Volume II: Pathway Models, vol. 469*
© 2008 Humana Press, a part of Springer Science+Business Media, New York, NY
Book doi: 10.1007/978-1-60327-469-2

describes cell biological techniques used to investigate how Wnts regulate these cellular processes in *C. elegans*.

2. Materials

1. 0.1% (w/v) Poly-L-lysine in water (Sigma).
2. 18 mm × 18 mm glass cover slips #1.5 (Corning).
3. Glass slides (Gold-Seal).
4. Sterile egg buffer (EB): 118 mM NaCl, 48 mM KCl, 2 mM CaCl$_2$, 2 mM MgCl$_2$, 25 mM Hepes, pH 7.3 (store at room temperature).
5. 3% (w/v) Agar in EB, melted and kept covered at 60°C.
6. Dissecting microscope.
7. Depression slides.
8. Needles: 25G 5/8.
9. *Caenorhabditis elegans* adult hermaphrodites.
10. Borosilicate capillaries pulled to narrow and wide bore sizes (1 mm inner diameter with filament, World Precision Instruments, *see* **Note 1**).
11. Mouth pipettes (Sigma Aspirator Tube Assembly).
12. Vaseline (melted in an eppendorf tube at 60°C).
13. Small paintbrush.
14. Polarized light microscope with DIC optics.
15. 8-well slides (MP Biomedicals).
16. Humid chamber (clear plastic box with damp paper and supports for slides).
17. Sodium hypochlorite (≥4% Cl, Sigma).
18. Chitinase/chymotrypsin mix: 3 U/mL chitinase (Sigma), 10 U/mL chymotrypsin (Sigma) in EB (store at –20°C).
19. Fix buffer (FB): 10 mM MES, pH 6.1, 125 mM KCl, 3 mM MgCl, 2 mM EGTA, 10% (w/v) sucrose, 4% (w/v) formaldehyde (store at 4°C for up to a month, but allow to warm to room temp before use).
20. 10 mg/mL L-α-lyso-phosphatidyl choline in water (store at –20°C, but allow to warm to room temp before use).
21. Agar/FB: 0.8% (w/v) Agar in FB without sucrose and formaldehyde
22. Agar/PBS: 0.8% (w/v) Agar in PBS.
23. 10% (v/v) Triton X-100 (Sigma) in PBS.

24. TRITC phalloidin 50 µg/mL in methanol (Sigma, store at –20°C).

25. Rabbit anti-P-Ser[19]-MLC (Abcam).

26. Tyramide Signal Amplification Kit (TSA, Molecular Probes).

27. DAPI 1 µg/mL in PBS (Sigma, store at –20°C).

28. Prolong mounting media (Molecular Probes).

29. Isolated blastomeres (method from Edgar and Goldstein *[9]*)

30. Edgar's embryonic growth media (EGM; store at 4°C; ref. *10*).

31. Red sepharose beads (Amersham Biosciences) in Wnt Buffer: 1X PBS, 1 M NaCl, 1% (w/v) CHAPS.

32. 22 mm × 22 mm glass cover slips #1.5 (Corning).

33. 22.81 µm glass beads (Whitehouse Scientific) in ddH$_2$0.

34. Purified Wnt protein in Wnt Buffer (see **item 31**; ref. *11*); mammalian Wnt proteins are also commercially available (R and D Systems) and the lyophilized protein can be resuspended in ddH$_2$0 to a concentration of 50 µg/mL.

3. Methods

In the following methods we describe techniques to assay downstream effects of Wnt signaling on cell fate and the cytoskeleton during *C. elegans* development. The methods include (1) mounting embryos for 4D-videomicroscopy and visualization of rhabditin granules, which are the easiest markers of endoderm specification *(12)*, (2) fixing and staining gastrulation stage embryos to detect myosin activation *(4)*, and (3) a new method we have developed for placing isolated embryonic cells in contact with a purified signaling molecule, for studying cell polarization in response to an intercellular signal.

3.1. Mounting Embryos for 4D Videomicroscopy and Visualization of Rhabditin Granules

1. Coat 18 × 18 mm cover slips with poly-L-lysine by placing cover slips on 7 µL of poly-L-lysine on a slide and heating on hotplate until all liquid is evaporated.

2. Allow cover slip to cool, then turn it over and place 10 µL of EB on coated side.

3. Make agar pad by placing a drop of 3% agar/EB on a slide, then place another slide on top (*see* **Note 2**).

4. Place young adult *C. elegans* in a depression slide with 200 µL of EB.

5. Cut adult worms in half by shearing across the worm's midline using two 25G needles.

6. Select early embryos under the dissecting microscope using a narrow bore capillary and mouth pipette, and transfer to the drop of EB on the coated cover slip. Embryos can be arranged by gently blowing EB out of mouth pipette before the embryos stick to the cover slip.

7. Invert the cover slip onto the 3% agar pad, and seal the preparation by painting around the edges of the slide with Vaseline. The Vaseline allows gas exchange but prevents desiccation.

8. Embryonic cell movements and division can now be visualized using DIC optics.

9. Rhabditin granules, which mark differentiated endoderm, can be visualized by looking at the embryos under polarized light. If analyzer and polarizer are set perpendicular to each other, to give maximum extinction, birefringent rhabditin granules will appear as bright spots (**Fig. 9.1**, compare wild-type with Wnt mutant embryos).

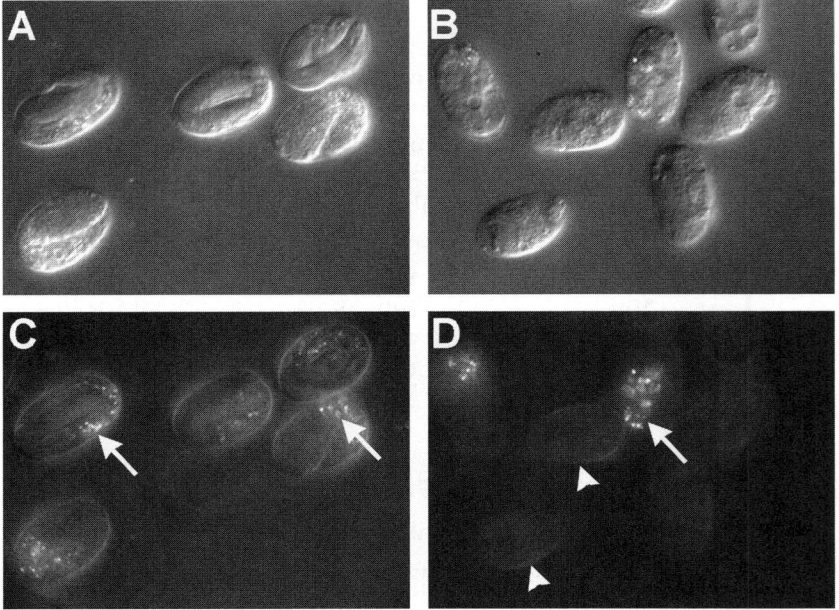

Fig. 9.1. Rhabditin granules under polarized light. Wild-type embryos (**A** and **C**) show the presence of birefringent rhabditin granules (arrows), marking the presence of endoderm. Compare with *mom-2*/Wnt mutant embryos **B** and **D**) where only 25 to 30% of embryos show the presence of rhabditin granules (arrows, arrowheads mark embryos with no granules). Figure shows embryos under DIC optics (**A** and **B**) and polarized light (**C** and **D**).

3.2. Fixing and Staining Gastrulation Stage Embryos for Active Myosin and Filamentous Actin

This protocol is adapted from reference *(13)*.

1. Coat 8-well slide by placing 3 µL of poly-L-lysine in each well and covering with a cover slip. Heat on hotplate until all liquid is evaporated, rinse well with ddH_2O, and allow to dry.

2. Place 10 young adult N2 worms in 200 µL of EB in a depression slide.

3. Cut adult worms in half by shearing across the worm's midline using two 25G needles.

4. Select embryos at four-cell stage using narrow bore capillary and mouth pipette, and transfer to a fresh depression slide well containing 200 µL of EB.

5. Leave to age at room temperature for 40 minutes in humidified chamber.

6. Use narrow bore capillary to move embryos to fresh well containing 10% sodium hypochlorite in EB and incubate for 3 minutes.

7. Wash embryos by moving through three fresh wells containing EB.

8. Place embryos in a fresh well and remove all excess EB. Add 15 µL chitinase/chymotrypsin mix and incubate at room temperature for 4 minutes (*see* **Note 3**).

9. Add 10 µL room temperature 10 mg/mL L- -lyso-phosphatidyl choline to 990 µL room temperature FB. Add 200 µL of this to embryos for 2 minutes to fix and permeabilize.

10. Transfer embryos, using a wide bore needle, to a fresh well that has been coated with agar/FB to prevent embryos from sticking, containing 200 µL of FB for 10 minutes.

11. Use wide bore capillary to move embryos to a fresh well coated with agar/PBS containing PBS and repeat three times.

12. Place embryos in 10 µL of PBS on 8-well slides coated with poly-L-lysine.

13. Extract embryos with PBS/0.2% Triton X-100 for 5 minutes.

14. Wash with PBS by removing all liquid and replacing with PBS. Repeat wash three times.

15. Block for 1 hour in humidified chamber using 1% block from TSA kit in PBS.

16. Add Phalloidin 1/400 and Rabbit anti-P-MLC 1/100 in 1% block in PBS and incubate in humidified chamber for 1 hour.

17. Wash three times, 5 minutes each with PBS.

18. Add goat anti-rabbit HRP (from TSA kit) 1/300 in 0.1% block in PBS and incubate for 1 hour in humidified chamber.

19. Wash six times, 5 minutes each with PBS.

20. Treat with tyramide reagent for 3 minutes as per TSA kit instructions.

21. Wash three times with PBS.

22. Add DAPI solution and incubate for 10 minutes.

23. Remove all liquid, add 7 µL of prolong mounting medium per well, cover with a cover slip, and allow to set overnight.

3.3. Method for Investigating Wnt/MOM-2-induced Cell Polarity In Vitro

3.3.1. Preparing Wnt-coated Beads

Wnt-coated beads can generally be used for 1 to 2 days. Store at 4°C.

1. Rock 50 µL of protein in Wnt Buffer with 10 µL of red sepharose beads at 4°C for at least 1 hour.

2. Remove unbound protein by washing the Wnt-coated beads twice with Wnt buffer. To do this, transfer 2 µL of coated beads to 98 µL of Wnt Buffer alone and invert the tube gently to wash the beads. Allow the beads to settle to the bottom of the tube and repeat wash by transferring 2 µL of coated beads into another 98 µL of Wnt buffer alone. Rock the tube gently again and allow the beads to settle (see **Note 4**).

3. Transfer the coated beads to EGM by again transferring 2 µL of beads to 98 µL of EGM. The Wnt Buffer contains a high amount of salt and detergent, so the amount of Wnt Buffer that is added to EGM is minimized.

3.3.2. Placing Isolated Cells in Contact with Wnt-coated Beads

1. Isolate P_1 cells with blastomere isolation techniques (9). Keep P_1 isolates in a humid chamber to prevent evaporation while preparing the cover slips. Cells that have not been isolated properly will bleb or will not continue to divide.

2. Resuspend the glass beads in water by flicking the tube and pipette 5 µL of the suspension onto a cover slip as five drops: one on each corner, and one in the middle (**Fig. 9.2A**, see **Note 5**).

3. Wick up all the excess water around the glass beads.

4. Pipette 2.5–3 µL of the Wnt-coated bead/EGM suspension onto the center of the cover slip (**Fig. 9.2A**).

5. Using the same needle as was used for blastomere isolation, mouth pipette the isolated P_1 cell (before it has divided into EMS and P_2) into the center of the EGM suspension and allow it to settle.

6. Gently place a glass slide over the face-up cover slip. Once the cover slip and slide come in contact, the media will spread

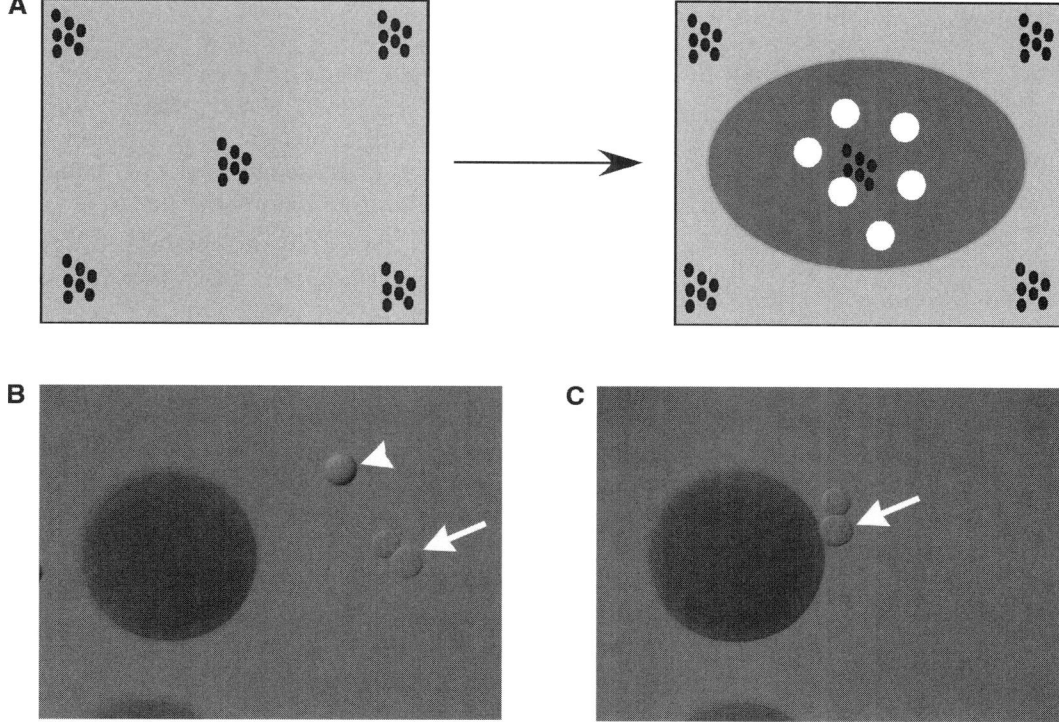

Fig. 9.2. Placing Wnt-coated beads in contact with isolated cells. (**A**) A cartoon depiction of the glass bead approach. Left: glass beads (black spots) in water are pipetted onto cover slips in 5 drops. Right: once the excess water is wicked away, the Wnt-coated beads (white circles) in EGM (dark gray) are pipetted onto the center of the cover slip. (**B**) Cells are placed in between a cover slip and slide, using glass beads (arrowhead) as a spacer. P_1 has divided into P_2 and EMS (arrow) near a Wnt-coated bead (large bead). (**C**) Using the glass beads as supporting rollers, the cover slip is manually pushed over the slide, and the cells (arrow) are placed in contact with the nearby Wnt-coated bead.

out, but it should not reach the edges of the cover slip (*see* **Note 6**).

7. The slides can now be flipped over such that the cover slip is on top of the slide.

8. Place the slides in a humid chamber until P_1 has divided into P_2 and EMS (**Fig. 9.2B**).

9. When EMS is born, gently roll the cover slip over the glass slide, using the glass beads as rollers: The glass beads will allow the cover slip to move along the glass slide, but prevent the cells from being compressed. The EMS cells must be in contact with the Wnt-coated beads within 5 minutes of being born (**Fig. 9.2C**, *see* **Note 7**).

10. Once EMS is in contact with a Wnt-coated bead, completely seal the slide by dipping a paintbrush into melted Vaseline and painting the Vaseline around the edges of the cover slip. Cells can now be visualized using DIC optics.

4. Notes

1. Borosilicate needles can be hand pulled over a flame or using a needle puller (e.g., the Sutter P-97 Flaming/Brown micropipette puller). The end is then broken using a razor blade or a microforge (Narishige) to give a bore size of ~50 μm for narrow needles or 80–100 μm for wide-bore needles.

2. Agar pads of the correct thickness are obtained by using slides coated with lab tape as spacers. Place two slides, to which two strips of tape have been applied to each, on either side of a clean slide. Add a drop of 60°C agar to the central slide, and immediately place another clean slide on top, resting on the tape strips and the liquid agar.

3. New batches of chitinase need to be tested. This is done by treating embryos with chitinase/chymotrypsin and determining how long it takes until three-cell embryos are found with the two anterior AB daughters appearing symmetrical, bulging the softened eggshell. This generally takes 3 to 8 minutes.

4. Be sure to be gentle with the suspension as the Wnt protein attaches to the red sepharose beads through non-covalent interactions and aggressive shaking and vortexing of the suspension may disturb these interactions.

5. Although most of the glass beads will be approximately 22 μm, there will also be bead sizes that range from ~19–25 μm. Larger glass beads make it harder to manipulate the position of the cells and the Wnt-coated beads, so to preferentially select for the smaller glass bead populations, allow the glass beads to settle for 5–10 seconds after flicking the tube. The larger beads will settle faster than the smaller beads. Then, pipette from the upper region of the solution, which should contain the smaller bead population.

6. It is best to minimize the volume pipetted onto the cover slip because if the liquid reaches the edges of the cover slip, it will be difficult to seal the slide without moving the cells and potentially disturbing the contact between the bead and the cells.

7. Although it is easier to remove the P_1 cell at the two-cell stage and allow P_1 to divide into EMS and P_2, EMS and P_2 cells can also be isolated at the four-cell stage after they are born. Isolating cells at the P_1 stage allows for more time to place EMS in contact with the Wnt-coated bead during the critical period, as EMS is only responsive to spindle positioning cues for approximately 5 minutes after EMS is born *(14)*.

Reference

1. Thorpe, C. J., et al. (1997) Wnt signaling polarizes an early C. elegans blastomere to distinguish endoderm from mesoderm. *Cell* 90, 695–705.

2. Rocheleau, C. E., et al. (1997) Wnt signaling and an APC-related gene specify endoderm in early C. elegans embryos. *Cell* 90, 707–716.

3. Korswagen, H. C. (2002) Canonical and non-canonical Wnt signaling pathways in Caenorhabditis elegans: variations on a common signaling theme. *Bioessays* 24, 801–810.

4. Lee, J. Y., et al. (2006) Wnt/Frizzled signaling controls C. elegans gastrulation by activating actomyosin contractility. *Curr Biol* 16, 1986–1997.

5. Walston, T., et al. (2004) Multiple Wnt signaling pathways converge to orient the mitotic spindle in early C. elegans embryos. *Dev Cell* 7, 831–841.

6. Thorpe, C. J., Schlesinger, A., Bowerman, B. (2000) Wnt signalling in Caenorhabditis elegans: regulating repressors and polarizing the cytoskeleton. *Trends Cell Biol* 10, 10–17.

7. Goldstein, B., et al. (2006) Wnt signals can function as positional cues in establishing cell polarity. *Dev Cell* 10, 391–396.

8. Schlesinger, A., et al. (1999) Wnt pathway components orient a mitotic spindle in the early Caenorhabditis elegans embryo without requiring gene transcription in the responding cell. *Genes Dev* 13, 2028–2038.

9. Edgar, L.G., Goldstein B. (2007) *Methods in Cell Biol, in press.*

10. Edgar, L. G. (1995) Blastomere culture and analysis. *Methods Cell Biol* 48, 303–321.

11. Willert, K., et al. (2003) Wnt proteins are lipid-modified and can act as stem cell growth factors. *Nature* 423, 448–452.

12. Laufer, J. S., Bazzicalupo, P., Wood, W. B. (1980) Segregation of developmental potential in early embryos of Caenorhabditis elegans. *Cell* 19, 569–577.

13. Costa, M., Draper, B. W., Priess, J.R. (1997) The role of actin filaments in patterning the Caenorhabditis elegans cuticle. *Dev Biol* 184, 373–384.

14. Goldstein, B. (1995) Cell contacts orient some cell division axes in the Caenorhabditis elegans embryo. *J Cell Biol* 129, 1071–1080.

Part V

Drosophila

Chapter 10

Function of the Wingless Signaling Pathway in *Drosophila*

Foster C. Gonsalves and Ramanuj DasGupta

Abstract

Signaling by the wingless pathway has been shown to govern numerous developmental processes. Much of our current understanding of wingless signaling mechanisms comes from studies conducted in *Drosophila melanogaster*, which offers superior experimental tractability for genetic and developmental studies. Wingless signaling is highly consequential during normal development and patterning of *Drosophila*. Its earliest identifiable role during development of *Drosophila* is in the embryonic segmentation cascade, wherein wingless functions as a segment polarity gene and serves to pattern each individual segment along the antero-posterior axis of the developing embryo. Subsequent developmental roles fulfilled by wingless include patterning the developing wings, legs, eyes, CNS, heart, and muscles. Each of these developmental contexts offers excellent systems to query mechanisms regulating different aspects of wingless signal transduction such as synthesis, secretion, reception, and transcription. This chapter presents a brief overview on the functions of wingless signaling during development of *Drosophila melanogaster*.

Key words: *Drosophila*, signaling pathways, Wingless, embryogenesis, organogenesis, patterning, polarity.

1. Introduction

The recent era of "-omes," heralded by the sequencing of the *Drosophila* genome has pervaded all aspects of scientific inquiry. *Drosophila melanogaster*, which to date remains one of the most established and tractable model organisms, has allowed vast elucidation of the molecular intricacies involved in the development of multicellular organisms. *Drosophila* as a model-system affords immense ease to the experimentalist and is thus extremely attractive for the study of wide variety of biological processes including development and behavior. Molecular details of several intercellular signaling pathways, including those elicited by Notch,

Elizabeth Vincan (ed.), *Wnt Signaling, Volume II: Pathway Models, vol. 469*
© 2008 Humana Press, a part of Springer Science+Business Media, New York, NY
Book doi: 10.1007/978-1-60327-469-2

Wingless, Hedgehog, JAK/STAT, and TGF-β family molecules, were initially unraveled due to the genetic simplicity of *Drosophila*. Moreover, decades of research has shown that an appreciable number of genes bear homologs in many other animals including humans; the establishment of various *in vivo* disease models in *Drosophila* has thus accelerated our understanding of a number of human diseases.

Pioneering work by Nusslein-Volhard and Wieschaus *(1)* led to the identification of a vast number of genes, which through subsequent research have addressed and answered several fundamental questions in developmental biology and gene regulation. Development of *Drosophila melanogaster* proceeds through three broad phases: embryonic, larval, and adult. Each of these stages of development allows one to query different questions pertaining to the biology of multicellular organisms. For example, patterning of the *Drosophila* embryo is a highly intricate and tightly regulated process that proceeds under the influence of a cascade of gene functions beginning in the oocyte *(2, 3)*. Study of the delineation of the antero-posterior (A/P) and dorso-ventral axis (D/V) in the embryo has led to elucidation of a complex interplay between various signaling pathways such as those elicited by wingless, hedgehog, and dpp *(4)*.

Drosophila, a holometabolous insect, undergoes metamorphosis through three larval instars. The larval stages have proven to be immensely useful in studying mechanisms of growth and patterning during organogenesis. Adult organs in the fly arise from groups of epithelial cells called imaginal discs, which are specified during late embryogenesis. Complex interplay between signaling pathways results in the growth, patterning, and differentiation of the adult organ. A recurring theme, both through embryonic and larval development of *Drosophila melanogaster*, like in many other multicellular organisms, is that a limited number of signaling pathways perform varied functions depending on the spatial and temporal context of development *(5, 6)*. The wingless pathway is one such highly conserved signaling pathway elicited by *wingless* (*wg*), which codes for secreted glycoproteins. The main intracellular outcome of the canonical Wg pathway is the stabilization of the cytoplasmic pool of a key mediator, *armadillo* (*arm*)/β-catenin (β-cat), which is otherwise degraded by a complex comprised of Axin (*axn*), GSK3β (*sgg*), and APC (*apc*). Initially identified as a key player in stabilizing cell–cell adherens junctions, *arm* is now known to also act as a transcription factor by forming a complex with the Pangolin/TCF (*pan*) family of HMG-box (High mobility group) transcription factors. Upon the binding of Wg to its receptor, Frizzled (*fz*), stabilized *arm* translocates to the nucleus, where, together with *pan*/TCF transcription factors, it activates downstream target genes. Wg signaling has been shown to regulate a plethora of fundamental developmental and cell biological processes such as cell proliferation, differentiation, and cell polarity *(7)*. In *Drosophila* these encompass development of the embryonic ectoderm, heart, muscles, CNS, wings, legs, and eyes *(7)*. This chapter presents a brief account of the signaling responses elicited by *wg*, so as to give the reader a flavor for the kinds of developmental roles *wg* family molecules fulfill in *Drosophila*.

2. Wingless and the Embryo

Segmental compartmentalization of the *Drosophila* embryo involves a hierarchical cascade of gene-functions initiated by maternal gene products such as Bicoid and Nanos *(8, 9)*. Subsequently, the so-called "gap genes" such as Hunchback, subdivide the embryo into broad overlapping domains along the antero-posterior axis, and ultimately delineate several contiguous segments via the action of the "pair rule" genes *(10)*. This genetic cascade culminates at the expression of "segment-polarity" genes such as *wingless* (*wg*), *hedgehog* (*hh*), and *engrailed* (*en*; *11–14*). The segment-polarity genes function to subdivide a given segment into anterior and posterior compartments *(15)*. During embryonic development, *wg* expression is first detected in the ectoderm at the cellular blastoderm stage (**Fig. 10.1**; ref. *16*). The expression is later resolved during germ-band extension stages into distinct parasegmetnal stripes, each of which is one cell wide *(16, 17)*. This expression pattern is dependent on pair-rule genes for its initial establishment; however subsequent maintenance of the striped *wg* expression relies on the expression and function of other zygotic-genes such as *gooseberry* (*gsb*) and *hh (17, 18)*. Interaction of the wingless signaling pathway with other conserved pathways is typified in the segmentation cascade of the *Drosophila* embryo. Wg signaling is required for maintaining *en* expression in adjoining cells, which in turn secrete Hh, which maintains *wg* in the neighboring stripe of cells via Patched (*ptc*) *(11, 17)*. Furthermore, combinatorial effects of *wg* and *hh* signaling results in a striped expression of *Serrate* (*Ser*); *wg* delimits *ser* expression by repressing *ser* from the anterior, while Hh exerts its repressive effect on *ser* from the posterior *(19)*. Segmental expression of *wg* in the embryo is critical in specification of cell fate in the embryonic ectoderm. Wingless expressing cells generate stretches of smooth cuticle interspersed between characteristic ventral denticular belts *(19, 20)*. The denticular belts are comprised of small tooth-like structures secreted by the ventral epidermal cells at the end of embryogenesis and are thought to play a crucial role in larval locomotion. *wg* mutant embryos exhibit a lawn of denticles due to the absence or

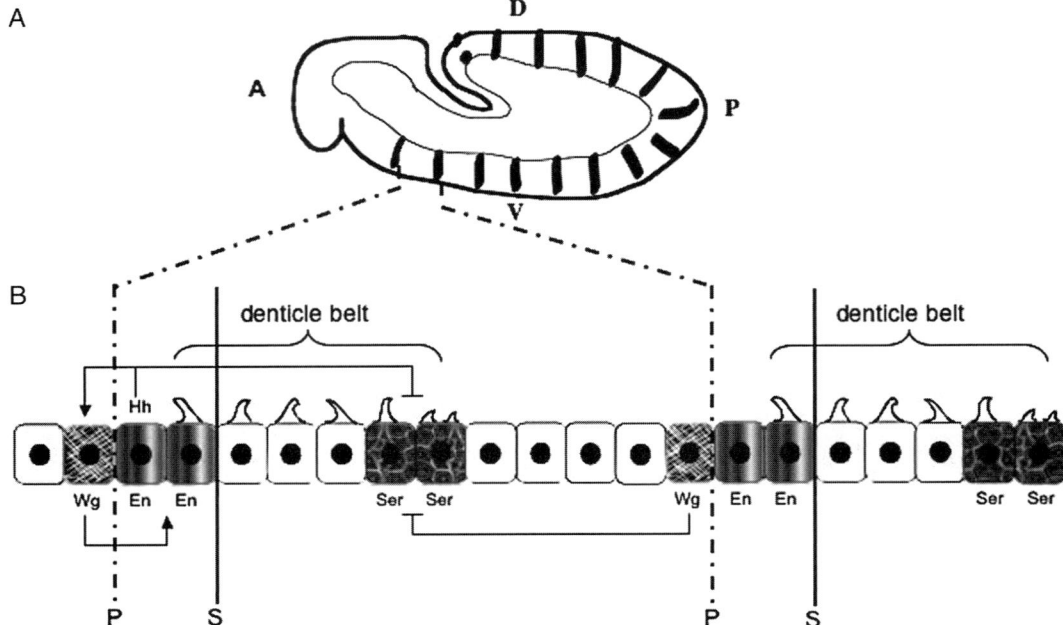

Fig. 10.1. Wingless and embryonic patterning. **(A)** By late embryogenesis, wingless is expressed in distinct stripes along the antero-posterior axis of the gastrulating embryo. Each stripe marks the parasegmental boundary (P). **(B)** *Wingless/wg* signaling activates expression of *Engrailed/en* in a stripe of cells posterior to the *wg* expressing cells. The *en* expressing cells mark the future segmanetal boundary (S). Expression of *wg* is initially dependent on *Hedgehog/Hh* signals emanating from stripes of cells located immediately posterior to the *wg*-expressing cells. Subsequently, *Hh* and *wg* signals function to delimit *Serrate/Ser* expression, resulting a *Ser* stripe, which is two to three cells wide. By the end of embryogenesis, *wg*-expressing cells secrete naked cuticle, while those devoid of *wg* signals secrete characteristic denticles, which comprise the ventral denticular belts. D, dorsal; V, ventral, A, anterior; P, posterior.

mis-specification of smooth-cuticle secreting epidermal cells; whereas ectopic expression of *wg* in the embryo results in all naked cuticle and a complete loss of denticular belts *(21, 22)*. The specification of smooth cuticle by *wg* is by virtue of its ability to repress shaven baby (*svb*), the gene that is necessary and sufficient to direct denticle formation *(23)*. Additional segment polarity mutants from various screens subsequently identified other components of the *wg* signal transduction pathway such as *arm, zeste-white 3* (*zw3*)/glycogen synthase kinase-3 (GSK-3), and *arrow* (*arr*). Furthermore, the fact that *wg* loss of function results in a lawn of denticles alluded to the possibility that *wg* is a secreted molecule; in a wild-type situation, it is thus able to exert its effect on cells which are several cell diameters away from the stripe of cells expressing *wg* and fate them as smooth cuticle *(24)*. More recently, a new wingless homolog, *wntD* has been identified, which is a target of the Dorsal/Snail/Twist gene network in the gastrulating *Drosophila* embryo *(25)*. WntD has been shown to block nuclear localization of Dorsal and serves to create a negative feedback loop to repress Dorsal activation *(26)*.

3. Wingless and Organogenesis

Much of the elucidation of wingless signaling has come from studies in the larval imaginal tissue, particularly the wing and eye disc. The adult wing develops from a group of epithelial cells called the wing imaginal disc, which are set aside during early embryogenesis. Expression of wingless along with *nubbin (nub)*, a POU domain protein, is responsible for defining the wing primordium during early patterning of the wing imaginal disc *(27)*. The expression of *wg* in the wing imaginal disc is complex and is under the control of distinct regulatory elements *(28)*. During the second larval instar, *wg* expression is detected in the ventral/anterior region of wing disc, which is later resolved into a narrow stripe of expression along the future dorso-ventral margin of the wing *(29)*. This results in a gradient of *wg* signaling activity and regulates downstream target genes in a concentration dependent manner. Furthermore, it has been proposed that the wingless gradient in developing imaginal tissue establishes a polar coordinate system encompassing the A/P, D/V, and P/D axes so that each cell attains a distinct fate depending on its position plane of the imaginal disc *(29)*. Thus, the formation of wg signaling gradient is thought to be crucial in providing positional cues to the cells in the wing pouch. Its establishment has been shown to be dependent on the combined action of the glypicans, *Division Abnomrally Delayed (Dally)*, and *Dally-like protein (Dlp; 30)*. Both *Dally* and *Dlp* code for heparan sulfate proteoglycans (HSPGs) and are linked to the plasma membrane via a glycosyl phosphatidyl linker. It is thought that secreted wingless is bound and stabilized by *Dally* and *Dlp* on the receiving cells; this binding appears to be crucial in delivering the *wg* ligand to cells distal from the source. Han et al. show that wingless is unable to move through a strip of cells mutant for these HSPGs, thus suggesting the extracellular diffusion of wingless is a highly regulated process *(30)*. Similar regulatory controls are proposed to function in the diffusion of other secreted signaling molecules such as Hh and Dpp *(30, 31)*.

Localized expression of *wg* in the third instar wing imaginal disc is dependent on Notch signaling elicited by the Serrate ligand *(32)*. At the D/V boundary, *wg* along with Notch signaling induces a zone of non-proliferating cells (ZNC), which later give rise to sensory bristles of the adult wing margin (**Fig. 10.2B**; refs. *33* and *34*). Johnston and Edgar showed that the ZNC is

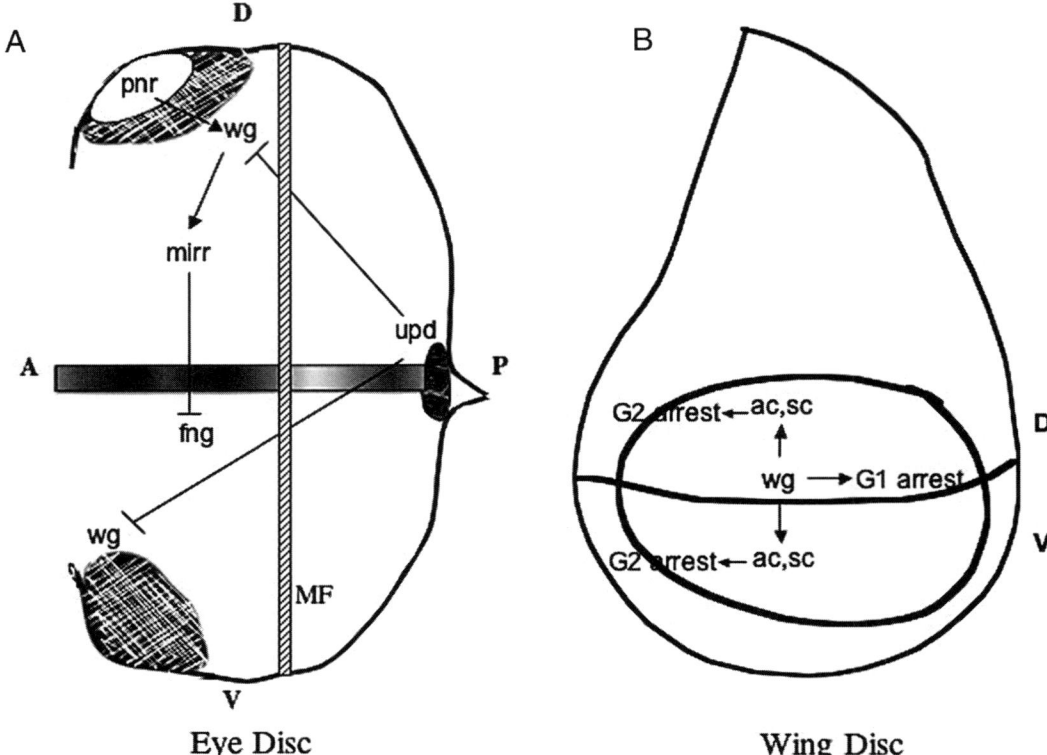

Eye Disc   Wing Disc

Fig. 10.2. Wingless and organ patterning. (**A**) The GATA family transcription factor, *pnr*, which is expressed along the dorsal margin, activates *wg* expression. *wg* in turn activates the Iroquois-complex gene, *mirror/mirr*, which represses *fng* expression, along the DV midline. This leads to Notch activation in this region and establishment of the DV midline. *Unpaired/upd*, expressed at the optic stalk represses *wg* activity at the dorsal and ventral anterior regions. (**B**) In the developing wing disc, *wg* emanating from the DV boundary establishes a gradient of signaling activity. In the dorsal and ventral anterior compartments, *wg* induces G2 arrest by activating *achaete-scute/asc* expression. On the other hand, in cells along the DV margin, *wg* signaling induces G1 arrest by the activation of *dE2F*. D, dorsal; V, ventral.

comprised of three subdomains, of which the cells in the central domain express *wg* *(34)*. In two of these subdomains, cells are arrested in the G2 phase of the cell cycle due to *wg* dependent induction of proneural genes, *achaete-scute* (*asc*), which in turn repress the mitotic inducer, *string* (Cdc25). In the third domain, *wg* emanating from the D/V boundary induces G1 arrest by inhibiting the activity of dE2F, which is required for activation of genes essential in DNA replication *(35)*. The ZNC is crucial in maintaining the D/V boundary of the developing wing, which in turn is critical for the restriction of dorsal vs. ventral lineage of cells in the wing pouch. The activity of wingless at the D/V boundary determines the final size of the adult wing; hence, such patterning activity of wingless is important in delimiting adult organ size within a narrow range for a given population. In addition to being expressed at the D/V boundary, *wg* is also expressed in two concentric rings along the periphery of the wing-pouch *(28, 29)*. Expression of *wg* in each of these rings is directed by distinct regulatory elements *(36)* and is directed by a complex interplay between *teashirt* (*tsh*), *nubbin* (*nub*), *rotund* (*rn*), and the *wg* target gene *vestigial* (*vg*). This *wg* expression at the wing-pouch border is key to the proper formation of the hinge (the region of the wing, adjoining the thorax). More recent studies have suggested that *wg* signaling, in addition to modulating whole organ patterning, also regulates cell shape changes in the distal wing epithelium. This is achieved by combinatorial *wg-* and *fat*-mediated regulation of DE-cad expression and cell-cell adhesion along the proximo-distal axis in the developing wing imaginal disc *(37)*. This may serve as an important node of interaction between signaling pathways regulating organ size and cell shape including planar polarity.

Polarity is a characteristic feature exhibited by cells arranged in epithelial layers; in *Drosophila*, Planar Cell Polarity (PCP) is reflected in the stereotyped orientation of wing hairs, body bristles, and the inter-ommatidial bristles of the adult eye. Detailed genetic analyses have implicated a core set of genes that function in the establishment of PCP. These include *frizzled* (*fz*), *dishevelled* (*Dsh*), *prickle* (*pk*), *flamingo* (*fmi*), and *diego* (*dgo*) *(38)*. Intriguingly, of these *fz* and *dsh* form key components of the *wg* signaling pathway; however, in the context of PCP, they are proposed to function independent of the canonical *wg* signaling. Wg, itself however does not appear to be directly involved in regulating PCP in the wing. In the embryonic epidermis however, *wg* has recently been shown to provide an instructive cue to orient the denticles in the larval cuticle in the correct direction *(39)*. Colosimo and Tolwinski demonstrate that ectopic expression of *wg* in the larval epidermis results in a reorientation of the denticles towards the source of *wg*, which is suggestive of a signaling role for the *wg* pathway in determining denticle polarity. Moreover, transcriptional readout from the *wg* pathway mediated by *arm* seems to be important in relaying this signal so as to result in correct orientation of the denticles *(39)*.

Another organ system in Drosophila where *wg* signaling is of paramount importance is its compound eye. The *Drosophila* eye is comprised of several hundred subunits called ommatidia, each of which is made up of eight photoreceptors and a few support cells, each arranged in an invariant pattern *(40)*. The development of each individual ommatidium is highly stereotyped and involves a complex interplay of various signaling pathways including Notch, RTK, Hh, JAK/STAT, and Wg *(40)*. The adult eye is derived from the embryonic primordia of epithelial cells, called the eye-antennal imaginal disc, which invaginates from the ectoderm during embryogenesis and grows in the larva (**Fig. 10.2A**). The eye-antennal disc gives rise to much of the adult head capsule as well as the compound eye and the antenna *(41)*. Specification of the retinal field depends on the activity of transcription factors, *eyeless* (*eye*), *eyes absent* (*eya*), *sine oculis* (*so*), and *dachshund* (*dac*; *42–45*). *Wg* signaling functions to restrict the area of the eye field, thereby defining the border between the retina and surrounding head-capsule *(46, 47)*. Clones of cells expressing a constitutively active form of *arm* (arm*) lose expression of the eye specification genes, *sine oculis, dachshund* and *eyes absent (47)*. Differentiation of the photoreceptors in the early eye imaginal disc of the third instar larvae begins at the posterior margin and progresses anteriorly in a sweeping wave across the eye primordia. This "wave of differentiation" is led by a depression in the disc known as the morphogenetic furrow. Wingless signaling from the dorsal and ventral anterior margins of the eye disc contributes to the correct spatial localization of morphogenetic furrow initiation *(48)*. Cells anterior to the furrow are proliferative and undifferentiated, whereas those posterior to the furrow exit the cell cycle and differentiate into proneural photoreceptor cells. As the furrow moves from the posterior to the anterior of the disc, the cells posterior to the furrow get recruited into clusters of photoreceptor cells which give rise to the adult compound eye of *Drosophila*. Furthermore, *wg* signaling plays a key role in patterning both the anterior-posterior as well as the dorso-ventral axes of the eye disc. This is achieved by the activation of *wg* expression in the dorsal margin of the eye disc by *pannier/pnr*, a GATA family transcription factor *(49, 50)*. *wg* in turn activates expression of the Iroquois complex genes in the dorsal compartment; whose activity in the ventral compartment is restricted by *unpaired/upd* emanating from the optic stalk *(51, 52)*. *Mirror/mirr*, an Iro-C gene represses fringe, forming a boundary of *fng* expression and activity at the DV midline, thus establishing the dorso-ventral axis in the developing eye disc *(53)*. Additionally in the third instar, *Hh* is expressed in the differentiating photoreceptors, while *Dpp* is expressed in a stripe along the progressing morphogenetic furrow. Each of these signaling molecules serve to downregulate anteriorly expressed genes, which are activated by *wg* emanating from the dorsal and ventral anterior regions of the eye disc. More recently, interaction of the *wg* signaling pathway with the JAK/STAT pathway in the eye disc has been reported to result in regional specification of the eye by promoting the specification of the eye field *(54)*.

The integration of combinatorial effects of the *wg* pathway with those from other signaling pathways are thus crucial in determining the final shape, form, and size of adult organs. Hence, the relative simplicity of *Drosophila melanogaster* has thus far proven highly advantageous in deciphering concepts underlying signaling mechanisms that govern development of multicellular organisms.

References

1. Nusslein-Volhard, C., Wieschaus, E. (1980) Mutations affecting segment number and polarity in Drosophila. *Nature* 287, 795–801.

2. Peri, F., Bokel, C., Roth, S. (1999) Local Gurken signaling and dynamic MAPK activation during Drosophila oogenesis. *Mech Dev* 81, 75–88.

3. Peri, F., Roth, S. (2000) Combined activities of Gurken and decapentaplegic specify dorsal chorion structures of the Drosophila egg. *Development* 127, 841–850.

4. Riechmann, V., Ephrussi, A. (2001) Axis formation during Drosophila oogenesis. *Curr Opin Genet Dev* 11, 374–383.

5. Gerhart, J. (1999) 1998 Warkany lecture: signaling pathways in development. *Teratology* 60, 226–239.

6. Affolter, M., Mann, R. (2001) Development. Legs, eyes, or wings—selectors and signals make the difference. *Science* 292, 1080–1081.

7. Wodarz, A., Nusse, R. (1998) Mechanisms of Wnt signaling in development. *Annu Rev Cell Dev Biol* 14, 59–88.

8. Hulskamp, M., Pfeifle, C., Tautz, D. (1990) A morphogenetic gradient of hunchback protein organizes the expression of the gap genes Kruppel and knirps in the early Drosophila embryo. *Nature* 346, 577–580.

9. Irish, V., Lehmann, R., Akam, M. (1989) The Drosophila posterior-group gene nanos functions by repressing hunchback activity. *Nature* 338, 646–648.

10. Sanson, B. (2001) Generating patterns from fields of cells. Examples from Drosophila segmentation. *EMBO Rep* 2, 1083–1088.

11. DiNardo, S., Sher, E., Heemskerk-Jongens, J., et al. (1988) Two-tiered regulation of spatially patterned engrailed gene expression during Drosophila embryogenesis. *Nature* 332, 604–609.

12. Lee, J. J., von Kessler, D.P., Parks, S., et al. (1992) Secretion and localized transcription suggest a role in positional signaling for products of the segmentation gene hedgehog. *Cell* 71, 33–50.

13. Muller, H., Samanta, R., Wieschaus, E. (1999) Wingless signaling in the Drosophila embryo: zygotic requirements and the role of the frizzled genes. *Development* 126, 577–586.

14. Lessing, D., Nusse, R. (1998) Expression of wingless in the Drosophila embryo: a conserved cis-acting element lacking conserved Ci-binding sites is required for patched-mediated repression. *Development* 125, 1469–1476.

15. DiNardo, S., Heemskerk, J., Dougan, S., et al. (1994) The making of a maggot: patterning the Drosophila embryonic epidermis. *Curr Opin Genet Dev* 4, 529–534.

16. van den Heuvel, M., Nusse, R., Johnston, P., et al. (1989) Distribution of the wingless gene product in Drosophila embryos: a protein involved in cell-cell communication. *Cell* 59, 739–749.

17. Ingham, P.W., Hidalgo, A. (1993) Regulation of wingless transcription in the Drosophila embryo. *Development* 117, 283–291.

18. Li, X., Noll, M. (1993) Role of the gooseberry gene in Drosophila embryos: maintenance of wingless expression by a wingless—gooseberry autoregulatory loop. *Embo J* 12, 4499–4509.

19. Alexandre, C., Lecourtois, M., Vincent, J. (1999) Wingless and Hedgehog pattern Drosophila denticle belts by regulating the production of short-range signals. *Development* 126, 5689–5698.

20. Lawrence, P. A., Sanson, B., Vincent, J. P. (1996) Compartments, wingless and engrailed: patterning the ventral epidermis of Drosophila embryos. *Development* 122, 4095–4103.

21. Sampedro, J., Johnston, P., Lawrence, P. A. (1993) A role for wingless in the segmental gradient of Drosophila? *Development* 117, 677–687.

22. Sanson, B., Alexandre, C., Fascetti, N., et al. (1999) Engrailed and hedgehog make the range of Wingless asymmetric in Drosophila embryos. *Cell* 98, 207–216.

23. Payre, F., A. Vincent, and S. Carreno, (1999) ovo/svb integrates Wingless and DER pathways to control epidermis differentiation. *Nature* 400, 271–275.

24. Howes, R., Bray, S. (2000) Pattern formation: Wingless on the move. *Curr Biol* 10, R222–R226.

25. Ganguly, A., Jiang, J., Ip, Y. T. (2005) Drosophila WntD is a target and an inhibitor of the Dorsal/Twist/Snail network in the gastrulating embryo. *Development* 132, 3419–3429.

26. Gordon, M. D., Dionne, M. S., Schneider, D. S., et al. (2005) WntD is a feedback inhibitor of Dorsal/NF-kappaB in Drosophila development and immunity. *Nature* 437, 746–749.

27. Ng, M., Diaz-Benjumea, F. J., Cohen, S. M. (1995) Nubbin encodes a POU-domain protein required for proximal-distal patterning in the Drosophila wing. *Development* 121, 589–599.

28. Rodriguez Ddel, A., Terriente, J., Galindo, M. I., et al. (2002) Different mechanisms initiate and maintain wingless expression in the Drosophila wing hinge. *Development* 129, 3995–4004.

29. Couso, J. P., Bate, M., Martinez-Arias, A. (1993) A wingless-dependent polar coordinate system in Drosophila imaginal discs. *Science* 259, 484–489.

30. Han, C., Yan, D., Belenkaya, T. Y., et al. (2005) Drosophila glypicans Dally and Dally-like shape the extracellular Wingless morphogen gradient in the wing disc. *Development* 132, 667–679.

31. Belenkaya, T. Y., Han, C., Yan, D., et al. (2004) Drosophila Dpp morphogen movement is independent of dynamin-mediated endocytosis but regulated by the glypican members of heparan sulfate proteoglycans. *Cell* 119, 231–244.

32. Diaz-Benjumea, F.J., Cohen, S. M. (1995) Serrate signals through Notch to establish a Wingless-dependent organizer at the dorsal/ventral compartment boundary of the Drosophila wing. *Development* 121, 4215–4225.

33. O'Brochta, D. A., Bryant, P. J. (1985) A zone of non-proliferating cells at a lineage restriction boundary in Drosophila. *Nature* 313, 138–141.

34. Johnston, L. A., Edgar, B. A. (1998) Wingless and Notch regulate cell-cycle arrest in the developing Drosophila wing. *Nature* 394, 82–84.

35. Duronio, R. J., O'Farrell, P. H., Xie, J. E., et al. (1995) The transcription factor E2F is required for S phase during Drosophila embryogenesis. *Genes Dev* 9, 1445–1455.

36. Neumann, C. J., Cohen, S. M. (1996) A hierarchy of cross-regulation involving Notch, wingless, vestigial and cut organizes the dorsal/ventral axis of the Drosophila wing. *Development* 122, 3477–3485.

37. Jaiswal, M., Agrawal, N., Sinha, P. (2006) Fat and Wingless signaling oppositely regulate epithelial cell-cell adhesion and distal wing development in Drosophila. *Development* 133, 925–935.

38. Fanto, M. McNeill, H. (2004) Planar polarity from flies to vertebrates. *J Cell Sci* 117, 527–533.

39. Colosimo, P. F., Tolwinski, N. S. (2006) Wnt, Hedgehog and junctional Armadillo/beta-catenin establish planar polarity in the Drosophila embryo. *PLoS ONE* 1, e9.

40. Voas, M. G., Rebay, I. (2004) Signal integration during development: insights from the Drosophila eye. *Dev Dyn* 229, 162–175.

41. Haynie, J. L., Bryant, P. J. (1986) Development of the eye-antenna imaginal disc and morphogenesis of the adult head in Drosophila melanogaster. *J Exp Zool* 237, 293–308.

42. Curtiss, J., Mlodzik, M. (2000) Morphogenetic furrow initiation and progression during eye development in Drosophila: the roles of decapentaplegic, hedgehog and eyes absent. *Development* 127, 1325–1336.

43. Halder, G., Callaerts, P., Flister, S., et al. (1998) Eyeless initiates the expression of both sine oculis and eyes absent during Drosophila compound eye development. *Development* 125, 2181–2191.

44. Halder, G., Callaerts, P., Gehring, W. J. (1995) Induction of ectopic eyes by targeted expression of the eyeless gene in Drosophila. *Science* 267, 1788–1792.

45. Pignoni, F., Zipursky, S. L. (1997) Induction of Drosophila eye development by decapentaplegic. *Development* 124, 271–278.

46. Lee, J. D., Treisman, J. E. (2001) The role of Wingless signaling in establishing the anteroposterior and dorsoventral axes of the eye disc. *Development* 128, 1519–1529.

47. Baonza, A., Freeman, M. (2002) Control of Drosophila eye specification by Wingless signalling. *Development* 129, 5313–5322.

48. Treisman, J. E., Rubin, G. M. (1995) wingless inhibits morphogenetic furrow movement in the Drosophila eye disc. *Development* 121, 3519–3527.

49. Heberlein, U., Borod, E. R., Chanut, F. A. (1998) Dorsoventral patterning in the Drosophila retina by wingless. *Development* 125, 567–577.

50. Maurel-Zaffran, C., Treisman, J. E. (2000) pannier acts upstream of wingless to direct dorsal eye disc development in Drosophila. *Development* 127, 1007–1016.

51. Cavodeassi, F., Diez Del Corral, R., Campuzano, S., et al. (1999) Compartments and organising boundaries in the Drosophila eye: the role of the homeodomain Iroquois proteins. *Development* 126, 4933–4942.

52. Zeidler, M. P., Perrimon, N. Strutt, D. I. (1999) Polarity determination in the

Drosophila eye: a novel role for unpaired and JAK/STAT signaling. *Genes Dev* 13, 1342–1353.

53. Yang, C. H., Simon, M. A., McNeill, H. (1999) mirror controls planar polarity and equator formation through repression of fringe expression and through

control of cell affinities. *Development* 126, 5857–5866.

54. Ekas, L. A., Baeg, G. H., Flaherty, M. S., et al. (2006) JAK/STAT signaling promotes regional specification by negatively regulating wingless expression in Drosophila. *Development* 133, 4721–4729.

Chapter 11

Visualization of PCP Defects in the Eye and Wing of *Drosophila melanogaster*

Natalia Arbouzova and Helen McNeill

Abstract

"Tissue" or "planar" polarity is a characteristic of many epithelial tissues and is not only required for proper cell alignment, but in many instances is absolutely essential for normal function. Planar cell polarity (PCP) is the polarization of cells within the plane of an epithelium in a direction perpendicular to the axis of the apico-basal polarity. The oriented hair alignment in mammalian skin (feather in birds or scales in fish) and highly organized stereocilia bundles in the vertebrate inner ear are examples of such tissue organization. PCP was first described in *Drosophila*, with non-canonical Wnt signaling (also called PCP signaling, see **Chapter 10**, Volume 1) shown to be critical for its establishment. Two of the best characterized PCP models in *Drosophila* are the developing wing and the eye, where the graded activity of the Frizzled (Fz) receptor determines proximo-distal orientation of the wing hairs and mirror-imaged patterning of ommatidia, respectively.

In this chapter, we describe simple methods to visualize PCP defects in the *Drosophila* eye and wing, in both developing and adult tissues. These methods include confocal immunofluorescent analysis of larval or pupal tissues, stained with antibodies specific to PCP components or cytoskeleton markers, and light microscopy of the adult eye and wing.

Key words: *Drosophila*, PCP, imaginal disc, microscopy.

1. Introduction

The *Drosophila* eye consists of 800 structural units, called ommatidia, each of which contains eight precisely arranged photoreceptors, termed R1 through R8. The orientation of photoreceptors is planar polarized, with ommatidia in the dorsal half of the eye having a mirror-image pattern to those in the ventral half of the eye.

Elizabeth Vincan (ed.), *Wnt Signaling, Volume II: Pathway Models, vol. 469*
© 2008 Humana Press, a part of Springer Science + Business Media, New York, NY
Book doi: 10.1007/978-1-60327-469-2

PCP organization first becomes visible in the larval imaginal disc. All stages of ommatidial differentiation are apparent in a single imaginal disc. The most mature ommatidia are located most posteriorly, behind the morphogenetic furrow (MF), a physical indentation that moves from posterior to anterior across the disc tissue (**Fig. 11.1A**), marking the start of photoreceptor differentiation. Undifferentiated cells are found in the anterior part of the disc, in front of the MF. First, the photoreceptor R8 differentiates, followed by subsequent recruitment of the pairs R2/R5, then R3/R4, R1/R6, and, eventually, R7 (reviewed in ref. *1)*. After recruitment of the photoreceptors R3/R4, ommatidia of the dorsal and ventral halves, though initially similarly oriented, undergo rotation in opposite directions, creating mirror-imaged eye patterning (**Fig. 11.1B** and ref. *2)*. This rotation is dependent on specification of R3 and R4 fates within the pair of the R3/4 precursor cells. R3/4 precursors have the potential to develop into either R3 or R4 and become determined upon PCP signaling. In PCP mutants, the specification of R3/R4 is randomized, with the equatorial cell often adopting R4 fate. This results in rotation in the opposite direction and, hence, reversed polarity. Also, symmetrical ommatidia may be observed, containing two R3 or two R4 photoreceptors. These phenotypes can be visualized by immunostaining of molecular markers at the third

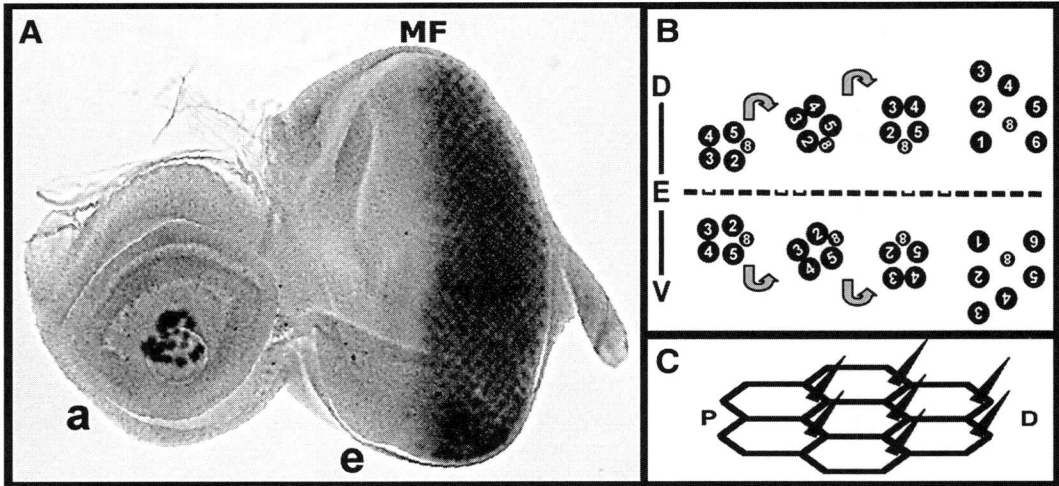

Fig. 11.1. PCP is evident in the *Drosophila* eye and wing. (**A**) Antenna (marked "a") and eye (marked "e") imaginal discs. Morphogenetic furrow is marked "MF." Differentiated photoreceptors are visualized by X-gal staining for elav-lacZ reporter. Unless otherwise specified, here and in other figures, discs and larvae are oriented anterior to the left and dorsal up. (**B**) A diagram showing rotation of ommatidial preclusters in the dorsal (D) and ventral (V) poles of the eye separated by equator (E). Upon recruitment of R3 and R4 into the precluster, the dorsal ommatidia rotate clockwise, while the ventral those rotate counter-clockwise (arrows). (**C**) A scheme depicting hexagonal-shaped epithelial cells of the pupa wing, with the trichomes pointing distally (D). On this scheme, proximal is to the left, distal is to the right.

instar stage, or scored in adult eye sections based on the morphology of ommatidia.

In the *Drosophila* wing, PCP is evident in the orientation of hairs (or trichomes) covering the wing cuticle, as each cell produces a single trichome at the distal cell edge that points distally (**Fig. 11.1C**). At early pupal stage of the wing development, cells that initially exhibit no asymmetry begin to accumulate actin filaments at the distal cell vertex at approximately 32–36 hours after prepupa formation *(3)*. This reorganization of the cytoskeleton is induced by asymmetric localization of the core PCP components: Fz, Dishevelled (Dsh), and Flamingo (Fmi) become enriched at the distal cell edge, while Van Gogh (Vang) and Prickle (Pk) are recruited to the proximal cell edge. In PCP mutants, this asymmetric localization is disrupted, resulting in alterations in the cytoskeleton and aberrant initiation of prehair formation. For example, in *fz* or *dsh* mutants, the trichome is formed in the center of the cell, and the direction it points is often randomized (reviewed in ref. *4)*. At pupal stage, the site of hair initiation, visualized by staining of actin filaments, as well as immunostaining of the core PCP proteins, can be used as a marker of PCP defects. In adult wings, hairs display characteristic swirling patterns visible under light microscope.

2. Materials

2.1. Imaginal Discs Staining, Mounting, and Confocal Fluorescence

1. Phosphate buffer saline (PBS): 130 mM NaCl, 7 mM Na_2HPO_4, 3 mM NaH_2PO_4.
2. Fixative: 4% formaldehyde (from 16% [w/v] paraformaldehyde, EMS Sciences) in PBS (*see* **Notes 1** and **2**).
3. Washing buffer: 0.1% (v/v) Triton X-100 in PBS (*see* **Note 3**).
4. Blocking solution: 5% (v/v) normal goat serum (Gibco®) in washing buffer.
5. High salt washing buffer: 530 mM NaCl, 7 mM Na_2HPO_4, 3 mM NaH_2PO_4 (5 M NaCl diluted 1:10 with PBS).
6. Primary antibodies: rat anti-Bar *(5, 6)*, rabbit anti-Spalt *(7, 8)*, mouse anti-βGal (Promega, 1:1000), mouse anti-Arm (DSHB, 1:20–1:100), mouse anti-Fmi (1:10; refs. *9* and *10)*, mouse or rabbit anti-GFP (Molecular Probes, 1:1000), rat anti-Elav (1:1000, DSHB).
7. Secondary antibodies: anti-rat and anti-mouse IgG conjugated to Cy2, Cy3 or Cy5 (Jackson Laboratory).
8. Mounting medium: 70% (v/v) glycerol in PBS or Vectashield® (Vector Labs).

9. Dumont #5 forceps (Fine Science Tools).

10. 35 mm Petri dishes (Nunc®) or 150 mm Petri dishes (Nunc) with a dark bottom covered with a layer of silicone (Sylgard®).

11. 1.5 mL microfuge tubes (Eppendorf®).

12. Microscope cover slips (18 × 18 mm, 20 × 20 mm or 22 × 22 mm) and 25 × 75 × 1.0 mm microscope slides (Fisher).

13. Nail polish (Maybelline) for sealing.

14. Leica M Stereomicroscope.

15. Leica TCS SP2 confocal microscope.

2.2. Adult Eye Embedding, Sectioning, and Microscopy

1. Washing buffer: 0.1 M sodium phosphate, pH 6.8 (to prepare, mix 0.2 M Na_2HPO4 and 0.2 M NaH_2PO4 in 72:28 ratio, respectively, and add an equal volume of water).

2. Fixative: 2% (v/v) glutaraldehyde (EMS Sciences) in 0.1 M sodium phosphate buffer (*see* **Note 4**).

3. 2% osmium in 0.1 M sodium phosphate buffer (*see* **Note 4**).

4. Ethanol dilution series (v/v): 30, 50, 70, 90, and 100%.

5. Propylene oxide (*see* **Note 4**).

6. Durcupan® ACM set for embedding (Fluka; *see* **Note 4**).

7. 0.5% (w/v) toluidine blue (Fluka) in water.

8. Single Edge Gem Teflon®-coated razor blades (EMS Sciences).

9. Flat embedding moulds (BEEM).

10. Microscope cover slips (18 × 18 mm, 20 × 20 mm or 22 × 22 mm) and 25 × 75 × 1.0 mm microscope slides (Fisher).

11. 70°C incubator.

12. Leica EM UC6 microtome equipped with a diamond knife (Diatome).

2.3. Pupal Wing Staining

1. Phosphate buffer saline (PBS): 130 mM NaCl, 7 mM Na_2HPO_4, 3 mM NaH_2PO_4.

2. Fixative: 4% formaldehyde (from 16% (w/v) paraformaldehyde, EMS Sciences) in PBS (*see* **Note 2**).

3. Washing buffer: 0.2% (v/v) Triton X-100 in PBS (*see* **Note 3**).

4. Blocking solution: 0.5–1% (v/v) BSA or 10% (v/v) normal goat serum (Gibco) in washing buffer (*see* **Note 5**).

5. Primary antibody: mouse anti-Fmi (1:10) *(9, 10)*, mouse or rabbit anti-GFP (Molecular Probes, 1:1000).

6. Secondary antibody: anti-mouse and anti-rabbit IgG conjugated to Cy2, Cy3 or Cy5 (Jackson Laboratory).

7. Nuclei markers: DAPI (Vectashield by Vector laboratories), DRAQ5™ (Biostatus).

8. Cytoskeleton marker: Texas Red-X-Phalloidin (Molecular Probes, 1:200) (*see* **Note 2** and **6**).

9. Methanol dilution series (v/v): 100, 90, and 50% (*see* **Note 2**).

10. Mounting medium: 10% (v/v) glycerol, 2.5% DABCO-antifade (Vectashield) in PBS (*see* **Note 7**).

11. Dumont #5 forceps (Fine Science Tools).

12. Scalpel blades #11 (EMS Sciences) or Kendall 21G syringe needle (Fisher).

13. 35 mm Petri dishes (Nunc) or 150-mm Petri dishes (Nunc) with dark-field bottom covered with silicone layer (Sylgard®).

14. Tissue culture 96-well plates (Nunc).

15. Microscope cover slips (18 × 18 mm, 20 × 20 mm or 22 × 22 mm) and 25 × 75 × 1.0 mm microscope slides (Fisher).

16. Leica TCS SP2 confocal microscope.

2.4. Adult Wing Embedding

1. Dehydration solution: 70% (v/v) ethanol, 30% (v/v) glycerol.

2. Mounting medium: DPX (EMS sciences) or Canada balsam (Merck).

3. Methods

3.1. Detection of PCP Defects in the Eye Imaginal Disc

Detection of PCP defects in the larval eye is based on immuno-histochemical visualization of certain photoreceptors, which are precisely localized within the ommatidium. Among commonly used markers are the homeobox protein BarH1, expressed in the R1 and R6 photoreceptors (**Fig. 11.2A-A"** and refs. *6* and *11*); Spalt transcription factors and *svp-lacZ* reporter mark the R3 and R4 pair *(12)*. The *mδ0.5-lacZ* reporter of N signaling shows R4-specific expression *(13)*. The cadherin Fmi, as well as *dsh-GFP* reporter, is initially expressed in R3 and R4, but at the onset of rotation becomes accumulated in R4 *(14)*. We also recommend co-staining with antibodies such as Elav, which marks neural cells behind the MF *(15)*, or Armadillo (Arm), expressed at the cell junctions *(16)*, to outline the photoreceptors and ommatidial preclusters. It is worth mentioning that the quality of antibodies is critical for obtaining best images, and commercially available anti-βGal antibodies proved to give finest pictures. However, the *Drosophila* lines carrying the reporters described earlier need to be crossed into the line of interest, which may take a while.

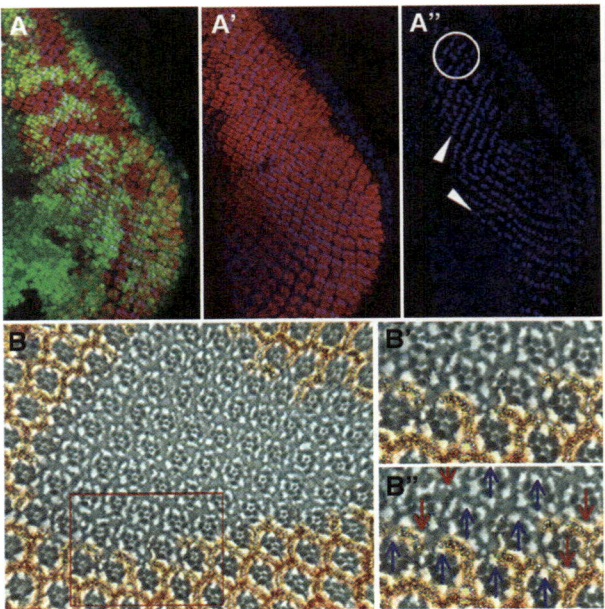

Fig. 11.2. Visualization of PCP defects in the eye. (**A**) Homozygous *sprouty* clones (marked by absence of GFP) on the heterozygous background (marked by GFP) are generated using Flp/FRT system *(18)*. Twin-spots are recognnized by more intense GFP. Eye imaginal discs are stained with anti-Elav (red) and anti-BarH1 (blue) antibodies. (**A'**) shows only red and blue channel from (**A**). (**A''**) shows blue channel alone. R1/R6 pairs marked by anti-BarH1 are aligned in more or less parallel lines within each pole (follow the direction of white arrowheads). Polarity reversals within the clones or in the neighboring ommatidia (non-autonomous effect) are recognized by R1/R6 pairs oriented perpendicular to these lines (encircled). (**B**) A tangential section through the adult eye of a *ds* mosaic animal. The clone is marked by absense of the pigment (brown color). Note that in wild-type tissues surrounding the clone, trapezoid-shaped ommatidia are pointing upward (dorsal pole shown, compare to Fig. 11.1B). In the clone, however, some ommatidia are reversed. (**B'** and **B''**) The region marked by red frame in (**C**) is shown at higher resolution, with blue arrows representing wild type orientation and red arrows marking reversed polarity ommatidia in (**C''**).

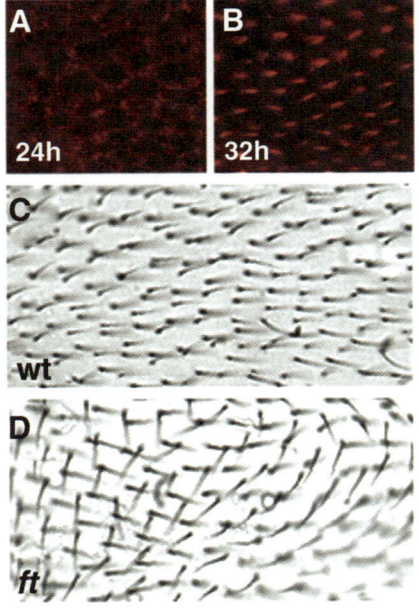

Fig 11.3. Visualization of PCP defects in the wing. (**A**) Wild-type pupal wing dissected at 24h APF. At this stage, actin cytoskeleton marked with phalloidin (red) demonstrates symmetrical submembrane localization. Here and in the following panels, proximal is to the left, distal is to the right. (**B**) Wild-type pupal wing dissected at 32 hours APF. Actin filaments form a single trichome on the distal edge of the cell. (**C**) Wild-type adult wing shows hairs pointing distalward. (**D**) Adult wing from a *ft^{GRV}* mosaic animal (clones are not marked) shows a swirling pattern typical for PCP mutants. Note that while some hairs retain proper wild-type position, others are misoriented, pointing proximally or perpendicularly to the wing surface.

1. Collect 3d instar wandering larvae (*see* **Note 8**) and place in a drop of PBS. Under stereomicroscope, dissect out imaginal discs as follows. To pull out eye-antennae disc complexes, hold the larva at the border between the thorax and the abdomen, using forceps (**Fig. 11.4A**). With a second pair of forceps, grasp the mouthparts and pull them out of the head. Ideally, only the eye-antennae complexes should stay connected to the mouthparts (**Fig. 11.4B,C**). If other tissues remain attached, still holding on to the mouthparts, remove the brain (**Fig. 11.4B**) and salivary glands. Then grab the *internal* part of the mouthparts and remove this and the eye-antennal discs from the *external* part of the mouthparts. All other tissues should be left behind at this step. While dissecting larvae, collect disc complexes in a microfuge tube in PBS on ice. Do not dissect longer than 1 to 2 hours.

2. Fix the complexes for 20 minutes at room temperature (*see* **Notes 1** and **2**).

3. Wash four times for 10 minutes in washing buffer (*see* **Note 3**).

4. Block with 5% (v/v) serum for 30 minutes at room temperature. Incubate with primary antibody 1 hour at room temperature or overnight at 4°C.

5. Wash eight times for 10 minutes with washing buffer. Some antibodies may give a high background. To minimize this background, the discs can be alternatively washed three times for 5 minutes in regular washing buffer, then in high-salt washing buffer for 20 minutes and again three times for 20 minutes in regular washing buffer.

6. Incubate with secondary fluorescent antibody for 1 to 2 hours at room temperature in the dark. Wash eight times for 10 minutes in the dark.

7. Equilibrate with 70% glycerol for 3 to 4 hours at room temperature or overnight at 4°C in the dark.

8. Mount in 70% glycerol on a microscope slide, cover with microscope cover slip, seal the edges of the slip with nail polish, and analyze under confocal microscope.

3.2. Detection of PCP in Adult Eyes

By the adult stage, ommatidia acquire a characteristic trapezoid shape, as apparent in tangential eye sections, and PCP defects can be assessed morphologically (**Fig. 11.2B-B″**).

1. Select flies of the appropriate genotype (*see* **Note 8**), anesthetize with CO_2.

2. Dissect heads away using #11 scalpel blade, gently cutting away one-half or one-third of the opposite eye bulge to expose the inside of the head.

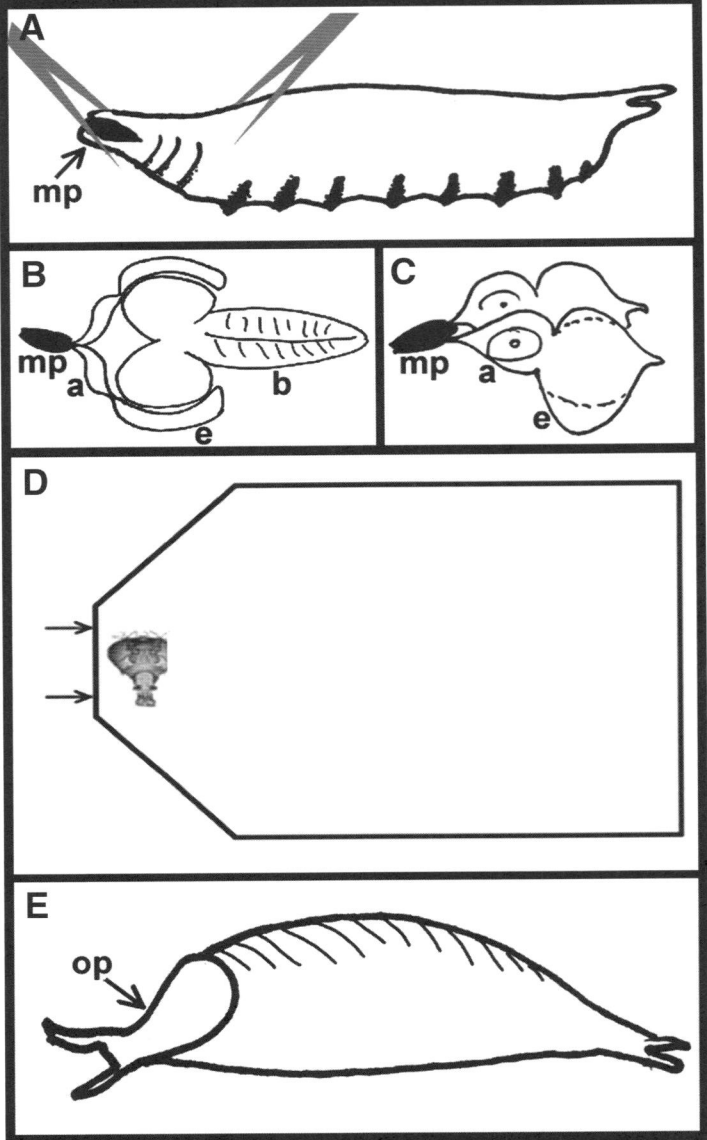

Fig. 11.4. Cartoons illustrating dissection and mounting. (**A**) For dissecting, hold the larva with a pair of forceps at the border between the thorax and the abdomen (approximately one forth of the body length) and pull the mouthpart (marked "mp") with the second pair of forceps. Lateral view is shown. (**B**) When pulled out, eye (marked "e") and antenna (marked "a") discs complex and, at times, the larval brain (marked "b") remain attached to the mouthpart. This cartoon shows dorsal view. (**C**) Ideally, only eye and antenna discs complex is attached to the mouthpart. This cartoon shows lateral view. (**D**) Place the head in a mould, with the eye to be analyzed as close as possible to the narrow end of the mould. The ventral (flat) surface of the head should be resting on the bottom of the mould. Arrows show the directions of trimming when the block is being prepared for sectioning. (**E**) Remove the operculum (marked "op"), which does not attach very close to the pupal body, with forceps and then remove the rest of the cuticle bit by bit.

3. Fix dissected tissues for at least 4 hours on ice. Add an equal volume of 2% osmium oxide and incubate on ice for 30 minutes (*see* **Note 4**).

4. Replace osmium/gluteraldehyde solution with fresh 2% osmium and incubate from 1 to 2 hours on ice to overnight at 4°C (*see* **Note 4**).

5. Rinse with washing buffer.

6. Dehydrate through 30 to 100% ethanol dilution series, incubating in each for 10 minutes on ice (each time remove as much liquid as possible, but leave enough to cover tissue). Wash once again in a fresh aliquot of 100% ethanol.

7. Replace ethanol with propylene oxide and leave for 10 minutes at room temperature. Wash in propylene oxide three more times (*see* **Note 4**).

8. Without removing the last wash of propylene oxide, add an equal volume of resin, mix well, and incubate overnight at room temperature.

9. Replace the resin/propylene mixture with pure resin (*see* **Note 4**). Incubate for at least 4 hours at room temperature.

10. Place resin in the moulds. Add one specimen per mould. Orientate the tissue, so that a flat surface is resting on the bottom of the mould, very close to the edge to be cut and in the right orientation for cutting (very important; **Fig. 11.4D**).

11. Bake the resin at 70°C for 15 hours. Do not leave for longer than 36 hours, as the resin may become brittle.

12. Remove the resin blocks from the moulds, mount onto microtome. Trim the block, leaving as little resin, surrounding the material, as possible. Tilt and rotate the specimen to orient the surface to be analyzed almost parallel to the sectioning plane. A knife angle is to be determined empirically (approximately 15°). Make series of 0.6-μm semithin sections, transfer onto a drop of distilled water on a microscope slide, dry.

13. Stain sections with 0.5% toluidine blue for 1 minutes, followed by several washes in distilled water. Analyze under light microscope.

3.3. Detection of PCP Defects in the Pupal Wing

Polarity proteins such as Fz and Fmi start to show clear asymmetric localization after about 24 hours APF (timings are given for culture at 25°C), with strongest patterns of localization between 28 to 32 hours APF. Trichomes begin to emerge at about 32 hours APF (although emergence is delayed in genotypes such as *fz⁻*) (*3, 17*). Polarity defects can be scored performing immunostaining of polarity markers such as Fmi or fz-GFP reporter, or staining for actin filaments, the major cytoskeletal component of the trichome (**Fig. 11.3 A,B**). To obtain satisfactory results, it is critically important to dissect pupae of the correct age. In particularly for trichomes, it is best to dissect when they are newly formed, because more mature, longer trichomes are easily subject to mechanical damage, which produces suggestive but meaningless swirling patterns. Note though, that in some genotypes, trichome emergence can be delayed until 34 to 36 hours APF.

Precise ageing is achieved by collecting white prepupae and incubating for the appropriate length of time at 25°C (note that the white prepupal stage is relatively brief). The white prepupal stage is the time following the point at which wandering third instar larvae evert their spiracles, stop moving, and harden their external cuticle. It lasts until the "tan" prepupal stage when the cuticle begins to pigment.

1. Collect prepupae with a blunt pair of forceps onto the walls of vials for ageing, trying to avoid cuticle damage.

2. Fix the pupae 31 to 32 hours later. Using blunt forceps, transfer a pupa into a drop of fix in a Petri dish (*see* **Note 2**). Remove the external (brown and crunchy) cuticle by cutting it open with a scalpel blade or the tip of a syringe needle. Alternatively, remove the operculum (the flap at the head end, which comes off easily, arrow in **Fig. 11.4E**) with forceps and then remove the cuticle bit by bit.

3. Transfer the pupa without its external cuticle into a large fresh drop of fix and leave for at least 30 minutes (preferably 1 hour or more).

4. Dissect the wings out from the thorax by holding the thorax of the pupa with forceps, grasping the wing around its most proximal region with a second pair of forceps, and pulling. Alternatively, tease away bits of the pupal cuticle (the cellophane wrapper) from each wing, until the whole wing is removed in one piece. When the cuticle is removed, fix the liberated wing for approximately 20 minutes.

5. Transfer the wings into a microwell plate with washing buffer (two to three wings per well in a total volume 15 µL). For immunostaining, block the tissues for 30 minutes at room temperature (*see* **Note 5**).

6. Replace the blocking solution with a primary antibody mix. Seal the plate to prevent drying up and incubate overnight at 4°C.

7. Wash 10 times with washing buffer, removing as much buffer as possible without touching the wings and leaving a little in the bottom of the well, so the wings do not dry out. For better results, leave the wings for 1 hour (or more) in the final wash at room temperature.

8. Remove as much liquid as possible, add 15 µL of secondary antibody mix, and incubate for 1 to 3 hours at room temperature or overnight at 4°C. For actin staining, add phalloidin to the secondary antibody mix and stain in parallel (*see* **Note 6**). If only phalloidin staining is needed, this can be done in washing buffer without blocking.

9. After incubation with the secondary antibody and/or phalloidin, wash 10 times at room temperature.

10. If to be imaged immediately, the wings can be mounted straightaway. However, the antibody interactions can be further stabilized by a post-fixation, in which case samples can be kept for weeks at 4°C (note, however, that some fluorochromes may fade away quickly). For post-fixation, remove the final wash, add fixative/wash buffer (1:1), and fix for 10 minutes at room temperature. Wash four times for 10 minutes with washing buffer and three times with PBS only (washing away detergent makes mounting easier).

11. Put a drop of mounting medium on a clean microscope slide (*see* **Note 7**). Transfer the wings, holding the proximal end with forceps, into the drop. Preferably, mount one wing on each slide (maximum three to six per slide). Let the slides dry at room temperature for about 1 hour, seal all edges with nail polish. Store at 4°C until imaging. Analyze under confocal microscope.

3.4. Detection of PCP Defects in the Adult Wing

At the adult stage, mutants for PCP genes show characteristic hair swirling phenotype in the wing (**Fig. 11.3C,D**). Note that swirling may also be an artefact caused by inaccurate handling or mounting of the wings. Therefore, we recommend including *fz*- or *dsh*− mutants as positive control and negative wild-type wings as a negative control in each experiment.

1. Select flies of the appropriate genotype (*see* **Note 8**), anesthetize with CO_2.

2. Detach the wings from the thorax, holding the most proximal part of the wing with forceps. Put in a drop of dehydration solution.

3. Mount in DPX or Canada balsam. Analyze under light microscope.

4. Notes

1. The composition of the fixative used may affect antibody specificity or affinity. The most common fixative is 4% formaldehyde in PBS. Alternatively, PLP or PEM fixatives can

be used. We suggest trying different fixation protocols for each newly used antibody.

PLP: 2% (w/v) formaldehyde, 0.01 M NaIO$_4$, 0.075 M lysine, 0.037 M sodium phosphate buffer, pH 7.2. To prepare PLP, dissolve 0.36 g lysine in 10 mL H$_2$O, 7.5 mL 0.1 M NaH$_2$PO$_4$, pH 7.2, 2.5 mL 0.1 M Na$_2$HPO$_4$ on ice. To 15 mL of the buffered lysine solution, add 2.5 mL 16% (w/v) formaldehyde and 50 mg NaIO$_4$. The fixative is stable for 1 to 2 weeks at 4°C.

PEM: 4% (w/v) formaldehyde, 0.1 M PIPES, pH7.0, 2 mM MgSO$_4$, 1 mM EGTA. PEM buffer is better prepared as 2x stock, with formaldehyde added immediately before use.

Note, that the 16% formaldehyde stock solution, once the glass vial opened, can be stored at 4°C for no longer than one month and should be added to PBS immediately before use.

2. Formaldehyde, phalloidin, and methanol are toxic! Wear protecting gloves and dispose in hazardous waste container.

3. Triton X-100 and Tween-20 are commonly used detergents for whole-mount antibody staining. Note, that Tween-20 permeabilizes only cellular membrane and cannot be used for detection of nuclear proteins. Triton X-100 is a stronger detergent and permeabilizes both cellular and nuclear membranes. Some antibodies may show low penetrance, in which case increasing Triton X-100 concentration to 0.2% (v/v) may help.

4. Osmium, glutaraldehyde, and propylene oxide are extremely toxic! Only handle them in the hood and wear gloves. To dispose of osmium, mix with 2 volumes vegetable oil and leave for 1 to 2 weeks when complete oxidation occurs.

To prepare resin for semi-thin sections, mix the following components of the Durcupan ACM set: 54 g resin (A), 44.5 g hardener (B), 2.5 g accelerator (C), and 10 g plasticiser (D). Use immediately or store in aliquots at –20°C. Do not subject to freeze–thaw more than three times. Handle resin with non-latex gloves as it is carcinogenic when unpolymerized. In order to protect working surfaces, cover them with aluminium foil (microscope, dissecting surface, etc). To dispose of waste resin, bake at 70°C overnight and discard only when fully polymerized.

5. Normal goat serum is commonly used to block tissues when crude serum antibody is used. However, it may interfere with phalloidin. If monoclonal antibody is to be used, we recommend blocking tissues with 0.5–1% (w/v) BSA.

6. A number of commercial phalloidins can be used to stain actin filaments, such as Texas Red-X-Phalloidin, FITC-, or TRITC-conjugated Phalloidins. However, we find that

Alexa-633-Phalloidin may be incompatible with detergent in washing buffer.

7. Due to low glycerol concentration, the mounting medium may not clear the wings very well. Higher glycerol concentrations or any viscous mountant (e.g., Mowiol) can be used instead. However, care should be taken as they may cause the wings to dehydrate and roll up on the slide. We recommend starting out with 10% glycerol.

8. A great number of recessive mutations display poor viability when homozygous or may be lethal, which complicates their study. In such case analysis of mosaic organisms proves to be very useful. In *Drosophila*, Flp/FRT system is commonly used to produce a clone of mutant cells on otherwise wild-type background *(18)*.

Acknowledgments

The authors would like to express their appreciation to Elizabeth Silva, Leslie Clayton, and Cedric Plachot for supplying data used in figures.

References

1. Mollereau, B., Domingos, P. M. (2005) Photoreceptor differentiation in Drosophila: from immature neurons to functional photoreceptors. *Dev Dyn* 232, 585–592.

2. Strutt, H., Strutt, D. (2003) EGF signaling and ommatidial rotation in the Drosophila eye. *Curr Biol* 13, 1451–1457.

3. Wong, L. L., Adler, P. N. (1993) Tissue polarity genes of Drosophila regulate the subcellular location for prehair initiation in pupal wing cells. *J Cell Biol* 123, 209–221.

4. Adler, P. N., Lee, H. (2001) Frizzled signaling and cell-cell interactions in planar polarity. *Curr Opin Cell Biol* 13, 635–640.

5. Fanto, M., Clayton, L., Meredith, J., et al. (2003) The tumor-suppressor and cell adhesion molecule Fat controls planar polarity via physical interactions with Atrophin, a transcriptional co-repressor. *Development* 130, 763–774.

6. Higashijima, S., Kojima, T., Michiue, T., et al. (1992) Dual Bar homeo box genes of Drosophila required in two photoreceptor cells, R1 and R6, and primary pigment cells for normal eye development. *Genes Dev* 6, 50–60.

7. Domingos, P. M., Brown, S., Barrio, R., et al. (2004) Regulation of R7 and R8 differentiation by the spalt genes. *Dev Biol* 273, 121–133.

8. Kuhnlein, R. P., Frommer, G., Friedrich, M., et al. (1994) spalt encodes an evolutionarily conserved zinc finger protein of novel structure which provides homeotic gene function in the head and tail region of the Drosophila embryo. *Embo J* 13, 168–179.

9. Shimada, Y., Usui, T., Yanagawa, S., et al. (2001) Asymmetric colocalization of Flamingo, a seven-pass transmembrane cadherin, and Dishevelled in planar cell polarization. *Curr Biol* 11, 859–863.

10. Usui, T., Shima, Y., Shimada, Y., et al. (1999) Flamingo, a seven-pass transmembrane cadherin, regulates planar cell polarity under the control of Frizzled. *Cell* 98, 585–595.

11. Hayashi, T., Kojima, T., Saigo, K. (1998) Specification of primary pigment cell and outer photoreceptor fates by BarH1 homeobox gene in the developing Drosophila eye. *Dev Biol* 200, 131–145.

12. Domingos, P. M., Mlodzik, M., Mendes, C. S., et al. (2004) Spalt transcription factors are required for R3/R4 specification and establishment of planar cell polarity in the Drosophila eye. *Development* 131, 5695–5702.

13. Cooper, M. T., Bray, S. J. (1999) Frizzled regulation of Notch signaling polarizes cell fate in the Drosophila eye. *Nature* 397, 526–530.

14. Yang, C. H., Axelrod, J. D., Simon, M. A. (2002) Regulation of Frizzled by fat-like cadherins during planar polarity signaling in the Drosophila compound eye. *Cell* 108, 675–688.

15. Robinow, S., White, K. (1988) The locus elav of Drosophila melanogaster is expressed in neurons at all developmental stages. *Dev Biol* 126, 294–303.

16. Peifer, M., Rauskolb, C., Williams, M., et al. (1991) The segment polarity gene armadillo interacts with the wingless signaling pathway in both embryonic and adult pattern formation. *Development* 111, 1029–1043.

17. Strutt, H., Strutt, D. (2002) Nonautonomous planar polarity patterning in Drosophila: dishevelled-independent functions of frizzled. *Dev Cell* 3, 851–863.

18. Xu, T., Rubin, G. M. (1993) Analysis of genetic mosaics in developing and adult Drosophila tissues. *Development* 117, 1223–1237.

Chapter 12

Wingless Signaling in *Drosophila* Eye Development

Kevin Legent and Jessica E. Treisman

Abstract

The secreted morphogen Wingless (Wg) has a variety of functions throughout *Drosophila* eye development, controlling tissue specification, growth, and patterning. Wg plays a critical role in subdividing the eye imaginal disc into separate primordia that will give rise to the adult retina and the surrounding head capsule. During larval development, *wg* is expressed in the anterior lateral margins of the eye disc, regions that will give rise to head cuticle; Wg signaling promotes the head fate and prevents these marginal regions from initiating ectopic photoreceptor differentiation. Expression of *wg* at the dorsal margin is earlier and stronger than at the ventral margin, allowing Wg to contribute to specifying the dorsal domain of the eye disc. Finally, during the pupal stages, *wg* expression surrounds the entire eye and a concentric gradient of Wg establishes several distinct peripheral retinal cell fates. This chapter reviews these aspects of Wg function and describes how to generate clones of cells mutant for genes encoding components of the Wg signaling pathway in the eye disc and examine their effects on photoreceptor differentiation by immunohistochemistry.

Key words: *Drosophila*, *wingless*, eye disc, photoreceptor, immunostaining, *dishevelled*, *axin*.

1. Introduction

The *Drosophila* compound eye consists of a hexagonal array of approximately 800 light-sensing units called ommatidia. Each ommatidium contains eight photoreceptor neurons, four lens-secreting cone cells, and two primary pigment cells; the ommatidia are surrounded by a lattice of secondary and tertiary pigment cells and mechanosensory bristles *(1)*. The adult eye and head capsule develop from an epithelial bilayer known as the eye-antennal imaginal disc. This structure derives from a primordial group of about 20 cells determined early during

Elizabeth Vincan (ed.), *Wnt Signaling, Volume II: Pathway Models, vol. 469*
© 2008 Humana Press, a part of Springer Science + Business Media, New York, NY
Book doi: 10.1007/978-1-60327-469-2

embryogenesis by expression of two Pax6 family transcription factors *(2–4)*. The disc invaginates from the embryonic ectoderm and grows by asynchronous cell divisions until the third larval instar *(1, 5)*. The retina arises from the columnar epithelial layer, which is partitioned into separate eye and antennal primordia during the second larval instar (**Fig. 12.1A**; ref. *6*). The eye disc epithelium is further subdivided into a central field that gives rise to the eye, and marginal regions that instead form the head capsule *(7)*. The overlying squamous peripodial epithelial layer gives rise to additional head structures and may also communicate with the columnar epithelium *(8, 9)*.

In the third larval instar, a wave of photoreceptor differentiation initiates at the posterior margin of the eye disc and progresses sequentially one row of ommatidia at a time toward the anterior. Differentiation is preceded by the morphogenetic furrow (MF), a transient vertical groove in the epithelium formed by contraction of the cells in the apical/basal dimension and constriction of their apical surfaces (**Fig. 12.1B**; ref. *10*). In the MF, cells are arrested in the G1 phase of the cell cycle *(11)*. Cells immediately posterior to the MF become organized into a regularly spaced array of five-cell preclusters containing the precursors of photoreceptors R8, 2, 5, 3, and 4 *(12)*. The remaining cells undergo a final round of cell division, the second mitotic wave, before R1, 6, 7 and the four cone cells are recruited to each ommatidium *(12)*. The precise lattice of pigment cells and bristles surrounding the ommatidia is formed by differentiation and cell death of the interommatidial cells during the pupal stages *(13)*.

The *wingless (wg)* gene is the founding member of the *Wnt* gene family and encodes a secreted signaling protein that acts as a morphogen *(14)*. In the eye disc, Wg acts primarily to promote head capsule differentiation and restrict eye development *(15–18)*. In addition, Wg signaling patterns the eye primordium along the dorsal-ventral and central-peripheral axes, and contributes to its growth. The following reviews these functions and provides a detailed protocol to generate and stain clones of cells mutant for Wg pathway components within the developing eye disc.

Fig. 12.1. Functions of Wg in eye development. Panels show schematic representations of second instar (**A**) and third instar (**B**) eye-antennal imaginal discs and a pupal eye (**C**). Dorsal is to the top and anterior is to the left. Wg–dependent genetic interactions governing D/V compartmentalization (**A**) and MF progression (**B**) are indicated. (**B**) shows *wg* expression in the third instar eye disc, which is restricted to small patches at the anterior dorsal and ventral margins and a thin stripe extending around the posterior margin. Expression of the retinal determination network proteins (Eya, Dac, and So) is encircled by a dashed black line. Note that Dac expression does not extend as far posteriorly as Eya and So. Expression of Ey in the anterior part of the eye disc is encircled by a dashed gray line. MF (morphogenetic furrow), PPN (preproneural domain). (**C**) indicates the concentric arrangement of head capsule, PR, DRO, and ommatidia lacking or containing bristles, which is established at pupal stages by Wg expressed in a ring surrounding the eye.

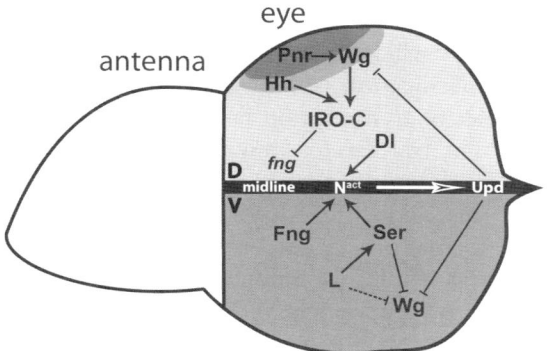

A. 2nd instar disc illustrating D/V patterning

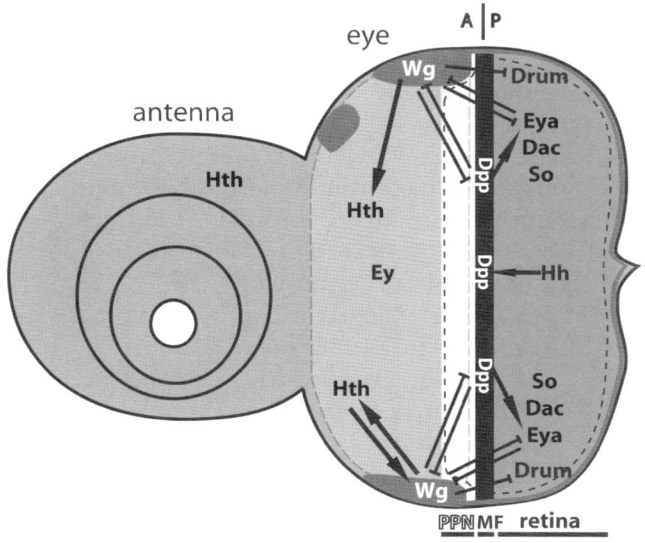

B. 3rd instar disc illustrating MF progression

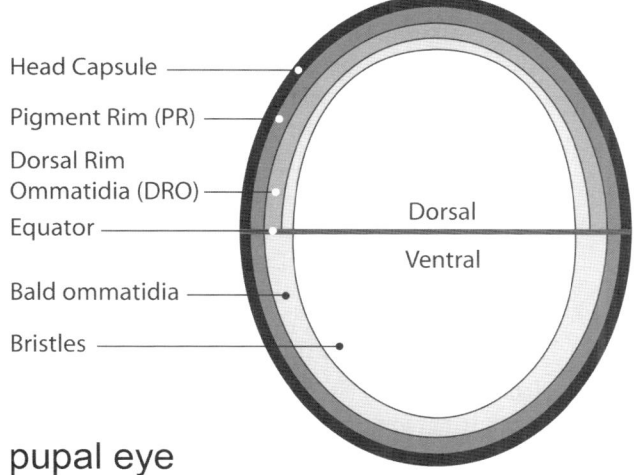

C. pupal eye

1.1. Wg Inhibits Eye Specification

A critical role for Wg is to define anterior regions of the eye disc that will give rise to the head capsule rather than the eye. Eye specification is controlled by the retinal determination genes, a set of transcription factors that function in a complex hierarchy. The two *Pax6* homologues *twin of eyeless (toy)* and *eyeless (ey)* act at the top of the cascade, with *toy* upstream of *ey*; both are necessary and sufficient for eye development but are also required for formation of the entire head, reflecting their early expression in the whole eye-antennal disc *(2–4, 19)*. Subdivision of the disc into separate eye and antennal primordia is first apparent in the early second instar, when Ey is downregulated in the anterior region that will give rise to the antenna *(6, 19, 20)*. Concomitantly with this change, the retinal determination protein Eyes absent (Eya) is specifically expressed at the posterior margin of the eye disc in response to signaling by the secreted molecules Hedgehog (Hh) and Decapentaplegic (Dpp) *(6, 19, 20)*. Ey is required for the expression of both Eya and Sine oculis (So), which interact with each other to form a compound transcription factor that promotes eye specification *(19, 21, 22)*. The final retinal determination protein, Dachshund (Dac), requires these proteins for its expression, but can also physically interact with Eya to promote eye development *(19, 23, 24)*. Positive feedback regulation stabilizes the expression of these four genes to lock in the retinal fate *(19, 21, 24–27)*.

Once the eye disc has been specified in the second instar, its cells express Ey together with two other transcription factors, Homothorax (Hth) and Teashirt (Tsh). When the MF initiates in the third instar, these factors become restricted to the most anterior region of the eye disc, where they promote proliferation *(28)*. A preproneural domain (PPN) anterior to the MF is defined by the loss of Hth, which is repressed by Dpp signaling, and the expression of Eya, So, and Dac *(28)*. Posterior to the MF, expression of Ey, Tsh and Dac is downregulated, but Eya and So are maintained and continuously required for photoreceptor differentiation (**Fig. 12.1B**; refs. *21, 28,* and *29*). The early function of the interactive retinal specification and determination network thus ensures that cells anterior to the MF are already committed to become retinal tissue before they differentiate.

wg expression is limited to anterior marginal regions of the eye disc that will give rise to head cuticle (**Fig. 12.1**), and many observations demonstrate that Wg signaling promotes head capsule formation at the expense of the retinal field *(15–18, 30)*. Reduction of Wg activity using a temperature-sensitive *wg* allele or removal of the positively acting downstream component *dishevelled (dsh)* from clones of cells results in expansion of the adult eyes into the dorsal head *(17, 18, 31)*. Conversely, removal of negative regulators of Wg signaling such as *shaggy (sgg)* or *axin (axn)* maximally activates the Wg pathway and

transforms eye tissue into head cuticle or other cuticular structures *(15, 31–33)*.

Specification of the head primordium by Wg probably occurs
early in eye disc development. Despite the limited domain of *wg*
transcription, Wg protein diffuses throughout the entire early
second instar eye disc, where it blocks Eya expression. Dpp
signaling is only able to induce Eya once the disc has grown large
enough that Wg can no longer reach the posterior cells *(6, 32,
34)*. Consistent with this model, the lack of eye development in
mutants for *eyegone (eyg)*, a *Pax6*-like gene that is required for
disc growth *(35)*, can be rescued by inhibiting Wg signaling in
posterior cells *(36)*. In the third instar eye disc, *wg* expressed at
the anterior dorsal and ventral margins *(8, 15, 17, 30, 31)* continues to repress the expression of the retinal determination genes
eya, *so* and *dac* in regions destined to form the dorsal head *(32)*.
Wg may also promote the head fate by enhancing expression of
the anteriorly expressed transcription factor Hth (**Fig. 12.1B**;
ref. *(37)*. These expression patterns are maintained by feedback
loops; Eya and Dac repress expression of *wg* at the posterior margin of the eye disc *(15, 36)*, while Hth maintains ventral Wg
expression *(37)*, contributing to the distinction between eye and
head fates. Another regulator of *wg* expression is the JAK/STAT
signaling pathway. The ligand Unpaired (Upd) is expressed at
the posterior margin of the early eye disc and promotes the
formation of the eye field through repression of *wg* transcription
(**Fig. 12.1A**; refs. *38* and *39*). The negative regulatory interactions between anteriorly expressed Wg and posteriorly expressed
signaling molecules including Hh, Dpp, and Upd thus define the
boundaries of the eye field.

Loss of *wg* activity transforms the head primordium into
eye tissue by initiation of ectopic morphogenetic furrows. This
occurs primarily at the dorsal margin of the eye disc, correlating
with the strong expression of *wg* in this region, but can also occur
at the ventral margin where *wg* is more weakly expressed (**Fig.
12.2B,D**; refs. *15, 17*, and *30*). These furrows then progress
inward toward the center of the disc *(15, 17)*. Conversely, ectopic
activation of the Wg pathway, using overexpression of Wg or an
activated form of Armadillo/β-catenin (Arm), or mutations in
sgg or *axn*, prevents normal MF initiation and progression (**Fig.
12.2C,E**; refs. *15, 16*, and *33*). Although high levels of Wg signaling can prevent the expression of *dpp*, a positive regulator of MF
movement *(31)*, Wg overexpression also has an inhibitory effect
downstream of Dpp receptor activation *(36)*. This might be due
to its ability to induce expression of Hth, a negative regulator of
photoreceptor differentiation that is repressed by Dpp signaling
(**Fig. 12.1B**; refs. *28, 37*, and *40*). Another target for Wg that is
likely to be relevant in this context is *drumstick (drum)*, a member
of the *odd-skipped* gene family, which normally contributes to MF

initiation at the posterior margin and is repressed by Wg at the anterior lateral margins *(41)*.

Taken together, these results suggest that division of the eye disc between an anterior head field and a posterior eye field relies on the balance between antagonistic Wg signaling in the anterior and Dpp signaling in the posterior. The ranges of these signals initially overlap, but are separated by growth of the disc anlage, which is driven by Notch activity *(6, 35)*.

1.2. Wg Patterns the Dorsal–Ventral Axis in the Eye Disc

In addition to establishing head primordia in anterior regions of the eye disc, Wg also contributes to distinguishing the dorsal and ventral domains. The GATA family transcription factor Pannier (Pnr) lies at the top of a regulatory cascade that leads to dorsal/ventral (D/V) compartmentalization of the eye disc and activation of the Notch receptor precisely along the D/V midline. It is not clear when dorsal-ventral differences are first established, as *pnr* transcripts are present in the dorsalmost region of the embryonic eye disc *(42)*, but a *pnr*-GAL4 line does not drive UAS-GFP expression in the dorsal eye disc until early second instar *(43)*. *wg-lacZ* is expressed in the dorsal peripodial membrane and dorsal margin cells beginning in the first instar *(8, 30, 44)*; at least at later stages, dorsal *wg* expression is dependent on *pnr* *(42)*. *wg* expression in the dorsal peripodial epithelium can be attributed to activation by Pnr of an eye-specific enhancer in the *wg* 3′ cis-regulatory region *(45)*.

Wg cooperates with Hh, which is also expressed dorsally in the late second instar eye disc, to activate the expression of the *Iroquois complex* (*Iro-C*) homeobox genes *araucan*, *caupolican* and *mirror* in the dorsal half of the disc *(8, 33, 42, 44, 46)*. The Iro-C proteins repress the expression of Fringe (Fng), a glycosyltransferase that makes the Notch receptor more responsive to one of its ligands, Delta (Dl), and less responsive to Serrate (Ser) *(47–49)*. The Fng expression border, in combination with the restriction of Ser to the ventral compartment and Dl to the dorsal compartment, results in activation of Notch precisely at the dorsal/ventral boundary of the disc, known as the equator *(47-49)* (**Fig. 12.1A**). Regulation of the *Dl* and *Ser* expression patterns is likely to involve both the dorsal Iro-C proteins and the ventral determinants Lobe and Sloppy-paired *(50, 51)*. Through its effect on the *Iro-C* genes, Wg thus contributes to the localized activation of Notch at the equator, which stimulates growth of the eye disc and later determines the initiation point of the MF.

An additional site of *wg* expression appears at the ventral margin of the early third instar eye disc *(8, 30, 44)*. Both Ser and its upstream activator Lobe are required for ventral eye development and growth *(43, 50)*. Lobe and Ser have been shown to repress ventral *wg* expression during the second instar; their absence results in ectopic Wg signaling that triggers cell death and loss of the ventral eye (**Fig. 12.1B**; ref. *52*). Formation of a normally patterned eye is thus critically dependent on the timing of *wg* expression in this region. Later in development, Wg diffuses from both the dorsal and ventral margins to establish inverse gradients of Dachsous and Four-jointed, two molecules that set up planar polarity in the eye disc *(53, 54)*. This function of Wg is discussed in more detail in **Chapter 11** in **Volume 2**.

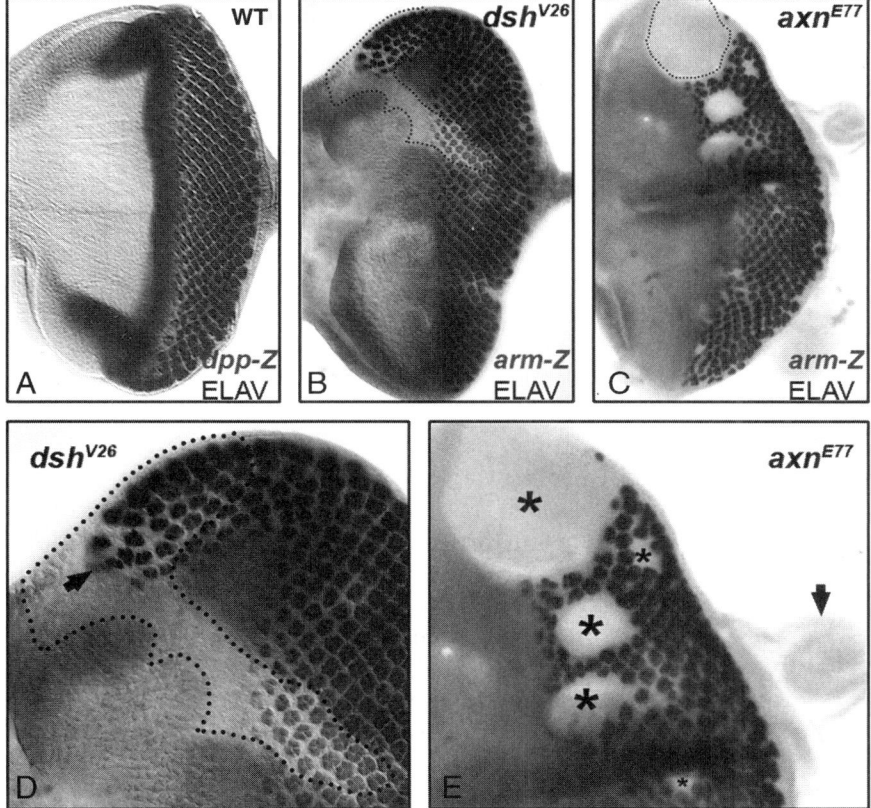

Fig. 12.2. Effects of genetic manipulation of the Wg pathway on photoreceptor differentiation. All pictures are third instar eye discs stained with X-gal to reveal β-galactosidase activity (gray) and with an antibody to the neuronal marker Elav (black). In each panel, anterior is to the left and dorsal to the top. (**A**) A wildtype disc expressing *dpp-lacZ*. Regularly spaced clusters of photoreceptors differentiate posterior to the MF, which is marked by *dpp-lacZ* activity. (**B–E**) show discs heterozygous for *dsh^{V26}* (**B, D**) or *axn^{E77}* (**C, E**) in which clones of homozygous mutant cells have been induced by the FLP/FRT technique and are visualized by the absence of *arm-lacZ* activity. (**B, D**) Removal of *dsh*, a positive intracellular effector of Wg signaling, in clones of cells abutting the anterior dorsal margin (clone boundaries indicated by dotted line), prevents Wg signal transduction and allows ectopic MF initiation resulting in photoreceptor differentiation anterior to the normal MF (black arrow in **D**, compare with **A**). (**C, E**) Conversely, removal of *axn*, a negative regulator of Wg signaling, from clones of cells within the differentiating retina (asterisks), allows ectopic activation of Wg signaling posterior to the MF and eliminates photoreceptor differentiation in these clones. Mutant clones at the posterior margin of the eye disc also grow beyond the normal boundaries of the disc (arrow in **E**). (**B**) is reprinted, with permission, from ref. *99*.

1.3. Wg Controls Growth of the Eye Disc

Both loss-of-function and gain-of-function studies suggest that Wg is a positive regulator of eye disc growth. Removal of *wg* early in larval development using a temperature-sensitive allele results in very small eye discs *(15, 17)*, although it is possible that this size reduction is due to premature differentiation of cells that would otherwise have continued to proliferate. Activation of the Wg signaling pathway by overexpression of Wg or removal of the negative regulator Axn from clones of cells results in dramatic overgrowth, especially in clones that contact the posterior margin *(15, 32, 33)*. Such clones do not respect the normal boundaries of the eye disc, but form large rounded projections beyond these boundaries (**Fig. 12.2C,E**; refs. *32* and *33*). A similar effect can be produced by misexpression of Tsh, which is not normally expressed in posterior margin cells, and this effect requires Wg signaling *(28, 55)*.

Final eye size must depend on the interplay between Wg and other growth regulators such as Dpp, Notch, and the Notch target Upd *(8, 47–49, 56–58)*. Expression of Wg or activated Arm throughout the eye disc reduces eye size in addition to blocking differentiation *(33, 36, 52)*, probably because it prevents the expression or action of these other factors. Because clones of cells in which Wg signaling is activated are still able to respond to molecules secreted by surrounding cells, Wg may have a mitogenic effect only in combination with additional growth regulatory signals.

1.4. Wg Surrounding the Eye Patterns the Peripheral Retina

As the MF progresses toward the anterior of the disc during the third instar, *wg* is transcribed in a thin stripe of cells along the posterior margin of the eye disc, adjacent to the retinal field *(17, 31)*. When the MF reaches the anterior of the eye disc at the end of the first day of pupal development, *wg* expression remains in a ring of presumptive head cuticle cells, immediately adjacent to and surrounding the entire developing eye. This is consistent with the transformation of *sgg* or *axn* mutant eye disc cells, in which Wg signaling is maximally active, into head cuticle *(31–33)*. This pattern of *wg* expression is maintained throughout pupation and in the adult head. Wg diffusion from this peripheral source establishes a gradient that patterns the peripheral retina *(31, 59)*.

The outermost region of the eye is organized into a series of concentric rings with different morphological features. At the periphery of the eye, abutting the head capsule, the pigment rim (PR) is a thick layer of pigment cells, devoid of photoreceptors, that insulates ommatidia from extraneous light rays *(59)*. On the dorsal side of the eye only, the ommatidia directly adjacent to the PR are specialized polarized light detectors called dorsal rim ommatidia (DRO; refs. *59* and *60*). Finally, bristles are absent from the outermost ommatidial rings, but present in the remainder of the eye (**Fig. 12.1C**). A series of gain- and loss-of-function experiments demonstrated that a gradient of Wg signaling organizes the differentiation of these concentric features *(59)*.

During the pupal phase, the most peripheral ring of ommatidia is eliminated by apoptosis *(61–63)*, while the surrounding secondary and tertiary pigment cells survive and contribute to the PR *(63)*. Wg signaling at mid-pupation is required for the death of these 80–100 ommatidia that often lack the full complement of cells and might compromise peripheral vision *(62–64)*. Consistently, *wg* overexpression or loss of the negative Wg pathway component Adenomatous polyposis coli (APC) 1 in the retinal lattice is sufficient to elicit apoptosis of all photoreceptors at mid-pupation *(63, 65, 66)*. Programmed cell death of the peripheral ommatidia also requires the Snail group transcription factors Worniu (Wor), Escargot (Esg), and Snail (Sna), which are targets of the Wg signaling pathway *(64)*.

While high levels of Wg activity result in photoreceptor death, allowing only pigment cells to differentiate, intermediate levels of Wg can induce the differentiation of DRO. A critical target of Wg for this activity is the transcription factor Hth, which acts in combination with the Iro-C homeodomain proteins expressed specifically in the dorsal eye *(59, 60)*. Finally, low levels of Wg signaling can prevent the formation of interommatidial bristles, at least in part through repression of the proneural gene *achaete* and its cofactor *daughterless* *(59, 67, 68)*. As endogenous *wg* is not essential for bristle repression, another *Drosophila* Wnt may contribute to this function *(67)*.

The multiple functions of Wg throughout eye development in part reflect its dynamic expression pattern, but probably also require spatial and temporal regulation of the responsiveness of surrounding cells. Combinatorial control of growth and differentiation by Wg and other secreted factors is also likely to play an important role in patterning the eye disc. Elucidating the functions of Wg in eye development has been critically dependent on the ability to analyze mosaic eye discs in which clones of cells are mutant for components of the Wg signaling pathway. This technique is described in the following sections.

2. Materials

	Drosophila strains	Comment	References
2.1. Useful Tools to Study Wg Signaling in the Drosophila Eye	wg^{CX4}	amorph	69
	wg^1	viable hypomorph	70
	wg^{IN}	not secreted	71
	wg^{IL114}	temperature sensitive : 17°C (P) → 25°C (R)	72
	wg^{I-12}	temperature sensitive : 17°C (P) → 25°C (R)	73

Drosophila strains	Comment	References
wg^{en-11}	enhancer trap, amorph	74
arr^{12-1}	loss of function	75
fz^{15}	amorph	76
$fz2^{C1}$	loss of function	77
dsh^{V26}	amorph	78
axn^{E77}	loss of function	33
sgg^{D127}	amorph	79
$dAPC^{Q8}$	loss of function	80
arm^{1}	amorph	81
$pan^{2}/dTCF^{2}$	amorph	82
$act5c>y^{+}>wg$	FLP-out	83
$tub>w^{+}>wg$	FLP-out	53
$tub>CD2, y+>$ $flu\text{-}wg$	FLP-out, HA-tagged	84
$GMR\text{-}wg$	expressed posterior to MF	53
$GMR\text{-}wg^{ts}$	temperature sensitive : $16.5°C\ (P) \rightarrow 25°C\ (R)$	59
$sev>w^{+}>arm^{*}$	activated form	53
$UAS\text{-}wg$	GAL4-responsive promoter	85
$UAS\text{-}wg^{ts}$	temperature sensitive : $16.5°C\ (P) \rightarrow 25°C\ (R)$	86
$UAS\text{-}GFP\text{-}wg$	GFP-tagged	87
$UAS\text{-}Nrt\text{-}$ $flu\text{-}wg$	tethered, HA-tagged	88
$UAS\text{-}dTCF^{\Delta N3}/$ $UAS\text{-}dTCF^{DN}$	dominant negative	82
$UAS\text{-}arm$	GAL4-responsive promoter	80
$UAS\text{-}arm^{K45}$	activated form	89
$UAS\text{-}arm^{S10}$	activated form	90
$UAS>CD2,$ $y^{+}>flu\text{-}\Delta arm$	N-terminal deletion, activated	88
$UAS\text{-}sgg$	GAL4-responsive promoter	91
$UAS\text{-}sgg^{S9A}/$ $UAS\text{-}sgg^{act}$	activated form	36
$UAS\text{-}axn$	GAL4-responsive promoter	92
$wg2.11\text{-}lacZ$	reporter	45
$arm\text{-}lacZ$	reporter	93

Drosophila strains	Comment	References
wg-GAL4	GAL4 expressed in *wg* pattern	*94*
arm-GAL4	GAL4 expressed in *arm* pattern	*95*
Antisera		
Anti-Wg	Developmental Studies Hybridoma Bank	
Anti-dAPC		*84*
Anti-Arm	Developmental Studies Hybridoma Bank	
Anti-Axn		*92*
Anti-Elav	Developmental Studies Hybridoma Bank (recognizes photoreceptor nuclei)	

2.2. Fly Stocks

1. *y,w, ey-FLP ; + ; FRT82B, arm-lacZ / TM6B.*
2. *y,w ; + ; FRT82B, axn^{E77} / TM6B.*
3. *y,w, arm-lacZ, FRT19A ; + ; ey-FLP / TM6B.*
4. *y,w, dsh^{V26}, FRT19A / FM7.*

2.3. Dissection

1. Two pairs of fine forceps (Dumont # 5).
2. A silicone dissection dish (Sylgard, Silicone Elastomer Kit).
3. A binocular microscope (Zeiss, Stemi SV 11).
4. A tungsten hook: ultra micro needle (Ted Pella, Inc.) bent into a hook using forceps.
5. Petri dishes (35 × 10 mm).
6. Polystyrene microwell mini trays, 60 wells (Nunc).
7. Microscope slides, cover slips (18 × 18 mm), and nail polish.
8. A flat metal plate that can fit in an ice bucket.

2.4. Solutions

1. 0.1 M phosphate buffer, pH 7.2: mix 1 M Na_2HPO_4 and 1 M NaH_2PO_4 in a 72:28 ratio, respectively, and add 9 volumes of water.
2. PEM: 0.1 M PIPES pH 7.0, 2 mM $MgSO_4$, 1 mM EGTA; it is usually kept as a 2X stock at 4°C.
3. Freshly prepared fixative: either 4% (w/v) formaldehyde in PEM: for 10 mL, mix 5 mL of 2X PEM, 1 mL of H_2O and 4 mL of 10% (w/v) methanol-free formaldehyde (Polysciences) OR 2% (w/v) formaldehyde in PLP: Prepare a phosphate buffer solution by mixing 75 mL of 0.1 M phosphate buffer, pH 7.2, 2.5 mL of 1 M Na_2HPO_4, and

122.5 mL of H$_2$O. To 15 mL of this, add 1 mL of water and 0.27 g lysine. Just before use, add 50 mg sodium periodate and 4 mL of 10% (w/v) methanol-free formaldehyde.

4. PTX: 0.1 M phosphate buffer, pH 7.2, 0.2% (v/v) Triton X-100. For some antibodies, other detergents such as saponin may be preferable.

5. Normal donkey serum (Jackson Immunoresearch) or goat serum if goat secondary antibodies are used.

6. Rat anti-Elav (Developmental Studies Hybridoma Bank) and HRP-conjugated donkey anti-rat (Jackson Immuno-research).

7. Freshly prepared diaminobenzidine (DAB) solution: for 250 μL staining solution, add 225 μL of 0.1 M phosphate buffer, 0.2% (v/v) Triton to 25 μL 5 mg/mL DAB; add 5 μL of 1% (w/v) cobalt chloride for intensification and 2.5 μL of 0.3% H$_2$O$_2$ (freshly diluted from a 30% stock). For double staining with X-gal, include 5 μL of 1% (w/v) nickel ammonium sulfate and 6 μL of 1% (w/v) cobalt chloride in the DAB solution. DAB is toxic; handle with gloves and deactivate used tips, etc., overnight in bleach before disposal. The 5 mg/mL solution should be stored in 25 μL aliquots at −80°C.

8. X-gal staining buffer: For 50 mL, mix 1.8 mL 0.2 M Na$_2$HPO4, 0.7 mL 0.2 M NaH$_2$PO4, 1.5 mL 5 M NaCl, 50 μL 1 M MgCl$_2$, 3.05 mL 50 mM K$_3$Fe(CN)$_6$, 3.05 mL 50mM K$_4$Fe(CN)$_6$, and H$_2$O to 50 mL.

9. 8% X-gal solution in dimethylformamide (store in aliquots at −80°C).

10. 80% glycerol (v/v; Roche molecular biology grade) in PBS.

3. Methods

3.1. Genetics

The FLP/FRT method is used to generate, during development, clones of homozygous mutant cells in the eye of a fly heterozygous for a mutation affecting Wg signaling *(96)*. The presence of Flipase recombination target (FRT) sites near the centromere allows exchange between the mutant and wild-type chromatids on homologous chromosomes. This recombination is catalyzed by the Flipase enzyme (FLP), which is specifically expressed in the developing eye under the control of the *eyeless* promoter *(97)*. Chromatid segregation at mitosis may result in a homozygous mutant cell, which will proliferate to form a mutant clone (**Fig. 12.3**). In the third instar eye disc, the presence of the ubiquitously expressed *armadillo-lacZ (arm-Z)* transgene

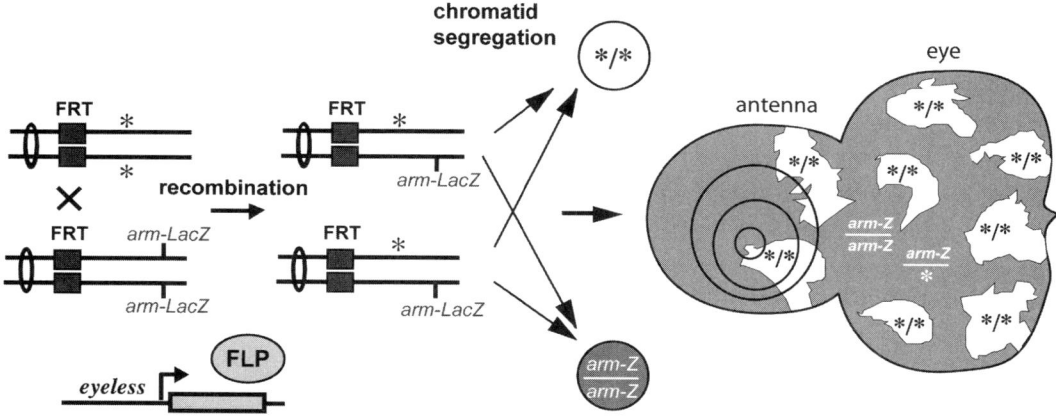

Fig. 12.3. Mitotic recombination using the FLP/FRT system. In cells heterozygous for a mutation (*/+), the FLP/FRT system enables mitotic recombination between homologous chromatids, resulting in the production of one homozygous mutant (*/*) daughter cell and one wild type (+/+) daughter cell. In third instar eye discs mosaic for the mutation, a P element carrying the *arm-lacZ* marker allows homozygous mutant clones (*/*) to be recognized by their lack of β-galactosidase activity (gray).

(93) distal to the FRT sites provides a way to discriminate clones of homozygous mutant cells, which lack β-galactosidase activity, from heterozygous or homozygous wild-type cells (**Fig. 12.3**). Photoreceptor differentiation within clones identified by the absence of X-gal staining can be monitored using an antibody to the neuronal marker Elav *(98)*.

In freshly yeasted food vials:

1. Cross 6-8 *y,w, ey-FLP; FRT82B, arm-lacZ / TM6B* females with 4-6 *y,w; FRT82B, axn^{E77} / TM6B* males.

2. Cross 6-8 *y,w dsh^{V26}, FRT19A/FM7* females with 4-6 *y,w, arm-lacZ, FRT19A/Y; eyFLP/TM6B* males.

Allow the flies to lay eggs for 2 to 3 days and transfer to a fresh vial. Approximately 1 week later, select non-*Tubby y,w, ey-FLP/y,w* or Y; *FRT82B, arm-lacZ / FRT82B, axn^{E77}* third instar larvae from **cross 1**. Select non-*Tubby y,w, arm-lacZ, FRT19A / y,w, dsh^{V26}, FRT19A ; eyFLP / +* third instar female larvae from **cross 2**.

3.2. Dissection

1. Remove third instar larvae from the food vial with forceps. Select wandering larvae that have left the food but have not yet pupariated. Transfer the larvae to a puddle of 0.1 M phosphate buffer (pH 7.2) on a dissection plate to wash off excess food.

2. Select a larva and move it to a second puddle of 0.1 M phosphate buffer (pH 7.2) on the same plate. Under the dissecting microscope, hold it gently about half way down the body with one pair of forceps. With a second pair of forceps, grasp the mouthparts and pull them out of the head (**Fig. 12.4A**). Usually the eye-antennal discs will remain

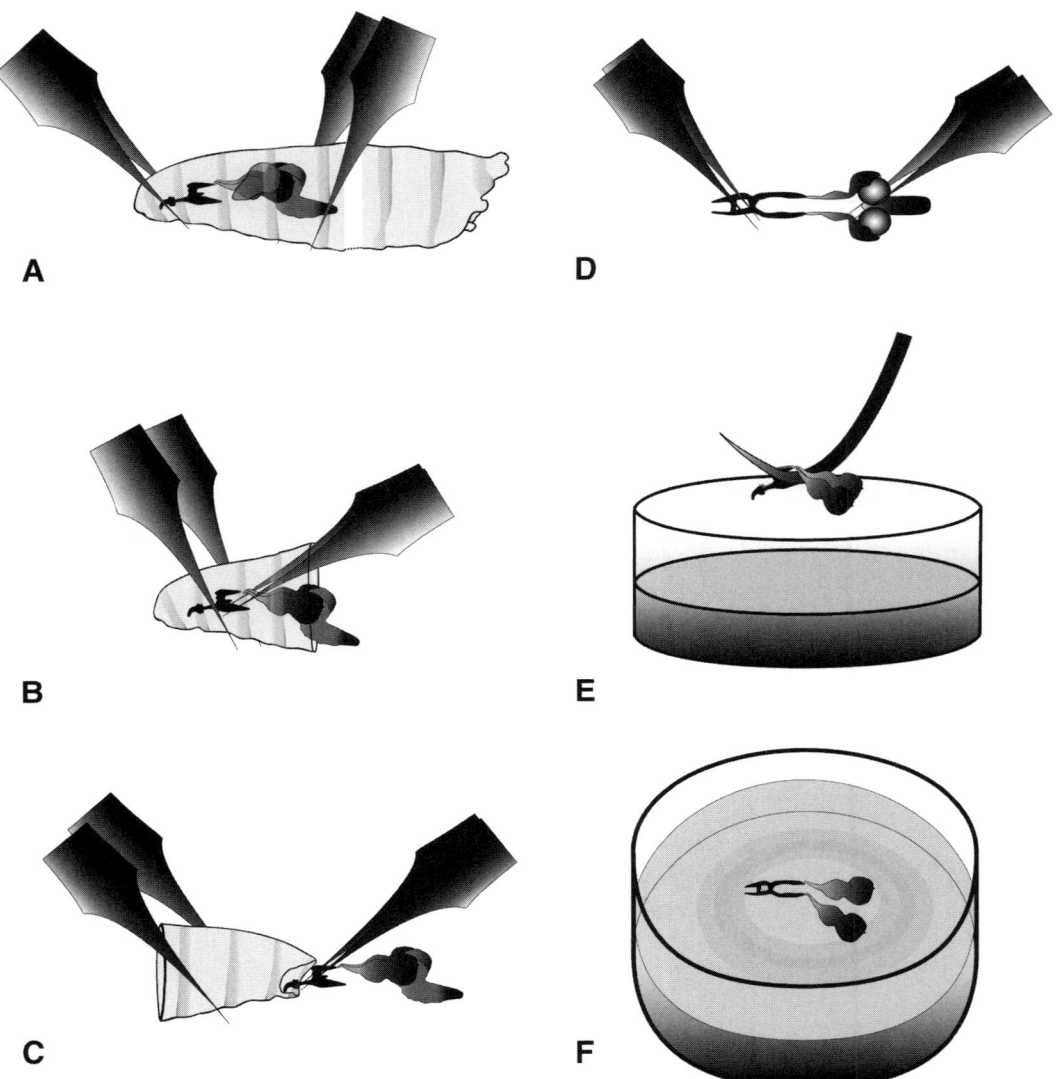

A

D

B

E

C

F

Fig. 12.4. Cartoon of the dissection protocol for eye disc staining. (**A**) Using forceps, open the larva by pulling on the mouthhooks (black). (**B–C**) Remove the larval cuticle and extract the eye-antennal discs attached to the mouthhooks. (**D**) Pin down the brain between the two hemispheres and pull on the mouthhooks to separate the eye discs from the brain. (**E**) Pick up the discs with a tungsten hook placed under the antennal discs, with the apical, convex side facing downward. (**F**) Deposit the discs on the surface of the fix, allowing the discs to flatten due to surface tension.

attached to the mouthhooks. Still holding the mouthhooks, remove other attached tissues such as the salivary glands and gut. Then hold the internal part of the mouthhooks and remove the external cuticle (**Fig. 12.4B,C**). To remove the brain, hold the mouthhooks with one pair of forceps and pin the brain to the dissection dish with one point of a second pair of forceps placed between the two hemispheres. Gently pull on the mouthhooks to break the optic stalk that links the eye disc to the brain (**Fig. 12.4D**). Leave the eye discs attached to the mouthhooks (*see* **Note 1**).

3. Using a tungsten hook, pick up the eye discs, placing the hook at the junction between the antennal imaginal disc and the mouthhooks (**Fig. 12.4E**). Deposit the discs on the surface of the fix solution (usually 4% formaldehyde [v/v] in PEM) in a 35 × 10 mm Petri dish placed on a flat metal plate in an ice bucket (**Fig. 12.4F**). Fix the discs for 25–35 minutes on ice (*see* **Note 2**).

4. Using the hook, transfer the eye discs attached to the mouthhooks into a 30 mm Petri dish filled with PTX. Wash the discs in this solution for 15 minutes on ice.

5. Transfer the discs to the antibody solution. A 60-well (10 × 6) microwell tray can be used for up to six different samples that will require 10 wells each (*see* **Note 3**). Up to 15 pairs of discs of one genotype can be stained together in a single well. Place a folded wet Kimwipe in the tray to maintain its humidity. Incubate the discs in primary antibody diluted in PTX, 10% (v/v) serum, overnight at 4°C. Rat anti-Elav can be used at 1:100.

6. In the following steps, use the hook to transfer the discs to the next well within the microwell tray at each step. Wash 3 × 5 min in PTX at room temperature. Incubate in secondary antibody diluted 1:200 in PTX, 10% (v/v) serum, for at least 2 hours at 4°C. Wash 3 times for 5 minutes in PTX at room temperature.

7. Transfer the discs to freshly prepared diaminobenzidine (DAB) solution. Staining will occur rapidly, so transfer only a few discs at a time and be prepared to stop the reaction before the discs are overstained.

8. Stop the reaction by transferring the discs to PTX.

9. Proceed with X-Gal staining if desired. Incubate in an Eppendorf tube containing prewarmed (65°C) X-gal staining buffer with a 1:40 dilution of 8% X-gal (7.5 µL in 300 µL) at 37°C (*see* **Note 4**). For *arm-lacZ* the staining time is usually 2 to 3 hours, but staining should be checked under a dissecting microscope. Wash in PBS, 0.1% (v/v) Triton X-100, rocking for 1 hour, to remove crystals of X-gal that may have become attached to the discs.

10. Mount discs in 80% (v/v) glycerol in PBS. Deposit a 20-µL drop of glycerol solution on a microscope slide. Transfer the discs into the glycerol. Remove the mouthhooks and cover the discs with a cover slip on the slide. Seal the cover slip with nail polish (*see* **Note 5**).

11. View discs on a compound microscope using Nomarski optics and a 20X or 40X objective lens (*see* **Note 6**).

4. Notes

1. Keeping the mouthhooks attached to the eye-antennal discs provides a safe way to handle both discs from one larva at the same time. One can then transfer the discs between wells with minimal damage or solution transfer by picking them up with a tungsten hook placed under the antennal discs, so that the mouthhooks fall to one side and the eye discs to the other.

2. The ventral and dorsal edges of the eye disc have a tendency to fold over, making it difficult to flatten the discs sufficiently when mounting them. When transferring the discs into the fix solution, pick them up by placing the tungsten hook under the antennal discs with the apical, convex, surface of the discs facing down. Gently deposit the discs apical side down on the surface of the fix, while moving the hook downward into the fix solution. Due to surface tension, the eye disc epithelium will unfold and flatten on the fix (**Fig. 12.3F**). PEM works well for antibodies to most nuclear and cytoplasmic proteins, but transmembrane proteins may be better preserved by fixing in PLP for 40–60 minutes on ice.

3. The use of a microwell mini tray allows each step of the staining protocol to take place in a 10–15 μL volume, conserving precious antibodies.

4. If the staining buffer cools down to room temperature, X-gal crystals will form and will be difficult to remove from the discs. To avoid this, prewarm the staining buffer at 65°C for at least 30 minutes. Add X-gal and transfer the tube to a 37°C water bath close to the microscope. Use the hook to place discs directly into the tube, removing it from the water bath for as short a time as possible.

5. To improve the quality of the mounting, drag each disc toward the same edge of the glycerol drop with its apical surface upward, until the disc immobilizes and flattens. Then gently place the cover slip on the drop beginning with the edge closest to the discs. The glycerol drop will enlarge but the flattened discs won't disperse or twist. Discs can be further flattened by gently pressing on the cover slip with forceps while observing the discs through the dissecting microscope to avoid using too much force.

6. Dorsal and ventral sides of the eye-antennal disc can be easily identified because the central circular fold within the antennal disc is closest to the ventral side (**Fig. 12.1B**).

Acknowledgments

We thank Jeffrey Lee and Esther Siegfried for fly stocks. The manuscript was improved by the critical comments of Inés Carrera, Kerstin Hofmeyer, Jean-Yves Roignant, and Josefa Steinhauer. This work was supported by the National Institutes of Health (grants EY13777 and GM56131 to J.E.T.). K.L. is a fellow of the Fondation pour la Recherche Médicale (F.R.M.).

References

1. Wolff, T., Ready, D. (1993) in (Bate, M., Martinez-Arias, A., eds.) *The Development of Drosophila melanogaster*, Cold Spring Harbor Laboratory Press, Cold Spring Harbor, NY, pp. 1277–1326.

2. Quiring, R., Walldorf, U., Kloter, U., et al. (1994) Homology of the *eyeless* gene of *Drosophila* to the *Small eye* gene in mice and *Aniridia* in humans. *Science* 265, 785–789.

3. Czerny, T., Halder, G., Kloter, U., et al. (1999) *Twin of eyeless*, a second *Pax-6* gene of *Drosophila*, acts upstream of *eyeless* in the control of eye development. *Mol Cell* 3, 297–307.

4. Kronhamn, J., Frei, E., Daube, M., et al. (2002) Headless flies produced by mutations in the paralogous *Pax6* genes *eyeless* and *twin of eyeless*. *Development* 129, 1015–1026.

5. Baker, N. E., Yu, S. Y. (2001) The EGF receptor defines domains of cell cycle progression and survival to regulate cell number in the developing *Drosophila* eye. *Cell* 104, 699–708.

6. Kenyon, K. L., Ranade, S. S., Curtiss, J., et al. (2003) Coordinating proliferation and tissue specification to promote regional identity in the *Drosophila* head. *Dev Cell* 5, 403–414.

7. Dominguez, M., Casares, F. (2005) Organ specification-growth control connection: new in-sights from the *Drosophila* eye-antennal disc. *Dev Dyn* 232, 673–684.

8. Cho, K. O., Chern, J., Izaddoost, S., et al. (2000) Novel signaling from the peripodial membrane is essential for eye disc patterning in *Drosophila*. *Cell* 103, 331–342.

9. Hallsson, J. H., Haflidadottir, B. S., Stivers, C., et al. (2004) The basic helix-loop-helix leucine zipper transcription factor Mitf is conserved in *Drosophila* and functions in eye development. *Genetics* 167, 233–241.

10. Ready, D. F., Hanson, T. E., Benzer, S. (1976) Development of the *Drosophila* retina, a neurocrystalline lattice. *Dev Biol* 53, 217–240.

11. Thomas, B. J., Gunning, D. A., Cho, J., et al. (1994) Cell cycle progression in the developing *Drosophila* eye: *roughex* encodes a novel protein required for the establishment of G1. *Cell* 77, 1003–1014.

12. Wolff, T., Ready, D. F. (1991) The beginning of pattern formation in the *Drosophila* compound eye: the morphogenetic furrow and the second mitotic wave. *Development* 113, 841–850.

13. Cagan, R. L., Ready, D. F. (1989) The emergence of order in the *Drosophila* pupal retina. *Dev Biol* 136, 346–362.

14. Baker, N. E. (1987) Molecular cloning of sequences from *wingless*, a segment polarity gene in *Drosophila*: the spatial distribution of a transcript in embryos. *EMBO J* 6, 1765–1773.

15. Treisman, J. E., Rubin, G. M. (1995) *wingless* inhibits morphogenetic furrow movement in the *Drosophila* eye disc. *Development* 121, 3519–3527.

16. Heslip, T. R., Theisen, H., Walker, H., et al. (1997) Shaggy and dishevelled exert opposite effects on Wingless and Decapentaplegic expression and on positional identity in imaginal discs. *Development* 124, 1069–1078.

17. Ma, C., Moses, K. (1995) *wingless* and *patched* are negative regulators of the morphogenetic furrow and can affect tissue polarity in the developing *Drosophila* compound eye. *Development* 121, 2279–2289.

18. Royet, J., Finkelstein, R. (1996) *hedgehog, wingless* and *orthodenticle* specify adult head development in *Drosophila*. *Development* 122, 1849–1858.

19. Halder, G., Callaerts, P., Flister, S., et al. (1998) Eyeless initiates the expression of both *sine oculis* and *eyes absent* during *Drosophila* compound eye development. *Development* 125, 2181–2191.

20. Pappu, K. S., Chen, R., Middlebrooks, B. W., et al. (2003) Mechanism of *hedgehog* signaling during *Drosophila* eye development. *Development* 130, 3053–3062.

21. Pignoni, F., Hu, B., Zavitz, K. H., et al. (1997) The eye-specification proteins So and Eya form a complex and regulate multiple steps in *Drosophila* eye development. *Cell* 91, 881–891.

22. Niimi, T., Seimiya, M., Kloter, U., et al. (1999) Direct regulatory interaction of the eyeless protein with an eye-specific enhancer in the *sine oculis* gene during eye induction in *Drosophila*. *Development* 126, 2253–2260.

23. Pappu, K. S., Ostrin, E. J., Middlebrooks, B. W., et al. (2005) Dual regulation and redundant function of two eye-specific enhancers of the *Drosophila* retinal determination gene *dachshund*. *Development* 132, 2895–2905.

24. Chen, R., Amoui, M., Zhang, Z., et al. (1997) Dachshund and eyes absent proteins form a complex and function synergistically to induce ectopic eye development in *Drosophila*. *Cell* 91, 893–903.

25. Curtiss, J., Mlodzik, M. (2000) Morphogenetic furrow initiation and progression during eye development in *Drosophila*: the roles of *decapentaplegic*, *hedgehog* and *eyes absent*. *Development* 127, 1325–1336.

26. Halder, G., Callaerts, P., Gehring, W. J. (1995) Induction of ectopic eyes by targeted expression of the *eyeless* gene in *Drosophila*. *Science* 267, 1788–1792.

27. Bonini, N. M., Leiserson, W. M., Benzer, S. (1993) The *eyes absent* gene: genetic control of cell survival and differentiation in the developing *Drosophila* eye. *Cell* 72, 379–395.

28. Bessa, J., Gebelein, B., Pichaud, F., et al. (2002) Combinatorial control of *Drosophila* eye development by *eyeless*, *homothorax*, and *teashirt*. *Genes Dev* 16, 2415–2427.

29. Mardon, G., Solomon, N. M., Rubin, G. M. (1994) *dachshund* encodes a nuclear protein required for normal eye and leg development in *Drosophila*. *Development* 120, 3473–3486.

30. Baker, N. E. (1988) Transcription of the segment-polarity gene *wingless* in the imaginal discs of *Drosophila*, and the phenotype of a pupal-lethal *wg* mutation. *Development* 102, 489–497.

31. Heslip, T. R., Theisen, H., Walker, H., et al. (1997) Shaggy and disheveled exert opposite effects on Wingless and Decapentaplegic expression and on positional identity in imaginal discs. *Development* 124, 1069–1078.

32. Baonza, A., Freeman, M. (2002) Control of *Drosophila* eye specification by Wingless signalling. *Development* 129, 5313–5322.

33. Lee, J. D., Treisman, J. E. (2001) The role of Wingless signaling in establishing the anteroposterior and dorsoventral axes of the eye disc. *Development* 128, 1519–1529.

34. Royet, J., Finkelstein, R. (1997) Establishing primordia in the *Drosophila* eye-antennal imaginal disc: the roles of *decapentaplegic*, *wingless* and *hedgehog*. *Development* 124, 4793–4800.

35. Dominguez, M., Ferres-Marco, D., Gutierrez-Avino, F. J., et al. (2004) Growth and specification of the eye are controlled independently by Eyegone and Eyeless in *Drosophila melanogaster*. *Nat Genet* 36, 31–39.

36. Hazelett, D. J., Bourouis, M., Walldorf, U., et al. (1998) *decapentaplegic* and *wingless* are regulated by *eyes absent* and *eyegone* and interact to direct the pattern of retinal differentiation in the eye disc. *Development* 125, 3741–3751.

37. Pichaud, F., Casares, F. (2000) *homothorax* and *iroquois-C* genes are required for the establishment of territories within the developing eye disc. *Mech Dev* 96, 15–25.

38. Ekas, L. A., Baeg, G. H., Flaherty, M. S., et al. (2006) JAK/STAT signaling promotes regional specification by negatively regulating *wingless* expression in *Drosophila*. *Development* 133, 4721–4729.

39. Bach, E. A., Ekas, L. A., Ayala-Camargo, A., et al. (2007) GFP reporters detect the activation of the *Drosophila* JAK/STAT pathway in vivo. *Gene Expr Patterns* 7, 323–331.

40. Pai, C. Y., Kuo, T. S., Jaw, T. J., et al. (1998) The Homothorax homeoprotein activates the nuclear localization of another homeoprotein, extradenticle, and suppresses eye development in *Drosophila*. *Genes Dev* 12, 435–446.

41. Bras-Pereira, C., Bessa, J., Casares, F. (2006) *odd-skipped* genes specify the signaling center that triggers retinogenesis in *Drosophila*. *Development* 133, 4145–4149.

42. Maurel-Zaffran, C., Treisman, J. E. (2000) *pannier* acts upstream of *wingless* to direct

dorsal eye disc development in *Drosophila. Development* 127, 1007–1016.

43. Singh, A., Choi, K. W. (2003) Initial state of the *Drosophila* eye before dorsoventral specification is equivalent to ventral. *Development* 130, 6351–6360.

44. Cavodeassi, F., Diez Del Corral, R., Campuzano, S., Dominguez, M. (1999) Compartments and organising boundaries in the *Drosophila* eye: the role of the homeodomain Iroquois proteins. *Development* 126, 4933–4942.

45. Pereira, P. S., Pinho, S., Johnson, K., et al. (2006) A 3' cis-regulatory region controls *wingless* expression in the *Drosophila* eye and leg primordia. *Dev Dyn* 235, 225–234.

46. Heberlein, U., Borod, E. R., Chanut, F. A. (1998) Dorsoventral patterning in the *Drosophila* retina by *wingless. Development* 125, 567–577.

47. Cho, K. O., Choi, K. W. (1998) Fringe is essential for mirror symmetry and morphogenesis in the *Drosophila* eye. *Nature* 396, 272–276.

48. Papayannopoulos, V., Tomlinson, A., Panin, V. M., et al. (1998) Dorsal-ventral signaling in the *Drosophila* eye. *Science* 281, 2031–2034.

49. Dominguez, M., de Celis, J. F. (1998) A dorsal/ventral boundary established by Notch controls growth and polarity in the *Drosophila* eye. *Nature* 396, 276–278.

50. Chern, J. J., Choi, K. W. (2002) Lobe mediates Notch signaling to control domain-specific growth in the *Drosophila* eye disc. *Development* 129, 4005–4013.

51. Sato, A., Tomlinson, A. (2007) Dorsal-ventral midline signaling in the developing *Drosophila* eye. *Development* 134, 659–667.

52. Singh, A., Shi, X., Choi, K. W. (2006) Lobe and Serrate are required for cell survival during early eye development in *Drosophila. Development* 133, 4771–4781.

53. Wehrli, M., Tomlinson, A. (1998) Independent regulation of anterior/posterior and equatorial/polar polarity in the *Drosophila* eye; evidence for the involvement of Wnt signaling in the equatorial/polar axis. *Development* 125, 1421–1432.

54. Yang, C. H., Axelrod, J. D., Simon, M. A. (2002) Regulation of Frizzled by fat-like cadherins during planar polarity signaling in the *Drosophila* compound eye. *Cell* 108, 675–688.

55. Singh, A., Kango-Singh, M., Sun, Y. H. (2002) Eye suppression, a novel function of *teashirt*, requires Wingless signaling. *Development* 129, 4271–4280.

56. Reynolds-Kenneally, J., Mlodzik, M. (2005) Notch signaling controls proliferation through cell-autonomous and non-autonomous mechanisms in the *Drosophila* eye. *Dev Biol* 285, 38–48.

57. Bach, E. A., Vincent, S., Zeidler, M. P., et al. (2003) A sensitized genetic screen to identify novel regulators and components of the *Drosophila* janus kinase/signal transducer and activator of transcription pathway. *Genetics* 165, 1149–1166.

58. Pignoni, F., Zipursky, S. L. (1997) Induction of *Drosophila* eye development by *decapentaplegic. Development* 124, 271–278.

59. Tomlinson, A. (2003) Patterning the peripheral retina of the fly: decoding a gradient. *Dev Cell* 5, 799–809.

60. Wernet, M. F., Labhart, T., Baumann, F., et al. (2003) Homothorax switches function of *Drosophila* photoreceptors from color to polarized light sensors. *Cell* 115, 267–279.

61. Hay, B. A., Wolff, T., Rubin, G. M. (1994) Expression of baculovirus P35 prevents cell death in *Drosophila. Development* 120, 2121–2129.

62. Cordero, J., Jassim, O., Bao, S., et al. (2004) A role for *wingless* in an early pupal cell death event that contributes to patterning the *Drosophila* eye. *Mech Dev* 121, 1523–1530.

63. Lin, H. V., Rogulja, A., Cadigan, K. M. (2004) Wingless eliminates ommatidia from the edge of the developing eye through activation of apoptosis. *Development* 131, 2409–2418.

64. Lim, H. Y., Tomlinson, A. (2006) Organization of the peripheral fly eye: the roles of Snail family transcription factors in peripheral retinal apoptosis. *Development* 133, 3529–3537.

65. Ahmed, Y., Hayashi, S., Levine, A., et al. (1998) Regulation of armadillo by a *Drosophila* APC inhibits neuronal apoptosis during retinal development. *Cell* 93, 1171–1182.

66. Ahmed, Y., Nouri, A., Wieschaus, E. (2002) *Drosophila* Apc1 and Apc2 regulate Wingless transduction throughout development. *Development* 129, 1751–1762.

67. Cadigan, K. M., Jou, A. D., Nusse, R. (2002) Wingless blocks bristle formation and morphogenetic furrow progression in the eye through repression of *daughterless. Development* 129, 3393–3402.

68. Cadigan, K. M., Nusse, R. (1996) *wingless* signaling in the *Drosophila* eye and embryonic epidermis. *Development* 122, 2801–2812.

69. Karim, F. D., Chang, H. C., Therrien, M., et al. (1996) A screen for genes that function downstream of Ras1 during *Drosophila* eye development. *Genetics* 143, 315–329.

70. Sharma, R. P., Chopra, V. L. (1976) Effect of the Wingless (*wg1*) mutation on wing and haltere development in *Drosophila melanogaster*. *Dev Biol* 48, 461–465.

71. van den Heuvel, M., Harryman-Samos, C., Klingensmith, J., et al. (1993) Mutations in the segment polarity genes *wingless* and *porcupine* impair secretion of the Wingless protein. *EMBO J* 12, 5293–5302.

72. Couso, J. P., Bishop, S. A., Martinez Arias, A. (1994) The Wingless signalling pathway and the patterning of the wing margin in *Drosophila*. *Development* 120, 621–636.

73. Bejsovec, A., Martinez Arias, A. (1991) Roles of *wingless* in patterning the larval epidermis of *Drosophila*. *Development* 113, 471–485.

74. Manoukian, A. S., Yoffe, K. B., Wilder, E. L., et al. (1995) The *porcupine* gene is required for *wingless* autoregulation in *Drosophila*. *Development* 121, 4037–4044.

75. Li, K., Kaufman, T. C. (1996) The homeotic target gene *centrosomin* encodes an essential centrosomal component. *Cell* 85, 585–596.

76. Tomlinson, A., Struhl, G. (1999) Decoding vectorial information from a gradient: sequential roles of the receptors Frizzled and Notch in establishing planar polarity in the *Drosophila* eye. *Development* 126, 5725–5738.

77. Chen, C. M., Struhl, G. (1999) Wingless transduction by the Frizzled and Frizzled2 proteins of *Drosophila*. *Development* 126, 5441–5452.

78. Yanagawa, S., van Leeuwen, F., Wodarz, A., et al. (1995) The dishevelled protein is modified by wingless signaling in *Drosophila*. *Genes Dev* 9, 1087–1097.

79. Bourouis, M., Moore, P., Ruel, L., et al. (1990) An early embryonic product of the gene *shaggy* encodes a serine/threonine protein kinase related to the CDC28/cdc2+ subfamily. *EMBO J* 9, 2877–2884.

80. Sanson, B., White, P., Vincent, J. P. (1996) Uncoupling cadherin-based adhesion from wingless signalling in *Drosophila*. *Nature* 383, 627–630.

81. Peifer, M., Rauskolb, C., Williams, M., et al. (1991) The segment polarity gene *armadillo* interacts with the wingless signaling pathway in both embryonic and adult pattern formation. *Development* 111, 1029–1043.

82. van de Wetering, M., Cavallo, R., Dooijes, D., et al. (1997) Armadillo coactivates transcription driven by the product of the *Drosophila* segment polarity gene dTCF. *Cell* 88, 789–799.

83. Struhl, G., Basler, K. (1993) Organizing activity of wingless protein in *Drosophila*. *Cell* 72, 527–540.

84. Hayashi, S., Rubinfeld, B., Souza, B., et al. (1997) A *Drosophila* homolog of the tumor suppressor gene *adenomatous polyposis coli* down-regulates beta-catenin but its zygotic expression is not essential for the regulation of Armadillo. *Proc Natl Acad Sci USA* 94, 242–247.

85. Azpiazu, N., Lawrence, P. A., Vincent, J. P., et al. (1996) Segmentation and specification of the *Drosophila* mesoderm. *Genes Dev* 10, 3183–3194.

86. Wilder, E. L., Perrimon, N. (1995) Dual functions of *wingless* in the *Drosophila* leg imaginal disc. *Development* 121, 477–488.

87. Pfeiffer, S., Ricardo, S., Manneville, J. B., et al. (2002) Producing cells retain and recycle Wingless in *Drosophila* embryos. *Curr Biol* 12, 957–962.

88. Zecca, M., Basler, K., Struhl, G. (1996) Direct and long-range action of a wingless morphogen gradient. *Cell* 87, 833–844.

89. Brunner, E., Peter, O., Schweizer, L., et al. (1997) *pangolin* encodes a Lef-1 homologue that acts downstream of Armadillo to transduce the Wingless signal in *Drosophila*. *Nature* 385, 829–833.

90. Pai, L. M., Orsulic, S., Bejsovec, A., et al. (1997) Negative regulation of Armadillo, a Wingless effector in *Drosophila*. *Development* 124, 2255–2266.

91. Steitz, M. C., Wickenheisser, J. K., Siegfried, E. (1998) Overexpression of *zeste white 3* blocks wingless signaling in the *Drosophila* embryonic midgut. *Dev Biol* 197, 218–233.

92. Willert, K., Logan, C. Y., Arora, A., et al. (1999) A *Drosophila* Axin homolog, Daxin, inhibits Wnt signaling. *Development* 126, 4165–4173.

93. Vincent, J. P., Girdham, C. H., O'Farrell, P. H. (1994) A cell-autonomous, ubiquitous marker for the analysis of *Drosophila* genetic mosaics. *Dev Biol* 164, 328–331.

94. Giraldez, A. J., Copley, R. R., Cohen, S. M. (2002) HSPG modification by the secreted enzyme Notum shapes the Wingless morphogen gradient. *Dev Cell* 2, 667–676.

95. Tolwinski, N. S., Wieschaus, E. (2001) Armadillo nuclear import is regulated by cytoplasmic anchor Axin and nuclear anchor dTCF/Pan. *Development* 128, 2107–2117.

96. Xu, T., Rubin, G. M. (1993) Analysis of genetic mosaics in developing and adult *Drosophila* tissues. *Development* 117, 1223–1237.

97. Newsome, T. P., Asling, B., Dickson, B. J. (2000) Analysis of *Drosophila* photoreceptor axon guidance in eye-specific mosaics. *Development* 127, 851–860.

98. Robinow, S., White, K. (1991) Characterization and spatial distribution of the ELAV protein during *Drosophila melanogaster* development. *J Neurobiol* 22, 443–461.

99. Janody, F., Lee, J. D., Jahren, N., et al. (2004) A mosaic genetic screen reveals distinct roles for *trithorax* and *polycomb* group genes in *Drosophila* eye development. *Genetics* 166, 187–200.

Chapter 13

High-Throughput RNAi Screen in *Drosophila*

Ramanuj DasGupta and Foster C. Gonsalves

Abstract

Genetic and biochemical analyses in model systems such as the fruitfly, *Drosophila melanogaster*, have successfully identified several genes that play key regulatory roles in fundamental cellular and developmental processes. However, the analyses of the complete genome sequences of *Drosophila*, as well as of humans, now reveal that traditional methods have ascribed functions to only a fraction of the total predicted genes. Thus, the roles for many, as yet unidentified genes, in normal development and cancer remain to be discovered. The challenge presented by the various large-scale genome projects is how to derive biologically relevant information from the raw sequences. The past few years have witnessed a rapid growth in the development and implementation high-throughput screening (HTS) technologies that researchers are now using to discover "gene-function" in an unbiased, systematic, and time-efficient manner. In fact one of the most promising functional genomic approach that has emerged in the past few years is based on RNA-interference (RNAi), in which the introduction of double-stranded RNA (dsRNA) into cells or whole organisms has been shown to be an effective tool to suppress endogenous gene expression. The RNAi technology has made it feasible to query the function of every gene in the genome for their potential function in a given cell-biological process using cell-based assays. This chapter discusses the application, advantages, and limitations of this powerful technology in the identification of novel modulators of cell-signaling pathways as well as its future scope and utility in designing more efficient genome-scale screens.

Key words: High-throughput screening (HTS), RNA-interference (RNAi), double-stranded RNA (dsRNA), Wnt/wg signaling, functional genomics, signaling networks.

1. A "Systems" Level Understanding of Cell Signaling Pathways

Over the past 25 years, research in cell and developmental biology has shown that only a handful of evolutionarily conserved cell signaling pathways are reiteratively used to generate an enormous diversity of cell types that comprise the organ systems and

Elizabeth Vincan (ed.), *Wnt Signaling, Volume II: Pathway Models, vol. 469*
© 2008 Humana Press, a part of Springer Science + Business Media, New York, NY
Book doi: 10.1007/978-1-60327-469-2

bodies of all multicellular organisms. The complexities of these signaling pathways however are only recently being appreciated. In a developing embryo, an individual cell receives signals from multiple signaling molecules at the same time; how the cell integrates these parallel signals and translates that into a transcriptional outcome (and often into a unitary developmental outcome such as cell fate specification), is a major unanswered question.

In the past few years, detailed investigations into the molecular mechanisms that regulate these cell-signaling pathways have altered our traditional notion of them being comprised of linear "cassettes" of genes. These cassettes were thought to transduce an extracellular signal from the plasma membrane to the nucleus through a variety of cytoplasmic intermediates, eventually switching on the expression of specific target genes. These signaling pathways are in fact highly complex. There is a rapidly growing list of genes that is being implicated in the control of multiple signaling pathways even though they were initially identified as "canonical" members of one specific pathway *(1–6)*. Classic examples of genes that might function as shared regulators are glycogen synthase kinase (GSK)-3 and casein kinase (CK)1α. Both GSK-3 and CK1α were initially identified as negative regulators of the Wnt/*wingless* *(wg)* pathway but now have been shown to function in the Sonic hedgehog (Shh)/*hedgehog (hh)* signaling pathway (refs. *4*, *5*, and *7*; *see also* Wnt homepage at www.stanford.edu/~rnusse/pathways/WntHH.html). It has been proposed that these shared regulators could serve as nodes of integration or "crosstalk" between the different signal transduction cascades. This cross-regulation between cell-signaling pathways could eventually determine the final outcome of a cell that is under simultaneous influence of multiple signaling cues.

Regulation of pathway components by members of other signal transduction cascades is especially evident for the Wnt/*wg* signaling pathway *(1–6)*. Wnts are a family of conserved signaling molecules that have been shown to regulate a plethora of fundamental developmental and cell biological processes such as cell proliferation, differentiation, and cell polarity *(8–10)*. Mutations in the Wnt genes or in those that encode regulators of the Wnt signaling pathway can cause devastating birth defects, including debilitating abnormalities of the central nervous system, axial skeleton, limbs, and occasionally other organs *(11–16)*. Aberrant Wnt signaling has also been linked to human disease such as retinal degeneration of the eye (FEVR; refs. *17* and *18)* and a variety of cancers including those of the liver, intestine, breast, and skin *(19, 20)*.

Wnts/*wg* encode secreted glycoproteins that activate receptor-mediated pathways leading to numerous transcriptional and cellular responses *(10, 21, 22)*. The main function of the canonical Wnt pathway is to stabilize the cytoplasmic pool of a key mediator, β-catenin (β-cat)/*armadillo* (*arm*), which is otherwise phosphorylated by GSK-3 within the inhibitory *axin* (*axn*) scaffold protein complex and degraded by the proteosome pathway (**Fig.13.1**). Initially identified as a key player in stabilizing cell–cell adherens junctions, β-cat is now known to also act as a transcription factor by forming a complex with the lymphoid enhancer factor/T cell factor (LEF/TCF) family of HMG-box (High mobility group) transcription factors *(8, 16, 22)*. Upon Wnt stimulation, stabilized β-cat/*arm* translocates to the nucleus, where, together with LEF/TCF transcription factors, it activates downstream target genes. The Wnt/*wg* pathway can also be activated by inhibiting negative regulators such as GSK-3, adenomatous polyposis coli (APC), and Axin that promote β-cat/*arm* degradation, or by introducing activating mutations in β-cat that renders it incapable of interacting with the degradation complex, thus stabilizing its cytosolic pool *(23, 24)*. In addition to the β-cat mediated "canonical" pathway, Wnt/*wg* signaling can also activate an alternative "non-canonical" pathway that may lead to protein kinase C (PKC) and c-Jun N-terminal kinase (JNK) activation, resulting in calcium release and cytoskeletal rearrangements *(8)*.

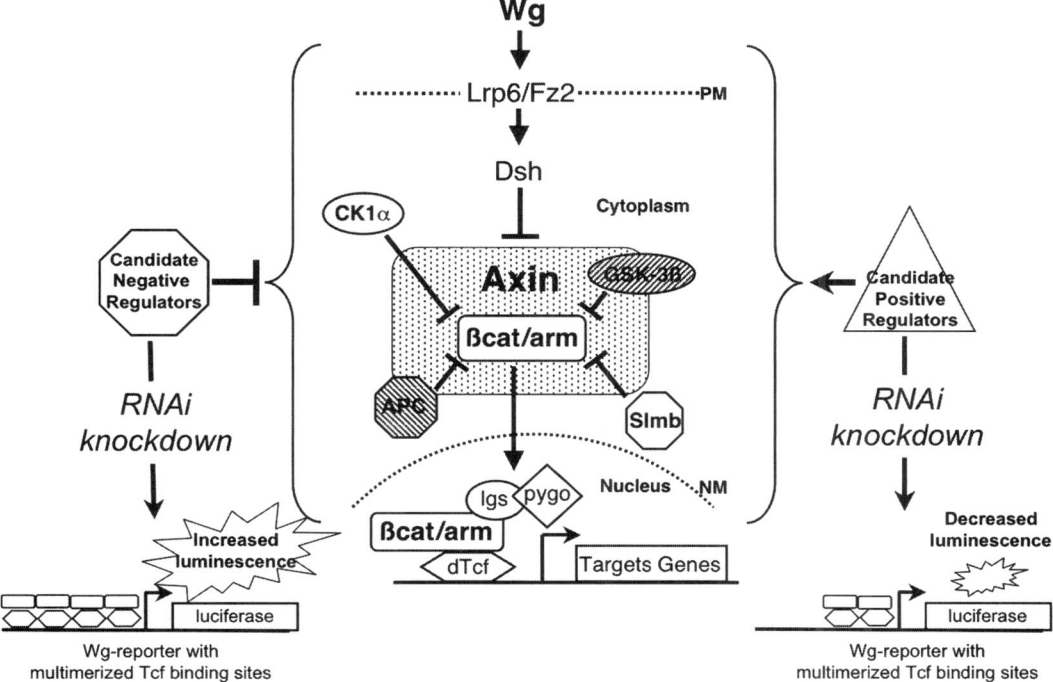

Fig. 13.1. Schematic diagram of the canonical (β-cat/arm-dependent) Wnt/*wingless* signaling pathway. Also shown is the rationale of RNAi screen setup. Knocking potential negative regulators should activate the wg-dependent luciferase reporter activity, whereas RNAi of positive regulators would result in a reduction of the reporter activity as measured by luminescence. It is important to note that the candidate modulators could be acting at any level in the signal transduction cascade, including regulating the transcription/expression of the known core components of the pathway.

Recent evidence from studies in a variety of model organisms suggests that there are several layers of regulation imposed on the Wnt pathway, thereby increasing the complexity of this already intricate network of controls. In addition to the canonical regulators that modulate Wnt/*wg* signaling activity, there are multiple nodes of crosstalk between the Wnt /*wg* and other signaling pathways. For example, *(1)* components of the Wnt/*wg* pathway can interact with specific mitogen-activated protein (MAP) kinase pathways involving the MAP kinase kinase kinase, TAK1, and MAP kinase NLK (Nemo-like kinase). TAK1 activates NLK, which in turn phosphorylates TCF and alters its binding to β-cat/*arm* and DNA *(25–29)*. Because TAK1 is activated by TGF-β and various other cytokines, it might provide an entry point for the regulation of the Wnt activity by other pathways *(30–32)*. *(2)* The Notch signaling pathway can antagonistically interact with the Wnt pathway. Notch modulates Wnt signaling activity by physical associating with β-cat/*arm* and preventing it from signaling in the nucleus *(33)*. *(3)* Studies in the *Drosophila* eye have suggested that an activated form of β-cat/*armadillo* (*arm**) is under specific negative regulation by epidermal growth factor receptor (EGFR)/MAP kinase signaling whereby MAPK signaling prevents *arm** from inducing apoptosis and early onset of cell death in the developing retina *(34, 35)*. *(4)* Recently, cross-regulation between cell-signaling pathways has also been shown to play a critical role in epithelial-mesenchymal transition (EMT) in a variety of mammalian cancer cell lines. For example, PDGF stimulation of a p68 family RNA helicase results in the displacement of β-cat from the Axin degradation complex in a Wnt independent manner. Such findings allude to the importance of interplay between different signaling pathways in determining/defining various disease states *(36, 37)*.

2. Functional Genomic Approaches to Study Cell Signaling Networks Using RNA-Interference (RNAi)

Classical forward genetic and biochemical approaches in model systems such as the fruitfly, *Drosophila melanogaster*, the nematode, *Caenorhabditis elegans* and more recently in zebrafish and mouse, have been enormously successful in identifying genes that play key regulatory roles in fundamental cellular and developmental processes. Understanding the normal function of these genes has provided significant insights into what goes awry in abnormal situations, such as tumorigenesis. Specifically, contributions made by studies in *Drosophila* are numerous, and many important discoveries involving oncogenes, tumor suppressors, and other crucial players involved in cell proliferation, differentiation, and death were made first in this organism.

Completion of the sequence of the *Drosophila* genome provides us with an unprecedented resource, as we can now fully evaluate the degree of conservation of this organism with others *(38, 39)*. The relevance of *Drosophila* to humans is best illustrated by the fact that more than 60% of the genes identified in human diseases (177 out of 289) have counterparts in *Drosophila (39)*. Interestingly, recent analyses of the complete genome sequences of model organisms such as *Drosophila* and *C. elegans*, as well as of humans reveal that traditional genetic and biochemical approaches have ascribed functions for only a fraction of the total number of predicted genes *(38, 40)*. For example, analysis of the *Drosophila* genome has led to the annotation of ~16,000 genes *(38, 41, 42)*, of which good functional annotation is available for only approximately 25% of the genes. Thus, the function of many as yet uncharacterized genes in normal development and cancer remain to be discovered. Although conventional genetic approaches will clearly continue to provide valuable information, new powerful methods are needed to systematically and rapidly analyze the functions of all ~16,000 predicted genes.

The full potential of the genome sequence can be realized by devising new technologies that efficiently and systematically bridge the gap between the genomic sequence of a predicted gene and its function. It is also increasingly clear that individual proteins, including Wnts and components of the Wnt pathway, are almost always found in a variety of complexes with numerous other molecules within a cell such as other proteins, DNAs or RNAs. Thus, it is the coordinate activity of these complex interactions that eventually determine the biological characteristics of a cell *(43–46)*. The same is also true of interaction/cross-talk between entire signaling pathways that regulate cell growth, proliferation and differentiation, as common effectors and/or integrators of multiple signaling pathways need to be coordinately regulated to determine cell behavior *(47, 48)*. A key challenge for the present day biologist is to devise ways of integrating information at the whole-genome scale in order to better understand the regulation and dynamics of complex molecular interactions and their function in determining cell biological and developmental events.

Although much is known about the Wnt/Wg signal transduction cascade, it is clear that neither have all components been identified, nor have the mechanisms of the existing players in different contexts been well defined *(37, 49–53)*. A comprehensive understanding of this signaling pathway by the identification of additional players, their mechanism of action and targets of the Wnt/Wg signal, is crucial for developing therapeutic intervention for the contribution of Wnt/Wg pathway to cancer. A key challenge in understanding the molecular network of regulation for the Wnt-signaling pathway is first to identify the genes that regulate or mediate crosstalk between the Wnt and other signaling pathways and, second, to investigate the molecular mechanisms of known and novel regulators/protein complexes by which they modulate the activity of the Wnt pathway. Although conventional genetic approaches have been enormously successful in identifying numerous components of the *Drosophila* Wnt/*wg* pathway, genetic redundancy, pleiotropic effects of mutations, selectivity of various mutagens and the masking of zygotic phenotypes due to expression of maternally provided gene products in the embryo (maternal effect), potentially limit the detection of additional regulatory factors. This is especially true for genes that might serve as nodes of interaction between the Wnt/*wg* and other signaling pathways, since mutations in them might lead to early lethality or pleiotropic phenotypes in the embryo or perhaps even in the oocyte, making it impossible to assign them as regulators of a specific pathway.

In this post-genomic era, the challenge presented by the various large-scale genome projects is how to derive biologically relevant information from the raw sequences. In the past few years, a number of approaches to mine this information have emerged, such as expression genomics, proteomics, bio-informatics, and functional genomics. Of these different approaches, functional genomics allows us to simultaneously test the function of all genes in a given genome in a rapid, systematic and unbiased manner in a variety of cell-based assays (in the case for *Drosophila* and mammalian cells) or in whole animals (as in *C. elegans*). One of the most promising functional genomic approaches that has emerged in the past few years is based on the use of double-stranded RNA (dsRNA) to knockdown gene function. In several organisms, introduction of a dsRNA has proven to be an effective tool to suppress gene expression through a process referred to as RNAi *(54–59)*. Importantly, the simple addition of dsRNA to *Drosophila* cells in culture reduces or eliminates efficiently the expression of target genes by RNAi, thus phenocopying loss-of-function (LOF) mutations *(60)*. The methodology in *Drosophila*, relies upon the use of long dsRNAs, which, following uptake by the cells, are processed by Dicer2 into a pool of 21–25bp small interfering RNAs (siRNAs) *(61)*. These siRNAs silence endogenous gene expression by triggering the cleavage of target mRNAs. In contrast to *Drosophila*, where long dsRNAs of more than 100bp are used as RNAi reagents, synthetic 21–23 bp siRNAs or vector-based short-hairpin RNAs (shRNAs) are used in mammalian cells to avoid the detrimental interferon response triggered by the cells in response to long dsRNAs *(62–65)*.

RNAi strategies offer several important advantages: (1) Both maternal and zygotic gene expression can be perturbed in early embryos; (2) phenotypic effects are immediately attributable to specific gene sequences; (3) functionally redundant members of a gene family can be blocked simultaneously; (4) phenotypic results can be acquired rapidly without extensive breeding; (5) large numbers of genes can be evaluated in a single screen; and

(6) by using appropriate dsRNA expression vectors, inhibitory effects can be targeted *in vivo* to particular tissues and stages of development. Importantly, the simple addition of dsRNA to *Drosophila* cells in culture reduces or efficiently eliminates the expression of target genes by RNAi, thus phenocopying LOF mutations *(60)*. Whole genome RNAi screens in *Drosophila* cell culture have now been used to screen for novel regulators of multiple signaling pathways, including the Wnt/*wingless* (*wg*), hedgehog (hh), JAK/STAT, and RTK/MAPK pathways. In addition, it is now possible to perform synthetic RNAi screens in *Drosophila* cells using multiple dsRNAs to uncover functions of genes that do not display a phenotype when mutated individually (see ref. *66*). Such screens would enable researchers to identify genes that are functionally redundant or act together in large protein complexes in the regulation of cell proliferation, growth, and apoptosis and also to study epistatic relationships, modifier effects, and synthetic phenotypes *(4, 51, 52, 66–70)*.

Although RNAi-mediated high-throughput screens (HTS) in mammalian cells are now being efficiently conducted *(71–75)*, there are still several advantages of conducting genome-wide RNAi screens in *Drosophila* cells. First, RNAi is extremely effective *(60, 76, 77)* and the excellent annotation of the genome allows almost full genome coverage *(38, 42)*. Second, the high conservation between the *Drosophila* and vertebrate genomes, and organization of important signaling pathways, makes the translation of the findings from flies to vertebrates obvious *(78–82)*. In fact it is likely to be more effective to perform such screens in *Drosophila* cells first and then look at the functions of their orthologs in the mammalian system *(83)*. Such a strategy overcomes the problem associated with the high degree of functional redundancy that exists in higher vertebrates. Third, the powerful genetics and the availability of large numbers of chemically and transposon-induced mutants and deficiency lines in *Drosophila* offers a unique opportunity to quickly validate *in vivo* the targets identified from the RNAi screens *(84–86)*. Further, a number of methodologies, such as targeted gene knockout and hairpin RNAi constructs, can also be employed to engineer loss-of-function mutations in specific genes and analyze their functions.

3. RNAi-Mediated HTS for Novel Modifiers of the Wnt/wg Signaling Pathway in *Drosophila* Cells

RNAi technology has now been used for several whole-genome screens in *Drosophila* using specific cell-based assays for a variety of cell signaling pathways *(4, 50, 51, 66, 67, 70, 83, 87, 88)*. This has been made possible by the availability of dsRNA libraries from a variety of sources, including the *Drosophila* RNAi Screening Center (DRSC), Boston, which was established by Dr. Norbert Perrimon's research group at the Harvard Medical School and commercial sources, such as Ambion and Eurogentec *(89)*. All *Drosophila* RNAi-screening libraries are based on long dsRNAs. These have been used in several screens that utilized either quantitative luciferase or fluorescent reporter-based assays or microscopy/image-based qualitative readouts.

3.1. Genome-wide RNAi Screens for Novel Modifiers of Wg Pathway Activity

Recently, several research laboratories have completed whole-genome or genome-scale screens in an effort to identify novel modulators of the Wnt/wg pathway in *Drosophila* cells. The first such RNAi-HTS for the Wnt pathway was designed by the Perrimon group at DRSC *(51)*. We developed a HT-assay based on the known ability of Wnt/β-cat signaling to activate transcription of luciferase reporter constructs in transfected cells. The screen was optimized for a HT-format using 384-well plates (detailed protocols available from DRSC Web site: http://flyrnai.org). Because the signal-to-noise ratio of normalized luciferase reporter activity is significantly reduced in small-volume reactions, we generated two improved versions of the existing TOP-Flash constructs *(90)*: dTF12, a fly cell-optimized Wg reporter with 12 multimerized TCF binding sites cloned upstream the *Drosophila* heat shock minimal promoter; and STF16, optimized for mammalian cells with 16 TCF binding sites upstream of a minimal TATA box from the thymidine kinase (TK) promoter. Both reporters are active in a variety of fly and mammalian cell lines *(51)*. Reporters with mutated Tcf-binding sites served as specificity controls. The co-transfection of a *Renilla luciferase* reporter, *PolIII-RL* was used as a control for transfection efficiency and cell viability. The ratio of the Wg-firefly luciferase (dTF12/STF16) and the renilla luciferse activity represented the "normalized" luciferase activity, which was used as a "true" measure of the pathway activity *(51)*.

A library of approximately 22,000 dsRNAs representing more than 95% of the genes in the *Drosophila* genome were tested for their effect on the activity of the Wg-reporters. HTS was achieved in a 384-well plate format, with individual dsRNA aliquoted in each well, together with a transfection mix of Wg-reporter, the control PolIII-RL and a *wg*-expressing plasmid (pMK33-*wg*). After 5 days of incubation to ensure complete knockdown of intended target genes, the normalized luciferase activity was measured in each well using ultra-sensitive HT-plate-readers with enhanced luminescence capabilities (**Figs.13.1** and **13.2**). A variety of statistical analyses were applied to the data to highlight those dsRNAs that had the most credible effect on the wg-reporter activity. Of these, 238 were identified and retested in secondary screens in 96-well plate format to ensure a more

Wg/Wnt Screen Outline

384 Well Plates with dsRNA's

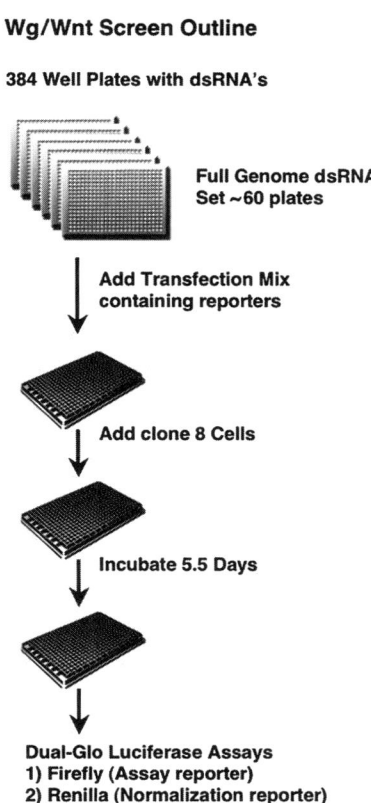

Full Genome dsRNA Set ~60 plates

Add Transfection Mix containing reporters

Add clone 8 Cells

Incubate 5.5 Days

Dual-Glo Luciferase Assays
1) Firefly (Assay reporter)
2) Renilla (Normalization reporter)

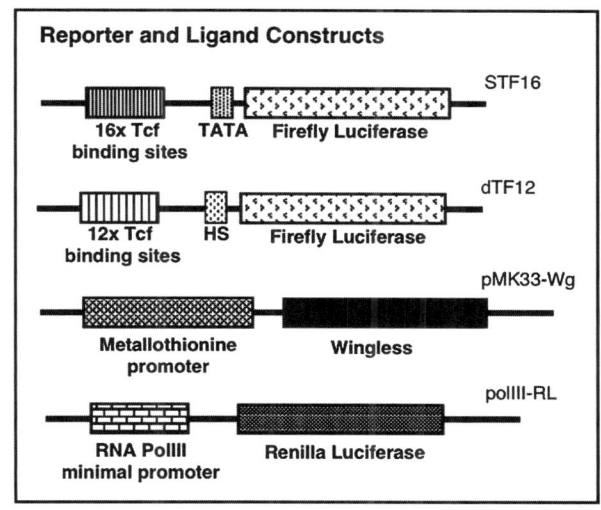

Reporter and Ligand Constructs

STF16 — 16x Tcf binding sites | TATA | Firefly Luciferase

dTF12 — 12x Tcf binding sites | HS | Firefly Luciferase

pMK33-Wg — Metallothionine promoter | Wingless

polIII-RL — RNA PolIII minimal promoter | Renilla Luciferase

Cell lines available for screening	Description of source	Wg signaling activity
Clone 8	Epithelial cells derived from imaginal discs, adherent cells, spindle shaped at low confluence, round at higher density	Robust
Schneider cell derivatives: SL2, S2R+, S2*, DL2	Embryonic, hemocyte lineage, phagocytic, round, flat cells semi-adherent	Robust in S2R+ cells
Kc167	Embryonic, phagocytic, round cells, semi-adherent	Robust

Fig. 13.2. Outline of an RNAi-mediated high-throughput screen (HTS) for the Wg-pathway. The dsRNA library targeting every annotated gene in the genome is aliquoted in "ready-to-screen" 384-well plate. Each dsRNA targeting a specific gene is housed in an individual well of a 384-well plate. On day 1, transfection mix containing the assay and control reporters, and wg-cDNA expressing vectors are added to the assay plates containing dsRNAs and "reverse-transfected" (cells added to the transfection mix) into clone eight cells. The dsRNA and reporters are incubated with the cells for 5 days to ensure complete knockdown of target genes after which luminescence/luciferase activity is measured with the Dual-Glo luciferase assay kit (from Promega Inc.) using a plate reader. The various assay reporters, controls, and cells lines that are available to assay activity of the Wg-signaling pathway, are also listed.

robust signal to noise ratio. Of them, 213/238 (90%) were re-confirmed in their ability to modify the wg-reporter activity.

A majority of the known core Wnt-pathway components were identified in the screen including the ligand receptor complexes such as *wg*, *arrow* (*arr*)/LDL-receptor-related protein-6 (Lrp6) and *frizzled* (*fz*); members of the β-cat/arm degradation complex or genes involved in negative regulation of β-cat activity, such as *axin*, *slmb*/β-TrCP, and *CK1* ; and genes involved in regulating the activity of the β-cat-transcriptional complex in the nucleus, such as *pangolin* (*pan*)/dTCF, *pygopus* (*pygo*) and *legless*(*lgs*)/Bcl-9; thus underscoring the robustness and validity of the Wnt screen in HT-format. Attention could then be confidently turned to 200 new sequences identified.

One of the critical challenges after a HTS is to find innovative ways to implement a variety of secondary screens in order to test if the "cherry-picked" candidate genes would pass through a series of "assay-filters," which would help prioritize genes of interest in the HTS-dataset. We approached this in several ways. First, because the *wg*-reporters can be activated in several *Drosophila* cell types (see table in **Fig.13.2**), additional secondary screens were performed to test the effect of dsRNA-mediated knockdown of the ~200 new sequences in S2R+ and Kc167 cells lines *(51)*. This could potentially weed out cell-type-specific factors. However, this analysis could be confounded by the difference in expression profiles of genes in the different cell lines. Second, candidate sequences were retested for their effect on Wg-reporter activity in the absence of the Wg ligand. As the positive sequences here could affect the baseline activity, or perhaps expression of key components of the wg- pathway, they may be of less interest than the ones that specifically modify Wg-induced reporter activity. Third, in several cases it was possible to effectively conduct a series of epistasis tests in *Drosophila* cells to obtain preliminary evidence for where in the pathway the new components act. The experiments tested whether the dsRNA knockdown of potential positive regulators could block signaling in the presence of positive stimuli (of known components) downstream of wg, such as expression of the constitutively activated co-receptor (ΔNLrp6), *Dishevelled* (*dsh*), or stabilized mutant β-cat. Fourth, using computational approaches such as predicting gene function of the candidate modulators based on functional annotation assigned by the Gene Ontology (GO) consortium and from the prediction of conserved protein domains identified by the InterPro database. In addition, reciprocal best blast (RBB) analysis was also conducted to identify putative conserved mammalian orthologs of the candidate regulators. This was successful in ~50% of the cases. Besides the obvious mammalian relevance, these genes could represent the more ubiquitous modulators of the Wnt signaling pathway that are not limited to species- or cell-type-specific regulators. To test the functional relevance, some of the putative mammalian orthologs were tested for their ability to modulate the Wnt-reporter activity using siRNA-mediated knockdown in human 293T cells. For example, the human homolog of a gene named CG4136, which encodes a paired-like homeobox protein, when expressed in zebrafish embryos by injection of RNA at the single-cell stage, gave a developmental phenotype similar to that of Wnt8 gain of function. This indicated that it could modulate Wnt signaling in a variety of systems, both invertebrate and vertebrate and also set a paradigm for systematically testing candidate genes for conserved function in a variety of cell types and model organisms *(51)*.

Interestingly, this initial screen for the Wg pathway also identified several genes that had been previously reported in genetic screens designed to find genes that could interact with the Wg-pathway, including: *lilli*, *brahma*, *osa*, *cdc2*, *string* (cdc-25), *N*, *mastermind* (*mam*), etc. These observations lend additional support to the putative function of novel regulators identified in forward genetic screens in the regulation of the Wnt/wg signaling pathway.

It is also encouraging to find independent reports of new regulators that are being discovered for the Wnt pathway in different model systems that were also identified in the initial Wg-screen. For example, a recent report described the function of p68 RNA helicase, an ortholog of the *Drosophila Rm62* gene identified in the Wg-screen *(36, 37)*. This protein was described to cause the dissociation of Axin from β-cat and promote the nuclear translocation of the latter, thereby causing epithelial to mesenchymal transformation in human colon cancer cell lines. We also isolated Tip60/CG6121, which had not been identified in prior genetic screens for the Wg pathway. However, recent studies in human cells and colorectal cancer cell lines have shown that the β-cat's C-terminal activation domain associates with TIP60/TRAPPP and a mixed-lineage-leukemia (MLL1/MLL2) SET1-type chromatin-modifying complex *in vitro*, and that this complex promotes H3K4 trimethylation at the c-Myc target gene *in vivo (91, 92)*.

In addition to the luciferase reporter-based assays, image/immuno-fluorescence-based assays can also be developed to monitor the activity of the Wnt signaling pathway in cells. Specifically, cell based assays can be developed for β-cat/Arm localization and/or stabilization of either the endogenous β-cat/Arm protein using antibodies (available from Developmental Studies Hybridoma Bank, DSHB), or Arm-GFP fusion protein, in stable cell lines expressing the fusion construct, in Wg-treated versus untreated cells. Furthermore, a cross-reactive phospho-specific β-cat antibody is available from Cell Signaling Technology that may be used to track changes in the phosphorylation status of β-cat upon the dsRNA-mediated knockdown of genes. In fact, this antibody has been successfully used in defining the molecular mechanisms by which CK1 and GSK-3 regulate phosphorylation and hence the activity of endogenous β-cat *(93)*. Similar assays can also be designed to study factors regulating the subcellular localization of other components of the Wnt/wg signaling pathway by making stable cell lines expressing GFP-fusion constructs of known factors. Finally, secretion and/or endocytosis of the Wg ligand or receptor complex can be assessed by tracking the movement of fusion constructs encoding fluorescently tagged versions of the Wg and its receptor(s) proteins.

While the definitive validation, physiological relevance and the function/molecular mechanisms of these candidate modulators in the Wnt signaling pathway remain to be elucidated by biochemical assays and *in vivo* testing in model organisms and different cell types, such screens offer a good starting point to tackle the complexities of such intricate signaling networks. However, because different cell types express distinct sets of genes, clearly not all modifiers would be found by screening in any given cell type. An interesting case in point is a similar RNAi screen that was performed by Bartscherer et al. in S2R+ and Kc167 cells *(50)*. In this screen, the authors surveyed the effect of knocking down ~2300 putative transmembrane proteins to identify genes that may be involved in the secretion of the Wingless ligand. The Wnt-responsive luciferase reporter and control renilla luciferase reporters were similar to the ones we used in our screen. In this screen, Bartscherer and colleagues identified a gene called *evenness interrupted* (*evi*; also called *wntless* because it was simultaneously identified as a recessive suppressor of a *wg* gain-of-function phenotype in the *Drosophila* eye *[49]*), and was shown to be required for secretion of Wg from the Wg-producing cells. *evi* knockdown in *clone8* (*cl8*) cells does not reduce the activity of the Wg-reporter and hence was not identified as a candidate regulator in our screen. This could be due to several reasons. For example, it is possible that evi is not expressed at high levels in *cl8* cells so its knockdown does not affect reporter activity; OR, the dsRNA used in the *cl8* cell screen was not as effective as the ones used in the S2R+/Kc167 cell screen.

Taken together, the whole-genome RNAi screens using cell-based assays provide a technology platform for efficient enrichment for potential modulators of a variety of cell signaling pathways. The availability of screen data as an open resource from DRSC (www.flyrnai.org) should allow researchers to directly investigate the function of some of these candidate genes that are most interesting to them. While some individual candidate genes will certainly gain prominence through further experimental analysis of their individual functions, the significance of others will be likely to emerge through bioinformatics/computational approaches of integrating these large data sets with those obtained from other similar functional screens.

3.2. Controlling the Targeting Specificity of dsRNAs in Whole-Genome Libraries: Off Target Effects (OTEs)

The development and application of genome-wide RNAi screens has occurred in parallel with a rapidly evolving understanding of the mechanism of RNAi, including the regulation and processing of dsRNAs, the factors that influence siRNA specificity and efficacy, as well as the biogenesis, expression, and function of microRNAs (miRNAs) in cells *(61, 94, 95)*. These recent developments have led to a much greater understanding of siRNAs and dsRNAs as RNAi reagents, especially with regard to their specificity in degrading the intended target gene *(89, 96)*.

The discovery of "off-target effects" (OTEs) which is the degradation of unintended or non-specific mRNA targets based on sequence homology of the dsRNA/siRNA used in any given RNAi experiment, has played a critical role in promoting a much greater appreciation of various rules dictating siRNA specificity. OTE were initially recognized as an important source of false positives in mammalian studies using single siRNAs for the knock-down of target genes *(63, 97)*. Subsequently, studies conducted with pools of siRNAs targeting the same transcript revealed that OTEs could be reduced (albeit not always eliminated), as undesirable effects of single siRNAs bearing perfect or partial homologies to other gene coding regions or their 3′-UTRs were diluted by the pooling method *(98, 99)*.

The protection against OTEs provided by pools of siRNAs was the main reason for arguing that OTEs would not be a significant issue in *Drosophila* or *C. elegans* screens, despite the fact that Dicer (RNase III ribonuclease)-mediated cleavage of long dsRNAs could give rise to siRNAs with partial (typically 19–21 bp) sequence complementarity to transcripts other than the intended target. Moreover, the failure to detect the existence of any member of the ubiquitous family of RNA-dependent RNA polymerase (RdRp) in *Drosophila* potentially eliminated the chances of any amplification step of target RNAs, hence limiting the effect of OTEs *(100)*. As such, OTEs arising from the knockdown of unintended target genes were not thought to be a significant source of cellular phenotypes, and thus were thought unlikely to contribute to the rate of false-positives in any HTS in these organisms. This line of reasoning, however, had not been rigorously tested experimentally and was questioned in a review article by Echeverri and Perrimon *(89)*. Shortly thereafter, two groups independently reported evidence for OTEs in *Drosophila* RNAi screens *(101, 102)*. Together these studies implicated identity stretches as short as 13nt for low complexity trinucleotide repeats (e.g., CAN repeats, Ma et al. 2006) or slightly longer (17nt-19nt and greater, Kulkarni et al. 2006) for more complex sequence homologies as contributing to false-positives in *Drosophila* RNAi screens. Although sequence homology can lead to OTEs, the mere presence of predicted-sequence homology to multiple transcripts does not necessarily translate into OTEs. For example, the Kulkarni et al. study revealed that 50 of 135 predicted off targets (OTs) for a dsRNA directed towards the PP2A-B′ gene did not reveal any changes in expression levels of the corresponding mRNAs. This may reflect the fact that the problematic siRNAs were not produced *in vivo* because of the processivity exhibited by Dicer when acting on dsRNAs *(61, 95, 103)*, or if they were, that they were not effective in knocking down their cognate targets. Thus *in silico* prediction of OTs will almost always over-estimate the incidence of OTEs that might occur with dsRNAs in an experimental setting.

However, it is important to immediately address the issue of OTs in the initial screens for cell signaling pathways. Because it is non-trivial to address all possible OTEs from dsRNAs targeting individual genes in the genome, perhaps the best solution is to use independent dsRNAs/siRNAs for re-validation of candidate sequences identified in the published screens for the Wnt and other cell signaling pathways. In order to address the issue of sequence-specific OTEs in genome-wide screens, the DRSC, Boston has recently generated a new version of whole-genome dsRNA screening collection, called the DRSC2.0. In addition, they have also assembled an independent collection, DRSC-validation (DRSC-v), for independent confirmation of hits identified in initial screens. These libraries are comprised of independent dsRNAs largely free of any predicted OTs. These new OT-free collections are being used to retest the "hits" obtained in the initial screens. It is also important to note that while cross-reacting sequences in dsRNAs can clearly lead to an increase in the rate of false-positives, there clearly are other factors, such as cell-type specificity, use of specific normalization vectors, and properties of the transcriptional reporters, which have a major impact on the outcome of luciferase reporter-based RNAi screens.

Undoubtedly the ultimate validation will be in determining the function of the candidate genes *in vivo* in animal model systems, which is underway for several candidate genes obtained in the Wg, Hh, and JAK/STAT pathway screens (addendum in ref. *87*; DasGupta, et al., unpublished). Additionally, our current understanding of OTEs associated with long dsRNAs is likely to be incomplete, and there may be other predictors (e.g., seed regions *[98]*) that, under the given circumstances, need to be avoided. Clearly the effort in designing better reagents are still evolving and they will continue to be a major focus of further investigation by groups interested in genome-wide screening approaches toward understanding the function of cell signaling pathways in development and cancer.

4. The Future Potential of Cell-Based RNAi Screens

4.1. RNAi-Based LOF Screens

In the future, the most obvious application of the RNAi technology would be RNAi-based direct LOF screens. These will continue to provide a powerful method to screen for additional factors that regulate the many aspects of Wnt signaling and cell-signaling pathways at large. The LOF screening involves identifying and characterizing the function of genes for any given cell biological process that can be qualitatively assayed by microscopy or quantified using plate-reader based luminescent or fluorescent reporters. However, it is important to realize that RNAi is a method of "knocking-down" gene function and is not a gene "knockout." Hence, the generation of an RNAi phenotype is highly dependent on the expression level, activity and half-life/stability of a target protein, especially in case of certain enzymes or signaling molecules where even low residual levels could be sufficient in fulfilling their cell biological functions. Therefore, it would be important to develop more innovative RNAi screening strategies, including modifier screens and synthetic screens, in order to realize the full potential of this screening technology.

In the near future, there will also be greater emphasis in performing RNAi LOF screens in primary cells, which offer the obvious advantage of assessing cell-type/context-specific phenotypes. Moreover the differentiation programs of primary cells often follow the *in vivo* differentiation patterns more closely, thereby making the outcome of such screens more physiologically relevant. While the scope and sophistication of the assays used in RNAi screens will increase over time, there is also an effort by some researchers to increase the throughput of the existing assay formats. At present most screens are being carried out in 96- or 384-well plates. However, dsRNA/siRNA libraries are now being printed on glass slides using microarrayers with each spot representing an individual dsRNA, cDNA or small molecule. This method of screening is known as solid-phase optimized transfection (SPOT)-RNAi *(104–106)*. Cells are simply plated on top of the slides that have dsRNAs or siRNAs spotted on known locations. The cells that land on the printed spots take up the dsRNA/siRNA, and form clusters of 80–200 cells where the targeted genes have been silenced. Although restricted to microscopy-based assays, this format vastly improves the throughput and significantly reduces the cost associated with such screens, while maintaining the ability to multiplex a variety of experiments.

4.2. Modifier Screens and Synthetic Screens

RNAi screens can be tailored using some of the same strategies that have already been developed and perfected in classical forward genetic screens. Of particular interest would be RNAi-modifier screens, which could be used to identify genes or new pathways that would either enhance or suppress a given phenotype of interest in a cell-based assay. The initial phenotype (to be modified) could be generated by the addition of dsRNA/siRNA targeting a known component of a particular cellular process under study. For example, for the Wnt signaling pathway, it should be possible to screen for "dsRNA-modifiers" of increased signaling activity (as measured by the Wnt-luciferase reporter) upon the RNAi-mediated knockdown of a known negative regulator, such as *axin*, which otherwise promotes the degradation of cytosolic β-cat. Such a screen should identify genes that are specifically required to regulate the activity of the stable/signaling pool of cytoplasmic β-cat, including those that may regulate β-cat's entry into the nucleus and components of the transcriptional complex, such as co-activators/repressors that mediate its activity as a transcriptional activator. This is similar to an "in-cell epistasis" assay, which was shown to work efficiently in *Drosophila cl8* cells *(51)*.

Another modification of RNAi screens could be the use of synthetic lethal screens, whereby lethal combinations of multiple non-lethal modifications/dsRNA treatments are sought. Such RNAi screens would involve the pre-treatment of cells with dsRNAs individually targeting genes that are otherwise mutated in naturally occurring genetic lesions (and have been linked to disease states such as cancer), but do not cause lethality by themselves. These pre-treated cells could then be transfected with a library of dsRNAs and scored for an RNAi-modified phenotype of cell-lethality, thereby offering a way of specifically killing cancerous cells while preserving the wild-type cells. Similar synthetic screens could also be used for the identification of functionally redundant genes or components of a pathway, which by their individual knockdown may not generate an RNAi-phenotype, but in combination, may collaborate to give a "score-able" cell-based phenotype. Such screens may be relevant to identify multifactorial/multi-genetic lesions that underlie the generation of human disease, including cancer.

4.3. Integration of Functional Genomic RNAi-HTS with Protein Interaction Screens

As a result of the variety of HTS for the Wnt/wg and other cell-signaling pathways, the field of signal transduction has witnessed a marked increase in complexity, is now being viewed as a network of cross-regulatory interactions between a variety of signaling molecules. Consequently, there are several new "candidate" components as potential contenders for a role in the Wnt or other cell signaling pathways. However, the upside of this increasingly complex picture is that bioinformatics and computational biology will take a closer look at this network at the "systems" level and perhaps offer a more realistic view of the regulation of signal transduction *in vivo*. This could be achieved by several ways. It is known that multiprotein complexes regulate the activity of most signal transduction pathways. The identification and analysis of their components through mass-spectrometry-based protein interaction screens are already providing important insights into how the ensemble of expressed proteins (proteome) maybe organized into functional units *(43, 44, 47, 48, 107, 108)*. However, the question still remains as to how these functional units coordinately regulate signaling pathways? This is where a systematic comparison of candidate genes (in a specific cell based assay) obtained from RNAi screens to that of known protein interaction databases would be immensely useful in understanding the "molecular context" of their activity. Moreover, mapping the RNAi functional network to that of the protein interaction networks could help identify important new regulators that were missed in RNAi screens and therefore generate some testable hypothesis regarding gene function. This strategy has already been implemented in *C. elegans* by Tewari et al., who used a systematic interactome mapping of the TGF-β signaling network in conjunction with functional analysis of the proteins found in the complex using RNAi *(109)*. For the Wnt pathway, the recent report of the protein "interactome" of β-cat from the Moon group should provide a template for similar comparative studies in the field of Wnt signal transduction.

Finally, a similar approach of cross-referencing the catalogue of cell-based phenotypes generated from RNAi screens with those of small molecule screens to identify inhibitors (or activators) of the Wnt or other signaling pathways *(68, 110, 111)*, could prove to be a powerful tool for identifying protein targets of candidate compounds that could modulate their oncogenic/tumor suppressor properties. Additionally, chemical genetic screens could be utilized to perform modifier screens to enhance or suppress specific RNAi phenotypes in cells.

5. Conclusion

In the future, the breakthroughs in developmental and cancer biology will rely on the efficient identification and functional characterization of a multitude of as yet unidentified, but important genes involved in oncogenesis. This information is essential for the generation of new therapeutic measures to prevent and/or treat cancer. The newly established RNAi technology together with the powerful forward genetic screens has the potential to revolutionize the field of signal transduction and cancer biology, because it provides an efficient method to identify and dissect the function of the majority of genes involved in specific signaling pathways that have been implicated in development and disease.

It is now widely recognized that whole-genome RNAi screens using cell-based assays provide a technology platform for efficient enrichment of potential modulators of cell signaling pathways. The early experience with RNAi reagents has led to a better understanding of their specificities and has already resulted in useful recommendations for best usage of the technology. However, while most of the individual screens are powerful by themselves, they cannot complete the big and often complex picture.

The long-term goal is to integrate a variety of HTS to understand and generate a global picture of how the Wnt signaling pathway is regulated at a molecular level and how it may interact with other signal transduction cascades. While RNAi screens may be used to assign new function to genes in the context of the Wnt pathway, we also need to query the multiprotein complexes in which the known regulators as well as candidate genes are found in the cell, to achieve a comprehensive understanding of their mechanism of action. The use of bioinformatics to integrate and compare the large data sets generated from multiple screens would be essential for extracting meaningful information from genome-scale screens. This approach will allow us to propose new testable hypotheses about the mechanisms of the regulators of the Wnt/wg signaling pathway. Finally, the combination of functional genomic screens with small molecule screens may hold the promise of drug discovery for the treatment of Wnt/wg pathway-related diseases. With this and future knowledge in hand, we expect to see many exciting applications in the next few years of this powerful technology.

References

1. De Toni, F., Racaud-Sultan, C. Chicanne, G., et al. (2006) A crosstalk between the Wnt and the adhesion-dependent signaling pathways governs the chemosensitivity of acute myeloid leukemia. *Oncogene* 25, 3113–3122.

2. Hasson, P., Egoz, N., Winkler, C., et al. (2005) EGFR signaling attenuates Groucho-dependent repression to antagonize Notch transcriptional output. *Nat Genet* 37, 101–105.

3. Letamendia, A., Labbe, E., Attisano, L. (2001) Transcriptional regulation by Smads: crosstalk between the TGF-beta and Wnt pathways. *J Bone Joint Surg Am* 83-A Suppl 1, S31–S39.

4. Lum, L., Yao, S., Mozer, B., et al. (2003) Identification of Hedgehog pathway components by RNAi in Drosophila cultured cells. *Science* 299, 2039–2045.

5. Musgrove, E. A. (2004) Wnt signalling via the epidermal growth factor receptor: a role in breast cancer? *Breast Cancer Res* 6, 65–68.

6. von Bubnoff, A., Cho, K. W. (2001) Intracellular BMP signaling regulation in vertebrates: pathway or network? *Dev Biol* 239, 1–14.

7. Nusse, R. (2003) Wnts and Hedgehogs: lipid-modified proteins and similarities in signaling mechanisms at the cell surface. *Development* 130, 5297–5305.

8. Miller, J. R., Hocking, A. M., Brown, J. D., et al. (1999) Mechanism and function of signal transduction by the Wnt/beta-catenin and Wnt/Ca2+ pathways. *Oncogene* 18, 7860–7872.

9. Polakis, P. (2000) Wnt signaling and cancer. *Genes Dev* 14, 1837–1851.

10. Wodarz, A., Nusse, R. (1998) Mechanisms of Wnt signaling in development. *Annu Rev Cell Dev Biol* 14, 59–88.

11. Ciruna, B., Jenny, A., Lee, D., et al. (2006) Planar cell polarity signalling couples cell division and morphogenesis during neurulation. *Nature* 439, 220–224.

12. Grove, E. A., Tole, S., Limon, J., et al. (1998) The hem of the embryonic cerebral cortex is defined by the expression of multiple Wnt genes and is compromised in Gli3-deficient mice. *Development* 125, 2315–2325.

13. Jiang, R., Bush, J. O., Lidral, A. C. (2006) Development of the upper lip: morphogenetic and molecular mechanisms. *Dev Dyn* 235, 1152–1166.

14. Kokubu, C., Heinzmann, U., Kokubu, T., et al. (2004) Skeletal defects in ringelschwanz mutant mice reveal that Lrp6 is required for proper somitogenesis and osteogenesis. *Development* 131, 5469–5480.

15. Shu, W., Jiang, Y. Q., Lu, M. M., et al. (2002) Wnt7b regulates mesenchymal

proliferation and vascular development in the lung. *Development* 129, 4831–4842.

16. Staal, F. J., Clevers, H. (2000) Tcf/Lef transcription factors during T-cell development: unique and overlapping functions. *Hematol J* 1, 3–6.

17. Kaykas, A., Yang-Snyder, J., Heroux, M., et al. (2004) Mutant Frizzled 4 associated with vitreoretinopathy traps wild-type Frizzled in the endoplasmic reticulum by oligomerization. *Nat Cell Biol* 6, 52–58.

18. Robitaille, J., MacDonald, M. L., Kaykas, A., et al. (2002) Mutant frizzled-4 disrupts retinal angiogenesis in familial exudative vitreoretinopathy. *Nat Genet* 32, 326–330.

19. Miyoshi, K., Hennighausen, L. (2003) Beta-catenin: a transforming actor on many stages. *Breast Cancer Res* 5, 63–68.

20. Miyoshi, K., Rosner, A., Nozawa, M., et al. (2002) Activation of different Wnt/beta-catenin signaling components in mammary epithelium induces transdifferentiation and the formation of pilar tumors. *Oncogene* 21, 5548–5556.

21. Moon, R. T., Kohn, A. D., De Ferrari, G. V., et al. (2004) WNT and beta-catenin signalling: diseases and therapies. *Nat Rev Genet* 5, 691–701.

22. Nusse, R. (1999) WNT targets. Repression and activation. *Trends Genet* 15, 1–3.

23. Logan, C. Y., Nusse, R. (2004) The Wnt signaling pathway in development and disease. *Annu Rev Cell Dev Biol* 20, 781–810.

24. Nusse, R. (2005) Wnt signaling in disease and in development. *Cell Res* 15, 28–32.

25. Behrens, J. (2000) Cross-regulation of the Wnt signalling pathway: a role of MAP kinases. *J Cell Sci* 113 (Pt 6), 911–919.

26. Ishitani, T., Kishida, S., Hyodo-Miura, J., et al. (2003) The TAK1-NLK mitogen-activated protein kinase cascade functions in the Wnt-5a/Ca(2+) pathway to antagonize Wnt/beta-catenin signaling. *Mol Cell Biol* 23, 131–139.

27. Ishitani, T., Ninomiya-Tsuji, J., Nagai, S., et al. (1999) The TAK1-NLK-MAPK-related pathway antagonizes signalling between beta-catenin and transcription factor TCF. *Nature* 399, 798–802.

28. Meneghini, M. D., Ishitani, T., Carter, J. C., et al. (1999) MAP kinase and Wnt pathways converge to downregulate an HMG-domain repressor in Caenorhabditis elegans. *Nature* 399, 793–797.

29. Smit, L., Baas, A., Kuipers, J., et al. (2004) Wnt activates the Tak1/Nemo-like kinase pathway. *J Biol Chem* 279, 17232–17240.

30. Arsura, M., Panta, G. R., Bilyeu, J. D., et al. (2003) Transient activation of NF-kappaB through a TAK1/IKK kinase pathway by TGF-beta1 inhibits AP-1/SMAD signaling and apoptosis: implications in liver tumor formation. *Oncogene* 22, 412–425.

31. Kaminska, B., Wesolowska, A., Danilkiewicz, M. (2005) TGF beta signalling and its role in tumour pathogenesis. *Acta Biochim Pol* 52, 329–337.

32. Matsumoto-Ida, M., Takimoto, Y., Aoyama, T., et al. (2006) Activation of TGF-beta1-TAK1-p38 MAPK pathway in spared cardiomyocytes is involved in left ventricular remodeling after myocardial infarction in rats. *Am J Physiol Heart Circ Physiol* 290, H709–H715.

33. Hayward, P., Brennan, K., Sanders, P., et al. (2005) Notch modulates Wnt signalling by associating with Armadillo/beta-catenin and regulating its transcriptional activity. *Development* 132, 1819–1830.

34. Freeman, M., Bienz, M. (2001) EGF receptor/Rolled MAP kinase signalling protects cells against activated Armadillo in the Drosophila eye. *EMBO Rep* 2, 157–162.

35. Voas, M. G., Rebay, I. (2004) Signal integration during development: insights from the Drosophila eye. *Dev Dyn* 229, 162–175.

36. He, X. (2006) Unwinding a Path to Nuclear beta-Catenin. *Cell* 127, 40–42.

37. Yang, L., Lin, C., Liu, Z. R. (2006) P68 RNA Helicase mediates PDGF-induced epithelial mesenchymal transition by displacing axin from beta-catenin. *Cell* 127, 139–155.

38. Adams, M. D., Celniker, S. E., Holt, R. A., et al. (2000) The genome sequence of Drosophila melanogaster. *Science* 287, 2185–2195.

39. Rubin, G. M., Yandell, M. D., Wortman, J. R., et al. (2000) Comparative genomics of the eukaryotes. *Science* 287, 2204–2215.

40. Venter, J. C., Adams, M. D., Sutton, G. G., et al. (1998) Shotgun sequencing of the human genome. *Science* 280, 1540–1542.

41. Hild, M., Beckmann, B., Haas, S. A., et al. (2003) An integrated gene annotation and transcriptional profiling approach towards the full gene content of the Drosophila genome. *Genome Biol* 5, R3.

42. Oliver, B., Leblanc, B. (2003) How many genes in a genome? *Genome Biol* 5, 204.

43. Bray, D. (2003) Molecular networks: the top-down view. *Science* 301, 1864–1865.

44. Hartwell, L. H., Hopfield, J. J., Leibler, S., et al. (1999) From molecular to modular cell biology. *Nature* 402, C47–C52.

45. Hucka, M., Finney, A., Sauro, H. M., et al. (2003) The systems biology markup language (SBML): a medium for representation and exchange of biochemical network models. *Bioinformatics* 19, 524–531.

46. Milo, R., Itzkovitz, S., Kashtan, N., et al. (2004) Superfamilies of evolved and designed networks. *Science* 303, 1538–1542.

47. Barabasi, A. L., Oltvai, Z. N. (2004) Network biology: understanding the cell's functional organization. *Nat Rev Genet* 5, 101–113.

48. Spirin, V., Mirny, L. A. (2003) Protein complexes and functional modules in molecular networks. *Proc Natl Acad Sci USA* 100, 12123–12128.

49. Banziger, C., Soldini, D., Schutt, C., et al. (2006) Wntless, a conserved membrane protein dedicated to the secretion of Wnt proteins from signaling cells. *Cell* 125, 509–522.

50. Bartscherer, K., Pelte, N., Ingelfinger, D., et al. (2006) Secretion of Wnt ligands requires Evi, a conserved transmembrane protein. *Cell* 125, 523–533.

51. DasGupta, R., Kaykas, A., Moon, R. T., et al. (2005) Functional genomic analysis of the Wnt-wingless signaling pathway. *Science* 308, 826–833.

52. Dasgupta, R., Perrimon, N. (2004) Using RNAi to catch Drosophila genes in a web of interactions: insights into cancer research. *Oncogene* 23, 8359–8365.

53. Hausmann, G., Banziger, C., Basler, K. (2007) Helping Wingless take flight: how WNT proteins are secreted. *Nat Rev Mol Cell Biol* 8, 331–336.

54. Sharp, P. A. (1999) RNAi and double-strand RNA. *Genes Dev* 13, 139–141.

55. Hutvagner, G., Simard, M. J., Mello, C. C., et al. (2004) Sequence-specific inhibition of small RNA function. *PLoS Biol* 2, E98.

56. Mello, C. C., Conte Jr., D. (2004) Revealing the world of RNA interference. *Nature* 431, 338–342.

57. Montgomery, M. K. (2006) RNA interference: unraveling a mystery. *Nat Struct Mol Biol* 13, 1039–1041.

58. Sen, C. K., Roy, S. (2007) miRNA: licensed to kill the messenger. *DNA Cell Biol* 26, 193–194.

59. Tabara, H., Grishok, A., Mello, C. C. (1998) RNAi in C. elegans: soaking in the genome sequence. *Science* 282, 430–431.

60. Clemens, J. C., Worby, C. A., Simonson-Leff, N., et al. (2000) Use of double-stranded RNA interference in Drosophila cell lines to dissect signal transduction pathways. *Proc Natl Acad Sci USA* 97, 6499–6503.

61. Zamore, P. D., Haley, B. (2005) Ribognome: the big world of small RNAs. *Science* 309, 1519–1524.

62. Bridge, A. J., Pebernard, S., Ducraux, A., et al. (2003) Induction of an interferon response by RNAi vectors in mammalian cells. *Nat Genet* 34, 263–264.

63. Scacheri, P. C., Rozenblatt-Rosen, O., Caplen, N. J., et al. (2004) Short interfering RNAs can induce unexpected and divergent changes in the levels of untargeted proteins in mammalian cells. *Proc Natl Acad Sci USA* 101, 1892–1897.

64. Sledz, C. A., Holko, M., de Veer, M. J., et al. (2003) Activation of the interferon system by short-interfering RNAs. *Nat Cell Biol* 5, 834–839.

65. Yang, S., Tutton, S., Pierce, E., et al. (2001) Specific double-stranded RNA interference in undifferentiated mouse embryonic stem cells. *Mol Cell Biol* 21, 7807–7816.

66. Kiger, A. A., Baum, B., Jones, S., et al. (2003) A functional genomic analysis of cell morphology using RNA interference. *J Biol* 2, 27.

67. Boutros, M., Kiger, A. A., Armknecht, S., et al. (2004) Genome-wide RNAi analysis of growth and viability in Drosophila cells. *Science* 303, 832–835.

68. Eggert, U. S., Kiger, A. A., Richter, C., et al. (2004) Parallel Chemical Genetic and Genome-Wide RNAi Screens Identify Cytokinesis Inhibitors and Targets. *PLoS Biol* 2, e379.

69. Friedman, A., Perrimon, N. (2004) Genome-wide high-throughput screens in functional genomics. *Curr Opin Genet Dev* 14, 470–476.

70. Nybakken, K., Vokes, S. A., Lin, T. Y., et al. (2005) A genome-wide RNA interference screen in Drosophila melanogaster cells for new components of the Hh signaling pathway. *Nat Genet* 37, 1323–1332.

71. Berns, K., Hijmans, E. M., Mullenders, J., et al. (2004) A large-scale RNAi screen in human cells identifies new components of the p53 pathway. *Nature* 428, 431–437.

72. Grimm, D., Pandey, K., Kay, M. A. (2005) Adeno-associated virus vectors for short hairpin RNA expression. *Methods Enzymol* 392, 381–405.

73. Milhavet, O., Gary, D. S., Mattson, M. P. (2003) RNA interference in biology and medicine. *Pharmacol Rev* 55, 629–648.

74. Paddison, P. J., Silva, J. M., Conklin, D. S., et al. (2004) A resource for large-scale

RNA-interference-based screens in mammals. *Nature* 428, 427–431.

75. Zheng, L., Liu, J., Batalov, S., et al. (2004) An approach to genomewide screens of expressed small interfering RNAs in mammalian cells. *Proc Natl Acad Sci USA* 101, 135–140.

76. Kennerdell, J. R., Carthew, R. W. (1998) Use of dsRNA-mediated genetic interference to demonstrate that frizzled and frizzled 2 act in the wingless pathway. *Cell* 95, 1017–1026.

77. Kennerdell, J. R., Carthew, R. W. (2000) Heritable gene silencing in Drosophila using double-stranded RNA. *Nat Biotechnol* 18, 896–898.

78. Bale, A. E. (2002) Hedgehog signaling and human disease. *Annu Rev Genomics Hum Genet* 3, 47–65.

79. Belvin, M. P., Anderson, K. V. (1996) A conserved signaling pathway: the Drosophila toll-dorsal pathway. *Annu Rev Cell Dev Biol* 12, 393–416.

80. Evans, C. J., Hartenstein, V., Banerjee U. (2003) Thicker than blood: conserved mechanisms in Drosophila and vertebrate hematopoiesis. *Dev Cell* 5, 673–690.

81. Pandur, P., Maurus, D., Kuhl, M. (2002) Increasingly complex: new players enter the Wnt signaling network. *Bioessays* 24, 881–884.

82. Wajant, H., Scheurich, P. (2004) Analogies between Drosophila and mammalian TRAF pathways. *Prog Mol Subcell Biol* 34, 47–72.

83. Gwack, Y., Sharma, S., Nardone, J., et al. (2006) A genome-wide Drosophila RNAi screen identifies DYRK-family kinases as regulators of NFAT. *Nature* 441, 646–650.

84. Adams, M. D., Sekelsky, J. J. (2002) From sequence to phenotype: reverse genetics in Drosophila melanogaster. *Nat Rev Genet* 3, 189–198.

85. Nagy, A., Perrimon, N., Sandmeyer, S., et al. (2003) Tailoring the genome: the power of genetic approaches. *Nat Genet* 33 Suppl, 276–284.

86. St Johnston, D. (2002) The art and design of genetic screens: Drosophila melanogaster. *Nat Rev Genet* 3, 176–188.

87. Baeg, G. H., Zhou, R., Perrimon, N. (2005) Genome-wide RNAi analysis of JAK/STAT signaling components in Drosophila. *Genes Dev* 19, 1861–1870.

88. Friedman, A., Perrimon, N. (2006) A functional RNAi screen for regulators of receptor tyrosine kinase and ERK signalling. *Nature* 444, 230–234.

89. Echeverri, C. J., Perrimon, N. (2006) High-throughput RNAi screening in cultured cells: a user's guide. *Nat Rev Genet* 7, 373–384.

90. Korinek, V., Barker, N., Willert, K., et al. (1998) Two members of the Tcf family implicated in Wnt/beta-catenin signaling during embryogenesis in the mouse. *Mol Cell Biol* 18, 1248–1256.

91. Kim, J. H., Kim, B., Cai, L., et al. (2005) Transcriptional regulation of a metastasis suppressor gene by Tip60 and beta-catenin complexes. *Nature* 434, 921–926.

92. Sierra, J., Yoshida, T., Joazeiro, C. A., et al. (2006) The APC tumor suppressor counteracts beta-catenin activation and H3K4 methylation at Wnt target genes. *Genes Dev* 20, 586–600.

93. Yanagawa, S., Matsuda, Y., Lee, J. S., et al. (2002) Casein kinase I phosphorylates the Armadillo protein and induces its degradation in Drosophila. *Embo J* 21, 1733–1742.

94. Bartel, D. P. (2004) MicroRNAs: genomics, biogenesis, mechanism, and function. *Cell* 116, 281–297.

95. Zamore, P. D., Tuschl, T., Sharp, P. A., et al. (2000) RNAi: double-stranded RNA directs the ATP-dependent cleavage of mRNA at 21 to 23 nucleotide intervals. *Cell* 101, 25–33.

96. Echeverri, C. J., Beachy, P. A., Baum, B., et al. (2006) Minimizing the risk of reporting false positives in large-scale RNAi screens. *Nat Methods* 3, 777–779.

97. Saxena, S., Jonsson, Z. O., Dutta, A. (2003) Small RNAs with imperfect match to endogenous mRNA repress translation. Implications for off-target activity of small inhibitory RNA in mammalian cells. *J Biol Chem* 278, 44312–44319.

98. Birmingham, A., Anderson, E. M., Reynolds, A., et al. (2006) 3' UTR seed matches, but not overall identity, are associated with RNAi off-targets. *Nat Methods* 3, 199–204.

99. Dharmacon. *Off-Target Effects: Disturbing the Silence of RNA Interference (RNAi).* Dharmacon RNA Technology 2006 [cited Dharmacon Technology Review].

100. Schwarz, D. S., Hutvagner, G., Haley, B., et al. (2002) Evidence that siRNAs function as guides, not primers, in the Drosophila and human RNAi pathways. *Mol Cell* 10, 537–548.

101. Kulkarni, M. M., Booker, M., Silver, S. J., et al. (2006) Evidence of off-target effects associated with long dsRNAs in Drosophila melanogaster cell-based assays. *Nat Methods* 3, 833–838.

102. Ma, Y., Creanga, A., Lum, L., et al. (2006) Prevalence of off-target effects in Drosophila

RNA interference screens. *Nature* 443, 359–363.

103. Carmell, M. A., Hannon, G. J. (2004) RNase III enzymes and the initiation of gene silencing. *Nat Struct Mol Biol* 11, 214–218.

104. Mousses, S., Caplen, N. J., Cornelison, R., et al. (2003) RNAi microarray analysis in cultured mammalian cells. *Genome Res* 13, 2341–2347.

105. Silva, J. M., Mizuno, H., Brady, A., et al. (2004) RNA interference microarrays: high-throughput loss-of-function genetics in mammalian cells. *Proc Natl Acad Sci USA* 101, 6548–6552.

106. Ziauddin, J., Sabatini, D. M. (2001) Microarrays of cells expressing defined cDNAs. *Nature* 411, 107–110.

107. Gunsalus, K. C., Ge, H., Schetter, A. J., et al. (2005) Predictive models of molecular machines involved in Caenorhabditis

elegans early embryogenesis. *Nature* 436, 861–865.

108. Veraksa, A., Bauer, A., Artavanis-Tsakonas, S. (2005) Analyzing protein complexes in Drosophila with tandem affinity purification-mass spectrometry. *Dev Dyn* 232, 827–834.

109. Tewari, M., Hu, P. J., Ahn, J. S., et al. (2004) Systematic interactome mapping and genetic perturbation analysis of a C. elegans TGF-beta signaling network. *Mol Cell* 13, 469–482.

110. Kapoor, T. M., Mayer, T. U., Coughlin, M. L., et al. (2000) Probing spindle assembly mechanisms with monastrol, a small molecule inhibitor of the mitotic kinesin, Eg5. *J Cell Biol* 150, 975–988.

111. Mayer, T. U., Kapoor, T. M., Haggarty, S. J., et al. (1999) Small molecule inhibitor of mitotic spindle bipolarity identified in a phenotype-based screen. *Science* 286, 971–974.

Part VI

Sea Urchin

Chapter 14

Wnt Signaling in the Early Sea Urchin Embryo

Shalika Kumburegama and Athula H. Wikramanayake

Abstract

Wnt signaling regulates a remarkably diverse array of cellular and developmental events during animal embryogenesis and homeostasis. The crucial role that Wnt signaling plays in regulating axial patterning in early embryos has been particularly striking. Recent work has highlighted the conserved role that canonical Wnt signaling plays in patterning the animal–vegetal (A–V) axis in sea urchin and sea anemone embryos. In sea urchin embryos, the canonical Wnt signaling pathway is selectively turned on in vegetal cells as early as the 16-cell stage embryo, and signaling through this pathway is required for activation of the endomesodermal gene regulatory network. Loss of nuclear β-catenin signaling animalizes the sea urchin embryo and blocks pattern formation along the entire A–V axis. Nuclear entry of β-catenin into vegetal cells is regulated cell autonomously by maternal information that is present at the vegetal pole of the unfertilized egg. Analysis of Dishevelled (Dsh) regulation along the A–V axis has revealed the presence of a cytoarchitectural domain at the vegetal pole of the unfertilized sea urchin egg. This vegetal cortical domain appears to be crucial for the localized activation of Dsh at the vegetal pole, but the precise mechanisms are unknown. The elucidation of how Dsh is selectively activated at the vegetal cortical domain is likely to provide important insight into how this enigmatic protein is regulated during canonical Wnt signaling. Additionally, this information will shed light on the origins of embryonic polarity during animal evolution. This chapter examines the roles played by the canonical Wnt signaling pathway in the specification and patterning of the A–V axis in the sea urchin. These studies have led to the identification of a novel role for canonical Wnt signaling in regulating protein stability, and continued studies of Wnt signaling in this model system are likely to reveal additional roles for this pathway in regulating early patterning events in embryos.

Key words: sea urchin, canonical Wnt signaling, A–V axis, β-catenin, endomesoderm.

1. Introduction

Early pattern formation in most animal embryos is strongly influenced by anisotropies present in the unfertilized oocyte. One asymmetry commonly seen in oocytes is the animal–vegetal

Elizabeth Vincan (ed.), *Wnt Signaling, Volume II: Pathway Models, vol. 469*
© 2008 Humana Press, a part of Springer Science+Business Media, New York, NY
Book doi: 10.1007/978-1-60327-469-2

(A–V) axis that establishes a fixed primary axis in oocytes of most triploblastic, bilaterally symmetric animals *(1)*. The molecular basis for the A–V axis is not well understood in most animals, but it is imparted by the asymmetric distribution of maternal determinants in the oocyte during oogenesis. Regardless of the mechanisms that establish this polarity, the A–V axis provides developmental information that is used as the first cues necessary for patterning a developing embryo *(1)*. Many species in the three bilaterian clades have oocytes that have a clear A–V axis, and in general, during embryogenesis in these animals, the endomesoderm arises at the vegetal pole and the animal pole gives rise to ectodermal tissues *(1, 2)*. These observations indicate that the developmental information along the A–V axis is important in germ layer segregation, one of the first subdivisions in the embryo. The vegetal pole also usually demarcates the site of gastrulation *(2)*. Although less well studied, an A–V polarity is also present in oocytes of diploblastic animals, the cnidarians and ctenophores *(1)*. Interestingly, studies have shown that unlike in bilaterians, the A–V axis in cnidarian eggs is not fixed, and the developmental potential for endoderm formation and gastrulation can be shifted to new locations using embryological approaches *(1)*. These observations suggest that during animal evolution the A–V axis may have become fixed in the common ancestor to the bilaterians *(3)*. A fixed A–V axis also correlates with the evolution of the third germ layer, the mesoderm. Thus, the molecular mechanisms that specify the A–V axis are of significant interest for developmental and evolutionary studies. Recent studies have revealed that the canonical Wnt pathway plays a critical role in the specification and patterning of the A–V axis in sea urchins and sea anemones *(4, 5)*, indicating that this pathway was likely used for this purpose in the common ancestor to cnidarians and bilaterians. A more complete understanding of how Wnt signaling is regulated in these organisms will provide insight into the evolution of pattern formation in animal embryos. The sea urchin embryo in particular has many attributes that make it a very useful model system for examining the regulation of the canonical Wnt pathway during early embryogenesis.

2. Cell Division and Cell Fate Specification Along the A–V Axis in the Early Sea Urchin Embryo

Theodor Boveri was the first embryologist to demonstrate that early pattern formation in the sea urchin embryo predictably followed a polarity present in the unfertilized egg *(6)*. This polarity now known as the A–V axis, not only provides developmental information for germ layer segregation, but also regulates the invariant mitotic patterns in the cleavage stage embryo. The first two cell divisions in the sea urchin embryo are along the A–V axis and are orthogonal to each other. The third cleavage is equatorial separating the embryo into four animal and four vegetal blastomeres of roughly the same size. At the fourth cleavage, the mitotic pattern changes in the animal and vegetal halves. The four animal-half blastomeres divide equally and perpendicular to the A–V axis to produce a ring of eight cells, while the four vegetal blastomeres divide unequally and obliquely to the A–V axis to produce four large macromeres and four small blastomeres called the micromeres that form at the most vegetal end of the 16-cell stage embryo. Cell lineage tracing has shown that each tier of blastomeres has very distinct fates in the embryo, with the animal half producing the ectoderm and the vegetal half producing all the endomesoderm and some ectoderm (7–9). The micromeres divide to produce four large micromeres and four small micromeres. The four small micromeres contribute to coelomic pouches in the embryo, while the four large micromeres serve an important endomesoderm-inducing function early in development, and later give rise to the primary mesenchyme cells (PMCs) that produce the skeletal rods in the larva (7, 9). The macromeres segregate into an upper veg1 and a lower veg2 tier by the 60-cell stage of development. Studies have shown that the veg2 tier is specified as endomesoderm by this stage of development. The blastomeres derived from the animal half are fated to form oral and aboral ectoderm (9). Thus, there is a striking segregation of cell fates along the A–V axis in the sea urchin embryo during the very early cleavage stages. It is now clear that this segregation of cell fates is mediated by the selective activation of the canonical Wnt pathway in the vegetal cells of the early embryo. Experimental evidence also strongly suggests that the initial specification of the A–V axis is mediated by unknown maternal factors that selectively activate canonical Wnt signaling in the vegetal blastomeres.

3. Activation of Canonical Wnt Signaling in the Early Sea Urchin Embryo and Regulation of Cell Fates Along the A–V Axis

The excellent optical clarity of sea urchin eggs and embryos, and the relative simplicity of the early embryonic stages have made sea urchins a useful embryonic model for studying the regulation of endogenous canonical Wnt signaling. Immunolocalization studies suggested that the initial activation of the canonical pathway is at the 16-cell stage when β-catenin, the mediator of canonical Wnt signaling, enters the nuclei of the micromeres. By the 32-cell stage, β-catenin is also seen in the macromere nuclei. At the 60-cell stage when the macromeres separate into veg1 and veg2 tiers, nuclear β-catenin is downregulated in the veg1 cells and remains at high concentrations in the veg2 and micromere cell nuclei *(10)*. Injection of a green fluorescent protein (GFP)-tagged β-catenin (β-*catenin::GFP*) mRNA showed a similar pattern of expression of the fusion protein, with nuclearization of β-catenin::GFP in vegetal blastomeres and the rapid clearing of the protein from animal half blastomeres via a GSK-3β-mediated degradation *(11)*. Consistent with its early nuclearization in vegetal blastomeres, functional studies have shown that nuclear β-catenin is essential for early endomesoderm cell fate specification and for pattern formation along the entire A–V axis *(4, 10)*. When a stable form of β-catenin, which cannot be targeted for GSK-3β mediated proteolysis is overexpressed in sea urchin eggs by mRNA microinjection, the resulting embryos are highly vegetalized *(4)*. In these vegetalized embryos, there is an expansion of endodermal cells as well as other vegetal cell derivatives such as pigment cells and muscle cells arising from SMCs. Conversely, embryos vegetalized by the overexpression of β-catenin have drastically reduced ectodermal cells *(4)*. In further experiments where animal halves were isolated from eight-cell stage embryos overexpressing the stable form of β-catenin, the resulting mesomere-derived embryoids formed ectopic endomesoderm *(4)*. This gain-of-function experiment showed that activated β-catenin is sufficient for activating endomesoderm cell fates in animal half-derived mesomeres that normally only follow ectodermal cell fates. A role for β-catenin in A–V axis patterning is also supported by loss-of-function experiments, in which nuclear β-catenin is depleted by the overexpression of the cell adhesion molecule cadherin *(4, 10)*. Overexpressed cadherin sequesters β-catenin at the cell surface, thereby depleting the signaling pool of β-catenin. When cadherin is overexpressed in early sea urchin embryos, β-catenin does not enter the nuclei of vegetal cells and these embryos are severely animalized *(4, 10)*. These animalized embryos lack endoderm and all mesodermal derivatives *(4, 10)*. Additionally, cadherin-overexpressing embryos do not express any aboral ectoderm markers, and are unable to form an oral–aboral axis, a process that requires vegetal signaling *(4, 12)*. Together, these results provide strong evidence that nuclear β-catenin is required for endomesoderm specification as well as for patterning along the A–V axis.

Nuclear β-catenin accumulates in different vegetal cell tiers in the early sea urchin embryo, and recent studies have shown that canonical Wnt signaling regulates specific developmental functions in each tier. Because β-catenin accumulates at high levels in micromere and macromere derivatives in the early embryo, global overexpression of mRNA cannot distinguish between the requirement and function of canonical Wnt signaling in each cell tier. To determine if nuclear β-catenin has specific functions in each vegetal cell tier, cadherin-overexpression has been combined with blastomere transplants. These experiments were done by either blocking nuclear β-catenin in the micromeres and then transplanting them to the animal half of a host embryo, or by transplanting wild-type micromeres to a micromere-deleted, cadherin-overexpressing embryo *(10, 13)*. When cadherin-overexpressing micromeres are transplanted to the animal pole of host embryos, these micromeres are unable to induce an ectopic gut in the mesomeres *(10)*, and when normal micromeres are transplanted to the vegetal pole of micromere-deleted, cadherin-overexpressing embryos, the transplanted micromeres are unable to induce endomesoderm *(13)*. These observations indicate that nuclear β-catenin is required in the micromeres for these cells to transmit the endomesoderm-inducing signal, and nuclear β-catenin is required in the macromere derivatives for these cells to respond to the micromere signal. Together, the aforementioned studies demonstrate that the nuclearization of β-catenin in vegetal cells of the early sea urchin embryo is the critical event in activation of pattern formation along the A–V axis.

4. Regulation of Nuclear β-catenin in the Vegetal Blastomeres of the Sea Urchin Embryo

There is increasing evidence that the A–V axis in the sea urchin egg is specified by maternal factors that locally activate canonical Wnt signaling at the vegetal pole. Because of this, considerable effort has been made in recent years to identify the maternally derived upstream regulators of nuclear β-catenin in early sea urchin embryos. One possible mechanism for A–V axis specification is that there is a localized Wnt ligand at the vegetal pole of the unfertilized egg that can selectively activate the canonical Wnt pathway in vegetal cells during early embryogenesis. In *Xenopus*, a maternal Wnt ligand (Wnt11) has been shown to be involved in the nuclearization of β-catenin in cells of the future dorsal side *(14, 15)*. *XWnt11* mRNA is initially localized at the vegetal pole and the mRNA is moved to the future dorsal side by the cortical rotation. Hence, it is possible that a similar mechanism is utilized in the sea urchin embryo, where a maternal Wnt localized to the vegetal pole is used to stabilize β-catenin in vegetal blastomeres. However, a genome-wide survey of the recently sequenced *Strongylocentrotus purpuratus* genome has failed to identify a *Wnt11* ortholog in this sea urchin. This genome-wide search has shown that there are 11 *Wnt* genes in the sea urchin embryo *(16)*. Preliminary studies have shown that although several of these *Wnt* genes are expressed maternally, none of them are localized at the vegetal pole, nor have they been shown to be able to nuclearize β-catenin in vegetal blastomeres in the developing embryo *(16)*. Another *Wnt* gene, *Wnt8*, plays an important role in pattern formation during early sea urchin development, but its expression starts in the 16-cell stage micromeres, after the initial nuclearization of β-catenin in these cells, and it is among the few direct target genes of β-catenin identified thus far *(17)*. Alternatively, a ligand-free mechanism may be used to drive β-catenin into vegetal cell nuclei. The seven-pass transmembrane cell-surface receptor, Frizzled (Fz) and its single-pass transmembrane co-receptor, Arrow/LDL receptor-related protein 5/6 (LRP 5/6) are important upstream regulators of the Wnt pathway. Although four Fz (Fz1/2/7, Fz4, Fz5/8, and Fz9/10) and at least one LRP receptor have been identified in the sea urchins, their functional roles, if any, in the activation of the canonical Wnt signaling pathway in early sea urchin development are yet to be determined *(16)*. To date, only the function of the Fz5/8 receptor has been characterized during sea urchin development. *Fz5/8* transcript expression is ubiquitous in the egg and early embryos *(18)*. But its expression narrows to a localized vegetal domain only at the mesenchyme blastula stage and thereinafter. Functional analysis of this receptor reveals that in fact Fz5/8 functions through the planar cell polarity (PCP) pathway, which typically regulates cell motility through a β-catenin-independent pathway. In sea urchins, Fz5/8 functions through the PCP pathway to initiate gastrulation and maintain late endodermal markers such as *Brachyury* in the archenteron of the gastrula stage sea urchin embryo *(18)*. This is one of the few examples of PCP signaling that has thus far been identified in sea urchins.

In contrast to the lack of knowledge of the cell surface ligands and receptors that regulate nuclear β-catenin in the early sea urchin embryo, there is a deeper understanding of the conserved cytoplasmic and nuclear components that regulate the stability and signaling activity of this protein along the A–V axis *(11, 19–22)*. β-catenin transcripts are found uniformly distributed in sea urchin eggs and early embryos *(23)*. Therefore, one mechanism for regulation of β-catenin protein in the early sea urchin embryo is potentially through its differential proteolysis along the A–V axis. Early studies identified a clear role for GSK-3β in regulating the differential stability of β-catenin along the A–V axis *(19)*. Unlike in *Xenopus* embryos where differential degradation of GSK-3β in the prospective dorsal blastomeres is partly responsible for stabilization of β-catenin *(24)*, in sea urchins, maternal GSK-3β transcripts and protein are maintained uniformly in the embryo *(11, 19)*. However, when a kinase-dead dominant-negative form of GSK-3β is coexpressed with β-catenin::GFP the GFP fusion protein is nuclearized in all blastomeres including the animal half-derived mesomeres *(11)*. Furthermore, overexpression of kinase-dead GSK-3β vegetalizes sea urchin embryos, while the overexpression of wild-type GSK-3β animalizes embryos *(11, 19)*. These results indicate that GSK-3β is necessary but this protein alone is not sufficient for the regulation of β-catenin in the vegetal blastomeres. These results also indicate that in order to selectively stabilize β-catenin in the endomesoderm there must be a Wnt pathway component upstream of GSK-3β and β-catenin that is localized, or locally activated in vegetal pole blastomeres at the time of endomesoderm specification *(25)*.

Recent work on the phosphoprotein Dishevelled (Dsh) has revealed an intriguing, previously unappreciated cytoarchitectural feature at the vegetal pole of the unfertilized sea urchin egg that appears to be critical for the localized activation of canonical Wnt signaling *(11, 25)*. The Dsh protein is a crucial regulator of the "destruction complex," consisting of GSK-3β, Axin, and APC in the canonical Wnt pathway. Dsh contains three highly conserved domains, the N-terminal DIX, the middle PDZ, and the C-terminal DEP domain *(26, 27)*. In addition to its role in the canonical Wnt pathway, Dsh also regulates signal transduction via the two non-canonical pathways *(26)* and has been shown to interact with a plethora of proteins making it challenging to study its function in a particular pathway *(26, 28)*. The flow of information to the canonical pathway or the non-canonical pathways appears to be mediated by the selective mobilization of distinct domains of the Dsh protein. For example, signaling via the canonical pathway requires the DIX and the PDZ domains while signaling via the non-canonical pathway requires the PDZ and the DEP domains *(26, 29)*. During canonical Wnt signaling, Dsh is believed to homodimerize and also to heterodimerize with Axin and form oligomers via its DIX domain *(30)*. Thus, Dsh function in the canonical Wnt pathway can be inhibited by the overexpression of the DIX domain of the molecule *(31, 32)*. It is believed that the DIX domain acts as a dominant-negative by dimerizing with the DIX domain of endogenous Dsh, thus preventing homodimerization *(32)*. Sea urchin embryos overexpressing this dominant-negative form of Dsh fail to nuclearize β-catenin, and as a result cannot specify endomesoderm, and become severely animalized *(11)*. Intriguingly, while blocking Dsh function in the canonical pathway results in an animalized phenotype, the overexpression of Dsh in the whole embryo does not lead to vegetalized embryos. In sea urchin embryos, vegetalization occurs due to the respecification of animal-half-derived mesomeres to endomesodermal cell fates *(4)*. Therefore, the observation that overexpressed Dsh cannot vegetalize sea urchin embryos strongly suggests that Dsh is tightly regulated or is not active in animal-half blastomeres. Whole-mount *in situ* hybridization studies show that Dsh mRNA is expressed ubiquitously throughout development *(11)*. But immunolocalization of endogenous Dsh and overexpression of a Dsh::GFP fusion protein by mRNA microinjection have shown that the protein is highly enriched in the vegetal cortex of the unfertilized egg, and in the uncleaved zygote *(11, 25, 33)*. This cortical enrichment is maintained throughout early development in the micromeres and in the overlying macromeres *(25, 33)*. The mechanism for this cortical vegetal enrichment of Dsh is unknown, but it points to a heretofore-unknown cytoarchitectural polarity that marks the vegetal pole. The domains that mediate Dsh targeting to the vegetal cortex have been characterized using a deletion analysis of a Dsh::GFP fusion protein, and it is believed that the targeting of Dsh to the vegetal cortex is essential for its activation *(11)*. Additionally, because overexpression of full-length Dsh has no effect on A–V patterning, these observations strongly suggest that the protein is not active in animal half blastomeres, and requires local activation at the vegetal pole *(11, 25)*. Therefore, the mechanisms that anchor Dsh to the vegetal cortical domain, and the mechanisms that regulate Dsh activity in this intriguing cytoplasmic structure(s) are outstanding questions. Elucidating these questions will provide significant insight into the mechanisms that activate Dsh in the canonical Wnt pathway. Additionally, this information will also shed significant light on the mechanisms that specify the A–V axis in the sea urchin embryo.

5. Nuclear β-catenin and Activation of Gene Regulatory Networks in the Sea Urchin Embryo

The activation of the canonical Wnt pathway through the nuclearization of β-catenin in the vegetal blastomeres starting at the 16-cell stage is the key upstream signal required for the activation of the endomesoderm gene regulatory network (EGRN; ref. *34*). As a result of this initial signal, a sequence of signaling events involving a large number of signaling molecules and transcription factors lead up to the specification and differentiation of the endoderm, the PMC and SMC derivatives. Extensive expression and functional studies carried out in recent years has resulted in the identification of a large number of essential components involved downstream of the initial β-catenin signal in the EGRN. The transcription factor *pmar1*, belonging to the paired class homeodomain family of transcription factors, is the earliest transcription factor to be identified downstream of nuclear β-catenin *(35)*. *pmar1* transcripts are expressed specifically in the micromeres at the onset of the 16-cell stage *(35)*. Molecular perturbation studies indicate that in the absence of nuclear β-catenin, the micromere function can be rescued by the overexpression of *pmar1*, thereby indicating that its function is downstream of the initial β-catenin signal. This rescue results in micromeres becoming PMCs, which form the skeletogenic spicules and a normal invagination of the archenteron *(35)*. The Delta ligand of the Notch-Delta signaling pathway is expressed in the micromeres between the eighth and the tenth cleavage stages *(36)*. Further studies show that when pmar1 is overexpressed it induces ectopic expression of *delta* in the micromeres between the eighth and the tenth cleavage stages, indicating that the Delta signal is downstream of pmar1 *(37)*. Subsequent activation of Notch-Delta signaling in vegetal plate cells has been shown to be necessary for segregation of the endomesoderm into endoderm and secondary mesoderm *(13, 38)*.

Wnt8 is another direct target gene of nuclear β-catenin in the early sea urchin embryo. Expression analysis data has shown that like *pmar1*, *Wnt8* transcripts are not maternally expressed. *Wnt8* expression is initiated at the 16-cell stage, immediately following the nuclear entry of β-catenin in these cells, and expression is tightly restricted to the micromeres *(17)*. Later on at the 32-cell stage, *Wnt8* expression expands to the macromeres, and similar to nuclear β-catenin at the 60-cell stage, its expression is seen in the veg2 but not the veg1 tier. Gain- and loss-of-function experiments have identified some important developmental roles of *Wnt8* in sea urchin embryos *(17)*. The overexpression of Wnt8 by mRNA injection results in embryos that have ectopic endoderm and ectopic spicules. Intriguingly, when Wnt8 is overexpressed in isolated animal halves, there is a strong induction of endoderm and the formation of spicules, but SMC-derived cell types are never seen in these explants. These observations suggest that even though Wnt8 is sufficient for endoderm and PMC specification, it is not sufficient for SMC specification. Knock down of Wnt8 function using morpholinos results in embryos lacking a gut, spicules, and reduced numbers of SMC cells, identifying a role for Wnt8 in endomesoderm formation *(17)*. Analysis of the cis-regulatory region of the *Wnt8* gene has revealed the presence of several Tcf-binding sites that are required for the correct spatial and temporal expression of *Wnt8 (39)*. These observations support the idea that *Wnt8* is an early target of nuclear β-catenin in the EGRN that regulates early events in endomesoderm specification. Additionally, it is believed that following the zygotic activation of *Wnt8*, the Wnt8 ligand maintains canonical Wnt signaling leading to the continuous nuclearization of zygotic β-catenin in the vegetal cells during early embryogenesis *(34, 39)*. The transcription factor Blimp1/Krox has been identified as another direct downstream target gene of β-catenin *(40, 41)*. Whole-mount *in situ* hybridization revealed that the spatial and temporal distribution of *blimp1/krox* transcripts closely resembles that of *Wnt8 (40)*. Furthermore a Blimp1/Krox DNA-binding site has been found in the regulatory region of the *Wnt8* gene adjacent to one of the Tcf1 sites. Mutational analysis of this site showed that Blimp1/Krox function is necessary for maintaining Wnt8 expression pattern in conjunction with Tcf1 *(39)*. Therefore, Blimp1/Krox functions as a positive regulator of *Wnt8* during later canonical Wnt signaling in sea urchin embryo *(39)*.

6. A Mutual Antagonism Between SoxB1 and Nuclear β-catenin Regulates A–V Axis Patterning in the Sea Urchin Embryo

Recent work done in the Angerer lab has identified an important interaction between the HMG-box transcription factor SoxB1 and nuclear β-catenin during A–V axis patterning in the early sea urchin embryo *(42, 43)*. SoxB1 protein is initially ubiquitously expressed in the egg and during the first three cleavage stages. However, at the fourth cleavage stage SoxB1 remains at high levels in the mesomeres and the macromeres, but becomes downregulated in the four micromeres that arise at the vegetal end of the embryo *(42)*. This downregulation correlates with the nuclear entry of β-catenin. Over the next several cleavage divisions, as nuclear β-catenin expression expands to additional cell tiers in the vegetal pole, SoxB1 protein is progressively downregulated in these same cells, such that by the blastula stage, nuclear β-catenin, and SoxB1 have mutually exclusive domains of expression *(43)*. Functional analysis has shown that nuclear β-catenin and SoxB1 have mutually antagonistic interactions that play a critical role in normal pattern formation in the embryo. Blocking the nuclear entry of β-catenin by cadherin overexpression results in the uniform expression of SoxB1 in all blastomeres in the embryo. Conversely, overexpression of an activated form of β-catenin that enters the nuclei of all cells in the embryo, results in the depletion of SoxB1 from the nuclei of all cells except at the extreme animal pole of the embryo. Interestingly, overexpression of SoxB1 by mRNA expression results in animalized embryos that fail to activate the EGRN, including those genes at the very top of the cascade *(43)*. These observations have suggested that SoxB1 functions at or near the level of β-catenin's transcriptional regulatory function *(43)*. Interestingly, mutually antagonistic interactions between β-catenin and other HMG-box domain-containing proteins have also been identified in other species including mammals, suggesting an ancient origin for this interaction in patterning embryos *(43)*.

Work in the Angerer lab has also identified a novel role for the canonical Wnt pathway in regulating protein turnover *(44)*. These studies were spurred by the observation that the rapid downregulation of SoxB1 protein from the micromeres was not readily explained by downregulation of SoxB1 transcription by nuclear β-catenin. This suggested that there is a post-transcriptional mechanism for downregulating SoxB1 protein in vegetal cells in the early embryo. When a GFP::SoxB1 fusion protein was expressed in embryos by mRNA injection into zygotes, it accumulated in a non-vegetal pattern, even though the *GFP::Sox1* mRNA was present in all blastomeres *(43, 44)*. Intriguingly, this fusion protein was uniformly expressed in the embryo when the canonical Wnt pathway was suppressed by cadherin overexpression. This result indicated that nuclear β-catenin is required for the turnover of SoxB1 protein in vegetal blastomeres, thus revealing a novel role for canonical Wnt signaling in regulated protein turnover *(43, 44)*. Future studies in sea urchins will likely continue to provide insight into the role of canonical Wnt signaling in novel cellular processes in embryos.

7. Conclusion

Sea urchin embryology has been an active area or research for more than 150 years, and many important discoveries have come from exploiting the advantages of this embryo for gaining insight into early developmental processes. One exciting area of recent investigation is the elucidation of the roles of canonical Wnt signaling in pattern formation along the sea urchin A–V axis. These observations were directly responsible for the work in sea anemones that led to the discovery of the crucial role of the canonical Wnt signaling pathway in regulating A–V axis patterning and germ layer segregation in these ancient animals *(4, 5)*. It is likely that future work in these simple embryos will continue to produce exciting and novel insights into the role of Wnt signaling in animal evolution and development.

Acknowledgments

This work was supported by NSF grant IOS 0446523, and the Ingeborg v. F. McKee Fund and the George F. Straub Trust of the Hawaii Community Foundation to AHW.

References

1. Goldstein, B., Freeman, G. (1996) Axis specification in animal development. *BioEssays* 19, 105–116.

2. Martindale, M. Q. (2005) The evolution of metazoan axial properties. *Nat Rev Genet* 6, 917–927.

3. Lee, P. N., Kumburegama, S., Marlow, H. Q., et al. (2007) Asymmetric developmental potential along the animal-vegetal axis in the anthozoan cnidarian, Nematostella vectensis, is mediated by Dishevelled. *Dev. Biol.* 310, 169–186.

4. Wikramanayake, A. H., Huang, L., Klein, W. H. (1998) β-catenin is essential for patterning the maternally specified animal-vegetal axis in the sea urchin embryo. *Proc Natl Acad Sci USA* 95, 9343–9348.

5. Wikramanayake, A. H., Hong, M., Lee, P. N., et al. (2003) An ancient role for β-catenin in the evolution of axial polarity and germ layer segregation. *Nature* 426, 446–450.

6. Boveri, T. (1901) Uber die Polaritat des Seeigel-Eies. *Verh Phys Med Ges Wurzburg* 34, 145–176.

7. Horstadius, S. (1939) The mechanisms of sea urchin development studies by operative methods. *Biol Rev Cambridge Phil Soc* 14, 132–179.

8. Horstadius, S. (1973) Experimental embryology of echinoderms. *Oxford: Clarendon Press.*

9. Davidson, E. H., Cameron, R. A., Ransick, A. (1998) Specification of cell fate in the sea urchin embryo: summary and some proposed mechanisms. *Development* 125, 3269–3290.

10. Logan, C. Y., Miller, J. R., Ferkowicz, M. J., et al. (1999) Nuclear β-catenin is required to specify vegetal cell fates in the sea urchin embryo. *Development* 126, 345–357.

11. Weitzel, H. E., Illies, M. R., Byrum, C. A., et al. (2004) Differential stability of b-catenin along the animal-vegetal axis of the sea urchin embryo mediated by dishevelled. *Development* 131, 2947–2956.

12. Duboc, V., Rottinger, E., Besnardeau, L., Lepage, T. (2004) Nodal and BMP2/4 signaling organizes the oral-aboral axis of the sea urchin embryo. *Dev Cell* 6, 397–410.

13. McClay, D. R., Peterson, R. E., Range, R. C., et al. (2000) A micromere induction signal is activated by β-catenin and acts through Notch to initiate specification of secondary mesenchyme cells in the sea urchin embryo. *Development* 127, 5113–5122.

14. Ku, M., Melton, D. (1993) Xwnt11: a maternally expressed Xenopus Wnt gene. *Development* 119, 1161–1173.

15. Tao, Q., Yokota, C., Puck, H., et al. (2005) Maternal Wnt11 activates the canonical Wnt signaling pathway required for axis formation in Xenopus embryos. *Cell* 120, 857–871.

16. Croce, J. C., Wu, S., Byrum, C., et al. (2006) A genome-wide survey of the evolutionarily conserved Wnt pathways in the sea urchin Strongylocentrotus purpuratus. *Dev Biol* 300, 121–131.

17. Wikramanayake, A. H., Peterson, R., Chen, J., et al. (2004) Nuclear βcatenin-dependent Wnt8 signaling in vegetal cells of the early sea urchin embryo regulates gastrulation and differentiation of endoderm and mesodermal cell lineages. *Genesis* 39, 194–205.

18. Croce, J., Duloquin, L., Lhomond, G., et al. (2005) Frizzled 5/8 is required in secondary mesenchyme cells to initiate archenteron invagination during sea urchin development. *Development* 133, 547–557.

19. Emily-Fenouil, F., Ghiglione, C., Lhomond, G., et al. (1998) GSK3β/shaggy mediates patterning along the animal-vegetal axis of the sea urchin embryo. *Development* 125, 2489–2498.

20. Huang, L., Li, X., El-Hodiri, H. M., et al. (2000) Involvement of Tcf/Lef in establishing cell types along the animal-vegetal axis of sea urchins. *Dev Genes Evol* 210, 73–81.

21. Vonica, A., Weng, W., Gumbiner, B. M., et al. (2000) TCF is the nuclear effector of the beta-catenin signal that patterns the sea urchin animal-vegetal axis. *Dev Biol* 217, 230–243.

22. Range, R. C., Venuti, J. M., McClay, D. R. (2005) LvGroucho and nuclear beta-catenin functionally compete for Tcf binding to influence activation of the endomesoderm gene regulatory network in the sea urchin embryo. *Dev Biol* 279, 252–267.

23. Miller, J. R., McClay, D. R. (1997) Changes of pattern adherence junction associated catenin accompany morphogenesis in the sea urchin embryo. *Dev Biol* 192, 323–339.

24. Dominguez, I., Green, J. B. A. (2000). Dorsal downregulation of GSK3β by a non-Wnt-like mechanism is an early consequence of cortical rotation in early Xenopus embryos. *Development* 127, 861–868.

25. Kumburegama, S., Wikramanayake, A. H. (2007) Specification and patterning of the animal-vegetal axis in sea urchins by the canonical Wnt signaling pathway. *Signal Trans* 7, 164–173.

26. Wallingford, J. B., Habas, R. (2005) The development biology of Disheveled: an enigmatic protein governing cell fate and cell polarity. *Development* 132, 4421–4436.

27. Wharton, K. A. (2003) Runnin' with the Dvl: proteins that associate with Dsh/Dvl and their significance to Wnt signal transduction. *Dev Biol* 253, 1–17.

28. Angers, S., Thorpe, C. J., Biechele, T. L., et al. (2006) The KLHL12-Cullin-3 ubiquitin ligase negatively regulates the Wnt-β-catenin

pathway by targeting Dishevelled for degradation. *Nature Cell Biol* 8, 348–357.

29. Malbon, C., Wang, H. (2006) Dishevelled: a mobile scaffold catalyzing development. *Curr Topics Dev Biol* 72, 153–166.

30. Kishida, S., Yamamoto, H., Hino, S., et al. (1999) DIX domains of Dvl and Axin are necessary for protein interactions and their ability to regulate β-catenin stability. *Mol Cell Biol* 19, 4414–4422.

31. Axelrod, J. D., Miller, R., Shulman, J. M., et al. (1998) Differential recruitment of Dishevelled provides signaling specificity in the planar cell polarity and wingless signaling pathways. *Genes Dev* 12, 2610– 2622.

32. Rothbacher, U., Laurent, M. N., Deardorff, M. A., et al. (2000) Dishevelled phosphorylation, subcellular localization and multimerization regulate its role in early embryogenesis. *EMBO J* 19, 1010–1022.

33. Wikramanayake, A., unpublished observations.

34. Davidson, E. H., et al. (2002) A genomic regulatory network for development. *Science* 295, 1669–1678.

35. Oliveri, P., Davidson, E. H., McClay, D. R. (2003) Activation of pmar1 controls specification of micromeres in the sea urchin embryo. *Dev Biol* 258, 32–43.

36. Sweet, H. C., Gehring, M., Ettensohn, C. A. (2002) LvDelta is a mesoderm-inducing signal in the sea urchin embryo and can endow blastomeres with organizer-like properties. *Development* 129, 1945–1955.

37. Oliveri, P., Carrick, D. M., Davidson, E. H. (2002) A regulatory gene network that directs micromere specification in the sea urchin embryo. *Dev Biol* 246, 209–228.

38. Sherwood, D. R., McClay, D. R. (1999) LvNotch signaling mediates secondary mesenchyme specification in the sea urchin embryo. *Development* 126, 1703–1713.

39. Minokawa, T., Wikramanayake, A. H., Davidson, E. H. (2005) Cis-regulatory inputs of the wnt8 gene in the sea urchin endomesoderm network. *Dev Biol* 288, 545–558.

40. Wang, W., Wikramanayake, A. H., Gonzalez-Rimbau, M., et al. (1996) Very early and transient vegetal-plate expression of SpKrox1, a Kruppel/Krox gene from Strongylocentrotus purpuratus. *Mech Dev* 60, 185–195.

41. Ransick, A., Rast, J. P., Minokawa, T., et al. (2002) New early zygotic regulators of endomesoderm specification in sea urchin embryos discovered by differential array hybridization. *Dev Biol* 246, 132–147.

42. Kenny, A. P., Oleksyn, D. W., Newman, L. A., et al. (2003) Tight regulation of SpSoxB factors is required for patterning and morphogenesis in sea urchin embryos. *Dev. Biol.* 261, 412–425.

43. Angerer, L. M., Kenny, A. P., Newman, L. A., et al. (2007) Mutual antagonism of SoxB1 and canonical Wnt sigrnaling in sea urchin embryo. *Signal Tran* 7, 174–180.

44. Angerer, L. M., Newman, L. A., Angerer, R. C. (2005) SoxB1 downregulation in vegetal lineages of sea urchin embryos is achieved by both transcriptional repression and selective protein turnover. *Development* 132, 999–1008.

Chapter 15

Detecting Expression Patterns of Wnt Pathway Components in Sea Urchin Embryos

Joanna M. Bince, Chieh-fu Peng, and Athula H. Wikramanayake

Abstract

The animal–vegetal (A–V) axis is a maternally established asymmetry that is present in most animal eggs, and it plays an important role in germ-layer segregation. Recent work has shown that the canonical Wnt signaling pathway plays an evolutionarily conserved role in specifying and patterning this axis. However, the precise mechanisms by which this pathway is activated in the early embryo to pattern the A–V axis are not known in most animals. The availability of the *Strongylocentrotus purpuratus* genome sequence, the ability to experimentally manipulate eggs and early embryos using embryological and molecular tools, and the superior optical clarity of sea urchin embryos makes them an important model for investigating the role of the canonical Wnt pathway in specifying and patterning the A–V axis. Here, we provide detailed protocols for determining the expression and localization of mRNA and proteins in early sea urchin embryos, which can be used in studies examining the regulation of Wnt signaling along the A–V axis.

Key words: Sea urchin, animal–vegetal axis, Wnt signaling, whole-mount *in situ* hybridization, immunoflourescence.

1. Introduction

The canonical Wnt signaling pathway plays an important role in axis specification and germ layer segregation in many animal embryos *(1, 2)*. Activation of signaling by this pathway occurs with some predictability along the animal–vegetal (A–V) axis in early embryos of animals as diverse as chordates and cnidarians, and it is marked by the asymmetric accumulation of nuclear β-catenin in a subset of blastomeres in the embryo *(1, 2)*. Where available, experimental evidence indicates that nuclearization of β-catenin in early embryos is regulated by maternal determinants

Elizabeth Vincan (ed.), *Wnt Signaling, Volume II: Pathway Models, vol. 469*
© 2008 Humana Press, a part of Springer Science + Business Media, New York, NY
Book doi: 10.1007/978-1-60327-469-2

localized along the A–V axis. However, except in *Xenopus*, where Wnt11 has been shown to be the upstream ligand regulating nuclearization of β-catenin during dorsal axis specification, information on the mechanisms regulating nuclear β-catenin along the A–V axis in embryos of other species is not well understood *(3)*. Additionally, searches of the available genome sequences of basal deuterostomes such as ascidians and sea urchins have failed to identify a Wnt11 gene *(4)*. A Wnt11 ortholog is present in the basal cnidarian *Nematostella vectensis*, but it is expressed too late in development to play a role in the nuclearization of β-catenin in early embryos in this animal *(5)*. Therefore, the mechanisms regulating the asymmetric activation of Wnt signaling along the A–V axis remains an outstanding question in most animal embryos. Identifying common mechanisms regulating the activation of canonical Wnt signaling in early embryos will provide important insights into the regulation of this pathway, and further insights into the role of this pathway in animal evolution.

The availability of the sea urchin *Strongylocentrotus purpuratus* genome now presents unprecedented opportunities for using this organism as a model for identifying maternal mechanisms regulating the Wnt pathway along the A–V axis *(4, 6)*. A genome-wide survey of Wnt pathway components in *S. purpuratus* has revealed that components of all three well-known Wnt pathways are highly conserved in sea urchins *(4)*. Of the 13 known Wnt subfamilies found in humans, 11 are present in the *S. purpuratus* genome. Comparison with the *Nematostella* genome indicates that the Wnt2 and Wnt11 genes have been lost from the sea urchin genome. Of the Wnt genes identified in *S. purpuratus*, five are expressed maternally, but their spatial expression pattern is currently unknown *(4)*. Here, we provide a detailed protocol for carrying out wholemount *in situ* hybridization that can be used to examine the spatial expression of the Wnt genes and other components of the Wnt signaling pathway in sea urchin eggs and embryos. Recent studies have increasingly highlighted the importance of subcellular compartments and protein localization in activation of the Wnt pathway *(7, 8)*. The superior optical clarity of sea urchin eggs and embryos will be useful in examining the subcellular distribution of components in the Wnt pathway during early embryogenesis. To facilitate such studies, we also provide a protocol for carrying out immunolocalization studies in sea urchin eggs and embryos.

2. Materials

2.1. Gamete Preparation

1. Sea urchins: *S. purpuratus* (Marinus Scientific, Garden Grove, CA).
2. 0.5 M potassium chloride (KCl).

3. 1 M amino-3-4-triazole (ATA; prepare fresh).

4. Instant Ocean artificial seawater (Marineland Labs/Aquarium Systems, Mentor, OH).

5. 70 µM Nytex filter (BD Biosciences, San Jose, CA).

6. 22-gauge needle and 10 mL syringe.

7. Embryological grade plasticware/glassware (*see* **Note 1**).

8. 250-mL sample cups (Fisher Scientific, Houston, TX).

9. 1-oz. disposable medicine cups (Fisher Scientific).

2.2. Dot Blot to Detect DIG-11-UTP Labeling

1. Nuclease-free water.

2. Nitrocellulose membrane (0.22 µM).

3. DIG-labeled control RNA (Roche Applied Science, Indianapolis, IN).

4. Ultraviolet Crosslinker (Amersham Biosciences, Piscataway, NJ).

5. Tris-buffered saline (TBS) with 0.5% (v/v) Tween-20 (TBST): Prepare 10X Stock TBS with 200 mM Tris-HCl, pH 7.5 and 1.5 M NaCl. Prior to use, dilute 100 mL TBS with 900 mL water and add 0.5% (v/v) Tween-20.

6. Blocking buffer: 5% (w/v) nonfat dry milk in TBST.

7. Primary antibody: 1:1000 dilution of anti-digoxigenin (DIG)-alkaline phosphatase (Roche).

8. Staining solution (1 mL): 0.1 M Tris-HCl, pH 9.5, 50 mM $MgCl_2$, 0.1 M NaCl, 1 mM levamisole, 4.5 µL of 18.75 mg/mL nitro blue tetrazolium chloride (NBT; Roche), 3.5 µL of 9.4 mg/mL 5-bromo-4-chloro-3-indolyl phosphate, toluidine salt (BCIP; Roche), sterile dH_2O.

2.3. Whole Mount In-Situ Hybridization (WMISH) Solutions (See Note 2)

Unless specified, water used for the following solutions should be first treated with 0.1% (v/v) diethylpyrocarbonate (DEPC) and autoclaved prior to use.

1. Fixative for unfertilized eggs and embryos up to the early pluteus stage: 4% (w/v) paraformaldehyde, 32.5% (v/v) filtered seawater, 32.5 mM MOPS, pH 7.0, 162.5 mM NaCl.

2. Fixative for larvae: 4% (w/v) paraformaldehyde, 0.1 M MOPS, pH 7.0, 0.5 M NaCl.

3. MOPS Buffer: 0.1 M MOPS, pH 7.0, 0.5 M NaCl, 0.1% (v/v) Tween-20.

4. Hybridization buffer I: 70% (v/v) formamide, 0.1 M MOPS, pH 7.0, 0.5 M NaCl, 0.1% (v/v) Tween-20, 1 mg/mL bovine serum albumin (BSA; Sigma-Aldrich, St. Louis, MO).

5. Hybridization buffer II: same as hybridization buffer but also includes the RNA probe at a final concentration of 0.1 ng/µL.

6. Blocking solution I: 0.1 M MOPS, pH 7.0, 0.5 M NaCl, 10 mg/mL BSA, 0.1% (v/v) Tween-20.

7. Blocking solution II: 0.1 M MOPS, pH 7.0, 0.5 M NaCl, 10% (v/v) sheep (goat) serum (*see* **Note 3**), 10 mg/mL BSA, 0.1% (v/v) Tween-20.

8. Antibody dilution solution: 0.1 M MOPS, pH 7.0, 0.5 M NaCl, 1% (v/v) sheep (goat) serum, 0.1 mg/mL BSA, 0.1% (v/v) Tween-20, anti-DIG antibody, sterile dH_2O.

9. Alkaline phosphatase buffer: 0.1 M Tris-HCl, pH 9.5, 50 mM $MgCl_2$, 0.1 M NaCl, 1 mM levamisole, sterile dH_2O.

10. Staining solution: 0.1 M Tris-HCl, pH 9.5, 50 mM $MgCl_2$, 0.1 M NaCl, 1 mM levamisole, 18.75 mg/mL nitro blue tetrazolium chloride (NBT; Roche), 9.4 mg/mL 5-bromo-4-chloro-3-indolyl phosphate, toluidine salt (BCIP; Roche), sterile dH_2O.

11. Mounting solution: 40% (v/v) glycerol in PBS (10 mM sodium phosphate, 150 mM NaCl, pH 7.4).

12. 1.5 mL extended tip transfer pipettes (Samco Scientific, San Fernando, CA).

13. Low-retention 0.6 mL microcentrifuge tubes (Fisher Scientific).

2.4. Immunofluorescence Solutions

1. Alexa Fluor® 568 goat anti-rabbit IgG (*see* **Note 3**; Invitrogen, Carlsbad, CA).

2. Phosphate buffered saline (PBS): 10 mM sodium phosphate, 150 mM NaCl, pH 7.4.

3. Immunofluorescence fixative: 4% paraformaldehyde (16% stock solution, Electron Microscopy Services, Hatfield, PA) in PBS (*see* **Note 4**).

4. 100% Methanol (chilled to −20°C).

5. Immunofluorescence blocking solution: 3 mg/mL BSA (Sigma Aldrich), PBS, 0.5% (v/v) Tween-20.

6. Mounting solution: 40% (v/v) glycerol in PBS (10 mM sodium phosphate, 150 mM NaCl, pH 7.4).

7. 4% sodium azide (w/v).

3. Methods

3.1. Whole-mount In-Situ *Hybridization (WMISH)*

PCR products or linearized template cDNA with a SP6, T7, or T3 promoter region can be transcribed with the corresponding RNA polymerases using ATP, CTP, GTP, UTP, and DIG-11-UTP. *In situ* hybridization to detect messenger RNA requires anti-sense

3.1.1. DIG-11-UTP Labeling of RNA Probe

RNA probes that will recognize complementary mRNA transcripts. As a negative control sense RNA probes should be prepared. To maximize yield of probe, *in vitro* reactions are prepared using the megaScript kit from Applied Biosystems (Chicago, IL).

1. To synthesize the riboprobe thaw, vortex and briefly centrifuge ($\geq$10,000g) all components of the Applied Biosystems megaScript kit.

2. For a 10 µL reaction (*see* **Note 5**), assemble the following components (at room temperature) in a microcentrifuge tube: 1 µg PCR products or linearized template DNA, 1 µL 10X reaction buffer, 1 µL 25 mM each ATP, CTP, GTP, 0.8 µL 75 mM UTP, 2.1 µL 10 mM DIG-labeled UTP, 1 µL enzyme mix, bring up volume to 10 µL with nuclease-free water.

3. Incubate at 37°C for 6 hours. To remove template DNA add 0.5 µL TURBO DNAse. Incubate at 37°C for 15 minutes.

4. To precipitate the riboprobe add 10 µL of nuclease-free water (*see* **Note 6**) and 10 µL lithium chloride precipitation solution to the reaction. If more than one reaction was prepared they can be combined at this point.

5. Vortex to mix contents and incubate at −20°C for a minimum of 30 minutes.

6. To pellet the RNA centrifuge at 4°C for 20 minutes at 13,000g.

7. Carefully remove the supernatant and wash the pellet by adding 0.5 mL 70% RNAse-free ethanol.

8. Repeat centrifugation at 4°C for 15 minutes at 13,000g.

9. Carefully decant and remove excess ethanol using a sterile pipette tip. Briefly dry pellet in a 65°C incubator. Resuspend in an appropriate volume of nuclease-free water.

10. To determine quality of the riboprobe remove aliquots for denaturing gel electrophoresis and spectrophotometric determination of RNA concentration.

11. RNA probes should be stored at −70°C until use (*see* **Note 7**).

3.1.2. Dot Blot to Detect DIG-11-UTP Labeling of RNA Probe

To determine the efficiency of the probe labeling reaction and to confirm probe concentration, a dot blot followed by immunochemical detection should be performed prior to beginning WMISH. DIG-labeled control RNA (Roche) can be used for comparison. For all immunoblotting steps nitrocellulose membranes should be placed on a rocking platform. Ensure that there is enough solution to completely cover the membrane.

1. Prepare a dilution series of DIG-labeled RNA probe using nuclease-free water (*see* **Note 8**). DIG-labeled control RNA should be prepared in the same dilutions for comparison.

2. Cut nitrocellulose membrane to desired size. Using a pencil draw circles (1 cm in size, 1 cm apart) on membrane corresponding to the number of dilution samples.

3. Add 1 μL of sample to each dot. Let dry at room temperature for 15 minutes or until nitrocellulose is completely dry.

4. To affix RNA to the nitrocellulose membrane, place it in an UV Crosslinker at 120 millijoules/cm^2 for the specified time. On most commercially available crosslinkers, the "auto-crosslink" feature, which produces a 1 minute exposure at a 254 nm wavelength, is sufficient.

5. The membrane can now be prepared for immunoblotting by incubation in 50 mL blocking buffer for 1 hour at room temperature.

6. Discard blocking buffer and wash the membrane twice for 10 minutes each in TBST.

7. The primary antibody, prepared fresh for each experiment, is used at a 1:1000 dilution using TBST. Incubate membrane in this solution for 1 hour at room temperature.

8. Discard the primary antibody solution and wash the membrane twice for 10 minutes each in TBST.

9. The staining solution, also prepared fresh for each experiment, is added to the membrane. During the detection step the membrane should be protected from light. The membrane should be replenished with fresh staining solution every 30 minutes or until a satisfactory signal is produced (*see* **Note 9**).

3.1.3. Gamete Preparation

Eggs and seawater should be kept at 15°C throughout the entire process. The following method can be used to obtain samples for WMISH and antibody staining.

1. Induce spawning of urchins by intracoelomic injection of 0.5 M KCl. To assess gamete quality, fertilize an aliquot of eggs with sperm.

2. If fertilization rates are higher than 95%, prepare eggs for culturing. Using a 250 mL sample cup wash the eggs with artificial seawater. Let eggs settle and remove excess seawater with a serological pipette. Add up to 100 mL of seawater and repeat for a total of three washes.

3. If early developmental stages are required for WMISH fertilization envelopes (FE) need to be removed prior to fixation (*see* **Note 10**). The hardened FE is not permeable to antibodies. For immunofluorescence samples the FE is removed after fixation.

4. Culture embryos at 15°C (*see* **Note 11**).

5. When embryos reach the desired stage collect an appropriate amount in a 50 mL conical tube. Pellet embryos in a clinical

centrifuge using a swing bucket rotor at 1000g for 1 minute. Decant supernatant removing as much seawater as possible and store embryos on ice.

3.1.4. WMISH Fixation

1. Add 10 volumes of WMISH fixative to the embryo pellet. For early stage to pluteus stage embryos fix overnight at 4 °C. For larval stages fix for 1 hour at 37°C.

2. Remove fixative and wash embryos using 10 volumes of MOPS buffer. Repeat for a total of five washes (*see* **Note 12**).

3. Fixed embryos can be stored in 70% ethanol indefinitely at −20°C.

3.1.5. Hybridization

Use 0.5 mL volumes for all incubation steps in the protocol. For wash steps allow embryos to settle before removing solutions. Centrifugation is not recommended.

1. Aliquot samples into 0.6 mL low retention tubes (*see* **Note 13**). A no probe control sample and a sample for the sense-strand RNA probe control should also be prepared.

2. If embryos are in 70% ethanol rehydrate them using MOPS buffer. Remove ethanol and wash embryos using MOPS buffer. Repeat three times (*see* **Note 14**). For embryos already in the MOPS buffer after fixation proceed to the following step.

3. Pre-hybridize samples in hybridization buffer I for 3 hours at 50°C. Rotating of samples is not necessary (*see* **Note 15**).

4. Remove buffer and add hybridization buffer II. Hybridize for 1 week at 50°C.

3.1.6. Wash Steps

All wash steps are performed at room temperature. Incubation time between washes is 10–15 minutes.

1. Wash embryos with MOPS buffer five times.

2. Incubate samples in hybridization buffer I for 3 hours at 50°C.

3. Wash the embryos with MOPS buffer three times.

3.1.7. Anti DIG-11-UTP Staining

1. Incubate samples in blocking solution I for 20 minutes at room temperature.

2. Remove solution and add blocking solution II. Incubate at 37°C for 30 minutes.

3. Dilute anti-DIG-AP antibody 1:1500 in antibody dilution buffer. Incubate samples overnight at room temperature.

4. Wash embryos with MOPS buffer four times.

5. Perform the final wash overnight at room temperature on a rotator.

3.1.8. Detection

1. Wash embryos with alkaline phosphatase buffer twice at room temperature. Increase incubation times between washes to 30 minutes.

2. Incubate embryos in staining solution (*see* **Note 16**).

3. To stop the reaction, remove staining solution and wash embryos with MOPS buffer five times (*see* **Note 17**).

3.1.9. Mounting and Viewing Embryos

1. To prepare embryos for observation remove an aliquot of the sample and place in new tube. Remove excess MOPS buffer and add mounting solution.

2. Prepare a shallow well on a cover slip using thin strips of cover slip glass (*see* **Note 18**).

3. Aspirate embryos and add sample directly into the well. Cover using a microscope slide and seal the cover slip using nail polish.

4. The stained embryos can be viewed using phase or differential interference contrast microscopy.

3.2. Whole-mount Antibody Staining Using Immunofluorescence

Immunohistochemistry is a powerful tool used to detect proteins and protein–protein interaction *in vivo*. It is widely used in research to determine the distribution and localization of biomarkers in different parts of a cell in order to gain insight to protein function. The method described in this section is recommended for use with immunofluorescence and confocal laser scanning microscopy studies on sea urchin embryos, but can be adapted for immunohistochemistry by using the appropriate secondary antibodies. When preparing embryo cultures it is not necessary to remove fertilization envelopes until after fixation. Unless stated otherwise, all steps are performed at room temperature and incubation times between washes is 20 minutes. Centrifugation of samples is not recommended.

1. Following gamete preparation, add 10 volumes of immunofluorescence fixative to the embryo pellet. Incubate sample on ice for 5 minutes and then at room temperature for an additional 20 minutes.

2. Remove as much fixative as possible and immediately add 10 volumes of ice-cold methanol. Incubate sample on ice for 10 minutes.

3. Remove methanol and wash embryos using 10 volumes of PBS. Repeat for a total of three washes.

4. Prior to antibody staining, fertilization envelopes need to be removed. Using a glass Pasteur pipette, slowly apply sample to appropriate cell strainer. Ensure that most of the envelopes are removed (*see* **Note 19**).

5. Transfer an appropriate amount of sample into 1.5 mL centrifuge tubes. Include a negative control (excluding primary antibody) sample as well.

6. Once embryos have settled, remove excess PBS and add blocking solution. Incubate for 15 minutes.

7. Remove blocking solution and add primary antibody solution (primary antibody diluted in PBS). Incubate for an hour at room temperature or overnight at 4°C.

8. To remove unbound primary antibody, wash embryos three times using blocking solution.

9. To detect primary antibody, add appropriate fluorophore-conjugated secondary antibody solution (secondary antibody diluted in PBS). Incubate samples for 45 minutes (*see* **Note 20**).

10. Wash embryos once in blocking solution and twice with PBS.

11. To view embryos, remove excess PBS and add mounting solution (*see* **Note 21**). Transfer to wells prepared on slides and cover slip as in **Subsection 3.9**.

4. Notes

1. Plasticware/glassware used to culture embryos should be thoroughly washed in hot water and rinsed with ultrapure water prior to use. Do not use detergents and chemicals to wash the plasticware/glassware. It is best to dedicate plasticware/glassware for embryo use only.

2. The protocol described in the chapter is adapted from published protocols from the Davidson Laboratory at CalTech *(9, 10)*.

3. We highly recommend this antibody. It produces brighter signals and undergoes less bleaching when observing embryos using confocal microscopy. For long term storage, aliquot small volumes of the antibody and store at −20°C. Avoid freeze–thawing of working stocks.

4. Fixative solutions should be prepared fresh. Previous experimentation suggests that stock 16% paraformaldehyde solutions should be used only on the day the vial was opened.

5. To maximize yield of riboprobe prepare more than one reaction. The reactions can be combined in a later step.

6. If low riboprobe yields are anticipated precipitate RNA with the lithium chloride solution only.

7. Previous experimentation suggests that probes stored in this way are good for 3–6 months.

8. Recommended dilutions range from 1:10, 1:100, 1:1000, 1:5000, 1:10000.

9. To determine probe concentration compare to DIG-labeled RNA control dilutions. Adjust probe concentration according to the concentration deduced from the dot-blots.

10. Fertilize eggs in seawater containing 0.5 mM ATA. At the completion of the first cell division embryos should be passed through a 70 µM Nytex filter (for *S. purpuratus*) or a 100 µM Nytex filter (for *Lytechinus pictus and L. variegatus)* to remove fertilization envelopes. This is done by pouring the culture through the filter in a slow but steady manner. When most of the envelopes have been removed allow embryos to settle. Remove supernatant and refresh with artificial seawater. Return cultures to 15°C.

11. Use Petri dishes when culturing smaller volumes. For larger volumes, embryos can be kept in beakers. To ensure proper aeration embryos raised in beakers need to be constantly stirred.

12. Incubation time between washes is 10–15 minutes.

13. Prior to hybridization fixed embryos at different stages can be mixed to reduce the number of samples.

14. For wash steps 1.5 mL extended tip transfer pipettes are used to remove supernatant.

15. While pre-hybridizing prepare Hybridization buffer II. A concentration of 0.1 ng/µL is used. Incubate this solution at 50°C for at least 30 minutes prior to use.

16. To accelerate substrate reaction replenish with new staining solution every hour. If a color reaction is not seen incubate samples in staining solution overnight.

17. Increase incubation times between washes to 20 minutes.

18. Thin strips of cover slip can be made by scoring a cover slip with a diamond knife, and then gently breaking it.

19. Observe samples under a dissecting microscope. Ensure that embryos are intact and that the envelopes are removed completely. Non-specific binding to the envelope can occur if it is not fully removed.

20. Use of fluorophore-conjugated secondary antibodies require that samples be kept away from direct light.

21. For long-term storage and to prevent bacterial contamination of samples, add sodium azide to a final concentration of 0.4% (w/v).

Acknowledgments

This work was supported by NSF grant IOS 0446523, and the Ingeborg v. F. McKee Fund and the George F. Straub Trust of the Hawaii Community Foundation to AHW.

References

1. Croce, J. C., McClay, D. R. (2006) The canonical Wnt pathway in embryonic axis polarity. *Sem Cell Dev Biol* 17, 168–174.

2. Kumburegama, S., Wikramanayake, A. H. (2007) Specification and patterning of the animal-vegetal axis in sea urchins by the canonical Wnt signaling pathway. *Signal Transduction* 7, 164–173.

3. Tao, Q., Yokota, C., Puck, H., et al. (2005) Maternal wnt11 activates the canonical wnt signaling pathway required for axis formation in *Xenopus* embryos. *Cell* 120, 857–871.

4. Croce, J., Wu, S., Byrum, C., et al. (2006) A genome-wide survey of the evolutionarily conserved Wnt pathways in the sea urchin *Strongylocentrotus purpuratus*. *Dev Biol* 300, 121–131.

5. Lee, P. N., Pang, K., Matus, D. Q., et al. (2006) A WNT of things to come: evolution of Wnt signaling and polarity in cnidarians. *Semin Cell Dev Biol* 17, 157–167.

6. Weinstock, G., The Sea Urchin Genome Sequencing Consortium. (2006) The genome of the sea urchin *Strongylocentrotus purpuratus*. *Science* 314, 941–952.

7. Bilic, J., Huang, Y. L., Davidson, G., et al. (2007) Wnt induces LRP6 signalosomes and promotes dishevelled-dependent LRP6 phosphorylation. *Science* 316, 1619–1622.

8. Yamamoto, H., Komekado, H., Kikuchi, A. (2006) Caveolin is necessary for Wnt-3a-dependent internalization of LRP6 and accumulation of beta-catenin. *Dev Cell 11*, 213–223.

9. Arenas-Mena, C., Cameron, A. R., Davidson, E. H. (2000) Spatial expression of Hox cluster genes in the ontogeny of a sea urchin. *Development* 127, 4631–4343.

10. Minokawa, T., Rast, J. P., Arenas-Mena, C., et al. (2004) Expression patterns of four different regulatory genes that function during sea urchin development. *Gene Expr Patterns* 4, 449–456.

Chapter 16

Functional Analysis of Wnt Signaling in the Early Sea Urchin Embryo Using mRNA Microinjection

Joanna M. Bince and Athula H. Wikramanayake

Abstract

The Wnt pathway is a highly conserved signal transduction pathway that plays many critical roles in early animal development. Recent studies have shown that this pathway plays a conserved role in the specification and patterning of the animal–vegetal (A–V) axis in sea urchins and sea anemones. These observations have suggested that the common ancestor to cnidarians and bilaterians used the Wnt signaling pathway for specifying and patterning this maternally established axis. Because the A–V axis plays a critical role in germ layer segregation, a better understanding of how the Wnt pathway is regulated along the A–V axis will provide key insight into the molecular mechanisms regulating germ layer segregation and germ layer evolution in animal embryos. Here, we provide a detailed protocol for using mRNA microinjection that can be used to analyze Wnt signaling in early sea urchin embryos. This protocol can also be adapted to introduce morpholino anti-sense oligonucleotides into sea urchin embryos.

Key words: Sea urchin, animal–vegetal axis, Wnt signaling, microinjection.

1. Introduction

The evolutionarily conserved Wnt signaling pathway regulates many important developmental events in animal embryos. The first studies that implicated Wnt signaling in early development of sea urchins were carried out by Emily-Fenouil et al. *(1)* and Wikramanayake et al. *(2)* who showed that GSK-3 and β-catenin signaling, respectively, were essential for pattern formation along

Elizabeth Vincan (ed.), *Wnt Signaling, Volume II: Pathway Models, vol. 469*
© 2008 Humana Press, a part of Springer Science + Business Media, New York, NY
Book doi: 10.1007/978-1-60327-469-2

the animal–vegetal (A–V) axis. Since these initial observations, other studies have confirmed that signaling through the canonical Wnt pathway is crucial for patterning the sea urchin A–V axis *(3, 4)*. In the early sea urchin embryo, canonical Wnt signaling is activated at around the 16-cell stage of development when the β-catenin protein specifically enters nuclei of endomesoderm cells at the vegetal pole of the embryo *(5)*. Experimental manipulations that prevent the nuclearization of β-catenin blocks endomesodermal cell fates and lead to animalized embryos that express only oral and neural cell fates *(1, 6)*. An outstanding question is how β-catenin is selectively nuclearized in the vegetal cells of the early sea urchin embryo. Nuclear entry of β-catenin in vegetal cells is clearly regulated by maternal factors, but their identity has remained elusive and the search for these factors is an active area of research. Because of the conserved role of the canonical Wnt pathway in regulating the A–V axis, understanding the molecular mechanisms that regulate nuclearization of β-catenin in the sea urchin embryo will likely provide insight into how this developmentally crucial anisotropy is specified in animal embryos *(7, 8)*.

Because of a relatively long generation time it has not been practical to develop genetic approaches for studying Wnt signaling in sea urchins. However, alternative manipulative tools such as mRNA overexpression for gain-of-function and dominant-negative strategies for blocking gene function, or knock down of gene function using morpholino anti-sense oligonucleotides have been used effectively to study the roles of Wnt signaling in early sea urchin development *(9, 10)*. Additionally, because of the ability to embryologically manipulate the early sea urchin embryo, these molecular methods have been coupled with embryological approaches to gain insight into the roles of Wnt signaling in different cell types during early development *(10)*. It is likely that this combined approach will continue to yield important insight into how maternally derived information is used to selectively activate nuclear β-catenin signaling during early embryogenesis. This chapter provides a detailed protocol for mRNA overexpression by microinjection that can be used to modify Wnt signaling during sea urchin embryogenesis. This protocol can also be adapted for introducing morpholino anti-sense oligonucleotides into sea urchin eggs and embryos.

2. Materials

2.1. Gamete Collection and Preparation

1. Sea urchins (*Strongylocentrotus purpuratus* & *Lytechinus pictus* [Marinus Scientific, Garden Grove, CA], *Lytechinus variegatus* [Duke University Marine Lab, Beaufort, NC]).

2. 0.5 M Potassium chloride (KCl).

3. Instant Ocean artificial seawater (Marineland Labs/Aquarium Systems, Mentor, OH).

4. Dejellying solutions: 1 M citric acid, pH 5.0, 1 M Tris-HCl, pH 8.0.

5. 22-gauge needle and 10-mL syringe.

6. 250-mL sample cups (Fisher Scientific, Houston, TX).

7. 1-oz. disposable medicine cups (Fisher Scientific).

2.2. Protamine Sulfate-Coated Injection Dishes

1. 60-mm plastic cell culture dish lids (Fisher Scientific).

2. 1% (w/v) protamine sulfate solution: 0.5 g protamine sulfate salt, ddH$_2$O to 50 mL.

2.3. Synthetic messenger RNA

1. mMessage mMachine High Yield Capped RNA Transcription Kit (Applied Biosystems).

2. Linearized cDNA that has been purified after restriction enzyme digestion (*see* **Note 1**) or PCR products. For synthetic mRNA synthesis, the final DNA concentration used should be approximately 1 µg/µL (*see* **Note 2**).

3. Quick Spin G-50 Sephadex Columns for RNA purification (Roche).

4. 25:24:1 phenol:chloroform:isoamyl alcohol (Roche).

5. Chloroform (Fisher Scientific).

6. Isopropanol (Sigma Aldrich).

7. 70% ethanol in 0.1% (v/v) diethylpyrocarbonate (DEPC)-treated water.

2.4. Injection Needles

1. Flaming/Brown Micropipette puller (Sutter Instrument CO., Novato, CA).

2. 1.0/0.75 mm (OD/ID) Thin-Wall Single-barrel Standard Borosilicate Glass Tubing (World Precision Instruments, Sarasota, FL).

2.5. Injection Apparatus

1. Stereo microscope with transmitted light base.

2. Inverted microscope (Axiovert 25, Carl Zeiss MicroImaging, Inc., Thornwood, NY; *see* **Note 3**).

3. Microinjector (IM-6 Narishige International USA, Inc., East Meadow, NY).

4. Micromanipulator (Leica Microsystems, Inc., Bannockburn, IL).

2.6. Mouthpipette

1. 1-mL Tuberculin Slip Tip Syringe (BD Biosciences).

2. Mouthpiece (HPI Hospital Products, Altamonte Springs, FL).

3. Rubber tubing, 1/8" ID, 3/16" OD (Cole Parmer Instrument Company, Vernon Hills, IL).

4. Rubber pipette bulbs.

3. Methods

3.1. Preparation of Synthetic messenger RNA

1. For a 20 µL reaction (*see* **Note 4**), assemble the following components in a microcentrifuge tube: 1 µg linearized template DNA, 2 µL 10X reaction buffer, 10 µL 2X NTP/CAP, 2 µL enzyme mix, up to 20 µL of nuclease-free water.

2. Incubate at 37°C for 2 hours (*see* **Note 5**). To remove template DNA, add 1 µL TURBO DNAse. Incubate at 37°C for 15 minutes.

3. Bring volume of reaction to 100 µL with nuclease-free water.

4. For RNA purification perform one 25:24:1 phenol:chloroform:isoamyl alcohol extraction followed by one 24:1 chloroform:isoamyl alcohol extraction.

5. Prepare the Quick Spin G-50 column (*see* **Note 6**). Keeping the column in an upright position, carefully apply sample to the center of the column bed. In a swing bucket rotor, centrifuge at 1100g for 4 minutes at 4°C.

6. Precipitate RNA by adding 1/10th volume ammonium acetate Stop Solution and an equal volume of isopropanol. Mix thoroughly and chill for at least 30 minutes at −20°C.

7. Centrifuge at 4°C for 20 minutes at 13,000g.

8. Carefully remove the supernatant. Wash the pellet by adding 1 mL of 70% ethanol (made with DEPC-treated water). Centrifuge at 4°C for 10 minutes at 13,000g.

9. Carefully decant and remove excess ethanol using a sterile pipette tip. Briefly dry pellet in a 65°C incubator. Resuspend in appropriate volume of nuclease-free water (*see* **Note 7**).

10. Remove aliquots for denaturing gel electrophoresis and spectrophotometric determination of RNA concentration.

3.2. Preparation of Protamine Sulfate-Coated Dishes

1. Assemble cell culture dish lids in rows. Add enough protamine sulfate solution to the first lid to cover the bottom with solution. Let sit for 1 minute and then transfer solution to another lid. Coated lids should washed by placing them in a beaker filled with ddH$_2$O. Continue down the row adding protamine sulfate solution to each lid. When all the dishes have been coated and washed, repeat with an additional wash step in ddH$_2$O. Lids can be left to dry at room temperature.

3.3. Preparation of Mouthpipettes

Mouthpipettes are used in the injection process to move eggs and embryos. They consist of rubber tubing connecting a mouthpiece to a glass micropipette.

1. To prepare the pipette adapter, start at the tip of a tuberculin slip tip syringe and cut a length of 3 cm. Cut a rubber pipette bulb approximately 2.5 cm from the open end. Insert the cut end of the syringe into the cut end of the rubber pipette bulb. Tape can be used to attach the end of the rubber bulb to the end of the syringe.

2. Insert the syringe tip of the pipette adapter to one end of rubber tubing. The mouthpiece is inserted into the opposite end. The glass micropipette is inserted into the open end of the pipette adapter. The rubber tubing should be between 2 to 3 feet in length.

3.4. Preparation of Injection Solutions and Micropipettes

1. mRNA injection solutions are prepared using nuclease-free water and 100% sterile glycerol. Final concentration of glycerol in injection solution is 40% (v/v).

2. To remove any particulate material that could clog the injection needles, the solution is filtered using a 0.2 μM, 4 mm syringe driven filter (Millipore, Billerica, MA). Filtered mRNA solutions are stable at –70°C for up to 1 week.

3. To fill injection needles with solutions two methods can be used: backfilling using a pipette or by micropipette (*see* **Note 8**).

4. Store filled needles at 4°C for at least 15 minutes prior to use. While injecting needles should be stored at –20°C. For longer storage (2 to 3 days) filled needles should be stored at –70°C.

5. Micropipettes are also used for lining the eggs in rows at the bottom of the dishes for injection. For this purpose the bore size of the pipette tip needs to be larger than the diameter of the eggs, but small enough to maintain control of the flow of eggs in and out of the tip during mouth pipetting (*see* **Note 9**).

3.5. Gamete Collection and Dejellying Eggs

When using cold-water species such as *S. purpuratus* and *L. pictus*, the seawater and collected eggs should be kept on ice (~4°C) throughout the entire injection process. For warmer water species such as *L. variegatus*, eggs and seawater should be kept at room temperature.

1. Induce spawning of urchins by intracoelomic injection (*see* **Note 10**) of 0.5 M KCl.

2. Sperm can be collected in a microcentrifuge tube using a pipette (*see* **Note 11**). Eggs are collected by inverting the female in a 1-oz. cup filled with seawater. Using the smaller cups prevents the urchins from crawling inside the cup.

To assess gamete quality, fertilize an aliquot of eggs with sperm (*see* **Note 12**).

3. Transfer eggs to a 250 mL sample cup filled up to 100 mL with seawater. Let eggs settle and remove excess seawater with a serological pipette. Repeat washes three times.

4. The jelly coat is stripped from eggs with an acidic seawater solution using the following protocol (*see* **Note 13**). When eggs are in the final wash remove as much seawater as possible. Add pH 5.0 seawater solution to eggs and gently swirl cup for 3:45 minutes (for *S. purpuratus)* or 1:20 minutes (for *L. pictus* and *L. variegatus)*. To ensure complete removal of jelly coat, eggs must be continuously swirled.

5. Using a glass Pasteur pipette, add 14 drops of 1.0 M Tris-HCl, pH 8.0, to the acidic seawater.

6. Add 100 mL of artificial seawater and allow eggs to settle. Repeat with two additional washes. After the last wash, resuspend eggs in 50 mL of seawater.

3.6. Microinjection of Sea Urchin Zygotes

1. To prevent hardening of fertilization envelopes a final concentration of 0.5 mM amino-3-4 triazole (ATA) is added to artificial seawater.

2. Score a line across a protamine-sulfate coated lid using the sharp edge of a glass Pasteur pipette tip. Fill dish with artificial seawater (~3 to 4 mL).

3. Gently swirl dejellied eggs until they are resuspended. Using a micropipette attached to a mouth pipette draw eggs into the pipette.

4. Place a protamine sulfate-coated injection dish on a stereomicroscope with a transmitted light base. Starting at the left side of the injection dish, slowly release dejellied eggs into the dish while arranging them in a row parallel to the score line. Move the dish from right to left while keeping the micropipette still. The goal is to have a single row of eggs one egg apart.

5. After completing the row do not disturb plate for at least 1 minute. This allows time for the eggs to adhere to the plate. Remove any eggs that have not adhered to the plate. Typically 200–300 eggs are injected in one dish.

6. Place the dish with the eggs on the stage of the inverted microscope and position it so that the row of eggs is perpendicular to the needle holder in the micromanipulator

7. Attach an mRNA-filled needle to the micromanipulator and apply enough oil pressure so that the oil is visible in the back of the needle. Using the lowest objective, position needle tip so that it comes close to but does not touch the score line.

8. Move the injection dish from right to left gently passing the needle across the score line to break the tip (*see* **Note 14**).

9. When a positive flow of the mRNA solution and a desirable needle tip opening is achieved, position the needle tip so that it comes close, but does not touch the eggs.

10. Fertilize eggs by adding 50 µL of a 1:1000 sperm dilution solution. In general more sperm are required to fertilize dejellied eggs.

11. Move the micromanipulator joystick forward in a steady rapid motion so that the needle tip passes through the egg surface and enters the cytoplasm. Deposit a small amount of solution inside the embryo and pull back (*see* **Note 15**).

12. Repeat the injection process using small, rapid hand movements. Continue down the row injecting all the fertilized eggs in the dish. Remove any uninjected eggs and any zygotes that rupture

13. Culture embryos at 15°C until desired stage (*see* **Note 16**).

14. Control embryos (uninjected or injected with a control mRNA) should be prepared alongside injected embryos at the beginning, middle, and end of the injection session.

4. Notes

1. Linearized expression vectors containing the cDNA of interest and PCR products with an RNA polymerase promoter site can be used for *in vitro* transcription using the mMessage mMachine kit. Typically restriction digests of cDNA are set up in 100 µL volumes. Twenty units of restriction enzyme are used to digest 20 µg of plasmid DNA. Digests are incubated at 37°C for 2 hours. Run an aliquot on an agar gel to confirm that the linearization is complete.

2. Following enzyme digestion to linearize the cDNA, purify the DNA using a 25:24:1 phenol:chloroform:isoamyl alcohol extraction. To precipitate DNA add 1/10th volume of 7.5M sodium acetate and 2 volumes of 100% ethanol to the supernatant. The sample is chilled for at least 30 min at –20°C. To recover the DNA, centrifuge at 13,000g for 10 minutes. The resulting pellet is washed with 70% ethanol and briefly dried at room temperature. The DNA pellet is suspended in nuclease-free water at a concentration of 0.5–1 µg/µL.

3. Inverted microscopes used for injections should have a stage wide enough to hold petri dish lids. A joystick control is used to move the stage.

4. Thaw reagents on ice. Be sure to vortex and centrifuge the 10X Reaction buffer and 2X NTP/CAP before use. 10X reaction buffer should be kept at room temperature while assembling the reaction.

5. For maximum yield, an additional 2-hour incubation is recommended following addition of 1 µL enzyme mix.

6. Gently invert column to ensure that beads are suspended. Remove the top of the column first and then remove the tip. Place column in collection tube provided. Allow the buffer to drain by gravity for 10 minutes. Discard excess buffer and put column back in collection tube. Place column and collection tube into a 15 mL conical tube. Using a swing bucket rotor centrifuge at 1100g for 2 minutes at 4°C. Remove column and place in a new collection tube. The column is now ready for the mRNA sample. Maximum load volume is 100 µL.

7. Store RNA at –70°C. If repeated freeze–thawing is avoided RNA is useable for a week when stored at –70°C.

8. To backfill needles using a P2 pipetter load needles by adding 0.5 µL of injection solution to the back of needle. The solution will gradually move from the back of the needle to the tip by capillary action. To backfill needles using a micropipette, start with a 9" glass Pasteur pipette. Heat the pipette over a Bunsen burner (at the point where the wider bore of the pipette narrows to a smaller bore) until it softens. Remove from flame and after a brief pause, quickly pull narrow end to elongate glass. The bore size of the pipette tip can be controlled by modulating the time between removing the hot glass from the flame and initiating the pulling motion to elongate the glass. If the pipette is pulled too early, it will break or will form very narrow unusable pipettes. If the pipette is allowed to cool too much, it will have a large bore size, which precludes its use for back-loading the injection needles. The micropipette tip needs to be small enough to fit into the injection needles. Cut excess glass off using a diamond knife. Using a mouth pipette, fill micropipette with injection solution, insert into the back of the glass needle, and carefully deposit solution into the area where the needle begins to taper.

9. Flame and pull a 9" Pasteur pipette as described in **Note 7**, but the bore size for pipettes used for handling eggs is larger than that used for back-loading needles. Some workers bend the pipettes used for transferring eggs at a 90° angle, but this is generally not necessary unless the injection dishes are deep. Using the lids of culture dishes will eliminate the need to angle the glass pipettes. Using a diamond knife, remove excess glass at the tip, leaving about 2 to 3 cm. To smoothen the tip, heat briefly at the base of the Bunsen flame.

10. Inject the soft tissue between the mouth and the shell.

11. If kept at 4°C, undiluted sperm are viable for 3 days.

12. Egg quality can be determined by examining the shape of the eggs and the fertilization success. Avoid using samples with oocytes or eggs that are misshapen. Immediately after fertilization it is important to examine the formation of the fertilization envelope. If the envelope doesn't fully form or is not round discard the eggs. Fertilization rates should be above 90%. In addition, egg quality decreases with time so dejellied eggs should be used as soon as possible.

13. 50 mL of room temperature artificial seawater is adjusted to pH 5.0 using 1 M citric acid. For cold-water sea urchins such as *S. purpuratus* and *L. pictus*, cool the seawater on ice for at least 10 minutes or until the temperature is at 4°C. For warm-water species such as *L. variegatus* the acidic seawater can be kept at room temperature.

14. The key is to pass the needle tip across the score line once to break the tip. Additional passes can result in the bore of the needle tip being too large or too blunt, precluding successful injections.

15. Because of the viscosity of the glycerol in the injection solution injected mRNA can be seen as a sphere inside the zygotes.

16. If embryos at different stages are needed for experiments the injected embryos have to be removed from the injection dishes using a mouth pipette and placed in 1% (w/v) noble agar coated dishes. Agar coated dishes are recommended because this prevents embryos from sticking to the bottom of the plastic culture dishes. These are prepared by dissolving 1% (w/v) noble agar in deionized water using a microwave. When the agar is fully dissolved, coat culture dishes with a thin layer, much in the same way as preparing the protamine sulfate coated dishes. To adhere agar to the dish, pass bottom of dish very briefly through a Bunsen flame. Dishes can be stored at 4–15°C.

Acknowledgments

This work was supported by NSF grant IOS 0446523 to AHW.

References

1. Emily-Fenouil, F., Ghiglione, C., Lhomond, G., et al. (1998) GSK3beta/shaggy mediates patterning along the animal-vegetal axis of the sea urchin embryo. *Development* 13, 2489–2498.

2. Wikramanayake, A. H., Huang, L., Klein, W. H. (1998) beta-catenin is essential for patterning the maternally specified animal-vegetal axis in the sea urchin embryo. *Proc Natl Acad Sci USA* 16, 9343–9348.

3. Angerer, L. M., Angerer, R. C. (2000) Animal-vegetal axis patterning mechanisms in the early sea urchin embryo. *Dev Biol* 1, 1–12.

4. Croce, J. C., McClay, D. R. (2006) The canonical Wnt pathway in embryonic axis polarity. *Semin Cell Dev Biol* 2, 168–174.

5. Logan, C. Y., Miller, J. R., Ferkowicz, M. J., et al. (1999) Nuclear beta-catenin is required to specify vegetal cell fates in the sea urchin embryo. *Development* 2, 345–357.

6. Yaguchi, S., Yaguchi, J., Burke, R. D. (2006) Specification of ectoderm restricts the size of the animal plate and patterns neurogenesis in sea urchin embryos. *Development* 133, 2337–2346.

7. Kumuregama, S., Wikramanayake, A. H. (2007) Specification and patterning of the animal-vegetal axis in sea urchins by the canonical Wnt signaling pathway. *Signal Transduction* 7, 164–173.

8. Lee, P. N., Kumuregama, S., Marlow, H. Q., et al. (2007) Asymmetric developmental potential along the animal-vegetal axis in the anthozoan cnidarian, *Nematostella vectensis*, is mediated by Dishevelled. *Dev Biol* doi:10.1016/j.ydbio.2007.05.040.

9. Howard, E. W., Newman, L. A., Oleksyn, D. W., et al. (2001) SpKrl: a direct target of beta catenin regulation required for endoderm differentiation in sea urchin embryos. *Development* 3, 365–375.

10. Wikramanayake, A. H., Peterson, R., Chen, J., et al. (2004) Nuclear beta-catenin-dependent Wnt8 signaling in vegetal cells of the early sea urchin embryo regulates gastrulation and differentiation of endoderm and mesodermal cell lineages. *Genesis* 3, 194–205.

Part VII

Zebrafish

Chapter 17

Wnt Signaling Mediates Diverse Developmental Processes in Zebrafish

Heather Verkade and Joan K. Heath

Abstract

A combination of forward and reverse genetic approaches in zebrafish has revealed novel roles for canonical Wnt and Wnt/PCP signaling during vertebrate development. Forward genetics in zebrafish provides an exceptionally powerful tool to assign roles in vertebrate developmental processes to novel genes, as well as elucidating novel roles played by known genes. This has indeed turned out to be the case for components of the canonical Wnt signaling pathway. Non-canonical Wnt signaling in the zebrafish is also currently a topic of great interest, due to the identified roles of this pathway in processes requiring the integration of cell polarity and cell movement, such as the directed migration movements that drive the narrowing and lengthening (convergence and extension) of the embryo during early development.

Key words: zebrafish, vertebrates, canonical Wnt signaling, β-catenin, Wnt/PCP, planar cell polarity, mutant, anti-sense morpholino oligonucleotides, Tilling.

1. Introduction

Wnt signaling is one of the most important pathways in nature, including plants, triggering events in the cell that bring about changes in gene expression, cell migration, and polarity (*1–3*). The pathway has been demonstrated to have a profound impact during normal development, in the maintenance of cellular homeostasis in renewing tissues and in the initiation and progression of disease, most particularly cancer (*4*). Wingless/Wnt signaling was first described in *Drosophila* and most of the components of the pathway, and the epistatic relationships between them, were first identified in this organism. Since then, there has been a rapidly growing appreciation of the two pathways that are triggered as a

Elizabeth Vincan (ed.), *Wnt Signaling, Volume II: Pathway Models, vol. 469*
© 2008 Humana Press, a part of Springer Science + Business Media, New York, NY
Book doi: 10.1007/978-1-60327-469-2

result of Wnt receptor (Frizzled) ligation by Wnt ligands, although in mammals most attention has been focused on the "canonical" Wnt signaling pathway. The activity of the canonical pathway depends on the cytoplasmic concentration of free β-catenin, which is normally kept at very low levels as a result of the combination of its tethering, through cadherin molecules, to the inner surface of the plasma membrane and its rapid ubiquitylation and degradation in proteasomes. A unique element of the canonical pathway is the critical regulatory role played by the so-called "destruction complex" in determining free β-catenin levels *(5)*. This complex comprises several components, including Apc, Axin, and GSK3, and in the absence of Wnt ligand this complex phosphorylates specific serine residues in the N-terminal region of β-catenin, thereby marking the molecules for ubiquitylation and subsequent degradation *(6)*. In response to activation by Wnt ligands such as Wnt1, 2, 3, 8, and 10, the destruction complex is disassembled, β-catenin molecules accumulate, translocate to the nucleus and bind to members of the Tcf/Lef family of transcription factors to activate new patterns of gene expression *(7)*. Pivotal to this process is the role of the canonical pathway-specific adaptor protein, LRP5/6. Upon docking of Wnt ligand onto Fzd receptors, the Wnt/Fzd complex interacts with the single span transmembrane protein LRP6, causing it to be phosphorylated by CK1γ *(8)*. Phosphorylated LRP6 then recruits the scaffold protein Axin to the plasma membrane, thereby disarming the destruction complex. All of the details of the canonical Wnt signaling pathway, and many more besides, have been established as a result of intense study in both invertebrate and vertebrate organisms, but not particularly in zebrafish. However, the value of zebrafish in this field has been to highlight various stages during development where canonical Wnt signaling was hitherto not known to be important.

Non-canonical Wnt signaling in the zebrafish is currently a topic of great interest, due to the identified roles of this pathway in processes requiring the integration of cell polarity and cell movement, such as the directed migration movements that drive the narrowing and lengthening (convergence and extension) of the embryo during early development. Non-canonical Wnt signaling is independent of β-catenin, and appears to trigger several independent downstream events, although all the details and probable cross-talk between the different pathways have yet to be revealed. Non-canonical Wnt signaling has been shown to affect release of Ca^{2+} from intracellular stores *(9)*, activate the JNK signaling pathway *(10, 11)*, activate the Rho family of small GTPases *(12–14)* and regulate E-cadherin recycling *(15)*. In vertebrate embryos, these signals ultimately affect cytoskeletal architecture and the establishment/modulation of cell polarity and cell behavior, and in some cases these functions have been examined most closely in zebrafish *(16)*.

2. Proteins That Mediate Wnt Signaling in Zebrafish

An attempt to illustrate the currently known complement of genes and encoded proteins that govern canonical and non-canonical Wnt signaling in zebrafish is shown in **Table 17.1**. In mammalian systems there are multiple Wnt ligands and Frizzled receptors, and the specificity of their interactions is currently under intense investigation. In zebrafish, the picture is complicated still further by the presence, for some of the players, of two highly related genes that arose as a result of a whole genome duplication event 320–404 million years ago, at the base of the teleost lineage *(17–20)*. Such highly related genes, for example *fzd7a* and *fzd7b*, are thought to have been retained during evolution as a result of their distinct, though partially overlapping, expression patterns *(21–23)*. Although the Wnt/β-catenin and Wnt/PCP pathways share some components, such as Fzd receptors and the cytoplasmic mediator Dishevelled (Dsh), the signals then go through divergent pathways. It is not clear yet precisely how the specificity of the two pathways is conferred. Although it is tempting to speculate that one Wnt ligand may partner with a specific Fzd, and thereby specify whether the canonical or non-canonical pathway imparts the signal, there is, however, no clear evidence so far to support this concept. Dsh may play a role in this decision, as it is well-known that different domains within the Dsh molecule are required to mediate the two different signaling pathways *(24)*. Also, recent evidence from zebrafish points to a pivotal role for the canonical Wnt signal inhibitor, Dickkopf (Dkk) in determining the downstream consequences of Fzd receptor engagement by various Wnt ligands (see later).

This chapter does not aim to provide a comprehensive account of Wnt signaling during zebrafish development. Rather, the goal is to draw attention to a few areas where Wnt studies in zebrafish have offered new insights, for example during head, heart and liver development. These examples have also been chosen to illustrate the different genetic approaches in fish that these studies have employed, including forward genetic screens, reverse genetics (anti-sense morpholino technology and TILLING), prior to the methods chapters that follow.

3. Canonical Wnt Signaling in Zebrafish Liver Development

Forward genetics in zebrafish provides an exceptionally powerful and unbiased tool to elucidate novel roles played by known genes in vertebrate developmental processes as well as assigning roles

Table 17.1
List of known genes in zebrafish that have been shown to play a role in Wnt signaling

Protein[1]	Gene[2]	Mutant/MO[3]	Wnt Pathway[4]	Role in Zebrafish Development[5]	References
Wnt ligand	wnt1	DF^{n5}, MO	Wnt/β-catenin	Neural patterning	107–110
	wnt2b	MO	Wnt/β-catenin	Pectoral fin / Mesoderm patterning	111, 112
	wnt2bb	prometheus (prt)	Wnt/β-catenin	Liver specification	25
	wnt3l	MO	Wnt/β-catenin?	Neural survival and patterning / Tailbud	110, 113–115
	wnt4a	MO	Wnt/PCP	Neuroepithelium / Convergence of midline organs	14, 48, 70
	wnt5b	pipetail (ppt), MO	Wnt/PCP	Convergence and Extension / Neuroepithelium / Pancreas / Tailbud	14, 39, 42, 46, 58, 70, 74, 116–119
	wnt8a	$Df(wnt8)^{w8}$	Wnt/β-catenin	Mesoderm patterning / Neural patterning / Tailbud	112, 120–125
	wnt8b	MO	Wnt/β-catenin	Neural patterning / Eye, hypothalamus	88, 110, 124, 126–128
	wnt10b	DF^{w5}, MO	Wnt/β-catenin	Neural patterning	109, 110
	wnt11	silberblick (slb)	Wnt/PCP	Convergence and extension / Eye / Convergence of midline organs	14, 15, 39, 47, 62, 71, 83, 88, 118
	wnt11r	TILLED, MO	Wnt/PCP	Convergence of midline organs	14, 129
	wnt16	TILLED			130

Frizzled receptor	fzd2	MO	Wnt/PCP (wnt5b)	Extension Pancreas Ciliogenesis	39, 119, 131
	fzd3	off-limits (olt)		Facial neuron migration	57
	fzd5 (fzd8c)	MO	Wnt/PCP (wnt11) Wnt/β-catenin (wnt8)	Eye Dorsal patterning	88, 123
	fzd7a	MO		Neural morphogenesis neural tube, pectoral fin, neuromasts	22, 132, 133
	fzd7b	TILLED, MO		Neural morphogenesis neural tube, pectoral fin, neuromasts, prechordal plate	22, 23, 132, 133
	fzd8a	MO	Wnt/β-catenin	Neural patterning: Wnt8b Eye: Wnt8b	88, 127
	fzd8b	Misexpression	Wnt/β-catenin (wnt8)	Dorsal patterning Anterior prechordal plate	134, 135
	fzd10	Misexpression	Wnt/β-catenin (wnt8)	Dorsal patterning Tail, neural tube, brain	123, 136
Soluble frizzled-related receptors	sfrp1a	Misexpression	Secreted inhibitor of Wnt/β-catenin	Forebrain Neural tube, gut, lateral line	100, 101, 137
	sfrp3/frzb	Misexpression	Wnt/β-catenin	Neural patterning Tail	100, 101, 120, 123, 138
	sfrp5	Misexpression		Forebrain development Eye	100, 139
	tlc	MO		Forebrain development Anterior neural border cells	126, 138, 140

(continued)

Table 17.1
continued

Protein[1]	Gene[2]	Mutant/MO[3]	Wnt Pathway[4]	Role in Zebrafish Development[5]	References
β-catenin	*ctnnb1*	MO	Wnt/β-catenin	Dorsal patterning Neurectodermal specification	*141, 142*
	ctnnb2	*ichabod (icb)*	Wnt/β-catenin	Dorsal patterning Neurectodermal specification	*141, 143, 144*
Destruction complex	*axin1*	Masterblind mbl	Wnt/β-catenin	Forebrain patterning Eye	*32, 138, 145*
	axin2/conductin/axil	Misexpression	Wnt/β-catenin	Dorsal patterning	*125, 146, 147*
	gsk3a	Misexpression	Wnt/β-catenin	Dorsal patterning	*147, 148*
	gsk3b	Misexpression	Wnt/β-catenin	Retinotectal projection	*147–149*
	apc	TILLED, MO	Wnt/β-catenin	Heart, gut Lens and retina	*36, 37, 150–154*
Putative transcriptional activator	*tcf4*	MO	Wnt/β-catenin	Brain and gut	*31, 155, 156*
	tcf7	MO	Wnt/β-catenin	Neural patterning Pectoral fin, notochord, brain, tailbud	*157, 158*
	tcf3a (tcf7l1a)	*headless (hdl)*, MO	Wnt/β-catenin	Dorsal patterning Neural patterning Eye	*28, 88, 155, 157, 159–161*
	tcf3b (tcf7l1b)	MO	Wnt/β-catenin	Neural patterning	*108, 110, 155, 157, 159, 161*
	lef1	MO, misexpression	Wnt/β-catenin	Ventral and posterior fate patterning Neural crest Hypothalamus	*125, 128, 151, 155, 157, 159, 161–164*

Putative positive regulator	*ccd1*	Misexpression	Wnt/β-catenin	Neural patterning	*165*
BTB protein Wnt/β-catenin antagonist	*KLHL12*	MO	Wnt/β-catenin	Dorsal patterning	*166*
Unknown	Gene not identified	*colgate col*	Wnt/β-catenin (*wnt8b*)	Dorsal patterning	*167*
Kinase targeting Tcf	*Nemo-like kinase (nlk)*	MO	Wnt/β-catenin (*wnt8, wnt 8b*)	Ventral and posterior fate patterning	*161*
Diego-related ankyrin repeat protein	*diversin*	MO	Wnt/β-catenin Wnt/JNK	Dorsal patterning Gastrulation movements Heart	*45, 146*
Secreted Wnt inhibitor	*dickkopf (dkk1)*	MO	Wnt/β-catenin and Wnt/PCP	Eye field morphogenesis	*36, 89, 91, 168*
Cytoplasmic signaling component	*dvl2*	MO	Wnt/β-catenin and Wnt/PCP		*40, 166*
	dvl3	MO	Wnt/β-catenin and Wnt/PCP		*166, 169*
Dishevelled-associated antagonist	*dapper1 (dact1)*	MO	Wnt/β-catenin	Ventral and posterior fate patterning	*40, 170, 171*
	dapper2 (dact2)	MO	Wnt/PCP	Convergence and extension	*40, 170, 171*
Kinase	*rock2a/rok2*	MO	Wnt/PCP	Convergence and extension	*13*
Small GTPase	*rhoab*	MO	Wnt/PCP	Gastrulation movements	*71, 74, 104*
	rhoad	MO	Wnt/PCP	Gastrulation movements	*71, 74, 104*

(continued)

Table 17.1
(continued)

Protein[1]	Gene[2]	Mutant/MO[3]	Wnt Pathway[4]	Role in Zebrafish Development[5]	References
Transmembrane atypical cadherin	fmi1a/celsr1a	MO	Wnt/PCP	Gastrulation movements	43, 57, 71
	fmi1b/celsr1b	MO	Wnt/PCP	Gastrulation movements	43, 57
	fmi2/celsr2	off-road (ord)	Wnt/PCP	Facial neuron migration	57, 71, 172
Four-pass transmembrane protein	trilobite/ strabismus1/ van gogh2	trilobite (tri)	Wnt/PCP	Convergence and extension Facial neuron migration	10, 41, 50, 63, 118, 173–177
Extracellular matrix proteoglycan	glypican/knypek	knypek (kny)	Wnt/PCP	Convergence and extension	44, 174–178
LIM and PET domain-containing protein	pk1/ prickle-like 1	MO	Wnt/PCP	Convergence extension Facial neuron migration	11, 42, 70
	Pk2/prickle-like 2	MO	Wnt/PCP	Minor role in convergence extension	11
Protein phosphatase	widerborst 2 wdb1/ ppp2r5e2	MO	Wnt/PCP	Convergence and extension	179
	widerborst wdb2/ ppp2r5e1	MO	Wnt/PCP	Convergence and extension	179
Scaffold protein	scribble1	landlocked (llk)	Wnt/PCP carriage return (also Apico-basal polarity)	Convergence and extension Facial neuron migration	180

[1]The general function of the protein product. [2]The gene name. Note that suffix -l indicating -like is used to denote a homolog that has not yet been assigned an ortholog. [3]The mutant name or MO to indicate that an antisense morpholino oligonucleotide MO has been used to knockdown expression. [4]The Wnt pathway in which the gene functions, if known. [5]The biological function of the gene in zebrafish development. Note that in some cases additional structures that express the gene, but as yet no function has been assigned, are indicated in parenthesis. The following genes have been identified by sequence, and for some expression patterns have been described, but as yet no biological function has been assigned: *wnt2 (93), wnt4b (94), wnt5a (93), wnt7b (93), wnt7 (95), wnt7a (96), wnt10a (97), fzd1 (98), fzd31 (57), fzd4 (98), fzd9 (99), fz12 (98), sfrp1b (100, 101), sfrp2 (100, 101), wif1 (29), ctnnbl1 (93), lrp5 (93), tcf3 (93), Groncho1 gro1 (102, 103), Groncho2 gro2 (103), daam1 (93), daam11 (93), rhoaa (104), rhoac (104), celsr21 (93), celsr3 (93), Strabismus2/van gogh 1 (105).* Recent synteny analysis has identified seven additional uncharacterized *wnt* homologs, identified as a *wnt11* ortholog, two *wnt7* orthologs, two *wnt9* orthologs and two *wnt6* orthologs *106.*

to novel genes. An excellent example of this, in the context of components of the canonical Wnt pathway, is the demonstration that the failure to form a liver in the zebrafish mutant, *prometheus (prt)*, is due to mutations (identified in three independent alleles) in a gene encoding a novel Wnt ligand, Wnt2bb/Prt *(25)*. *prt* was identified in a forward genetic screen that was designed to exploit a transgenic zebrafish line *Tg(gutGFP)^{s854}*, which expresses GFP specifically in the developing liver, pancreas and intestine from 30hpf *(26)*. The line provided an ideal background to screen for mutants with abnormalities in the size and morphology of the digestive organs, and illustrates the amenity of tissue-specific fluorescent lines for screens focused on the identification of genetic lesions playing a role in the development of particular tissues of interest. In this study, *prt* expression in two discrete bilateral stripes in the lateral plate mesoderm was found to be indispensable for liver induction in the adjacent endodermal cells of the primitive gut tube. The defect was apparent at 22 hours post-fertilization (hpf) due to the absence of a liver anlagen, which in wild-type embryos is marked by *prox1* and *hhex1* expression. That Wnt2bb/Prt activates the canonical pathway was demonstrated by experiments in which the mutant phenotype was recapitulated (phenocopied) in a heat shock-inducible transgenic line, *Tg(dnTcf3-GFP) (25)*. The application of a single heat shock, which causes production of a dominant negative form of the transcription factor, T-cell factor 3 (Tcf3), capable of blocking Wnt/β-catenin signaling, at any stage between 16hpf and 21hpf resulted in a strong decrease in the formation of hepatic tissue which was not observed if the heat shock was applied at 25hpf. This paper provides the first demonstration of a positive role for canonical Wnt signaling in a specific time window during liver specification, aside from its already described roles in liver growth and regeneration (for review, see ref. *27)*.

4. Canonical Wnt Signaling in Zebrafish Head Development

Zebrafish *headless (hdl)* was also identified in an ENU-mutagenesis screen, this time for genes that control early neurogenesis *(28)*. Mutant embryos show a mild phenotype (including slightly smaller eyes) that is not lethal and the animals reach sexual maturity. Progeny of homozygous *hdl* mutants, which are devoid of both maternal and zygotic *hdl* (MZ*hdl* embryos), were discovered to lack eyes, forebrain, and part of the midbrain. When it was shown that *hdl* mutants harbor a splice-site mutation in the gene encoding Tcf3, this advanced previous work in *Xenopus* that had established the importance of the vertebrate head organizer as a source of Wnt signaling antagonists, such as Dkk, in the formation of a reverse rostrocaudal gradient of Wnt signaling activity that induces head formation *(29, 30)*. Various experiments established that in this scenario, the role of Tcf3 is to act as a transcriptional repressor of Wnt target gene expression, and that inhibition of canonical Wnt signaling is a primary requirement for head induction. The authors proposed that the mechanism through which secreted Wnt antagonists such as Dkk-1 work is by binding to Wnt ligands, thereby preventing them from de-repressing Wnt target genes. Interestingly, alternative splicing of zebrafish transcripts encoding the highly related protein, Tcf4, generates isoforms lacking a domain encoded by exon 4 and 5 sequences. This domain is required for the interaction with Groucho-like co-repressors, thereby providing a mechanism for a switch between activating and repressor activities of Tcf proteins *(31)*.

Another zebrafish mutant harboring a mutation in a gene that encodes a component of the canonical Wnt signaling pathway is *masterblind*; in these embryos a fate transformation results in the expansion of the midbrain at the expense of the forebrain and eyes *(32)*. The mutation resides in the GSK3-binding domain of the gene encoding Axin, a key component in the β-catenin destruction complex. Analysis of this mutant has emphasized the indispensable role of canonical Wnt signaling in rostrocaudal patterning and head formation, and introduced the concept that thresholds of Wnt activity may determine different posterior to anterior fates within the neural plate, analogous to the role played by a gradient of Bmp activity in allocating neural fate in the dorsoventral axis.

5. Canonical Wnt Signaling in Zebrafish Heart Valve Formation and Intestinal Cancer

Reverse genetic approaches in zebrafish have been limited so far by the failure to achieve targeted genome manipulation by homologous recombination in pluripotent zebrafish stem cells, although cultured cell lines analogous to embryonic stem (ES) cells in mice have recently been established *(33)*. While this deficiency is likely to be addressed in the future, the current approach to establishing lines of zebrafish carrying inherited mutations in genes of interest is through TILLING (Targeting Induced Local Lesions IN Genomes), which combines chemical mutagenesis with screening for mutations in specific regions of genes of interest in pooled PCR products (using direct sequencing or methods based on heteroduplex detection), resulting in the identification of missense and nonsense mutant alleles *(34, 35)*. Using this methodology, a line of zebrafish carrying a mutation in the *apc* gene (*apc^mcr^*) was isolated *(36)*. The mutation produced a stop codon that mapped to a region homologous to the mutation cluster region (mcr) of the human colon tumor suppressor gene *APC*, encoding a severely truncated protein likely to be functionally inactive. In contrast to mice, zebrafish embryos containing two copies of the mutated *apc* allele complete gastrulation (presumably due to the presence of maternally-deposited *apc* mRNA) but die later in development due to multiple defects, including abnormal cardiac valve formation, revealing a previously undiscovered role for canonical Wnt signaling in heart development *(36)*. Interestingly, when the otherwise phenotypically wild-type heterozygous *apc* carriers were grown to adulthood, they developed intestinal, hepatic, and pancreatic cancers that were apparent from 4 months of age *(37)*. The intestinal lesions closely resembled human adenomatous polyps, including the characteristic features of nuclear pseudostratification, loss of goblet cells and increased nuclear-to-cytoplasmic ratios consistent with dysplastic epithelium, emphasising that cancer pathways are highly conserved between mammals and fish. Unlike in mammals, however, the wild-type *apc* allele was retained, indicating that in zebrafish, *apc* haploinsufficiency alone is sufficient to predispose to intestinal tumors.

6. The Zebrafish Wnt/PCP Pathway

The zebrafish non-canonical Wnt pathway (which we will refer to as Wnt/PCP) shares components with the *Drosophila* planar cell polarity (PCP) pathway, such as the Wnt receptor Frizzled (encoded by *fzd* genes; refs. *38* and *39*), the cytoplasmic mediator Dishevelled (encoded by *dvl2* and *dvl3*; ref. *40*), the four-pass transmembrane protein Strabismus (encoded by *tri*; refs. *10* and *41*), the LIM and PET domain-containing protein Prickle (encoded by *pk1*; ref. *11* and *42*), the seven-pass transmembrane atypical cadherin Flamingo (encoded by the *fmi1a/celsr1a* and *fmi1b/celsr1b* genes; ref. *43*) in addition to an extracellular matrix heparan sulphate proteoglycan, Glypican 4/6 Knypek (encoded by *kny*; ref. *44*). A zebrafish ortholog of the Drosophila *diego* gene, known as *diversin*, also appears to function in the Wnt/PCP pathway *(45)*. The Wnt ligands Wnt4a, Wnt5b, Wnt11, and Wnt11r are implicated as ligands signaling through the Wnt/PCP pathway, although research in zebrafish has largely focused on the two genes for which mutant alleles have been identified, *wnt5b/ppt* (*pipetail*; refs. *39* and *46*) and *wnt11/slb* (*silberblick*; refs. *47* and *48*). These two *wnt* mutants, and the mutant alleles of *tri/stbm* and *kny*, were identified in large-scale genetic screens by their defective gastrulation movements *(49–51)*. The roles of the other Wnt/PCP genes have been determined through analysis of candidate genes using anti-sense morpholino oligonucleotides, the favored reverse genetic technology for zebrafish. Embryos at the one-cell stage are injected with a highly stable morpholino oligonucleotide complementary to either the translation start site of the targeted gene to block protein synthesis, or to a splice junction to block authentic mRNA processing *(52–56)*. Both of these strategies provide effective "knock down" of protein expression in the ensuing "morphant" embryos, but may have different outcomes, depending on whether the targeted gene is maternally expressed or not. More severe phenotypes may result from morpholinos directed to the translation start-site, since these are equally effective at blocking both maternal and zygotic mRNA translation. In contrast, splice-site directed morpholinos are only effective against zygotic transcripts.

Several of the Wnt/PCP proteins also have Wnt/PCP-independent roles, for example the migration of the nVII facial neurons requires Tri/Stbm and Pk1, along with a third Flamingo homolog, Fmi2/Celsr2 (encoded by *ord*; refs. *41*, *42*, and *57*). These Wnt-independent roles are not considered in this chapter.

Genetic interactions between *wnt5b/ppt*, *wnt11/slb* and *kny* place these components together in the Wnt/PCP pathway *(39, 44, 58)*. Indeed, based on the interaction of the Drosophila heparan sulphate proteoglycans Dally and Dally-like with the Wnt homolog Wingless, it is possible that the *kny*-encoded Glypican4/6 may stabilize or support Fzd function at the cell membrane *(44, 59–61)*. Epistasis analysis has placed *tri/stbm* and *pk1* in a parallel pathway, and so these may modulate the activity of Wnt/PCP rather than being positioned within the same linear pathway *(41, 42, 50, 62)*. Wnt11/Slb appears to act predominantly in the morphogenesis of anterior tissues such as formation of the axial structures of the head, while Wnt5b/Ppt, Kny, Pk1, and Tri/Stbm play a more significant role in posterior tissues: for example, Wnt5b/Ppt regulates the morphogenetic movements required for tail formation *(41, 42, 44, 46, 47, 49, 51, 58, 63)*. Indeed, double mutants lacking both *wnt11/slb* and *wnt5b/ppt* show far more dramatic morphogenetic phenotypes than either mutant alone *(39)*, indicating a lack of redundancy between the actions of these two ligands. In contrast, the *fmi1a/celsr1a* and *fmi1b/celsr1b* genes redundantly control cell movements along the entire anterior–posterior axis *(43)*. Consistent with the concept that each Wnt ligand pairs with a specific Fzd receptor, there is evidence that Fzd2 may be the receptor for Wnt5b/Ppt signaling *(39, 64)* and, although their function is as yet unassigned, there are two Fzd7a orthologs with expression overlapping that of Wnt11/Slb *(21, 22)*.

7. Comparison of Zebrafish Wnt/PCP and *Drosophila* PCP

One fascinating difference between the *Drosophila* PCP pathway and vertebrate Wnt/PCP is that there is no known involvement of Wnt ligands in *Drosophila* PCP *(65)*. The core *Drosophila* PCP components (*fz, dsh, fmi, strb/vang, pk and dgo*) show polarized subcellular localization in response to global polarity patterning signals, refined by cell–cell feedback mechanisms, but the identity of the instructive global signals are not yet clear (prime candidates are the atypical cadherins fat, daschous, and four-jointed; refs. *65* and *66*). In vertebrates, it is not clear that orthologs of these cadherins play any role in Wnt/PCP, but instead, a family of Wnt ligands activates the Wnt/PCP pathway *(67)*. Despite the immediately obvious hypothesis that a gradient of Wnt protein or function would set up a global polarity pattern in zebrafish, it appears that the zebrafish Wnt ligands play a permissive role in Wnt/PCP, rather than relaying such directional instructions. This is evidenced by the broad expression patterns of the Wnt/PCP ligands, and the key observations that *wnt11/slb* and *wnt5b/ppt* mutant embryos can be completely rescued by ubiquitous expression of *wnt11/slb* and *wnt5b/ppt* mRNA respectively *(46, 47)*. *slb* mutant embryos can also be rescued by ubiquitous expression of a truncated form of Dsh, DshΔDIX, (able to mediate Wnt/PCP signaling without inhibiting endogenous canonical Wnt/β-catenin signaling) suggesting that downstream components of Wnt/PCP can also act permissively *(47, 68)*.

The *Drosophila* PCP pathway specifies polarity of cells within the plane of an epithelium *(65)*. The core PCP proteins show complex and interdependent patterns of membrane localization. In zebrafish, Wnt/PCP is required for regulation of polarity in dynamic epithelial structures such as the neuroepithelium, and mesenchymal cell movements such as mesodermal movements during gastrulation. In non-epithelial tissues, it is not yet clear if vertebrate Wnt/PCP is regulating cell structure via the same mechanisms as in *Drosophila* epithelia *(16, 69)*. Indeed, it has been difficult to analyze the subcellular localization of endogenous Wnt/PCP pathway components in dynamically moving tissues, but a few studies have addressed this in detail. Thus, the localization of Pk1 to the anterior (lateral) side of developing neuroepithelial cells has been demonstrated *(70)*, and in gastrulating mesoderm the recruitment of Fzd7, Dsh, and Wnt11/Slb to the cell membrane in response to Wnt11/Slb signaling has been shown to proceed in a *fmi* dependent way *(71)*. Rok2 has also been shown to undergo Wnt/PCP-dependent changes in cellular localization *(13)*. The particular amenability of zebrafish to live fluorescent analysis will surely result in many more such insights.

Zebrafish Wnt/PCP pathway does share the mix of cell-autonomous and cell non-autonomous functions that have been observed for the core *Drosophila* PCP genes *(65, 72)*. Tri/Stbm, Rok2, Pk1, Wnt5b/Ppt, and Wnt11/Slb all show cell non-autonomous functions indicating that, in keeping with the Drosophila PCP pathway, cell–cell interactions play a role in maintaining or modulating these functions, possibly through subcellular protein localization *(11, 13, 39, 41, 47)*.

8. Wnt/PCP Genes Regulate Vertebrate CE Movements

Zebrafish Wnt/PCP signaling regulates the morphogenetic movements underlying convergence and extension (CE), the mechanism by which the embryo narrows in the medio-lateral axis, while lengthening the anterior–posterior axis *(69)*. Disruption of the functions of Wnt11/Slb, Wnt5a/Ppt, Kny, Tri/Stbm, Pk1, or Rho-associated kinase 2 (Rok2) results in embryos with a shorter anterior–posterior axis, and broader notochord and somites *(11, 13, 39, 41, 42, 44, 47)*. The individual mesodermal cells show a loss of the cell elongation and mediolateral orientation that is usually seen in these cells during zebrafish gastrulation *(73)*. Rescue experiments suggest that *wnt11/slb* mediates its morphogenetic role through both Rok2-dependent and Rok2-independent mechanisms *(13)*. There is also evidence for the small GTPase RhoA functioning downstream of Wnt5b/Ppt and Wnt11/Slb signaling in CE, but the cell biological outcomes have yet to be investigated *(14, 74)*.

The cellular mechanisms of CE in *Xenopus laevis* have been analyzed in detail, and the role of the Wnt/PCP pathway is well established. Interestingly, however, there are several key differences between mesendodermal CE in *Xenopus* and zebrafish, suggesting that the same pathway may regulate somewhat different cell biological mechanisms for the same developmental result. During *Xenopus* CE, the mesodermal cells move as a sheet of adherent mesenchymal cells, using bipolar or unipolar protrusive activity to drive medio-lateral cell intercalation *(75)*. In teleost fish such as the zebrafish and *Fundulus heteroclitus*, ectodermal cells form a sheet and may use some of the same mechanisms as *Xenopus* mesodermal CE but mesodermal cells migrate as individuals or in small clusters *(73, 76, 77)*. Moreover, the cellular behaviors of mesodermal cells in the dorsal, lateral, and ventral parts of the zebrafish embryo are quite different, and these differences are thought to be primarily mediated by the dorsoventral BMP activity gradient *(78)*. Thus, only the mesodermal cells on the very dorsal side of the zebrafish embryo show the high level of mediolateral alignment and elongation that is typical of *Xenopus* mesodermal cell intercalation *(78–80)*. In late gastrulation, there is mediolateral elongation of ectodermal and mesodermal cells undergoing CE movements and this is regulated by components of the Wnt/PCP pathway *(69, 73)*.

9. Direct Regulation of Cell Behavior by Non-canonical Wnt Ligands

In addition to the conserved functions of vertebrate Wnt/PCP in regulating medio-lateral cell orientation, recent research has shown that zebrafish non-canonical Wnt ligands directly mediate cell biological outcomes required for morphogenesis and cell migration. Detailed analysis of two roles of Wnt11/Slb indicate that the downstream effects (or even direct effects) of Wnt/PCP activation impact not only on cell polarization, but also protrusion activity and cell adhesion. The prechordal plate is a cohesive group of axial mesendodermal progenitor cells that precedes the axial mesoderm (notochord progenitors) in a migration movement away from the dorsal margin, and later plays a central role in forebrain patterning *(81)*. These cells do not exhibit the CE movements of other mesodermal cells, but rather undergo directed cell migration on a substrate of the overlying epiblast *(77,82)*.

E-cadherin-mediated cell cohesion is required for co-ordinated prechordal plate cell migration at the onset of gastrulation *(15, 82)*. Wnt11/Slb regulates this migration movement independently of the core PCP genes and in a cell non-autonomous manner *(47, 62)*. To do this, it modulates E-cadherin recycling and the subcellular localization of E-cadherin by regulating the GTPase Rab5c, which is a key component in the regulation of intracellular trafficking such as receptor-mediated endocytosis *(15, 82)*. Wnt11/Slb appears to control the velocity and persistence of the cell movements by directing the orientation of cellular processes. Rather than affecting the number, size, and shape of protrusions, or the overall direction of movement, it enhances the stability of those protrusions that are oriented in the direction of migration *(62)*. The role of Wnt signaling in this process is therefore to facilitate and stabilize the cellular processes, allowing directional cell movement without providing the directional signal itself *(15, 83)*. Indeed, this research also provides a link between the dorsoventral BMP activity gradient and the cell behaviors that correlate with this gradient, as the BMP activity has been shown to affect cadherin-dependent cell adhesion *(84)*.

Wnt11/Slb also regulates cell contact persistence in gastrulating cells *(71)*. *In vivo* assays, in which mRNAs encoding fluorescently labeled Wnt/PCP proteins are injected into 1-cell zebrafish embryos, and the protein localization examined in the animal pole region (where endogenous Wnt11/Slb and Fzd7a are not present), or in the gastrulating mesoderm, have suggested that Wnt11/Slb signaling can induce Fzd7a, Dsh, and Wnt11/Slb to co-accumulate at the cell membrane in a *fmi1a/celrs1a*, *fmi1b/celsr1b*, *fmi2/celsr2/ord* dependent manner *(71)*. These sites of Wnt11/Slb-induced Fzd7a accumulations show stronger persistence of cell–cell contact, suggesting these are more strongly adhesive subdomains. The atypical cadherin Fmi2/Celsr2/Ord accumulates to these Fzd7a foci independently of Wnt11/Slb signaling, and contributes to this cell–cell contact persistence, although the physiological relevance of this interaction is unclear as *fmi2/celsr2/ord* null mutants are viable and show no CE defects *(57, 71)*. It is not clear if the Wnt11/Slb-mediated induction of Fzd7a accumulation and regulation of E-cadherin recycling are independent, as there is precedent from Drosophila studies for the two processes to be linked *(85)*. Studies such as these are starting to identify the key cellular behaviors that are modulated by the Wnt/PCP or non-canonical Wnt signaling pathways, and link them together with other events that mediate the global patterning of the embryo.

10. Wnt/PCP Regulates the Re-establishment of Epithelial Structure

Wnt/PCP also plays a key role in neural tube morphogenesis. This is an epithelial tissue, but again it is not clear if the cell biological mechanisms are similar to those in *Drosophila* PCP. During zebrafish neurulation, dividing cells round up, losing their polarized appearance, and move their cell body to the midline apical surface (retaining an attachment to the basement membrane) to divide. One daughter cell re-establishes its polarized shape and position in the epithelium, while the other intercalates into the contralateral epithelium, possibly regulating the symmetry of cell number in the neural tube *(73, 86, 87)*. The ability to insert into this epithelium cell-autonomously requires the *tri/stbm* gene of the Wnt/PCP pathway *(70)*. Embryos lacking all maternal and zygotic Wnt11/Slb, Wnt5b/Ppt, and Wnt4 protein (*MZspt*; *MZppt* double mutants injected with *wnt4*-directed MO) showed similar neurulation defects *(70)*. This study demonstrates that a Pk1-GFP fusion protein exhibits PCP-dependent localization specifically to the anterior lateral membrane of the neural keel cells, suggesting that planar polarity may mirror anterior–posterior patterning, and thus integrate this with the medio-lateral apical-basal polarity of the neural epithelium.

11. Integration of Wnt/β-catenin and Wnt/PCP Pathways

There is also evidence that both branches of Wnt signaling may interact to control the morphogenesis of some tissues. For example, in early forebrain development, coordination of eye field specification and morphogenesis is a result of the combined regulation by Wnt11/Slb, Fzd5, and Wnt/β-catenin signaling *(88)*. Indeed, Wnt11 carries out two roles in eye field development, modulating cell coherence to direct eye field morphogenesis, and antagonizing Wnt/β-catenin signals to consolidate the eye field territory *(88)*.

Some very interesting recent work has proposed that, during zebrafish gastrulation, a secreted Wnt antagonist, Dickkopf1 (Dkk1), may modulate both canonical and noncanonical Wnt pathways, to effect a balance between Wnt control of gene expression and Wnt/PCP regulated cell behaviors *(89;* **Fig. 17.1**). Dkk1 was first identified as a negative regulator of the Wnt/β-catenin pathway through its role as a "head inducer" during gastrulation in *Xenopus* and mice *(29, 90)*. In zebrafish, Dkk1 is induced by maternally-derived Wnt signaling, and is secreted from the prechordal plate to pattern the anterior neural plate *(91)*, which it achieves by binding to the Frizzled co-receptors LRP5/6 *(92)*. As a result of this interaction, Dkk causes crosslinking of LRP5/6 to a further class of transmembrane proteins, the Kremens, which promotes the internalization of the complex and inactivation of LRP5/6 (reviewed in ref. *4)*. New insight into the role of Dkk1 was achieved using two non-overlapping antisense morpholino oligonucleotides targeted to the translation start site of *dkk1* mRNA, which were injected into one-cell stage zebrafish embryos to block Dkk1 protein synthesis from both maternally deposited and zygotic mRNAs *(89)*. In addition to the expected dramatic reduction in the size of the eye and telencephalon, it was also noted in the morphant embryos that the rostral most limit of the axial mesendoderm was shifted further rostral in Dkk1 morphants, suggesting that the anterior movement of this tissue during gastrulation had been accelerated. This effect was independent of β-catenin activity, but dependent on Kny, a component of the Wnt/PCP pathway. Indeed, ectopic expression of *dkk1* mRNA in *Xenopus* embryos was found to activate Wnt/PCP signaling, resulting in elevated levels of phosphorylated JNK activity. These and other analyses led the authors to propose that Dkk1 acts at the level of Wnt receptors, concomitantly repressing Wnt/β-catenin signaling and activating Wnt/PCP through interactions with LRP5/6 and Kny. Such a pivotal role for Dkk1 could be achieved by dissociation of the LRP/Fzd complex in response to Dkk1, thereby liberating Fzd to interact with Wnt ligands defined as PCP activators, such as Wnts 4 and 5. Modulation of both

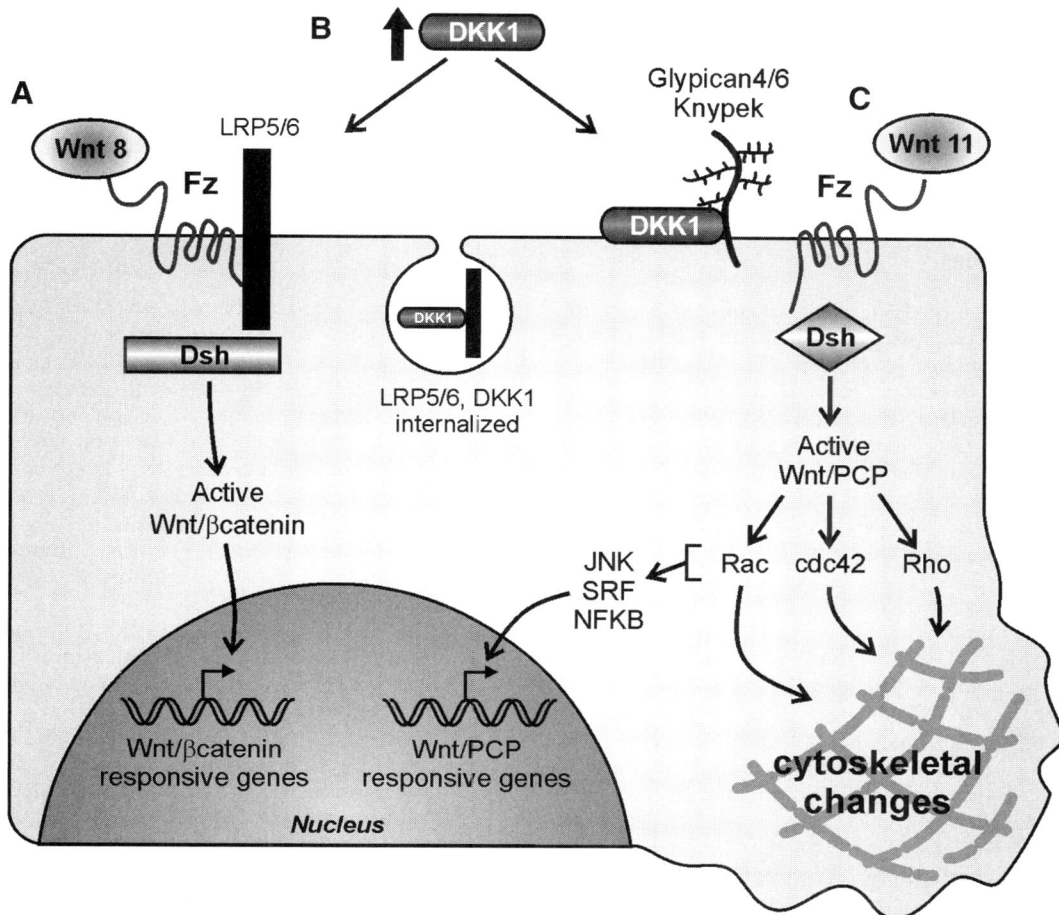

Fig. 17.1. Dickkopf plays a pivotal role in regulating the activities of the Wnt/β-catenin and Wnt/PCP pathways. Dickkopf1 (Dkk1) levels determine the availability of Frizzled (Fzd) receptors for interaction with ligands that activate the Wnt/β-catenin and the Wnt/PCP pathways. **(A)** In the presence of low levels of Dkk, Wnt ligands that activate Wnt/β-catenin signaling (such as Wnt 8; shaded oval) interact with Fzd receptors which in turn engage with LRP5/6 (solid black bar). Dishevelled (Dsh; shaded rectangle) is recruited to mediate the Wnt/β-catenin signal. **(B)** Upon elevation of the Dkk concentration, Dkk interacts with LRP5/6 and the complex is internalized and inactivated by endocytosis. **(C)** Liberated Fzd receptors are available to interact with Wnt ligands of the Wnt/PCP pathway (such as Wnt 11; shaded oval); concomitantly, Dkk binds to Glypican 4/6 Knypek (Kny). The switch in the class of Wnt ligand engagement causes Dsh to adopt an alternative conformation (shaded diamond) that triggers Wnt/PCP signaling.

pathways allows Dkk1 to integrate anterior-posterior patterning and gastrulation movements to regulate formation and morphogenesis of the eye field.

In summary, a combination of forward genetic and reverse genetic approaches in zebrafish has revealed novel roles for canonical Wnt and Wnt/PCP signaling during vertebrate development. The overall amenity, developmental manipulability, and genetic tractability of zebrafish ensure that future studies in this model organism are likely to continue to yield insights in this exciting and burgeoning field.

Acknowledgments

Work on zebrafish intestinal development in the Heath laboratory is supported by the National Health and Medical Research Council (NHMRC), Australia.

References

1. Haubrick, L. L., Assmann, S. M. (2006) Brassinosteroids and plant function: some clues, more puzzles. *Plant Cell Environ* 29, 446–457.

2. Plickert, G., Jacoby, V., Frank, U., et al. (2006) Wnt signaling in hydroid development: formation of the primary body axis in embryogenesis and its subsequent patterning. *Dev Biol* 298, 368–378.

3. Sansom, O. J., Reed, K. R., Hayes, A. J., et al. (2004) Loss of Apc in vivo immediately perturbs Wnt signaling, differentiation, and migration. *Genes Dev* 18, 1385–1390.

4. Clevers, H. (2006) Wnt/beta-catenin signaling in development and disease. *Cell* 127, 469–480.

5. van Noort, M., Clevers, H. (2002) TCF transcription factors, mediators of Wnt-signaling in development and cancer. *Dev Biol* 244, 1–8.

6. van Noort, M., Meeldijk, J., van der Zee, R., et al. (2002) Wnt signaling controls the phosphorylation status of beta-catenin. *J Biol Chem* 277, 17901–17905.

7. Barker, N., Morin, P. J., Clevers, H. (2000) The Yin-Yang of TCF/beta-catenin signaling. *Adv Cancer Res* 77, 1–24.

8. Davidson, G., Wu, W., Shen, J., et al. (2005) Casein kinase 1 gamma couples Wnt receptor activation to cytoplasmic signal transduction. *Nature* 438, 867–872.

9. Westfall, T. A., Brimeyer, R., Twedt, J., et al. (2003) Wnt-5/pipetail functions in vertebrate axis formation as a negative regulator of Wnt/beta-catenin activity. *J Cell Biol* 162, 889–898.

10. Park, M., Moon, R. T. (2002) The planar cell-polarity gene stbm regulates cell behaviour and cell fate in vertebrate embryos. *Nat Cell Biol* 4, 20–25.

11. Veeman, M. T., Slusarski, D. C., Kaykas, A., et al. (2003) Zebrafish prickle, a modulator of noncanonical Wnt/Fz signaling, regulates gastrulation movements. *Curr Biol* 13, 680–685.

12. Jopling, C., den Hertog, J. (2005) Fyn/Yes and non-canonical Wnt signalling converge on RhoA in vertebrate gastrulation cell movements. *EMBO Rep* 6, 426–431.

13. Marlow, F., Topczewski, J., Sepich, D., et al. (2002) Zebrafish Rho kinase 2 acts downstream of Wnt11 to mediate cell polarity and effective convergence and extension movements. *Curr Biol* 12, 876–884.

14. Matsui, T., Raya, A., Kawakami, Y., et al. (2005) Noncanonical Wnt signaling regulates midline convergence of organ primordia during zebrafish development. *Genes Dev* 19, 164–175.

15. Ulrich, F., Krieg, M., Schotz, E. M., et al. (2005) Wnt11 functions in gastrulation by controlling cell cohesion through Rab5c and E-cadherin. *Dev Cell* 9, 555–564.

16. Veeman, M. T., Axelrod, J. D., Moon, R. T. (2003) A second canon. Functions and mechanisms of beta-catenin-independent Wnt signaling. *Dev Cell* 5, 367–377.

17. Christoffels, A., Koh, E. G., Chia, J. M., et al. (2004) Fugu genome analysis provides evidence for a whole-genome duplication early during the evolution of ray-finned fishes. *Mol Biol Evol* 21, 1146–1151.

18. Vandepoele, K., De Vos, W., Taylor, J. S., et al. (2004) Major events in the genome evolution of vertebrates: paranome age and size differ considerably between ray-finned fishes and land vertebrates. *Proc Natl Acad Sci USA* 101, 1638–1643.

19. Crow, K. D., Stadler, P. F., Lynch, V. J., et al. (2006) The "fish-specific" Hox cluster duplication is coincident with the origin of teleosts. *Mol Biol Evol* 23, 121–136.

20. Hoegg, S., Brinkmann, H., Taylor, J. S., et al. (2004) Phylogenetic timing of the fish-specific genome duplication correlates with the diversification of teleost fish. *J Mol Evol* 59, 190–203.

21. El-Messaoudi, S., Renucci, A. (2001) Expression pattern of the frizzled 7 gene during zebrafish embryonic development. *Mech Dev* 102, 231–234.

22. Sumanas, S., Kim, H. J., Hermanson, S. B., et al. (2002) Lateral line, nervous system,

and maternal expression of Frizzled 7a during zebrafish embryogenesis. *Mech Dev* 115, 107–111.

23. Ungar, A. R., Calvey, C. R. (2002) Zebrafish frizzled7b is expressed in prechordal mesoderm, brain and paraxial mesoderm. *Mech Dev* 118, 165–169.

24. Wharton, K. A., Jr. (2003) Runnin' with the Dvl: proteins that associate with Dsh/Dvl and their significance to Wnt signal transduction. *Dev Biol* 253, 1–17.

25. Ober, E. A., Verkade, H., Field, H. A., et al. (2006) Mesodermal Wnt2b signalling positively regulates liver specification. *Nature* 442, 688–691.

26. Field, H. A., Ober, E. A., Roeser, T., et al. (2003) Formation of the digestive system in zebrafish. I. Liver morphogenesis. *Dev Biol* 253, 279–290.

27. Thompson, M. D., Monga, S. P. (2007) WNT/beta-catenin signaling in liver health and disease. *Hepatology* 45, 1298–1305.

28. Kim, C. H., Oda, T., Itoh, M., et al. (2000) Repressor activity of Headless/Tcf3 is essential for vertebrate head formation. *Nature* 407, 913–916.

29. Hsieh, J. C., Kodjabachian, L., Rebbert, M. L., et al. (1999) A new secreted protein that binds to Wnt proteins and inhibits their activities. *Nature* 398, 431–436.

30. Kiecker, C., Niehrs, C. (2001) A morphogen gradient of Wnt/beta-catenin signalling regulates anteroposterior neural patterning in Xenopus. *Development* 128, 4189–4201.

31. Young, R. M., Reyes, A. E., Allende, M. L. (2002) Expression and splice variant analysis of the zebrafish tcf4 transcription factor. *Mech Dev* 117, 269–273.

32. Heisenberg, C. P., Houart, C., Take-Uchi, M., et al. (2001) A mutation in the Gsk3-binding domain of zebrafish Masterblind/Axin1 leads to a fate transformation of telencephalon and eyes to diencephalon. *Genes Dev* 15, 1427–1434.

33. Fan, L., Collodi, P. (2006) Zebrafish embryonic stem cells. *Methods Enzymol* 418, 64–77.

34. McCallum, C. M., Comai, L., Greene, E. A., et al. (2000) Targeted screening for induced mutations. *Nat Biotechnol* 18, 455–457.

35. Wienholds, E., van Eeden, F., Kosters, M., et al. (2003) Efficient target-selected mutagenesis in zebrafish. *Genome Res* 13, 2700–2707.

36. Hurlstone, A. F., Haramis, A. P., Wienholds, E., et al. (2003) The Wnt/beta-catenin pathway regulates cardiac valve formation. *Nature* 425, 633–637.

37. Haramis, A. P., Hurlstone, A., van der Velden, Y., et al. (2006) Adenomatous polyposis coli-deficient zebrafish are susceptible to digestive tract neoplasia. *EMBO Rep* 7, 444–449.

38. Sumanas, S., Ekker, S. C. (2001) Xenopus frizzled-7 morphant displays defects in dorsoventral patterning and convergent extension movements during gastrulation. *Genesis* 30, 119–122.

39. Kilian, B., Mansukoski, H., Barbosa, F. C., et al. (2003) The role of Ppt/Wnt5 in regulating cell shape and movement during zebrafish gastrulation. *Mech Dev* 120, 467–476.

40. Waxman, J. S., Hocking, A. M., Stoick, C. L., et al. (2004) Zebrafish Dapper1 and Dapper2 play distinct roles in Wnt-mediated developmental processes. *Development* 131, 5909–5921.

41. Jessen, J. R., Topczewski, J., Bingham, S., et al. (2002) Zebrafish trilobite identifies new roles for Strabismus in gastrulation and neuronal movements. *Nat Cell Biol* 4, 610–615.

42. Carreira-Barbosa, F., Concha, M. L., Takeuchi, M., et al. (2003) Prickle 1 regulates cell movements during gastrulation and neuronal migration in zebrafish. *Development* 130, 4037–4046.

43. Formstone, C. J., Mason, I. (2005) Combinatorial activity of Flamingo proteins directs convergence and extension within the early zebrafish embryo via the planar cell polarity pathway. *Dev Biol* 282, 320–335.

44. Topczewski, J., Sepich, D. S., Myers, D. C., et al. (2001) The zebrafish glypican knypek controls cell polarity during gastrulation movements of convergent extension. *Dev Cell* 1, 251–264.

45. Moeller, H., Jenny, A., Schaeffer, H. J., et al. (2006) Diversin regulates heart formation and gastrulation movements in development. *Proc Natl Acad Sci USA* 103, 15900–15905.

46. Rauch, G. J., Hammerschmidt, M., Blader, P., et al. (1997) Wnt5 is required for tail formation in the zebrafish embryo. *Cold Spring Harb Symp Quant Biol* 62, 227–234.

47. Heisenberg, C. P., Tada, M., Rauch, G. J., et al. (2000) Silberblick/Wnt11 mediates convergent extension movements during zebrafish gastrulation. *Nature* 405, 76–81.

48. Ungar, A. R., Kelly, G. M., Moon, R. T. (1995) Wnt4 affects morphogenesis when misexpressed in the zebrafish embryo. *Mech Dev* 52, 153–164.

49. Hammerschmidt, M., Pelegri, F., Mullins, M. C., et al. (1996) Mutations affecting morphogenesis during gastrulation and tail formation in the zebrafish, Danio rerio. *Development* 123, 143–151.

50. Heisenberg, C. P., Nusslein-Volhard, C. (1997) The function of silberblick in the positioning of the eye anlage in the zebrafish embryo. *Dev Biol* 184, 85–94.

51. Solnica-Krezel, L., Stemple, D. L., Mountcastle-Shah, E., et al. (1996) Mutations affecting cell fates and cellular rearrangements during gastrulation in zebrafish. *Development* 123, 67–80.

52. Ekker, S. C. (2000) Morphants: a new systematic vertebrate functional genomics approach. *Yeast* 17, 302–306.

53. Nasevicius, A., Ekker, S. C. (2000) Effective targeted gene 'knockdown' in zebrafish. *Nat Genet* 26, 216–220.

54. Corey, D. R., Abrams, J. M. (2001) Morpholino antisense oligonucleotides: tools for investigating vertebrate development. *Genome Biol* 2.

55. Draper, B. W., Morcos, P. A., Kimmel, C. B. (2001) Inhibition of zebrafish fgf8 pre-mRNA splicing with morpholino oligos: a quantifiable method for gene knockdown. *Genesis* 30, 154–156.

56. Sumanas, S., Larson, J. D. (2002) Morpholino phosphorodiamidate oligonucleotides in zebrafish: a recipe for functional genomics? *Brief Funct Genomic Proteomic* 1, 239–256.

57. Wada, H., Tanaka, H., Nakayama, S., et al. (2006) Frizzled3a and Celsr2 function in the neuroepithelium to regulate migration of facial motor neurons in the developing zebrafish hindbrain. *Development.* 133, 4749–4759.

58. Marlow, F., Gonzalez, E. M., Yin, C., et al. (2004) Notail co-operates with non-canonical Wnt signaling to regulate posterior body morphogenesis in zebrafish. *Development* 131, 203–216.

59. Baeg, G. H., Lin, X., Khare, N., et al. (2001) Heparan sulfate proteoglycans are critical for the organization of the extracellular distribution of Wingless. *Development* 128, 87–94.

60. Tsuda, M., Kamimura, K., Nakato, H., et al. (1999) The cell-surface proteoglycan Dally regulates Wingless signalling in Drosophila. *Nature* 400, 276–280.

61. Lin, X., Perrimon, N. (1999) Dally cooperates with Drosophila Frizzled 2 to transduce Wingless signalling. *Nature* 400, 281–284.

62. Ulrich, F., Concha, M. L., Heid, P. J., et al. (2003) Slb/Wnt11 controls hypoblast cell migration and morphogenesis at the onset of zebrafish gastrulation. *Development* 130, 5375–5384.

63. Sepich, D. S., Myers, D. C., Short, R., et al. (2000) Role of the zebrafish trilobite locus in gastrulation movements of convergence and extension. *Genesis* 27, 159–173.

64. Sumanas, S., Kim, H. J., Hermanson, S., et al. (2001) Zebrafish frizzled-2 morphant displays defects in body axis elongation. *Genesis* 30, 114–118.

65. Klein, T. J., Mlodzik, M. (2005) Planar cell polarization: an emerging model points in the right direction. *Annu Rev Cell Dev Biol* 21, 155–176.

66. Strutt, H., Price, M. A., Strutt, D. (2006) Planar polarity is positively regulated by casein kinase Iepsilon in Drosophila. *Curr Biol* 16, 1329–1336.

67. Fanto, M., McNeill, H. (2004) Planar polarity from flies to vertebrates. *J Cell Sci* 117, 527–533.

68. Sheldahl, L. C., Slusarski, D. C., Pandur, P., et al. (2003) Dishevelled activates Ca2+ flux, PKC, and CamKII in vertebrate embryos. *J Cell Biol* 161, 769–777.

69. Tada, M., Concha, M. L., Heisenberg, C. P. (2002) Non-canonical Wnt signalling and regulation of gastrulation movements. *Semin Cell Dev Biol* 13, 251–260.

70. Ciruna, B., Jenny, A., Lee, D., et al. (2006) Planar cell polarity signalling couples cell division and morphogenesis during neurulation. *Nature* 439, 220–224.

71. Witzel, S., Zimyanin, V., Carreira-Barbosa, F., et al. (2006) Wnt11 controls cell contact persistence by local accumulation of Frizzled 7 at the plasma membrane. *J Cell Biol* 175, 791–802.

72. Strutt, H., Strutt, D. (2005) Long-range coordination of planar polarity in Drosophila. *Bioessays* 27, 1218–1227.

73. Concha, M. L., Adams, R. J. (1998) Oriented cell divisions and cellular morphogenesis in the zebrafish gastrula and neurula: a time-lapse analysis. *Development* 125, 983–994.

74. Zhu, S., Liu, L., Korzh, V., et al. (2006) RhoA acts downstream of Wnt5 and Wnt11 to regulate convergence and extension movements by involving effectors Rho kinase and Diaphanous: use of zebrafish as an in vivo model for GTPase signaling. *Cell Signal* 18, 359–372.

75. Keller, R., Shih, J., Domingo, C. (1992) The patterning and functioning of protru-

sive activity during convergence and extension of the Xenopus organiser. *Dev Suppl*, 81–91.

76. Trinkaus, J. P., Trinkaus, M., Fink, R. D. (1992) On the convergent cell movements of gastrulation in Fundulus. *J Exp Zool* 261, 40–61.

77. Warga, R. M., Kimmel, C. B. (1990) Cell movements during epiboly and gastrulation in zebrafish. *Development* 108, 569–580.

78. Myers, D. C., Sepich, D. S., Solnica-Krezel, L. (2002) Bmp activity gradient regulates convergent extension during zebrafish gastrulation. *Dev Biol* 243, 81–98.

79. Wallingford, J. B., Rowning, B. A., Vogeli, K. M., et al. (2000) Dishevelled controls cell polarity during Xenopus gastrulation. *Nature* 405, 81–85.

80. Keller, R. E., Danilchik, M., Gimlich, R., et al. (1985) The function and mechanism of convergent extension during gastrulation of Xenopus laevis. *J Embryol Exp Morphol* 89 Suppl, 185–209.

81. Kiecker, C., Niehrs, C. (2001) The role of prechordal mesendoderm in neural patterning. *Curr Opin Neurobiol* 11, 27–33.

82. Montero, J. A., Carvalho, L., Wilsch-Brauninger, M., et al. (2005) Shield formation at the onset of zebrafish gastrulation. *Development*. 132, 1187–1198.

83. Puech, P. H., Taubenberger, A., Ulrich, F., et al. (2005) Measuring cell adhesion forces of primary gastrulating cells from zebrafish using atomic force microscopy. *J Cell Sci* 118, 4199–4206.

84. von der Hardt, S., Bakkers, J., Inbal, A., et al. (2007) The Bmp gradient of the zebrafish gastrula guides migrating lateral cells by regulating cell-cell adhesion. *Curr Biol* 17, 475–487.

85. Classen, A. K., Anderson, K. I., Marois, E., et al. (2005) Hexagonal packing of Drosophila wing epithelial cells by the planar cell polarity pathway. *Dev Cell* 9, 805–817.

86. Kimmel, C. B., Warga, R. M., Kane, D. A. (1994) Cell cycles and clonal strings during formation of the zebrafish central nervous system. *Development* 120, 265–276.

87. Geldmacher-Voss, B., Reugels, A. M., Pauls, S., et al. (2003) A 90-degree rotation of the mitotic spindle changes the orientation of mitoses of zebrafish neuroepithelial cells. *Development* 130, 3767–3780.

88. Cavodeassi, F., Carreira-Barbosa, F., Young, R. M., et al. (2005) Early stages of zebrafish eye formation require the coordinated activity of Wnt11, Fz5, and the Wnt/beta-catenin pathway. *Neuron* 47, 43–56.

89. Caneparo, L., Huang, Y. L., Staudt, N., et al. (2007) Dickkopf-1 regulates gastrulation movements by coordinated modulation of Wnt/beta catenin and Wnt/PCP activities, through interaction with the Dally-like homolog Knypek. *Genes Dev* 21, 465–480.

90. Mukhopadhyay, M., Shtrom, S., Rodriguez-Esteban, C., et al. (2001) Dickkopf1 is required for embryonic head induction and limb morphogenesis in the mouse. *Dev Cell* 1, 423–434.

91. Shinya, M., Eschbach, C., Clark, M., et al. (2000) Zebrafish Dkk1, induced by the pre-MBT Wnt signaling, is secreted from the prechordal plate and patterns the anterior neural plate. *Mech Dev* 98, 3–17.

92. Mao, B., Wu, W., Li, Y., et al. (2001) LDL-receptor-related protein 6 is a receptor for Dickkopf proteins. *Nature* 411, 321–325.

93. Sprague, J., Bayraktaroglu, L., Clements, D., et al. (2006) The Zebrafish Infromation network (ZFIN): the zebrafish model organism database. *Nucleic Acids Res* 34, D581–D585.

94. Liu, A., Majumdar, A., Schauerte, H. E., et al. (2000) Zebrafish wnt4b expression in the floor plate is altered in sonic hedgehog and gli-2 mutants. *Mech Dev* 91, 409–413.

95. Krauss, S., Korzh, V., Fjose, A., et al. (1992) Expression of four zebrafish wnt-related genes during embryogenesis. *Development* 116, 249–259.

96. Norton, W. H., Mangoli, M., Lele, Z., et al. (2005) Monorail/Foxa2 regulates floorplate differentiation and specification of oligodendrocytes, serotonergic raphe neurones and cranial motoneurones. *Development* 132, 645–658.

97. Kelly, G. M., Lai, C. J., Moon, R. T. (1993) Expression of wnt10a in the central nervous system of developing zebrafish. *Dev Biol* 158, 113–121.

98. Wang, Y., Macke, J. P., Abella, B. S., et al. (1996) A large family of putative transmembrane receptors homologous to the product of the Drosophila tissue polarity gene frizzled. *J Biol Chem* 271, 4468–4476.

99. Van Raay, T. J., Wang, Y. K., Stark, M. R., et al. (2001) frizzled 9 is expressed in neural precursor cells in the developing neural tube. *Dev Genes Evol* 211, 453–457.

100. Peng, G., Westerfield, M. (2006) Lhx5 promotes forebrain development and activates transcription of secreted Wnt antagonists. *Development* 133, 3191–3200.

101. Tendeng, C., Houart, C. (2006) Cloning and embryonic expression of five distinct

sfrp genes in the zebrafish Danio rerio. *Gene Expr Patterns* 6, 761–771.

102. Kobayashi, M., Nishikawa, K., Suzuki, T., et al. (2001) The homeobox protein Six3 interacts with the Groucho corepressor and acts as a transcriptional repressor in eye and forebrain formation. *Dev Biol* 232, 315–326.

103. Wülbeck, C., Campos-Ortega, J. A. (1997) Two zebrafish homologues of the Drosophila neurogenic gene groucho and their pattern of transcription during early embryogenesis. *Dev Genes Evol* 207, 156–166.

104. Salas-Vidal, E., Meijer, A. H., Cheng, X., et al. (2005) Genomic annotation and expression analysis of the zebrafish Rho small GTPase family during development and bacterial infection. *Genomics* 86, 25–37.

105. Jessen, J. R., Solnica-Krezel, L. (2004) Identification and developmental expression pattern of van gogh-like 1, a second zebrafish strabismus homologue. *Gene Expr Patterns* 4, 339–344.

106. Molven, A., Njolstad, P. R., Fjose, A. (1991) Genomic structure and restricted neural expression of the zebrafish wnt-1 (int-1) gene. *Embo J* 10, 799–807.

107. Amoyel, M., Cheng, Y. C., Jiang, Y. J., et al. (2005) Wnt1 regulates neurogenesis and mediates lateral inhibition of boundary cell specification in the zebrafish hindbrain. *Development* 132, 775–785.

108. Lekven, A. C., Buckles, G. R., Kostakis, N., et al. (2003) Wnt1 and wnt10b function redundantly at the zebrafish midbrain-hindbrain boundary. *Dev Biol* 254, 172–187.

109. Riley, B. B., Chiang, M. Y., Storch, E. M., et al. (2004) Rhombomere boundaries are Wnt signaling centers that regulate metameric patterning in the zebrafish hindbrain. *Dev Dyn* 231, 278–291.

110. Ng, J. K., Kawakami, Y., Buscher, D., et al. (2002) The limb identity gene Tbx5 promotes limb initiation by interacting with Wnt2b and Fgf10. *Development* 129, 5161–5170.

111. Ramel, M. C., Buckles, G. R., Baker, K. D., et al. (2005) WNT8 and BMP2B co-regulate non-axial mesoderm patterning during zebrafish gastrulation. *Dev Biol* 287, 237–248.

112. Buckles, G. R., Thorpe, C. J., Ramel, M. C., et al. (2004) Combinatorial Wnt control of zebrafish midbrain-hindbrain boundary formation. *Mech Dev* 121, 437–447.

113. Thorpe, C. J., Weidinger, G., Moon, R. T. (2005) Wnt/beta-catenin regulation of the Sp1-related transcription factor sp5l promotes tail development in zebrafish. *Development* 132, 1763–1772.

114. Poss, K. D., Shen, J., Keating, M. T. (2000) Induction of lef1 during zebrafish fin regeneration. *Dev Dyn* 219, 282–286.

115. Amsterdam, A., Sadler, K. C., Lai, K., et al. (2004) Many ribosomal protein genes are cancer genes in zebrafish. *PLoS Biol* 2, E139.

116. Golling, G., Amsterdam, A., Sun, Z., et al. (2002) Insertional mutagenesis in zebrafish rapidly identifies genes essential for early vertebrate development. *Nat Genet* 31, 135–140.

117. Gong, Y., Mo, C., Fraser, S. E. (2004) Planar cell polarity signalling controls cell division orientation during zebrafish gastrulation. *Nature* 430, 689–693.

118. Kim, H. J., Schleiffarth, J. R., Jessurun, J., et al. (2005) Wnt5 signaling in vertebrate pancreas development. *BMC Biol* 3, 23.

119. Agathon, A., Thisse, C., Thisse, B. (2003) The molecular nature of the zebrafish tail organizer. *Nature* 424, 448–452.

120. Erter, C. E., Wilm, T. P., Basler, N., et al. (2001) Wnt8 is required in lateral mesendodermal precursors for neural posteriorization in vivo. *Development* 128, 3571–3583.

121. Lekven, A. C., Thorpe, C. J., Waxman, J. S., et al. (2001) Zebrafish wnt8 encodes two wnt8 proteins on a bicistronic transcript and is required for mesoderm and neurectoderm patterning. *Dev Cell* 1, 103–114.

122. Momoi, A., Yoda, H., Steinbeisser, H., et al. (2003) Analysis of Wnt8 for neural posteriorizing factor by identifying Frizzled 8c and Frizzled 9 as functional receptors for Wnt8. *Mech Dev* 120, 477–489.

123. Ramel, M. C., Buckles, G. R., Lekven, A. C. (2004) Conservation of structure and functional divergence of duplicated Wnt8s in pufferfish. *Dev Dyn* 231, 441–448.

124. Weidinger, G., Thorpe, C. J., Wuennenberg-Stapleton, K., et al. (2005) The Sp1-related transcription factors sp5 and sp5-like act downstream of Wnt/beta-catenin signaling in mesoderm and neuroectoderm patterning. *Curr Biol* 15, 489–500.

125. Houart, C., Caneparo, L., Heisenberg, C., et al. (2002) Establishment of the telencephalon during gastrulation by local antagonism of Wnt signaling. *Neuron* 35, 255–265.

126. Kim, S. H., Shin, J., Park, H. C., et al. (2002) Specification of an anterior neuroectoderm patterning by Frizzled8a-mediated

Wnt8b signalling during late gastrulation in zebrafish. *Development* 129, 4443–4455.

127. Lee, J. E., Wu, S. F., Goering, L. M., et al. (2006) Canonical Wnt signaling through Lef1 is required for hypothalamic neurogenesis. *Development* 133, 4451–4461.

128. Groves, J. A., Hammond, C. L., Hughes, S. M. (2005) Fgf8 drives myogenic progression of a novel lateral fast muscle fibre population in zebrafish. *Development* 132, 4211–4222.

129. Katoh, M. (2005) WNT/PCP signaling pathway and human cancer (review). *Oncol Rep* 14, 1583–1588.

130. Oishi, I., Kawakami, Y., Raya, A., et al. (2006) Regulation of primary cilia formation and left-right patterning in zebrafish by a noncanonical Wnt signaling mediator, duboraya. *Nat Genet* 38, 1316–1322.

131. Fong, S. H., Emelyanov, A., Teh, C., et al. (2005) Wnt signalling mediated by Tbx2b regulates cell migration during formation of the neural plate. *Development* 132, 3587–3596.

132. Knowlton, M. N., Chan, B. M., Kelly, G. M. (2003) The zebrafish band 4.1 member Mir is involved in cell movements associated with gastrulation. *Dev Biol* 264, 407–429.

133. Kim, S. H., Park, H. C., Yeo, S. Y., et al. (1998) Characterization of two frizzled8 homologues expressed in the embryonic shield and prechordal plate of zebrafish embryos. *Mech Dev* 78, 193–201.

134. Nasevicius, A., Hyatt, T., Kim, H., et al. (1998) Evidence for a frizzled-mediated wnt pathway required for zebrafish dorsal mesoderm formation. *Development* 125, 4283–4292.

135. Nasevicius, A., Hyatt, T. M., Hermanson, S. B., et al. (2000) Sequence, expression, and location of zebrafish frizzled 10. *Mech Dev* 92, 311–314.

136. Pezeron, G., Anselme, I., Laplante, M., et al. (2006) Duplicate sfrp1 genes in zebrafish: sfrp1a is dynamically expressed in the developing central nervous system, gut and lateral line. *Gene Expr Patterns* 6, 835–842.

137. Kapsimali, M., Caneparo, L., Houart, C., et al. (2004) Inhibition of Wnt/Axin/beta-catenin pathway activity promotes ventral CNS midline tissue to adopt hypothalamic rather than floorplate identity. *Development* 131, 5923–5933.

138. Kim, H. S., Shin, J., Kim, S. H., et al. (2007) Eye field requires the function of Sfrp1 as a Wnt antagonist. *Neurosci Lett* 414, 26–29.

139. Stigloher, C., Ninkovic, J., Laplante, M., et al. (2006) Segregation of telencephalic and eye-field identities inside the zebrafish forebrain territory is controlled by Rx3. *Development* 133, 2925–2935.

140. Bellipanni, G., Varga, M., Maegawa, S., et al. (2006) Essential and opposing roles of zebrafish beta-catenins in the formation of dorsal axial structures and neurectoderm. *Development* 133, 1299–1309.

141. Kelly, G. M., Erezyilmaz, D. F., Moon, R. T. (1995) Induction of a secondary embryonic axis in zebrafish occurs following the overexpression of beta-catenin. *Mech Dev* 53, 261–273.

142. Kelly, C., Chin, A. J., Leatherman, J. L., et al. (2000) Maternally controlled b-catenin-mediated signaling is required for organizer formation in the zebrafish. *Development* 127, 3899–3911.

143. Maegawa, S., Varga, M., Weinberg, E. S. (2006) FGF signaling is required for {beta}-catenin-mediated induction of the zebrafish organizer. *Development* 133, 3265–3276.

144. van de Water, S., van de Wetering, M., Joore, J., et al. (2001) Ectopic Wnt signal determines the eyeless phenotype of zebrafish masterblind mutant. *Development* 128, 3877–3888.

145. Schwarz-Romond, T., Asbrand, C., Bakkers, J., et al. (2002) The ankyrin repeat protein Diversin recruits Casein kinase Iepsilon to the beta-catenin degradation complex and acts in both canonical Wnt and Wnt/JNK signaling. *Genes Dev* 16, 2073–2084.

146. Shimizu, T., Yamanaka, Y., Ryu, S. L., et al. (2000) Cooperative roles of Bozozok/Dharma and Nodal-related proteins in the formation of the dorsal organizer in zebrafish. *Mech Dev* 91, 293–303.

147. Tsai, J. N., Lee, C. H., Jeng, H., et al. (2000) Differential expression of glycogen synthase kinase 3 genes during zebrafish embryogenesis. *Mech Dev* 91, 387–391.

148. Tokuoka, H., Yoshida, T., Matsuda, N., et al. (2002) Regulation by glycogen synthase kinase-3beta of the arborization field and maturation of retinotectal projection in zebrafish. *J Neurosci* 22, 10324–10332.

149. Eisinger, A. L., Nadauld, L. D., Shelton, D. N., et al. (2006) The adenomatous polyposis coli tumor suppressor gene regulates expression of cyclooxygenase-2 by a mechanism that involves retinoic acid. *J Biol Chem* 281, 20474–20482.

150. Nadauld, L. D., Chidester, S., Shelton, D. N., et al. (2006) Dual roles for adenomatous

polyposis coli in regulating retinoic acid biosynthesis and Wnt during ocular development. *Proc Natl Acad Sci USA* 103, 13409–13414.

151. Nadauld, L. D., Phelps, R., Moore, B. C., et al. (2006) Adenomatous polyposis coli control of C-terminal binding protein-1 stability regulates expression of intestinal retinol dehydrogenases. *J Biol Chem* 281, 37828–37835.

152. Nadauld, L. D., Sandoval, I. T., Chidester, S., et al. (2004) Adenomatous polyposis coli control of retinoic acid biosynthesis is critical for zebrafish intestinal development and differentiation. *J Biol Chem* 279, 51581–51589.

153. Nadauld, L. D., Shelton, D. N., Chidester, S., et al. (2005) The zebrafish retinol dehydrogenase, rdh11, is essential for intestinal development and is regulated by the tumor suppressor adenomatous polyposis coli. *J Biol Chem* 280, 30490–30495.

154. Dorsky, R. I., Snyder, A., Cretekos, C. J., et al. (1999) Maternal and embryonic expression of zebrafish lef1. *Mech Dev* 86, 147–150.

155. Shelton, D. N., Sandoval, I. T., Eisinger, A., et al. (2006) Up-regulation of CYP26A1 in adenomatous polyposis coli-deficient vertebrates via a WNT-dependent mechanism: implications for intestinal cell differentiation and colon tumor development. *Cancer Res* 66, 7571–7577.

156. Nyholm, M. K., Wu, S. F., Dorsky, R. I., et al. (2007) The zebrafish zic2a-zic5 gene pair acts downstream of canonical Wnt signaling to control cell proliferation in the developing tectum. *Development* 134, 735–746.

157. Veien, E. S., Grierson, M. J., Saund, R. S., et al. (2005) Expression pattern of zebrafish tcf7 suggests unexplored domains of Wnt/beta-catenin activity. *Dev Dyn* 233, 233–239.

158. Dorsky, R. I., Itoh, M., Moon, R. T., et al. (2003) Two tcf3 genes cooperate to pattern the zebrafish brain. *Development* 130, 1937–1947.

159. Pelegri, F., Maischein, H. M. (1998) Function of zebrafish beta-catenin and TCF-3 in dorsoventral patterning. *Mech Dev* 77, 63–74.

160. Thorpe, C. J., Moon, R. T. (2004) nemo-like kinase is an essential co-activator of Wnt signaling during early zebrafish development. *Development* 131, 2899–2909.

161. Dorsky, R. I., Sheldahl, L. C., Moon, R. T. (2002) A transgenic Lef1/beta-catenin-dependent reporter is expressed in spatially restricted domains throughout zebrafish development. *Dev Biol* 241, 229–237.

162. Ishitani, T., Matsumoto, K., Chitnis, A. B., et al. (2005) Nrarp functions to modulate neural-crest-cell differentiation by regulating LEF1 protein stability. *Nat Cell Biol* 7, 1106–1112.

163. Leung, T., Soll, I., Arnold, S. J., et al. (2003) Direct binding of Lef1 to sites in the boz promoter may mediate pre-midblastula-transition activation of boz expression. *Dev Dyn* 228, 424–432.

164. Angers, S., Thorpe, C. J., Biechele, T. L., et al. (2006) The KLHL12-Cullin-3 ubiquitin ligase negatively regulates the Wnt-beta-catenin pathway by targeting Dishevelled for degradation. *Nat Cell Biol* 8, 348–357.

165. Nambiar, R. M., Henion, P. D. (2004) Sequential antagonism of early and late Wnt-signaling by zebrafish colgate promotes dorsal and anterior fates. *Dev Biol* 267, 165–180.

166. Hashimoto, H., Itoh, M., Yamanaka, Y., et al. (2000) Zebrafish Dkk1 functions in forebrain specification and axial mesendoderm formation. *Dev Biol* 217, 138–152.

167. Ryu, S. L., Fujii, R., Yamanaka, Y., et al. (2001) Regulation of dharma/bozozok by the Wnt pathway. *Dev Biol* 231, 397–409.

168. Gillhouse, M., Wagner Nyholm, M., Hikasa, H., et al. (2004) Two Frodo/Dapper homologs are expressed in the developing brain and mesoderm of zebrafish. *Dev Dyn* 230, 403–409.

169. Waxman, J. S. (2005) Regulation of the early expression patterns of the zebrafish Dishevelled-interacting proteins Dapper1 and Dapper2. *Dev Dyn* 233, 194–200.

170. Wada, N., Javidan, Y., Nelson, S., et al. (2005) Hedgehog signaling is required for cranial neural crest morphogenesis and chondrogenesis at the midline in the zebrafish skull. *Development* 132, 3977–3988.

171. Bingham, S., Higashijima, S., Okamoto, H., et al. (2002) The Zebrafish trilobite gene is essential for tangential migration of branchiomotor neurons. *Dev Biol* 242, 149–160.

172. Henry, C. A., Hall, L. A., Burr Hille, M., et al. (2000) Somites in zebrafish doubly mutant for knypek and trilobite form without internal mesenchymal cells or compaction. *Curr Biol* 10, 1063–1066.

173. Lopez-Schier, H., Hudspeth, A. J. (2006) A two-step mechanism underlies the planar polarization of regenerating sensory hair cells. *Proc Natl Acad Sci USA* 103, 18615–18620.

174. Marlow, F., Zwartkruis, F., Malicki, J., et al. (1998) Functional interactions of genes mediating convergent extension, knypek and trilobite, during the partitioning of the eye primordium in zebrafish. *Dev Biol* 203, 382–399.

175. Sepich, D. S., Calmelet, C., Kiskowski, M., et al. (2005) Initiation of convergence and extension movements of lateral mesoderm during zebrafish gastrulation. *Dev Dyn* 234, 279–292.

176. Biemar, F., Argenton, F., Schmidtke, R., et al. (2001) Pancreas Development in Zebrafish: Early Dispersed Appearance of Endocrine Hormone Expressing Cells and Their Convergence to Form the Definitive Islet. *Dev Biol* 230, 189–203.

177. Hannus, M., Feiguin, F., Heisenberg, C. P., et al. (2002) Planar cell polarization requires Widerborst, a B' regulatory subunit of protein phosphatase 2A. *Development* 129, 3493–3503.

178. Wada, H., Iwasaki, M., Sato, T., et al. (2005) Dual roles of zygotic and maternal Scribble1 in neural migration and convergent extension movements in zebrafish embryos. *Development* 132, 2273–2285.

Chapter 18

Determination of mRNA and Protein Expression Patterns in Zebrafish

Elizabeth L. Christie, Adam C. Parslow, and Joan K. Heath

Abstract

Optically transparent zebrafish embryos provide an excellent vertebrate model system in which to reveal specific mRNA and protein expression patterns during development. Whole-mount preparations can be used to generate three-dimensional color or fluorescent readouts of the expression pattern of a given gene (or genes), matched with a bright-field image of all the tissues in the developing embryo. Whole-mount mRNA *in situ* hybridization (WISH) has long been the method of choice for revealing gene expression patterns in zebrafish because this method depends only on being able to identify a relatively short region of nucleotide sequence unique for the gene of interest. In contrast, the scarcity of antibodies that are specific to or cross-react with zebrafish proteins has limited the widespread use of immunocyto-chemical applications, though this situation will improve in the future. The elucidation of the specific expression patterns of Wnt pathway genes in zebrafish has made a major contribution to our current understanding of their roles in vertebrate development.

Key words: whole-mount *in situ* hybridization, riboprobe, whole-mount immunocytochemistry, antibody, vibrating microtome, sections.

1. Introduction

The embryonic development of zebrafish is very similar to that of higher vertebrate organisms and several advantages, including the optical transparency of zebrafish embryos and their rapid, external development, make the zebrafish an ideal model organism in which to study the genetic basis of vertebrate development. Moreover, the large number of offspring (a female typically produces >200 eggs per week) allows for easy collection of ample embryos early in development for experimentation. Zebrafish embryos can be easily stained to reveal the pattern of mRNA and protein expression during development.

Elizabeth Vincan (ed.), *Wnt Signaling, Volume II: Pathway Models, vol. 469*
© 2008 Humana Press, a part of Springer Science+Business Media, New York, NY
Book doi: 10.1007/978-1-60327-469-2

Whole-mount mRNA *in situ* hybridization (WISH) is the method of choice for revealing gene expression patterns in zebrafish. This technique has the advantage that it requires the identification of only a relatively short region (preferably 300–600 bp) of nucleotide sequence specific for the gene of interest. Current data indicates that there are approximately 21,000 genes in the zebrafish genome, and in the most recent release of the zebrafish genome sequence over half the genes have been manually annotated and can be accessed from the Sanger Centre Web site (www.sanger.ac.uk/Projects/D_rerio/).

WISH requires the generation of an anti-sense RNA probe (riboprobe) specific and complementary to the mRNA of interest, binding of the riboprobe to its complementary mRNA in fixed whole embryos, and detection of the riboprobe. A sense riboprobe prepared from the same template is the usual and most acceptable control. It is generally quite straightforward to identify a suitable region for the template, as the 3′ untranslated regions (UTRs) of translated genes are usually unique and of the appropriate length, although every sequence selected should be first BLASTed against the zebrafish genome to test its specificity. In contrast, although robust immunocytochemical methods have been developed, these remain severely limited by the lack of availability of antibodies that cross-react with zebrafish proteins. While a number of cross-reacting antibodies raised against highly conserved mammalian proteins have proven to be extremely useful for zebrafish studies, there remains a very large number of less well-conserved zebrafish proteins for which no antibodies exist. This problem is well-recognized by the zebrafish community and is being addressed in a coordinated fashion. This chapter describes in detail methods used in our laboratory for WISH and immunocytochemistry of whole mounts and thick (200-µm) sections, though it should be stated at the outset that these are entirely derivative of methods that were established by other workers in the field *(1, 2)* more than a decade ago.

2. Materials

2.1. Whole-mount mRNA In Situ Hybridization

2.1.1. Preparation of the Riboprobe

1. 3 *M* Sodium acetate.
2. 1X TE buffer: 10 m*M* Tris-HCl, pH 7.5, 1 m*M* ethylene diamine tetraacetic acid (EDTA).
3. Transcription mix: 200–400 ng DNA template (either linearized plasmid DNA or a PCR product containing an RNA polymerase promoter site), 2 µL 10X transcription buffer*, 2 µL NTP Labeling mix (usually containing

digoxygenin (DIG)-labeled UTP but fluoroscein-labeled UTP is a suitable alternative and is frequently used in double *in situ* hybridization applications)*, 1 µL RNase inhibitor*, 1 µL T3 or T7 or SP6 RNA polymerase*, RNase-free water (either provided in the kit or DEPC-treated water) to 20 µL (*see* **Note 1**). Keep individual components, except the transcription buffer, on ice until ready to use. (* Items provided in the Roche RNA labeling kit.)

4. Precipitation solution: 1 µL 0.5 *M* EDTA, pH 8, 2.5 µL 4 *M* LiCl, 75 µL absolute ethanol. This volume of precipitation solution is required for each riboprobe synthesis.

2.1.2. Embryo Collection

1. Embryo medium: 1.2 g Ocean Nature Sea Salt (Aquasonic or equivalent) in 20 L deionized water, final concentration is 60 µg/L.

2. Embryo medium with PTU: 0.003% N-phenylthiourea (PTU) (Sigma) in embryo medium.

2.1.3. Embryo Fixation

1. Benzocaine/tricaine (Sigma; final concentration 200 mg/L) to anaesthetize embryos prior to fixation if the embryos are older than 48 hours post fertilization (hpf) at the time of fixation (refer to local animal ethics committee regulations).

2. 2 mg/mL Pronase (Sigma) in embryo medium.

3. 1X PBS: 0.8% NaCl, 0.02% KCl, 0.02 *M* phosphate buffer, pH 7.4.

4. 4% (w/v) paraformaldehyde (PFA) in 1X PBS. This is prepared from a 16% (w/v) PFA stock solution (Electron Microscopy Services). The 16% stock solution is provided in sealed ampoules and is stored at room temperature. It should be freshly diluted prior to use. Once diluted, the 4% solution can be kept at 4°C and used within the following week or stored frozen at –20°C and thawed once for use.

5. PBST: PBS containing 0.1% (w/v) Tween-20 (Sigma); the Tween-20 prevents the fixed embryos from sticking together.

6. 33% and 66% (v/v) methanol in PBS or PBST.

2.1.4. In Situ Hybridization

1. Proteinase K (Sigma) stock solution: 10 mg/mL in PBST.

2. 20X SSC stock solution: 175.3 g NaCl, 88.22 g sodium citrate, ddH$_2$O to 1000 mL.

3. HYB$^-$: 5X SSC, 50% deionized formamide, 0.1% (w/v) Tween-20. Filter HYB$^-$ solution, wrap in foil and store at –20°C.

4. HYB$^+$: HYB$^-$, 50 μg/mL heparin, 5 mg/mL ribonucleic acid Type VI Torula Yeast (Sigma). Filter HYB$^+$ solution, wrap in foil, and store at −20°C.

5. HYB$^+$ / riboprobe: HYB$^+$, 1 ng/μL riboprobe.

6. HYB wash solutions: 75% HYB$^+$, 25% 2X SSC; 50% HYB$^+$, 50% 2X SSC; 25% HYB$^+$, 75% 2X SSC.

7. 20 μg/mL RNase A in PBST (optional, see **Table 18.1**).

8. 2X SSCT, 0.1% (w/v) Tween-20.

9. 0.2X SSCT, 0.1% (w/v) Tween-20.

10. Blocking solution: 5% (v/v) fetal calf serum (*see* **Note 2**) and 0.2% (w/v) bovine serum albumin (BSA) in PBST.

11. Antibody solution: 5% (v/v) fetal calf serum (*see* **Note 2**), 0.2% (w/v) BSA, 1:4000 dilution anti-DIG-AP Fab fragment (anti-DIG antibody) (Roche) in PBST.

12. Coloration buffer: 100 mM Tris-NaOH, pH 9.5, 50 mM MgCl$_2$, 100 mM NaCl, 0.1% (v/v) Tween-20, deionized water. Prepare fresh on the day of use to avoid precipitation.

13. 4-Nitro Blue Tetrazolium (NBT) solution (as supplied by Roche).

14. 5-Bromo 4-Chloro 3-Indolyl Phosphate (BCIP) solution (Roche).

15. Staining solution: 3.3 μL NBT stock solution and 3.3 μL BCIP stock solution per 1 mL coloration buffer. Container must be wrapped in foil to protect from light. Prepare just before use.

16. Low melting temperature agarose (such as SeaPlaque Agarose, Cambrex Bio Science) in PBS.

2.1.5. Modifications for Double In Situ Hybridization Experiments

1. Blocking solution: 5% (v/v) fetal calf serum (*see* **Note 2**) and 0.2% (w/v) BSA in PBST.

2. Antibody solution 2: 5% (v/v) fetal calf serum (*see* **Note 2**), 0.2% (w/v) BSA and anti-fluorescein-AP Fab fragments (Roche) in PBST.

3. Fast Red buffer: 0.1 M Tris, pH 8.2, 0.1% (v/v) Tween20, 300 mM NaCl.

4. Fast Red staining solution: Dissolve ½ Fast-Red tablet (Roche) per mL in Fast-Red buffer. Before using, spin the solution for 1 minute at full speed (16,000–18,000 g) in a microcentrifuge to pellet any unwanted precipitate.

2.2. Immunocytochemistry for Protein Expression

1. Embryo medium: 1.2 g Ocean Nature Sea Salt (Aquasonic) in 20 L deionized water, final concentration is 60 μg/L.

2.2.1. Embryo Collection

Table 18.1
Troubleshooting: Whole-mount mRNA *in situ* hybridization

Problem	Remedy	Modification to protocol
Weak staining	If the signal is consistently weak, a cold acetone step may be included after **Subsection 3.1.5.1.3**. Acetone will improve the penetration of the riboprobe by increasing the membrane permeability of the embryo.	After **Subsection 3.1.5.1, step 3**, remove PBST and add cold acetone to embryos, keep at −20°C for 8 minutes. Then continue with the protocol.
High background staining	If the protocol gives a high level of background staining there are a few additional steps that may reduce the background signal. Try each step separately to see which gives the best results.	1. *RNase Treatment* RNase treatment reduces background signals by degrading any unhybridized probe. Perform these steps after embryos have been hybridized with the riboprobe (after **Subsection 3.1.5.2., step 5,** then continue the protocol as normal). 1. Incubate embryos in 20 µl/mL RNase A for 30 minutes at 37°C. 2. Wash embryos in 2X SSCT for 10 minutes at 37°C. 3. Wash embryos in 50% formamide, 2X SSCT at 60–70°C for 1 hour. 4. Wash embryos in 2X SSCT at 60–70°C for 15 minutes, with gentle rocking. 5. Wash embryos twice in 0.2X SSCT for 15 min at 60–70°C, with gentle rocking. 2. *Preadsorption of anti-DIG antibody* In order to reduce background staining, Thisse and Thisse (1) recommend the following steps, which are designed to block the nonspecific binding sites on the anti-DIG antibody. 1. Add antibody to blocking solution (1:1000 dilution). 2. Add 10 mL of antibody in blocking solution to ~500 previously fixed embryos and rock gently for several hours at room temperature. 3. Remove supernatant, the preadsorbed antibody is ready to use or can be stored at 4°C.

t1.1
t1.2

t1.3

t1.4
t1.5
t1.6
t1.7
t1.8
t1.9
t1.10
t1.11
t1.12

t1.13
t1.14
t1.15
t1.16
t1.17
t1.18
t1.19
t1.20
t1.21
t1.22
t1.23
t1.24
t1.25
t1.26
t1.27
t1.28
t1.29
t1.30

2.2.2. Embryo Fixation	1. Benzocaine/tricaine (200 µg/mL) to anaesthetize embryos prior to fixation if they are older than 48 hpf (refer to local animal ethics committee regulations).
	2. 2 mg/mL pronase in embryo medium.
	3. 4% paraformaldehyde (PFA) in 1X PBS (16% PFA, Electron Microscopy Services), *see* **Subsection 2.1.3.**, **step 4**.
	4. PBST: PBS containing 0.1% (w/v) Tween-20 (Sigma).
2.2.3. Embryo Permeabilization and Antibody Staining	1. PBD: PBST + 0.5% Triton-X.
	2. Blocking solution: 1% (w/v) BSA, 1% (v/v) serum (*see* **Note 2**) in PBD.
	2. Antibody dilution solution: 0.2% (w/v) BSA in PBD.
2.2.4. Preparation of Thick Sections for Determination of Cellular Protein Localization Patterns	1. Low melting temperature (LMT) agarose (Cambrex or equivalent).
	2. Vibrating microtome (Leica VT 1000S or equivalent).

3. Methods

3.1. *Whole-mount mRNA* In Situ Hybridization

*3.1.1. Preparation of the Riboprobe (See **Notes 3** and **4**)*

This protocol is based on the widely used method of Thisse and Thisse *(1)*.

1. Linearize 5 µg of plasmid DNA containing the riboprobe template with a restriction enzyme that cleaves at a site flanking the insert to be transcribed. Digest for 2 hours with the appropriate restriction enzyme to generate the sense or anti-sense template (*see* **Note 5**). Alternatively, the riboprobe template can be generated by RT-PCR (*see* **Subsection 3.1.2**).

2. To obtain high yields from *in vitro* transcription, DNA should be free of contaminating proteins, RNA and RNases. At the end of the reaction, vortex with an equal volume of a 1:1 mixture of phenol:chloroform. Take the upper, aqueous phase and mix and vortex with an equal volume of chloroform. Harvest the upper, aqueous phase containing the linearized DNA. Alternatively, purify the reaction with a PCR purification kit (Qiagen or equivalent).

3. Precipitate the DNA with 1/10 volume 3 *M* Sodium Acetate and 2.5 volumes of absolute ethanol, chill at −20°C for 15 minutes, and centrifuge for at least 15 minutes at full speed in a microcentrifuge. Wash the pellet with 0.5 mL RNase-free 70% ethanol (ethanol made to 70% with DEPC-treated water) at room temperature. Remove the

supernatant, re-spin the tube for a few seconds, and remove the residual fluid with a fine pipette tip.

4. Resuspend the DNA in 1X TE to 0.5–1.0 µg/mL (*see* **Note 6**).

5. Visualize 1 µL of the linearized DNA on a 1% agarose gel to confirm that digestion is complete.

6. Keep the components of the transcription mix on ice, except the transcription buffer, which should be kept at room temperature (*see* **Note 7**). Assemble the transcription mix at room temperature, pipette the mixture up and down to combine components, spin briefly, and incubate for 3 hours at 37°C.

7. Digest the template DNA by adding 1 µL RNase-free DNase I (provided in the Roche RNA labeling kit) for 15 minutes at 37°C.

8. Add the precipitation solution to stop the synthesis reaction and precipitate the RNA by incubating for 30 minutes at –70°C (*see* **Note 8**).

9. Centrifuge for 30 minutes at 4°C at full speed in a microcentrifuge (16,000–18,000g).

10. Remove the supernatant carefully and wash the pellet with 70% ethanol, centrifuge and remove ethanol with a fine pipette tip, air dry for 1 minute and resuspend in 20 µL of sterile RNase-free water.

11. Check the integrity, size and yield of a 1 µL aliquot of the RNA probe on a 1% denaturing (formaldehyde) agarose gel alongside an RNA mass ladder (Invitrogen 0.16–1.77kb RNA ladder or equivalent). A good yield of riboprobe should be in the range of 100–300 ng/µL (*see* **Note 9**).

12. The riboprobe can be stored in aliquots at –70°C in this state or at a concentration of 1 ng/µL in fresh HYB⁻.

3.1.2. Alternative Method for Riboprobe Preparation: PCR-generation of a DNA Template with a Terminal Phage Promoter

1. Design primer pairs to amplify a 300–600 bp region of interest, ensuring that they are specific to the gene of interest.

2. For the sense probe add either a SP6 or T3 or T7 binding site to the 5′ end of the left primer, while for the anti-sense probe add either a SP6 or T3 or T7 binding site to 3′ end of the right primer, ensuring that at least 11 bases of the primer are specific for the gene of interest. Where possible aim for the final two bases of the SP6 or T3/7 binding sites to overlap with the sequence specific for the target mRNA.

3. Perform a PCR using zebrafish cDNA from an RT-PCR reaction, or a plasmid containing the cDNA insert, as the template. Check the authenticity of the PCR product by running 5 µL on an agarose gel or by nucleotide sequencing.

4. If there are multiple bands, extract the dsDNA band of the correct size using a gel extraction kit (Qiagen QIAquick or equivalent).

5. If there appears to be a low concentration of the desired PCR product, perform multiple PCR reactions and combine the products. Then follow **steps 6 to 11** in **Subsection 3.1.1**.

3.1.3. Embryo Collection

1. Collect embryos from pairs of breeding fish within 15–30 minutes of spawning (*see* **Note 10**).

2. Incubate embryos in 9 cm diameter bacterial grade Petri dishes containing approximately 30 mL of embryo medium with PTU at 28.5°C (in an incubator or on a warming plate) until the embryos are at the required stage *(3)* for the experiment. To investigate the expression pattern of a novel gene, a range of staged embryos should be investigated.

3.1.4. Embryo Fixation

1. Anaesthetize embryos that are older than 48 hpf with 200 μg/mL benzocaine or tricaine.

2. To avoid fixing embryos in a curled-up state (i.e., embryos aged between 18 hpf and ~40 hpf that have not hatched), dechorionate embryos manually with fine sharp watchmaker's forceps (Dumont No. 5, NeoLab or equivalent). Alternatively, treat embryos with 2 mg/mL pronase in embryo medium for 1 minute at 28.5°C (*see* **Note 11**). For embryos younger than 18 hpf, dechorionation is easier after fixation and rinsing (**step 3**).

3. Fix embryos in 5 mL tubes such as Technoplas tubes (Crown Scientific) in 4% PFA overnight at 4°C (*see* **Note 12**).

4. Remove the PFA and wash embryos twice in PBST for 5 minutes.

5. Dehydrate embryos through 33 and 66% methanol for 5 minutes each at room temperature.

6. Wash embryos in 100% methanol for 10 minutes at room temperature.

7. Transfer embryos to fresh 100% methanol and store at −20°C for at least 2 hours until required (*see* **Note 13**). This step is required to assist in the penetration of the riboprobe.

3.1.5. In Situ Hybridization

*3.1.5.1. Day 1 (See **Notes 3** and **14**)*

1. *Rehydration*: Rehydrate embryos through 66% methanol and 33% methanol for 5 minutes each. Then rinse the embryos 2–4X in PBST for 5 minutes.

2. *Proteinase K digestion*: Transfer embryos to Eppendorf tubes or 2 mL NUNC cryovials (*in vitro* or equivalent). Embryos will remain in these tubes until **step 3.5.1.3**. If embryos are 24 hpf or younger move straight to **step 3**.

Otherwise, undertake a limited digestion of the embryos with proteinase K to make them more permeable to the riboprobe as follows:

24–36 hpf embryos – 20 minutes in 1 μL 10 mg/mL proteinase K / 1 mL PBST

48 hpf embryos – 25 minutes in 2 μL 10 mg/mL proteinase K / 1 mL PBST

72 hpf embryos – 30 minutes in 3 μL 10 mg/mL proteinase K / 1 mL PBST

96 hpf embryos – 35 minutes in 4 μL 10 mg/mL proteinase K / 1 mL PBST

120 hpf embryos – 40 minutes in 5 μL 10 mg/mL proteinase K / 1 mL PBST

144 hpf embryos – 45 minutes in 5 μL 10 mg/mL proteinase K / 1 mL PBST

3. Rinse embryos immediately in PBST.

4. Re-fix the embryos in 4% PFA for 30 minutes.

5. Preheat HYB⁻ and HYB⁺ to 65°C.

6. Rinse the embryos 5X in PBST for 5 minutes each rinse.

7. *Pre-hybridization:* Incubate embryos in HYB⁻ for 15 minutes at 65°C in a hybridization oven with rocking. Then incubate embryos for 3–5 hours in HYB⁺ at 65°C with rocking.

8. *Hybridization:* If not already stored in HYB⁺, dilute the riboprobe to a concentration of 1 ng/μL in 300 μL fresh HYB⁺ and pre-heat at 65°C for 5 minutes. Remove the HYB⁺ from the embryos and quickly add 200 μL of the heated riboprobe mix to the embryos and incubate at 65°C overnight with rocking.

3.1.5.2. Day 2 (and 3, if Incubating Primary Antibody Overnight)

1. Prepare the following solutions and pre-warm the first six to 65°C: 75% HYB⁺/ 25% 2X SSC; 50% HYB⁺/ 50% 2X SSC; 25% HYB⁺/ 75% 2X SSC; 100% 2X SSC; 0.2X SSC + 0.1% Tween-20; 0.1X SSC + 0.1% Tween-20 (X2); 66% 0.2X SSC/33% PBST; 33% 0.2X SSC/66% PBST; PBST. The required volume of each solution is 2 mL per Netwell insert to be used.

2. Remove the HYB⁺ riboprobe mix and store at –70°C for re-use (*see* **Note 15**).

3. Transfer embryos to Netwell inserts, membrane diameter 15 mm, mesh size 74 μm (Corning) in a 12-well plate (*see* **Note 16** and **Fig. 18.1**; all steps until 11 can be performed in Netwells in a 12-well plate) and wash embryos in pre-warmed solutions at 65°C with rocking as follows:

Fig. 18.1. Netwell inserts assist in the transfer of embryos from one solution to another during the staining and washing steps of whole-mount *in situ* hybridization and immunocytochemistry. (**A**) Netwell inserts are short polystyrene tubes fitted with polyester mesh bottoms. The ones shown here (15-mm diameter) fit into the wells of a 12-well plate. (**B**) Embryos rest on the mesh and can easily be transferred from one solution to another.

75% HYB$^+$ / 25% 2X SSC	10 minutes
50% HYB$^+$ / 50% 2X SSC	10 minutes
25% HYB$^+$ / 75% 2X SSC	10 minutes
100% 2X SSC	10 minutes
0.2X SSC + 0.1% Tween-20	20 minutes
0.1X SSC + 0.1% Tween-20	2X 20 minutes

4. Meanwhile, prepare the blocking solution.

5. Wash embryos on a rotating platform at room temperature.

66% 0.2X SSC/33% PBST	5 minutes
33% 0.2X SSC/66% PBST	5 minutes
PBST	5 minutes

6. Incubate embryos in blocking solution for 3–5 hours on a rotating platform.

7. *Antibody incubation:* Prepare fresh antibody solution 1 hour before incubation is complete, adding anti-DIG antibody (at 1:3000 or 1:4000 dilution) just prior to the next step. Incubate embryos in antibody solution on a rocking platform overnight at 4°C or for 4 hours at room temperature.

8. Remove antibody solution and rinse embryos in PBST.

9. Wash embryos 6X in PBST for 15 minutes with gentle rocking. Meanwhile, prepare coloration buffer.

10. Wash embryos 3X in coloration buffer for 5 minutes. Meanwhile, prepare staining solution (*see* **Note 17**).

11. *Signal development:* Transfer embryos out of Netwells and into the wells of a new 12- or 24-well plate. Incubate embryos in staining solution: shield from the light by wrapping the plate in foil or by placing in a dark box for as long as is required for appropriate staining (can be minutes for highly expressed genes, such as *myoD* at 16hpf, or hours or days depending on the level of gene expression). For weakly expressed genes a proportion of the embryos can be left overnight at 4°C, which will slow down the staining reaction and avoid overstaining. The embryos can be warmed up to room temperature again if staining is still not sufficient. Monitor embryos intermittently with a stereomicroscope, changing the staining solution every few hours.

12. Stop the staining reaction when staining is judged to be sufficient, or when background staining starts to appear, by washing the embryos in PBST. Refer to **Table 18.1** for troubleshooting advice if the staining is too weak or the background is too high.

13. Incubate embryos in fresh coloration buffer for 5 minutes (*see* **Note 18**).

14. Wash embryos 3X in PBST for 5 minutes.

15. Embryos are ready for imaging or can be stored in PBST at 4°C for several months.

16. *Imaging:* Place embryos in 80% glycerol/PBST or 2% methyl cellulose and visualize using a brightfield or fluorescence stereomicroscope.

3.1.6. Sectioning of WISH-embryos (if desired)

1. Fix embryos in 4% PFA for either 1 hour at room temperature or overnight at 4°C.

2. Rinse embryos 3X in PBST for 5 minutes.

3. Rinse embryos twice in 70% ethanol for 5 minutes.

4. Embed embryos within 3 days in low melting temperature agarose in PBS (1–3%: higher concentrations set quicker, consequently providing less time to arrange embryos).

5. Trim the block around the embryos, infiltrate with melted paraffin wax and section as for histology. Generally 5–10 μm sections are optimal (10 μm sections are better if the staining is quite weak).

3.1.7. Modifications for Double Labeling by In Situ *Hybridization*

1. Follow the above protocol from **Subsection 3.1.1.** until the end of **Subsection 3.1.5.2**, **step 12**, including both riboprobes at **Subsection 3.1.5.1.**, **step 8**. When generating the riboprobes use the DIG labeling mix in the transcription mix

of one riboprobe and the Fluoroscein labeling mix in the transcription mix of the other riboprobe (*see* **Note 19**).

2. Following **Subsection 3.1.5.2.12,** inactivate the anti-DIG antibody by incubating the embryos in 10 mM EDTA at 65°C for 30 minutes.

3. Wash embryos for 30 minutes in 100% methanol.

4. Incubate embryos in 75% methanol in blocking solution, 50% methanol in blocking solution and 25% methanol in blocking solution for 10 minutes each.

5. Block embryos in blocking solution for 2 hours at room temperature.

6. Incubate embryos in antibody solution 2 overnight at 4°C.

7. Wash embryos as described in **Subsection 3.1.5.2., steps 8–10.** Prepare Fast-Red buffer and staining solution.

8. Stain embryos by following **Subsection 3.1.5.2., step 11,** using Fast-Red staining solution. As the Fast-Red application does not provide as intense staining as NBT/BCIP, the probe that gives the stronger signal should be selected for Fast-Red labeling. (This may be the riboprobe that targets the mRNA with the highest expression level, or be related to the properties of the riboprobe itself.)

9. Rinse embryos in PBST.

10. Wash embryos 3X in 100% methanol for 5–10 minutes.

11. Rinse embryos in PBST. Embryos are ready for imaging or sectioning and can be stored in PBST at 4°C for several months.

3.2. Determination of Specific Protein Expression Patterns Using Whole-mount Immunocytochemistry

3.2.1. Embryo Collection

1. Collect embryos from pairs of breeding fish within 15–30 minutes of spawning (*see* **Note 10**).

2. Incubate embryos in a 9-cm diameter bacterial grade Petri dishes containing approximately 30 mL of embryo medium at 28.5°C (in an incubator or on a warming plate; *see* **Note 20**). Incubate at this temperature until the embryos are at the required stage *(3)*. To investigate the expression pattern of a novel protein, a range of staged embryos should be investigated.

3.2.2. Embryo Fixation

1. Dechorionate embryos manually with sharp, fine watchmaker's forceps or by treatment with 2 mg/mL pronase in embryo medium for 1 minutes at 28.5°C (*see* **Note 11**). For embryos younger than 10 hpf, dechorionation is easier after fixation and rinsing (**step 4**; *see* **Note 21**).

2. Anaesthetize embryos with 200 μg/mL benzocaine or tricaine.

3. Fix the embryos in 5 mL tubes such as Technoplas tubes (Crown Scientific) in 4% PFA made up in PBS overnight

at 4 °C (*see* **Note 22**). For embryos that will eventually be sectioned on a vibrating microtome, lightly fix the embryos either in 4% PFA/PBS at room temperature for 2–3 hours or in 2% PFA/PBS overnight at 4 °C (it is important not to over-fix the embryos, otherwise they will become too brittle to section properly).

4. Remove the PFA and rinse the embryos three times in PBST for 5 minutes at room temperature.

3.2.3. Immunocytochemistry

1. *Primary antibody blocking step:* Transfer the embryos to Netwells in a 12-well plate, (*see* **Note 23**; **Fig. 18.1**). Permeabilize the embryos in PBD at room temperature with gentle rotation for 2–6 hours. Remove this solution and completely cover the embryos with blocking solution (this is approximately 400 μL of solution/well). Incubate at room temperature for approximately 2 hours on a rocking platform.

2. *Primary antibody incubation:* Dilute primary antibody in antibody dilution solution according to the manufacturer's recommendations. Common dilutions include: 1:200, 1:500, 1:1000, 1:2000, although this has to be determined empirically for each antibody. Remove blocking solution and incubate embryos with the primary antibody dilution solution overnight at 4°C on a rocking platform. To prevent evaporation of the antibody dilution solution, apply masking tape around the seal between the lid and plate.

3. *Secondary antibody blocking:* Wash the embryos three times in PBD for 20 minutes at room temperature on a rocking platform. Remove PBD and incubate embryos in blocking solution for 1 hour at room temperature.

4. *Secondary antibody incubation:* Dilute secondary antibody in blocking solution according to the manufacturer's recommendations (*see* **Note 24**). Common dilutions for secondary antibodies are in the range of 1:200–1:2000. Remove PBD and incubate embryos in secondary antibody dilution solution overnight at 4°C on a rocking platform.

5. *Secondary antibody removal:* Wash the embryos three times in PBD for 20 minutes at room temperature on a rocking platform.

6. *Imaging of embryos stained with secondary antibodies:* If the secondary antibody is conjugated to a fluorophore such as AlexaFluor 488 (Molecular Probes), mount the embryos on a cover slip in low melting temperature agarose for visualization in a confocal microscope (such as a Nikon C1 laser scanner attached to a Nikon TE-2000-E inverted microscope configured with filters suitable for FITC fluorescence). Images can be collected in a Z-series and projected using a

maximum intensity method. Alternatively, a solution of 80% glycerol/PBST solution or 2% methyl cellulose may be used to hold the embryos in position for visualization with a fluorescence dissecting microscope (such as a Leica MZFLIII). Meanwhile, keep the remainder of the embryos in complete darkness and/or stored at 4°C. This can be achieved by wrapping the plate in aluminum foil. If the embryos are going to be stored for an extended period of time, apply masking tape around the seal between the lid and plate to prevent evaporation. If the secondary antibody is conjugated to alkaline phosphatase (AP), develop the embryos as described in **Subsection 3.1.5.2.**, **steps 8–15**, place in a solution of 80% glycerol/PBST solution or 2% methyl cellulose and image in a stereo- or compound microscope.

3.3. Immunocytochemistry on Thick Sections for Determination of Specific Protein Localization Patterns

3.3.1. Embedding Media Preparation

1. Maintain a beaker of 600 mL of boiling water on a hot plate (*see* **Note 25**).

2. Place 25 mL of PBS and 1 g of LMT agarose into a 50 mL Falcon tube to make up a solution of 4% LMT agarose (*see* **Note 26**).

3. Loosen the lid of the Falcon tube and place into the beaker of boiling water.

4. Every 5 minutes release and retighten the lid of the Falcon tube to release the build up of pressure, carefully avoiding being burnt by the hot steam or agarose. Swirl the solution gently to prevent the agarose from forming clumps.

5. Boil the agarose until it has melted completely (*see* **Note 27**).

3.3.2. Embryo Sectioning

Note that the following instructions relate to the preparation of transverse sections, but can be adjusted for other planes of section.

1. Wash the embryos three times in PBST for 5 minutes during the agarose preparation.

2. To assist in the embedding process, pierce the yolk of the embryos with sharp fine watchmaker's forceps or insect pins (*see* **Note 28**).

3. Pipette the embryos into disposable cryomolds (intermediate-size = 15 × 15 × 5 mm; Sakura) and remove the excess PBS with a 200 µL pipette tip so only the embryos remain (*see* **Note 29**).

4. Cut approximately 1 cm off a number of standard 1000 µL pipette tips (1 per block to be embedded).

5. For transverse sections of embryos >48 hpf, add approximately 1.2 mL of 4% low melting temperature agarose into the cryomold with the modified pipette tips (*see* **Note 30**). Modified tips can be used once only.

6. Immediately following the addition of the agarose to the cryomold, remove any air bubbles that have formed on the surface (*see* **Note 31**).

7. Using a fine manipulating device such as a fine metal probe or a paintbrush hair attached to a yellow tip, drag the embryos away from the cryomold edges into the middle of the mould. Push down the embryos until they are lying horizontally close to the bottom surface of the dish.

8. For transverse sections, manipulate the embryos so that the head of the embryo is resting on the bottom surface with the tail pointing vertically upwards (**Fig. 18.2A**). Evenly space the embryos around the centre of mold (*see* **Note 32**).

9. Do not move the mold until the agarose solution has started to set (agarose becomes cloudy) otherwise the embryos may move out of position (*see* **Note 33**).

10. To prevent the agarose drying-out, place a small amount of PBS onto the surface of the agarose (*see* **Notes 33–35**).

3.3.3. Sectioning

1. Snap a double-sided razor blade into two and screw one half tightly into the blade holder (*see* **Note 36**).

2. Place the blade holder into position and fill the vibrating microtome buffer tray with PBS.

3. Flip the agarose out of the cryomold with a scalpel blade (**Fig. 18.2B**).

4. Trim all four edges of the block to reduce the amount of agarose to cut away during sectioning. Aim to remove any meniscus in the agarose to allow for an even cut during sectioning (*see* **Notes 37 and 38**).

5. Superglue the agarose mold to the vibrating microtome stage so that the embryo heads are facing up (**Fig. 18.2C,D**).

6. Begin sectioning on the vibrating microtome according to the manufacturer's instructions. Suitable parameters to begin with on the Leica VT1000S are: speed and frequency offset at 4, section thickness 150–200 μm. Collect single sections onto a paintbrush and transfer to the wells of a 24-well plate containing 300 μL PBS (*see* **Note 39** and **Fig. 18.2E**).

3.3.4. Immunohistochemistry on Thick Sections

1. *Primary antibody blocking step:* Remove PBS solution and incubate sections with enough blocking solution to completely cover the sections (approximately 400 μL) at room temperature for approximately 2 hours on a rocking platform.

2. *Primary antibody incubation:* Dilute the primary antibody in antibody dilution solution according to the manufacturer's recommendations or determined empirically. Remove the

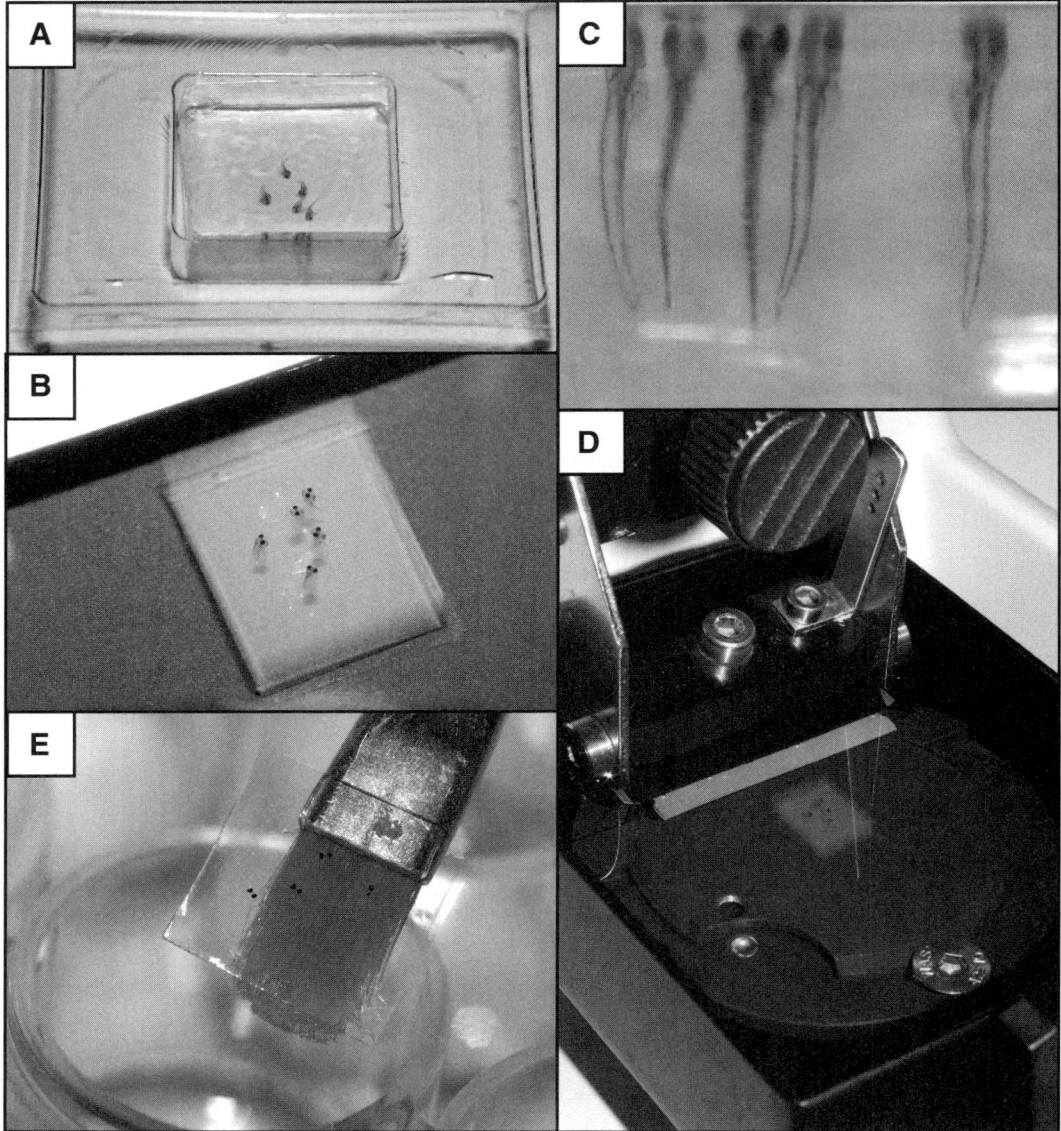

Fig. 18.2. Preparation of fixed, stained embryos for sectioning on a vibrating microtome. (**A**) For transverse sections, the embryos are arrayed vertically in a cryomold tray containing 4% low melting temperature agarose. The heads of the embryos are pointing downward. (**B**) When set, the agarose block is flipped carefully out of the tray. (**C**) The embryos are aligned in parallel in roughly the same plane. (**D**) The block is mounted onto the plate of the vibrating microtome (in this case a Leica VT1000S). (**E**) Horizontal sections are collected on a paintbrush and transferred to the wells of a 12-well plate for the subsequent immunocytochemical detection of proteins.

blocking solution and incubate the embryos with primary antibody dilution solution overnight at 4°C on a rocking platform. To prevent evaporation of the antibody solution, apply masking tape around the seal between the lid and the plate.

3. Wash the embryos three times in PBST for 20 minutes at 4°C on a rocker.

4. Remove PBST and incubate embryos in blocking solution for 2 hours at 4°C.

5. *Secondary antibody incubation:* Dilute the secondary antibody in antibody dilution solution according to the manufacturer's recommendations (*see* **Note 24**). Common dilutions of secondary antibodies for this application are 1:500, 1:1000, and 1:2000. Remove PBST and incubate embryos in secondary antibody solution overnight at 4°C on a rocking platform.

6. *Secondary antibody removal:* Wash the embryos three times in PBST for 20 minutes at 4°C on a rocking platform.

7. *Imaging:* Remove sections required for imaging with a paintbrush and place onto microscope slide for fluorescence analysis (*see* **Note 39**). In order to image the other side of the section, a slide can be placed on top of the section and carefully inverted to release the section from the first slide. The plate containing the remainder of the sections for imaging should be kept in complete darkness and stored at 4°C. This can be achieved by covering it in aluminum foil.

4. Notes

1. DEPC-treatment inactivates RNase in solutions. To make DEPC-treated water, add 0.1% (v/v) DEPC to H_2O (eg 0.5 mL DEPC to 500 mL H_2O) and shake. Incubate at 37°C for 1 hour or at room temperature for several hours, then autoclave for 15 minutes at 15 psi on liquid cycle.

2. Fetal calf serum can be replaced with any serum (e.g., goat or horse serum) except fish serum.

3. Wear gloves and use RNase-free reagents.

4. The preparation of the riboprobe can be undertaken in advance and the riboprobe(s) stored at −70°C for at least a year and probably indefinitely.

5. Restriction enzymes that create 5` overhangs or blunt ends should be used. There is one report that restriction enzymes that create 3` overhangs such as *Kpn*I and *Pst*I may permit template synthesis from the inappropriate strand.

6. The final concentration of the DNA must be at least 0.1 µg/µL.

7. Once thawed, the spermidine in the transcription buffer can co-precipitate the DNA template if stored on ice. If this occurs, incubate at 37°C to re-dissolve.

8. Alternatively, add 0.2 µL 0.5 M EDTA to stop the reaction and remove unincorporated nucleotides by adjusting the volume to 50 µL and purifying by passing twice through a RNase-free Sephadex G-25 or G-50 spin column such as a Probe Quant G 50 spin column (GE Healthcare) or equivalent, then move directly to **step 11**.

9. Riboprobe yields substantially less than this may give poor results. If the yield is less than 100 ng/µL repeat **steps 6 to 11** in **Subsection 3.1.1.**

10. This ensures that all embryos are synchronized at the same developmental stage.

11. Rinse the pronase-treated embryos 4X in embryo medium to remove all the pronase. Embryos usually fall out of their chorions with gentle swishing during the rinse. Older embryos may require longer in the pronase.

12. Fixation can also be performed for 5 hours at room temperature.

13. Embryos can be stored in methanol at −20°C for several months.

14. All further steps are to be performed at room temperature unless otherwise stated.

15. Probes can be used up to ten times depending on their quality. Using them more often can result in weaker signals. The probe in HYB⁺ is stable for at least a year and probably indefinitely when stored at −20 or −70°C.

16. **Steps 3 to 10** in **Subsection 3.1.5.2.** can also be performed in Eppendorf tubes or in the wells of a 12-well plate without Netwells. Netwell inserts are short polystyrene tubes fitted with polyester mesh bottoms that allow easy transfer of embryos between solutions (*see* **Fig. 18.1**).

17. When working with NBT and BCIP stock solutions wear neoprene gloves (Ansell or equivalent).

18. This step increases the intensity of the color.

19. When generating the riboprobes, it is best to use the Fluoroscein labeling mix for the riboprobe that gives the stronger signal.

20. Do not stack Petri dishes containing embryos on a warming plate as the temperature of the embryo medium in the stacked dishes will fall below 28.5°C and embryo development will slow down. Embryos can be incubated in stacks in a 28.5°C incubator.

21. Embryos less than 10 hpf should be fixed prior to dechorionation. For older embryos, dechorionation is easier on unfixed (live) embryos. Removing the skin from fixed embryos 3dpf or older assists in the penetration of antibodies.

22. PFA is volatile and carcinogenic and thus the fixation step should be conducted in a fume hood. Only use a small volume of PFA to just cover the embryos.

23. **Subsection 3.2.2.** to **3.2.3.**, **step 5** can also be performed in 5-mL Technoplas tubes (Crown Scientific) or 12-well plates without Netwells. Netwell inserts are short polystyrene tubes fitted with polyester mesh bottoms that allow easy transfer of embryos between solutions (*see* **Fig. 18.1**).

24. If a fluorescent secondary antibody is used, all solutions containing aliquots of the secondary antibody must be maintained in complete darkness. This can be achieved by covering the plate with aluminum foil.

25. Place a warning sign on the hot plate.

26. Do not allow the agarose to clump in the bottom of the Falcon tube as this will increase the length of time required to dissolve the solution.

27. During the embedding process the agarose needs to be maintained in a molten state. This can be achieved by keeping the agarose in the boiling water during embedding.

28. The snipping of the yolk allows the agarose to penetrate the embryos. This increases the likelihood that the embryo sections will stay in the agarose following sectioning, rather than falling out.

29. 4% low-melting-temperature agarose sets relatively quickly (1–3 minutes) so a maximum of six embryos should be embedded at a time. If more embryos are required, the percentage of low melting temperature agarose can be dropped to no lower than 3%.

30. Embryos older than 48 hpf require 1.2 mL of agarose solution due to their length. If performing transverse sections on embryos less than 48 hpf, a reduced amount of agarose solution can be used (~800 µL).

31. This may be performed using a plastic disposable transfer pipette, or a 200-µL pipette tip.

32. View the mold from the side to determine if all embryos are in the correct position with embryo heads on the bottom surface, and tails completely vertical. All embryos need to be at the same level in the mold so that comparable tissue structures can be observed in the same section.

33. The embryos can still be manipulated somewhat once the surface of the agarose has hardened a little.

34. If PBS is placed on the embryos before the agarose has set, an impression will form on the surface which will cause problems during sectioning.

35. If sectioning is to be done on another day, place the block in a wet box (closed plastic container with wet tissues in it) at 4°C in the refrigerator.

36. Keep razor blade in wrapping during breaking process, or use a wire cutter to reduce risk of injury.

37. Do not cut the block too close to the embryos as this will increase the risk of losing embryos during sectioning.

38. Cut one corner off the block so that the orientation of the sections can be followed during the imaging of the embryos.

39. This step requires gentle manipulation as the embryo sections can be easily lost if roughly handled.

Acknowledgments

The authors thank Heather Verkade, Andrew Trotter, and Felix Ellett for comments on the manuscript. Elizabeth Christie and Adam Parslow are Ludwig institute PhD students in the Department of Surgery at the Royal Melbourne and Western Hospitals, University of Melbourne, and are recipients of a Melbourne Research Scholarship and an Australian Postgraduate Award, respectively.

References

1. Thisse, C., Thisse, B. (1998). High resolution whole-mount in situ hybridization. Zebrafish Science Monitor, Vol 5. Eugene: University of Oregon Press, available at: http://zfin.org/zf_info/monitor/vol5.1/vol5.1.html#High%20Resolution%20Whole-Mount%20in%20situ%20Hybridization.

2. Schenborn, E. T., Mierendorf, R. C. (1985) A novel transcription property of SP6 and T7 RNA polymerases: dependence on template structure. *Nucleic Acids Res* 13(17), 6223–6236.

3. Kimmel, C. B., et al. (1995) Stages of embryonic development of the zebrafish. *Dev Dyn* 203(3), 253–310.

Chapter 19

Manipulation of Gene Expression During Zebrafish Embryonic Development Using Transient Approaches

Benjamin M. Hogan, Heather Verkade, Graham J. Lieschke, and Joan K. Heath

Abstract

The rapid embryonic development and high fecundity of zebrafish contribute to the great advantages of this model for the study of developmental genetics. Transient disruption of the normal function of a gene during development can be achieved by microinjecting mRNA, DNA or short chemically stabilized anti-sense oligomers, called morpholinos (MOs), into early zebrafish embryos. The ensuing development of the microinjected embryos is observed over the following hours and days to analyze the impact of the microinjected products on embryogenesis. Compared to stable reverse genetic approaches (stable transgenesis, targeted mutants recovered by TILLING), these transient reverse genetic approaches are vastly quicker, relatively affordable, and require little animal facility space. Common applications of these methodologies allow analysis of gain-of-function (gene overexpression or dominant active), loss-of-function (gene knock down or dominant negative), mosaic analysis, lineage-restricted studies and cell tracing experiments. The use of these transient approaches for the manipulation of gene expression has improved our understanding of many key developmental pathways including both the Wnt/β-catenin and Wnt/PCP pathways, as covered in some detail in **Chapter 17** of this book. This chapter describes the most common and versatile approaches: gain of function and loss of function using DNA and mRNA injections and loss of function using MOs.

Key words: zebrafish embryos, microinjection, DNA, RNA, morpholino, mutant, loss-of-function phenotype, gain-of-function phenotype.

1. Introduction

Zebrafish embryonic development proceeds rapidly from a single cell embryo to an embryo containing most major organs within the first 2 to 3 days of life *(1)*. The speed of this process allows for the analysis of gene function during embryonic development

Elizabeth Vincan (ed.), *Wnt Signaling, Volume II: Pathway Models, vol. 469*
© 2008 Humana Press, a part of Springer Science+Business Media, New York, NY
Book doi: 10.1007/978-1-60327-469-2

using transient gene overexpression and gene knock down technologies. These approaches involve the direct microinjection of DNA, mRNA, or anti-sense oligomers into early embryos, leading to expression or activity of microinjected products within developing somatic tissues shortly afterwards (for reviews, *see* refs. *2* and *3*). Microinjecting nucleic acids provides transient gain- or loss-of-function analysis of genes of interest throughout the first 3 days of development (approximately) and given that the most studied developmental processes occur within this period the method has become perhaps the most widely used approach for genetic manipulation in the zebrafish research community. In practice, a single operator can microinject multiple samples into large numbers of embryos (200–500 embryos in several hours), taking advantage of the high fecundity of the zebrafish. These experiments therefore permit statistically powerful *in vivo* genetic analysis within a short timeframe.

Techniques for gene misexpression by RNA microinjection, originally developed for *Xenopus* studies *(4, 5)*, utilize the microinjection of synthetically produced mRNA to provide highly uniform transient expression in developing zebrafish embryos. Uniform expression results from RNA microinjection into the streaming yolk of a one- to four-cell embryo, or into the cytoplasm of a single cell embryo. In order to confer temporal and/or spatial restriction on outcomes from RNA injections, microinjecting into single cells of early cleavage-stage embryos *(6)* and the use of caged, photoactivatable modified RNAs have also been described *(7)*.

DNA microinjection results in a mosaic expression pattern because unlike mRNA, which must simply access the cytoplasm, the DNA must penetrate each nucleus for expression *(8–10)*. Taking mosaic expression into consideration, the microinjection and transient analysis of DNA constructs has been widely used, particularly to dissect the function of promoter regions and *cis*-acting regulatory domains (for a description of this application, *see* ref. *11)*, or to take advantage of the randomized compartmentalization of expression.

The use of short interfering RNAs (siRNA) to knockdown target mRNAs by RNA interference (RNAi) has proven invaluable for loss-of-function studies in mammalian cell culture, *C. elegans*, *Drosophila*, and *Arabidopsis*. However, despite promising recent advances, the development of reliable RNAi technology in the zebrafish has not yet become widespread, in part because of initial demonstrations of non-specificity, and will not be discussed as a method here *(12–17)*.

The use of morpholino (MO) anti-sense oligomers for transient gene "knock down" provides a workable tool for the study of gene loss-of-function phenotypes during zebrafish embryogenesis *(18, 19)*. Although other types of chemically modified, stabilized, anti-sense oligomers are available (e.g., peptide nucleic acids; PNAs) (20, 21), their use has not become widespread in the zebrafish community and hence they will not be discussed further here. MO oligomers consist of the same bases found in DNA, and complementary base-pairing is the basis of their target specificity, but the backbone is built of chemically stable phosphorodiamidate intersubunit linkages (22, 23). As a result, MO oligomers are uncharged and (in general) non-toxic at concentrations capable of inhibiting mRNA function (18, 24). They are also resistant to chemical degradation by DNA nucleases, increasing their stability over developmental time (22, 23). They are generally targeted to bind either the translation start site, the immediately adjacent 5'UTR (thereby blocking ribosome binding and function of mature mRNA), or pre-mRNA processing sites. As such, their ability to effectively knock down gene function is a result of blocking mRNA translation or its accurate splicing. MOs have become the most widely used tool for loss-of-function studies in zebrafish. However, even the very earliest reports of the use of MOs recognized experimental risks associated with this anti-sense technology as well as the need for careful controls (18, 25–33). The potential for non-specific toxicity and off-target effects must be anticipated when designing experiments with MOs and hence MO experiments must be rigorously controlled (see reviews in refs. 24 and 34). In **Section 4**, we provide a discussion of the considerations faced when designing and interpreting MO experiments and some general guidelines for appropriate controls.

Here we provide the following series of methods: preparation of microinjection needles and arraying of zebrafish embryos for microinjection; microinjection of DNA, mRNA and MOs; calculation of molar amounts delivered and a guide for the design of MO controls, which we hope will have use when designing loss-of-function experiments in zebrafish. All these approaches are largely derivative of methods established over the past decade or more by numerous zebrafish researchers. In addition, we have included a reference list and table of reagents which have been used successfully to inhibit components of the Wnt/β-catenin and Wnt/PCP pathways (see **Note 1** and **Table 19.1**).

2. Materials

2.1. Preparation and Loading of Needles for Microinjection

1. Injection needle capillaries: thin-walled borosilicate glass capillaries with filament (1.0 mm OD, 0.78 mm ID, 75 mm L, Harvard Apparatus or equivalent). Note that some microinjection apparatus require capillaries without filaments.

Table 19.1
Methods to analyze and disrupt zebrafish Wnt signaling

Application	Name	Biochemical effect	Biological effect
Chemical inhibitors	Lithium chloride	Ectopically activates Wnt/β-catenin pathway by inhibiting Gsk3 activity	Dorsalization and loss of anterior neural structures (forebrain and eyes; *46, 96–98*)
Dominant negative mRNA	DN-*fzd3a* (C-terminal truncation)	Dominant negative form of Fzd3a	Loss of facial neural migration. Phenocopy of *fzd3a* mutant (*ord*) (*50*)
	DN-gsk3b (*Xenopus laevis*)	Ectopically activates Wnt/β-catenin pathway by disrupting Gsk3β activity	Phenocopies *mbl/axin1* mutant phenotype (*98*)
	DN-*gsk3b*	Ectopically activates Wnt/β-catenin pathway by inhibiting Gsk3β activity	Dorsalization (*99*)
	DN-axin (*Xenopus laevis*) (GSK3β binding motif alone)	Ectopically activates Wnt/β-catenin pathway by inhibiting Axin activity	Dorsalization (*99*)
	ΔN *β-catenin*	Constitutively active form of β-catenin	Dorsalization (*100, 101, 108*)
	DN-*fzd8a*-CRDTM	Dominant negative form of Fzd8a (encodes CRD domain and first transmembrane domain)	Expansion of eye and anterior head structures (*46*)
	DN-*rock2a* (Rho kinase 2)	Dominant negative form of Rock2a	Convergence extension defects (*102*)
	DN-*fzd8b* (called *fzAΔT*) (C-terminal truncation)	Dominant negative form of Fzd8b	Ventralization (*103*)
	*dsh*ΔDIX (*Xenopus laevis*) (deletion of N-terminus)	Disrupts Wnt/β-catenin signaling with no effect on Wnt/PCP pathway	Convergence extension defects (*100,104–107*)
	*dsh*ΔDEP (*Xenopus laevis*) (deletion of C-terminus)	Disrupts Wnt/PCP pathway with no effect on Wnt/β-catenin signaling.	Rescues *wnt11/slb* mutant (*100, 107*)
	ΔN-*tcf3* (deletion of β-catenin binding domain)	Dominant negative form of Tcf3	Defective gastrulation (*108*)

(continued)

**Table 19.1
(continued)**

Application	Name	Biochemical effect	Biological effect
	tcf3-BD (β-catenin binding domain alone)	Dominant negative that depletes the pool of β-catenin	Ventralization *(108)*
Fluorescent mRNA	EGFP-tagged Prickle (*Xenopus laevis*)	To visualize the dynamic asymmetric localization of Prickle during neural tube morphogenesis	No phenotype at low doses *(109, 110)*
	venus-YFP/ cerulean-ECFP tagged *wnt11*, *fzd7a*, *fzd7b*, *dsh*, *celsr2/fmi2*	To visualize dynamic sub-cellular localization	No phenotypes at low doses *(66)*
	venus-YFP tagged *fzd3a*	To visualize dynamic and inter-dependent sub-cellular localization	*(50)*
Transgenic lines for misexpression	Tg(*hsp70-wnt11-HA*)	Heat shock promoter driving HA-tagged *wnt11/slb* expresses HA-tagged *wnt 11/slb* upon heat incubation or targeted induction using laser heat	Rescues *wnt11/slb* mutant *(111, 112)*
	Tg(*hsp70: dkk1-GFP*)	Heat shock promoter driving GFP fused *dkk1* (secreted inhibitor of Wnt/β-catenin pathway)	Phenocopies *wnt8* mutant *(113, 114)*
	Tg(*hsp70: wnt5b-GFP*)	Heat shock promoter driving GFP fused *wnt5b/ppt* (activates Wnt/PCP)	Convergence extension defects *(114)*
	Tg(*hsp70:Δtcf3-GFP*) (N-terminal truncation)	Heat shock promoter driving GFP fused Δ*tcf3* (Dominant negative form of Tcf3)	Dorsalization *(113, 114)*
	Tg(*hsp70: wnt8a-GFP*)	Heat shock promoter driving GFP fused Wnt8 (activated Wnt//β-catenin signaling)	Loss of anterior neural structures *(114–116)*
	Tg(*TOPdGFP*)	A β-catenin responsive promoter expressing destabilized GFP (to assay Wnt/β-catenin signaling)	Faithfully recapitulates the known patterns of β-catenin activity during development *(117)*

2. Micropipette pulling apparatus (P-97 Flaming/Brown Micropipette Puller, Sutter Instruments or equivalent).

3. Sequencing gel pipette tips (Eppendorf microloader tips or equivalent).

4. Fine sharpened forceps (Dumont No. 5, NeoLab or equivalent).

5. Microinjection set up: Stereomicroscope with a large working distance (Leica M28Z or equivalent), Micro-manipulator (World Precision Instruments M3301 or equivalent), Pneumatic microinjection apparatus (World Precision Instruments, Pneumatic Picopump, PV820 or equivalent; *see* **Fig. 19.1**).

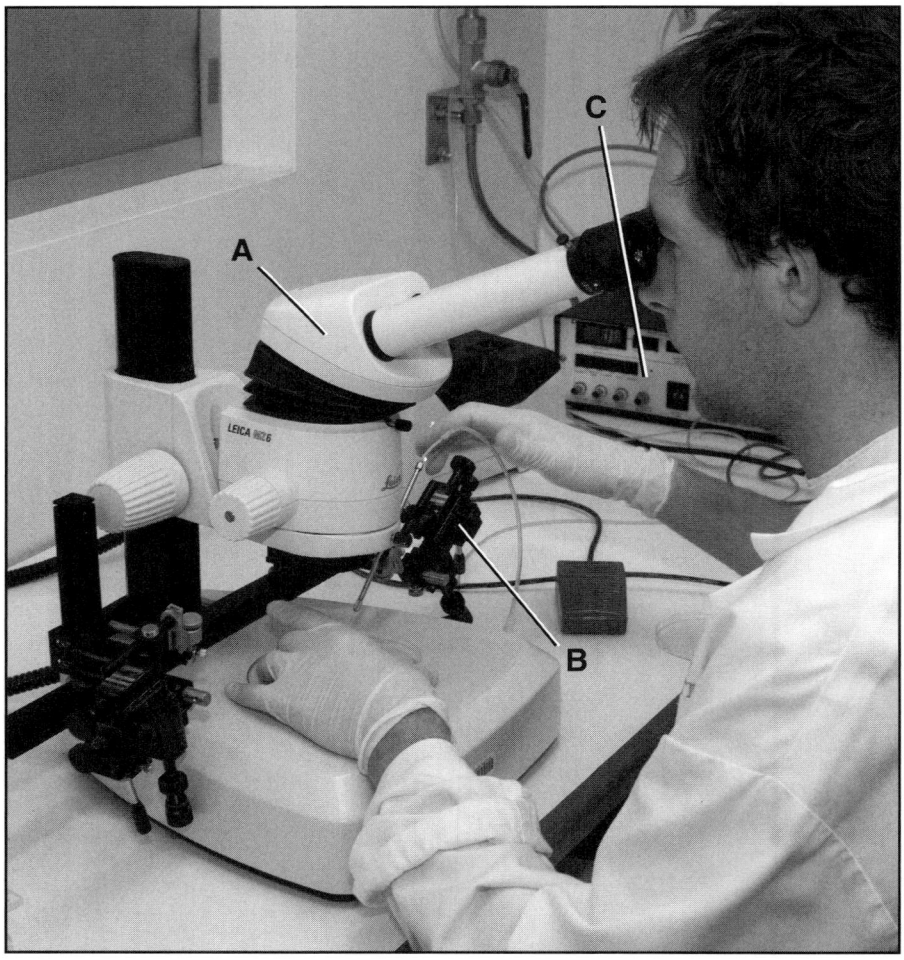

Fig. 19.1. A typical microinjection set-up for the microinjection of zebrafish embryos. (**A**) Stereomicroscope. (**B**) Micro-manipulator. (**C**) Pneumatic microinjection apparatus.

2.2. Preparation of Agarose Mounting Tray

1. Egg water: 1.2 g Ocean Nature Sea Salt (Aquasonic) in 20 L deionized water, final concentration = 60 μg/L.

2. 1–3% (weight/volume) agarose in egg water.

3. Custom-made mould for microinjection tray production. Typically such a mould will be made of plastic and have overall dimensions of 65 × 40 mm with up to six parallel ridges, 1.5 mm apart and 50 mm long, to create grooves 3 mm apart for positioning of the zebrafish embryos (e.g., see **Fig. 19.2**).

2.3. Preparation of DNA

1. Restriction enzyme digestion mix: 5–10 μg of plasmid DNA encoding the construct of interest, 5 μL restriction enzyme that restricts the plasmid at a single site, 10 μL 10X enzyme buffer, 1 μL 100X BSA, RNase-free (DEPC-treated or commercially available) water to 100 μL total volume.

2. DEPC-treated water. DEPC-treatment inactivates RNase in solutions. To make DEPC-treated water, add 0.1% (v/v) DEPC to H_2O (e.g., 0.5 mL DEPC to 500 mL H_2O) and shake. Incubate at 37°C for 1 hour or at room temperature for several hours, then autoclave for 15 minutes at 15 psi on liquid cycle.

3. 3 M sodium acetate.

4. 1X TE buffer: 10 mM Tris-HCl pH 7.5, 1 mM ethylene diamine tetraacetic acid [EDTA].

5. Ethanol: 100% and 70% in water.

6. 1.5–2.0% (w/v) agarose in 1X TAE buffer. To prepare a 50X TAE Buffer stock solution, in a final volume of 1 L, add 242g Tris, 100 mL of 0.5 M EDTA, pH 8.0, 57 mL of glacial acetic acid and adjust the pH to 7.6–7.8 with HCl.

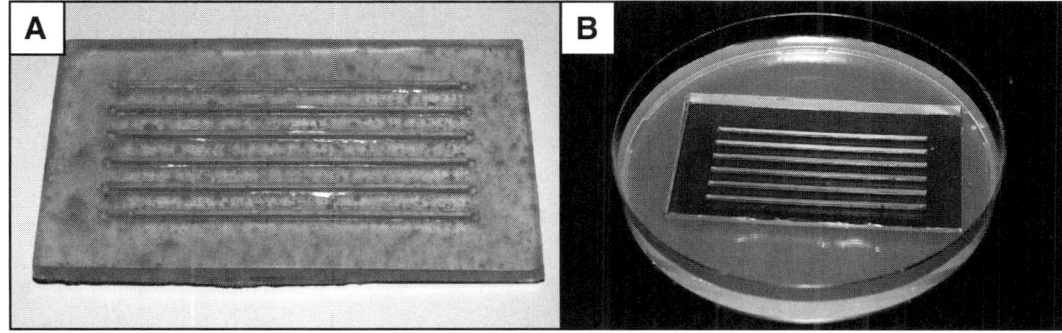

Fig. 19.2. A microinjection tray mould and agarose microinjection tray. (**A**) A suitable mould for generating a microinjection tray. The overall dimensions of this one are 65 mm wide × 40 mm long. The grooves are formed by ridges that are 50 mm long and 3 mm apart. (**B**) An agarose microinjection tray.

7. Injection tracer. Two commonly used injection tracers are Phenol red and rhodamine dextran, although other tracers (e.g., Texas red, fluorescein isothiocyanate [FITC]) can also be used. Phenol red has the advantage that it does not interfere with other reporter molecules such as dsRed, cherry, or green fluorescent protein (GFP), whereas rhodamine dextran is detected over broader wavelengths. Phenol red working solution: 0.5% Phenol red (Sigma) in RNase-free water. Use at a 1:10 or 1:20 ratio according to the desired strength. Rhodamine dextran working solution: 5% (w/v) Rhodamine B isothiocyanate-Dextran (Sigma) in 0.2 M RNase-free KCl. Prepare as a 100 mg/2mL stock solution, filter to remove particles using a 0.22 μm centrifuge spin column, and store in 5–10 μL aliquots at –70°C. Use at a 1:1 ratio (or less) according to the desired fluorescence strength.

8. DNA microinjection solution in KCl: 10–200 ng/μL linearized purified DNA, 0.1 M KCl, 1/10–1/20 Phenol red working solution.

9. DNA solution for an I-SceI co-microinjection: 50–100 ng/μL DNA, 5 μL 10X I-SceI buffer*, 2.5 μL I-SceI enzyme*, 2 μL Phenol red working solution, water to 50 μL. (*Items provided in Roche I-SceI Meganuclease kit.)

10. DNA/RNA solution for transposase-mediated injections: DNA construct in transposon backbone vector (Tol-2 or Sleeping Beauty) 25 ng/μL, mRNA encoding transposase 25 ng/μL, 1/10–1/20 Phenol red working solution.

2.4. Preparation of RNA

1. Restriction enzyme digestion mix: 5–10 μg of plasmid DNA encoding the cDNA of interest, 5 μL restriction enzyme that restricts the plasmid at a single site, 10 μL 10X enzyme buffer, 1 μL 100X BSA, RNase-free water to 100 μL total volume.

2. Transcription mix: 1 μg of linearized DNA template*, 10 μL 2X NTP/CAP labeling mix*, 2 μL 10X transcription buffer*, 2 μL 10X SP6 RNA polymerase enzyme mix*, the total volume is 20 μL. (*Items provided in Ambion SP6 mMESSAGE MACHINE kit.)

3. Micronjection mix: RNA to desired concentration, RNase-free water, 1/10–1/20 Phenol red working solution.

2.5. Preparation of Morpholinos (MOs) for Microinjection

1. Microinjection mix: MO at desired concentration with a 1/10–1/20 dilution of the Phenol red working solution.

2.6. Calculation of amount of MO injected

1. Paraffin oil (Relative density 0.865, Viscosity coefficient 171 mPa.s).

2. Standard metric laboratory hemocytometer.

3. Methods

3.1. Preparation and Loading of Needles for Microinjection

1. Store and handle glass capillaries in an RNase-free environment.

2. Remove glass capillaries with gloves and carefully load into the micropipette puller maintaining RNase-free conditions.

3. Pull microinjection needles using the micropipette puller.

4. Adjust settings of the micropipette puller during needle production until the needles produced are "stubby" with a tip that is not overly tapered (*see* **Note 2**).

5. Produce multiple needles with the same settings and store by arraying in a row on either a strip of double-sided tape or common application putty (Blu-Tac or equivalent) in an RNase-free environment (e.g., sterile 13-cm Petri dish).

6. Load needle with microinjection solution using a pipette fitted with a tip designed for nucleotide sequencing gels (Eppendorf microloader tips or equivalent).

7. Fit loaded needles to the working arm of the microinjection apparatus.

8. Break the tip of the needle with forceps immediately prior to the microinjection procedure to produce an opening of approximately 0.1 mm when the needle is loaded and fitted to the microinjection apparatus (*see* **Note 2**). The needle is now ready for microinjection.

3.2. Preparation of Agarose Mounting Tray and Arraying of Embryos

1. Prepare an agarose solution (in the range of 1–3%) in egg water. The optimal concentration of agarose depends on the experience of the operator. The higher the % agarose, the better the mould floats (*see* **item 2** below); however this is offset to some extent by the faster speed at which it sets, which may preclude accurate positioning of the mould.

2. Pour warm agarose into a plastic 9-cm Petri dish and gently lower in the plastic microinjection tray mould allowing it to float on the surface of the agarose (*see* **Note 3**). Allow the agarose microinjection tray to set before carefully removing the mould (*see* **Fig. 19.2**).

3. Store agarose microinjection trays at 4°C for up to several weeks prior to use (seal the plastic Petri dish with parafilm to prevent it from drying out). Microinjection trays can be re-used multiple times.

4. Fill the agarose microinjection tray with egg water, and pre-warm to ~20–28°C. Pipette a large number of synchronous one- to two-cell stage embryos for microinjection into the egg water in the agarose microinjection tray. For a

description of embryo collection and developmental stages, *see* refs. *1* and *35.*

5. Push the embryos into alignment along the grooves in the agarose microinjection tray so that they sit side by side in single rows (*see* **Fig. 19.3**).

6. Remove the majority of the egg water after alignment so that the surface tension is increased and holds the arrayed embryos firmly in position for microinjection. Embryos are now aligned and ready for the microinjection process.

3.3. Preparation and Microinjection of DNA for Transient Expression Analysis

1. Prepare plasmid DNA for microinjection from a clean, high-quality preparation. To avoid toxic affects; midi-prep or maxi-prep DNA is recommended.

2. Linearize DNA for microinjection by incubating with an appropriate restriction enzyme until the reaction has gone to completion (usually 1–2 hours). If co-microinjecting the

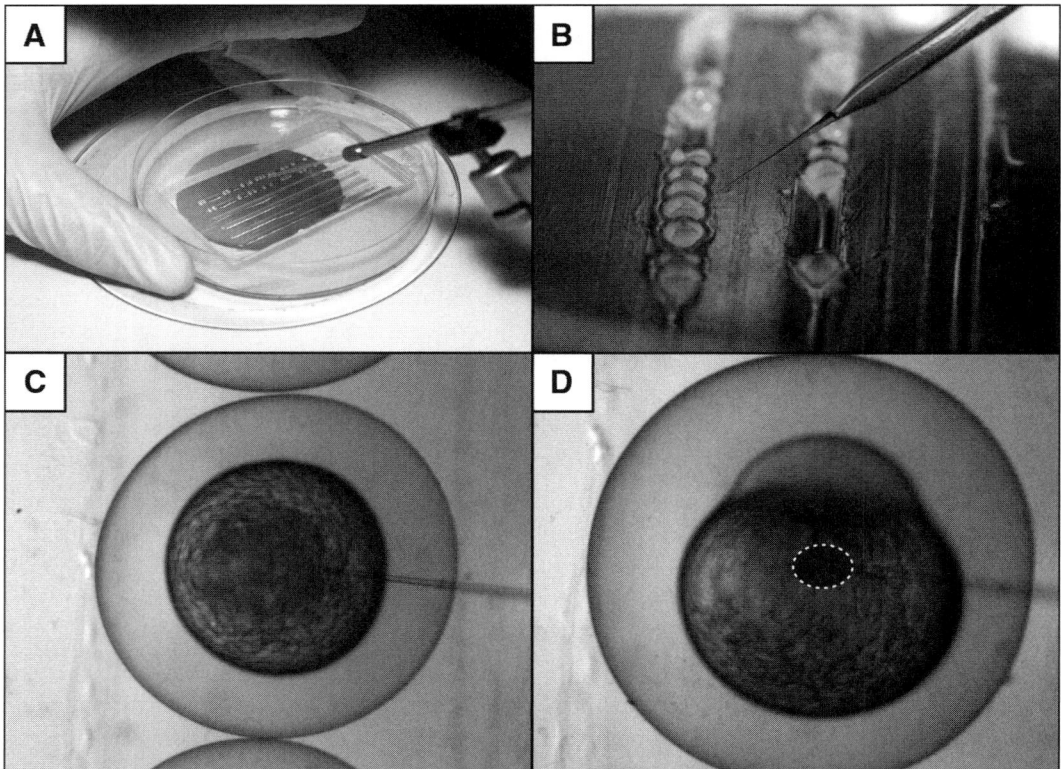

Fig. 19.3. Microinjection of zebrafish embryos. (**A, B**) Embryos arrayed for microinjection. (**C**) Microinjection of a one-cell embryo, animal pole uppermost, displaying the needle inserted into the yolk as required for mRNA or MO microinjections. (**D**) Lateral view of a one-cell embryo being microinjected. The white dashed line depicts the target bolus size, which corresponds to a volume of approximately 1.0 nL. Positioning the bolus adjacent to the fertilized egg as shown facilitates the injected RNA/MO being swept into the cell(s) by ooplasmic streaming (*see* **Note 9**).

DNA with either I-*Sce*I or transposase mRNA, this step is not required (*see* **Note 4**).

3. Purify the linearized DNA by extracting the solution with phenol/chloroform followed by ethanol precipitation or by using a commercially available purification column (Qiagen QIAquick PCR purification or equivalent).

3.3.1. Phenol/chloroform Extraction and Precipitation of DNA

1. Vortex digestion mixture with an equal volume of phenol/chloroform (1:1 ratio).

2. Centrifuge at top speed (16,000–18,000 g) in a bench microcentrifuge for 30 minutes at 4°C.

3. Remove the upper, aqueous phase and precipitate DNA by adding 1/10 volume of 3 *M* Sodium Acetate and 2.5 volumes of absolute ethanol. Incubate at room temperature for 15 minutes.

4. Centrifuge the solution at top speed (16,000–18,000g) in a bench microcentrifuge for 30 minutes at 4°C.

5. Remove the supernatant and wash the pellet with 500 μL 70% ethanol by inverting the tube several times.

6. Centrifuge at full speed (16,000–18,000 g) for 5 minutes at 4°C.

7. Remove the supernatant and all residual fluid with a pipette tip.

8. Dry the pellet briefly (5 minutes at room temperature) and resuspend the DNA in a suitable volume of 1X TE buffer (10 m*M* Tris-HCl, pH 7.5, 1 m*M* EDTA) or water.

9. Determine the concentration of the purified DNA with a spectrophotometer and visualize (1 μL) of the DNA using agarose gel electrophoresis to confirm that the DNA is completely linearized.

3.3.2. Microinjection of Purified, Linearized DNA

1. Prepare the DNA solution for microinjection by adding a tracer dye (e.g., Phenol red, rhodamine-dextran) to allow the microinjection bolus to be visualized clearly, and load the solution into a microinjection needle as described above (*see* **Note 5**).

2. Array the embryos on the agarose microinjection tray so that the cytoplasm of the single cell (or blastodisc) is visible and easily accessible.

3. Pass the prepared microinjection needle though the chorion and into the embryo. For DNA microinjections the bolus of DNA solution must be delivered into the cytoplasm of the single cell of one-cell stage embryos. The needle can be passed directly into the single cell or through the yolk, *see* **Fig. 19.3**.

For a consideration of controls for DNA microinjection experiments, *see* **Note 6**.

3.4. Preparation and Microinjection of RNA

3.4.1. RNA Preparation

1. Linearize the plasmid DNA containing the mRNA template of interest using a clean, high-quality DNA preparation (midi-prep or maxi-prep). Use a restriction enzyme that cuts the plasmid in a position downstream (3′) of the polyadenylation signal sequence and does not cut in the polymerase binding site or construct. Digest the plasmid for 2 to 3 hours at 37°C.

2. Purify digested DNA using either phenol/chloroform extraction followed by ethanol precipitation (*see* **Subsection 3.3.1.**) or a commercially available purification column (Qiagen QIAquick PCR purification or equivalent).

3. Dissolve the DNA pellet in RNase-free (DEPC-treated or commercially available) water or elute the DNA from the column in RNase-free water.

4. Determine the concentration of the purified DNA with a spectrophotometer and visualize (1 µL) of the DNA using agarose gel electrophoresis to confirm that the DNA is completely linearized.

5. Prepare the mRNA for microinjection using a commercially available *in vitro* transcription kit for the synthesis of capped, polyA-tailed RNA from a DNA template (Ambion mMESSAGE mMACHINE kit or equivalent; *see* **Note 7**).

6. Thaw and vortex the components of the transcription mix prior to use.

7. Use 1 µg of purified template DNA for the transcription reaction.

8. Prepare the transcription mixture on ice. (The transcription buffer should be warmed first to room temperature because this solution contains spermidine, which may co-precipitate the DNA template if kept on ice).

9. Incubate the mixture at 37°C for 2 to 3 hours.

10. Add 2 µL of RNase-free Turbo DNase (available in the mMESSAGE mMACHINE kit) to the mixture to digest the template upon completion of the reaction. Incubate for 15 minutes at 37°C, maintaining RNase-free conditions.

11. Purify the RNA using a commercially available RNA cleanup kit (Qiagen RNAeasy purification column, or equivalent). It is important that the RNA for microinjection is of high purity to avoid toxicity and nonspecific phenotypes. Lithium chloride cannot be used to precipitate mRNA for microinjection, as lithium ions induce developmental defects.

12. Elute the RNA from the column using RNase-free water.

13. Test the integrity, size and yield of RNA on a 1% denaturing agarose gel next to a commercially available RNA ladder

(Invitrogen 0.24–9.5 kb RNA ladder, or equivalent). Alternatively, denature the RNA sample and ladder in RNA loading buffer at 70°C for 5 minutes and run on a 1% TAE agarose gel. High-quality microinjection RNA (free of template DNA) runs as a single high yield band of the predicted size.

14. Confirm the specific yield of RNA and estimate purity using a spectrophotometer. A yield in the range of 0.2–0.8 µg/µL is expected from this method, and an absorbance ratio (OD_{260}/OD_{280}) that is close to 2.0 indicates a pure preparation.

15. Store the prepared RNA in small aliquots (2–5 µL dependent on yield) at –70 to –80°C and avoid repetitive freeze/thawing of individual aliquots (*see* **Note 8**).

3.4.2. RNA Microinjection

1. Prepare the mRNA for microinjection by adding a tracer dye (e.g., Phenol red, rhodamine-dextran) to allow the microinjection bolus to be visualized clearly and load the solution into a microinjection needle as described in **Subsection 3.1.**

2. The microinjection mix should be prepared with RNase-free solutions; gloves should be worn at all times. Keep the microinjection mixture on ice prior to loading the microinjection needle.

3. Load the microinjection needle with clean RNase-free tips as described earlier.

4. When arraying embryos on the agarose microinjection tray for RNA microinjection it is not essential for the cytoplasm of the cell/s to be accessible to the needle. RNA can be injected directly into the yolk of one- to four-cell embryos (*see* **Note 9**).

5. Pass the loaded microinjection needle though the chorion and into the yolk of the embryo and inject, *see* **Fig. 19.3**.

For a consideration of controls for RNA microinjection experiments, *see* **Note 10**.

3.5. Preparation and Microinjection of Morpholino Oligomers

1. Dissolve lyophilized MO stock in water to either the manufacturer's suggested concentration (add 300 µL for a concentration of 300 nM, approximately 8.3 µg/µL) or to a higher concentration if desired (many researchers add approximately 50 µL for a concentration of ~50 µg/µL; *see* **Note 11**).

2. Store stock solutions of MOs in small aliquots at –20°C for extended periods of time and avoid repeated freeze thawing (*see* **Note 11**).

3. Store working solutions at 4°C or room temperature for shorter periods of time (*see* **Note 11**).

4. Prepare microinjection mix containing the desired concentration of MO diluted in water with a tracer dye (e.g., Phenol red, rhodamine-dextran; *see* **Note 12**). MOs are stable and therefore do not require RNase-free conditions for the preparation of the microinjection mix or the needle for microinjection.

5. Proceed with needle loading and microinjection as described in **Subsection 3.1.**

3.6. Calculation of Amount Delivered

1. Prepare a hemocytometer by applying a droplet of paraffin oil to its gridded surface.

2. Using a pre-prepared needle inject 5–10 droplets (boluses) into the oil immediately above the grid of the hemocytometer. Make sure the boluses are of consistent size.

3. Measure the diameter and radius of each bolus by determining the number of hemocytometer grids spanned by the bolus.

4. Calculate the volume of each bolus using the volume of a sphere equation (volume = $4/3\pi r^3$, where r = the measured radius of your bolus).

5. Calculate the amount in nanograms of DNA, RNA or MO oligomer based on the concentration of the microinjection solution and the average bolus volume.

6. Standardize this process for multiple microinjections by calculating the required number of gridlines spanned to give a bolus volume of 1 nL. Adjust your bolus in each microinjection to this diameter using a hemocytometer (or a gridded eyepiece if your microscope has one). This method permits standardization of all microinjections within a laboratory, even when performed by different operators.

3.7. Treatment of Embryos Post-microinjection

1. After microinjection, remove embryos from agarose microinjection tray using a plastic disposable transfer pipette (Samco; 3 mL large bulb or equivalent). Carefully flood the microinjection tray with warmed egg water and pipette the injected embryos to a clean Petri dish in egg water. Incubate at 28.5°C.

2. Examine embryos on a stereomicroscope at the 8- to 32-cell stages to confirm that the tracer dye has been taken up by the cells.

3. Incubate the embryos until they have progressed through gastrulation (or ~8–10 hours after microinjection) and remove dead, unfertilized and abnormal embryos, recording

numbers. Repeat the removal of abnormal embryos 24 hours after microinjection.

4. Record the number of severely abnormal embryos post-microinjection to provide a record of the potential toxicity of the microinjected DNA, mRNA or MO, and/or flaws in microinjection technique such as mechanical trauma from an inferior needle.

For a consideration of controls for MO microinjection experiments, *see* **Notes 13–16**.

4. Notes

1. The following Wnt pathway genes have been knocked down by targeted anti-sense morpholino oligomers (this list is not meant to be exhaustive): *wnt1 (36, 37), wnt2b (38), wnt2bb (39), wnt3l (40), wnt4a (41), wnt5b (42, 43), wnt8a (44, 45), wnt8b (46, 47), wnt10b (37), wnt11 (41, 43, 48), wnt11r (41), fzd2 (42, 49), fzd3 (50), fzd5 (called fzd8c) (51), fzd7a (49), fzd7b (49), fzd8a (46), ctnnb1 (52, 53), ctnnb2 (53), axin (49), apc (54), tcf4 (55, 56), tcf7 (57), tcf3a/tcf7l1a (57,58), tcf3b/tcf7l1b (36, 57–59), lef1 (60), klhl12 (61), diversin (62), dkk1 (59), dvl2 (61, 63), dvl3 (61), dact1 (64), dact2 (64, 65), rhoab (66, 67), rhoad (66), fmi1/celsr1a (50, 68), fmi1b/celsr1b (50, 68), fmi2/celsr2 (50, 66), tri (69), pk1 (70), scribble (71).* (Information from ref. *72*.)

   In addition, various DNA/RNA constructs have been used to interfere with Wnt pathway gene function. For example, RNAs encoding dominant negative forms of zebrafish Wnt/ β-catenin pathway proteins have been produced by *in vitro* transcription and injected into one-cell embryos. Other dominant negative constructs encode *Xenopus* homologs that have been used successfully in zebrafish embryos to disrupt Wnt signaling. **Table 19.1** provides a list of many of the constructs that have been used.

2. Microinjection needles should not be produced with a very fine tip. This leads to a bendy needle which does not easily pierce the chorion (the protective protein coat around the embryo). When breaking the tip prior to microinjection, it is important that the hole produced is not too wide. A wide diameter will physically damage injected embryos, while a hole which is too small in diameter will become blocked during microinjection. The amount of solution delivered can be adjusted using the microinjection time and pressure setting

on the microinjection apparatus, to allow for variation in tip diameter. Sufficient hold pressure is always required to prevent backflow of egg water into the needle.

3. The mould should be lowered onto the agarose and allowed to float in position. Do not submerge the mould beneath the agarose. Wetting the mould first with egg water helps prevent bubbles in the agarose. *See* **Fig. 19.2** for the final result.

4. There are several ways to inject DNA to achieve transient mosaic expression. DNA for transient analysis can be injected as a circular plasmid directly. However, we suggest linearizing the plasmid for optimal transient expression; in our experience a greater proportion of injected embryos will show mosaic expression if the plasmid is linearized. Note that for stable transgenesis, not covered in this chapter, a linearized cDNA insert (with the vector backbone removed) should be used for microinjection. There are a number of ways to optimize the efficiency of DNA microinjections for transient or stable transgenesis. A vector containing I-*Sce*I meganuclease sites either side of the insert of interest is now commonly used *(73–75)*. Co-microinjection of DNA with I-*Sce*I enzyme leads to higher numbers of expressing embryos and often less mosaicism. This approach also greatly improves the chances of germline insertions for the generation of stable transgenic lines. Alternatively, transposase-mediated DNA microinjections also greatly improve the efficiency of post microinjection expression and transgenesis. These approaches require the cDNA of interest to be cloned into a transposon backbone (either Tol-2 or Sleeping Beauty based vectors are used) and co-injected with mRNA encoding transposase, which induces transposition of the DNA into the host genome. For reviews and methods for the use of these approaches, *see* refs. *76–78*.

5. The required DNA concentration will vary dependent on the application for which the method is being used. As a guide, 10–200 ng/μL is a good starting point.

6. All microinjections should include controls for technique and construct toxicity. In particular, DNA microinjections are commonly complicated by toxicity. Some useful technical controls include diluent or empty vector microinjections. Interpreting DNA microinjection experiments in the face of the mosaicism that occurs is often facilitated by following the construct distribution and expression with a tracer. This may be achieved by building an IRES-containing construct that drives co-expression of the cDNA of interest with a cDNA encoding GFP, β-galactosidase or a similar reporter molecule.

7. The cDNA template for transcription must be cloned into an appropriate vector containing a specific RNA polymerase binding site (SP6, T7 or T3) upstream of the insert of interest (in a sense orientation) and a polyadenylation signal downstream of the insert (a commonly used vector is pCS2+ *[79]*). Kits for the generation of capped mRNA are available for each of these common RNA polymerase binding sites.

8. The amount of mRNA to inject is highly dependent on the gene being studied. Reports have used ranges as broad as 0.1–500 pg. For an uncharacterized mRNA, it is best to start by injecting a series of concentrations and then optimize the concentration for later experiments based on the phenotypes observed.

9. As the early embryo develops, maternally deposited RNA is taken up from the yolk (where it has been earlier deposited by the mother) by ooplasmic streaming *(80, 81)*. mRNA that has been injected into the yolk will also be taken up efficiently into the developing embryo. However, as this process only occurs during the first few cell divisions, microinjection of early embryos is recommended for consistent ubiquitous expression. Injections are best carried out using one- to four-cell stage embryos.

10. The approaches used to control mRNA microinjection will vary depending on the biological process being studied, but RNA microinjections should also include controls that cover technique and construct toxicity. Some generally applicable controls that can be considered are: the microinjection of mRNA encoding GFP (or equivalent reporter) only, demonstrating a concentration-dependent response to the injected mRNA, and a correlation between gain-of-function and loss-of-function phenotypes. Although not widely used as yet, the microinjection of an inactive, mutant RNA that encodes a biologically inert but nonetheless translated protein (i.e., with a premature stop or point mutation) is an ideal RNA control if available or deemed necessary in the experimental context. However, such a control requires considerable understanding of the biology of the protein to ensure that the resulting product is truly inert (some truncating mutations or amino acid variations may inadvertently produce dominant negative forms). A common issue in RNA injections is establishing with certainty whether a very subtle or no phenotypic outcome is the real result. It is necessary in this case to confirm that intact RNA was injected and translated. The best method to achieve this is to demonstrate that the encoded protein is produced by Western blotting, either using a specific antibody if available, or by using a commercial antibody to a tag, such as *Myc*, encoded in the mRNA

construct. Alternatively, co-injecting the test RNA with RNA encoding a reporter molecule such as a fluorescent protein or β-galactosidase will demonstrate whether the reporter mRNA was injected properly and translated *in vivo*. Co-injection of the test RNA with a fluorescent reporter RNA also permits the removal of embryos that did not, for miscellaneous reasons, receive translatable RNA, by screening for reporter expression 8–24 hours after injection, thereby increasing the experimental signal:noise ratio.

11. MOs are currently exclusively produced and supplied to the research community by Genetools LLC. The Genetools Web site (www.gene-tools.com) provides detailed recommendations for the use of MOs. For storage, the following key points can be taken from the manufacturer's recommendations:

   a) MOs are highly stable and can be stored at room temperature when in regular use as a stock or microinjection mix.

   b) For long-term storage it is recommended to freeze MO stocks in aliquots (–20°C). However, it is worth noting that cooling or freezing of MOs may lead to their precipitation and so it is recommended that they should be heated to 65°C for 5 minutes after frozen storage.

   c) Genetools recommends storing MOs at a concentration of 300 n*M*. MOs can be stored at higher concentration (e.g., 50 µg/µL) but this may increase the chance of precipitation.

12. The concentration of MO which inhibits gene function but avoids toxic affects seems to be highly variable between MOs. In general, a bolus containing 1–15 ng is a good starting point. However, the concentrations at which toxicity is seen are not predictable. Ideally, every individual MO should be injected in a concentration series to determine the optimal dose.

13. The literature contains numerous examples of the use of MOs to inhibit translation or processing of a targeted mRNA, confirming their utility as a transient reverse genetics tool. However, an important requirement for their use as a reverse genetics tool for gene function discovery is that they specifically inhibit only the targeted mRNA, without inhibiting other mRNAs (off-target effects). It is thus essential when performing MO experiments that this fact is firmly established for the MOs in question; this requires the use and reporting of appropriate controls. The initial report of the use of MOs in zebrafish by Nasevicius and Ekker *(18)* demonstrated that MOs can effectively and efficiently knock

down mRNA function and can produce a specific phenocopy of known mutants. They also demonstrated that different MOs induce phenotypes at different concentrations, that MOs can induce phenotypes with varying penetrance, and that MOs can induce non-specific phenotypes (such as widespread cell death) not associated with the gene being targeted *(18)*. After 7 years of their use, such non-specific effects of MOs are considered common in the zebrafish community. Recently some of the non-specific effects associated with MOs have been shown to be mediated by p53, providing a plausible mechanism explaining non-specific MO effects *(82)*.

14. The best control for MO efficacy, when an antibody to the gene being targeted is available, is a Western blot showing definitively that the MO results in reduced protein levels. However, as antibodies are only infrequently available for zebrafish proteins, several other controls are commonly used to demonstrate the effectiveness of a MO to affect its target. The capability of a MO to target its intended sequence *in vivo* can be demonstrated artificially using a fluorescent reporter mRNA containing the MO binding site fused to the translational start site of GFP or similar fluorescent protein. Injecting such a reporter mRNA along with a targeting MO should prevent translation of the GFP protein, resulting in reduced fluorescence compared to embryos injected with the MO and an equivalent dose of GFP mRNA lacking the MO target site (examples of such controls can be found in references *27, 83–85)*. Although not applicable for translation start-site targeting MOs, a commonly used control to check the efficacy of splice-site MOs is reverse transcriptase (RT)-PCR. MOs targeted to splice donor or acceptor sites of pre-mRNAs can cause the production of incorrectly processed mRNA transcripts by exon skipping or by intron retention. This leads to alterations in processed mRNA size, which can be observed using RT-PCR of RNA preparations from injected embryos (some examples of the use of this control can be found in refs. *86–91)*. These controls all confirm the efficacy of a MO in targeting the correct binding site *in vivo*, but they do not tackle the important question of MO specificity because they do not address the possibility that off-target or non-specific effects are contributing to the observed MO-induced phenotype.

15. One method to control for the specificity of the phenotype is to demonstrate that two (or more), preferably non-overlapping, MOs targeted to different regions of the same mRNA, are capable of inducing the identical phenotype. For example, it is advisable to compare the phenotypes

achieved by microinjection of any two MOs directed to the translation start site, 5′ untranslated region or any of the pre-mRNA splice acceptor or donor sites. This control attests to the specificity of the induced phenotype to the mRNA being targeted but is inadequate if the phenotype demonstrated is similar to common MO off-target or nonspecific effects such as widespread cell death or developmental delay *(82)*. Probably the best method for assessing specificity is to rescue the MO-induced phenotype(s) by microinjection of mRNA encoding the gene being inhibited. An mRNA engineered not to contain the MO binding site is co-injected into the same embryo as the MO to determine whether this can rescue the MO induced phenotypes, thus demonstrating specificity of the phenotype to this specific mRNA (some examples can be found in references *30, 92–94)*. However, if the mRNA itself results in an overexpression phenotype, this rescue experiment may not be achievable. A commonly used control for microinjection toxicity, nonspecific effects and MO specificity has been the use of random sequence or mismatched control MOs. However, the variability in effect of any two MOs, especially when considering active concentration and toxicity differences, does not make the value of this control for demonstrating biological specificity intuitively obvious. Although microinjection of these control MOs tests for microinjection-induced defects, such control MOs are of limited value alone and are inferior to the specificity and efficacy controls described earlier. Finally, the biological context of MO-induced phenotypes can provide convincing argument for the specificity of MO-induced phenotypes. Correlating the existing knowledge about the gene, its expression pattern, any phenotypes known in other species or models and the MO-induced phenotype observed down the microscope is an important aspect of interpreting these experiments. Ultimately, the "gold standard" for MO specificity is the degree to which the MO phenotype accurately phenocopies hypomorphic and/or null mutant alleles of the targeted gene. While this control is of course not available when MOs are being employed to study the function of novel genes, as the number of cloned zebrafish mutants increases, the number of retrospective morphant/mutant correlations and validations will grow. Of some concern at the present time, is the growing number of examples of discordant descriptions of morphant and subsequent mutant phenotypes.

16. MOs targeting the translation start will inhibit all transcripts, whether maternally derived or produced by the zygote, from shortly after the time of injection. In contrast, splice-site directed MOs will only inhibit protein produc-

tion from newly synthesised zygotic transcripts. No MOs will inhibit only maternally deposited mRNA in wildtype embryos. Photoactivatable caged MOs have recently been described, offering some flexibility to control the onset and location of MO action *(95)*. Although MOs are stable, their effect is transient. They are diluted at every cell division throughout embryonic growth and lose the ability to efficiently block translation or splicing at late embryonic or larval stages. To provide one example, Yan et al. (2002) targeted the *sox9b* gene with splice-site directed MOs and used RNase protection assays to determine the amount of unspliced versus spliced RNA present over time. They observed an increase from 20% of mRNA being correctly spliced at 28 hours post-fertilization (hpf) to 55% of mRNA being correctly spliced by 4 days post-fertilization (dpf), indicating a progressive loss of efficacy over time *(87)*. Some late MO phenotypes have been reported (e.g., at 5 dpf), but for late phenotypes consideration should be given to the possibility that the phenotype represents the cumulative outcome of earlier events, rather than a specific action of the MO at the late time-point at which the phenotype is observed. The duration of MO efficacy is likely to be target gene and MO-specific, and may be influenced by factors such as microinjection concentration and levels of endogenous gene expression. It is therefore important that controls for MO efficacy be demonstrated at or before the time-point at which the phenotype is analyzed, and it must be remembered that later roles of the gene may not be uncovered.

Although the selection of MO controls will vary between projects, we recommend as a baseline, three distinct types of control be used. Taken together, these three controls will lead to a high level of confidence that the phenotypes observed are due to knock down of the target gene:

a) A control demonstrating the efficacy of the MO in vivo, at the molecular level (e.g., Western blot for start-codon targeting MOs and/or RT-PCR for splice-targeting MOs, or a synthetic fluorescent construct inhibition experiment).

b) A control demonstrating the specificity of the targeting event at the phenotypic level (e.g., rescue with an untargetable mRNA, or the use of multiple, non-overlapping, MOs).

c) Correlation between the expression pattern or known biology of the gene and the phenotype.

Table 19.2 provides an easy reference guide to the general advantages and disadvantages of the various different MO controls.

Table 19.2
Advantages and disadvantages of specific MO controls

Aim of control	Control	Advantages	Disadvantages
Recognition of non-specific or off-target MO effects	Random sequence control MO*	Controls against non-specific MO and microinjection-induced defects	Targeting MO and control MO may not have identical chemistry post-synthesis
	Mismatched base pair control MO*	As for random sequence MO	As for random sequence MO; possible residual MO activity despite mismatches
	Correlation of phenotype with expression pattern and biology	Direct correlation of specific expression and specific phenotype gives confidence in analysis	Does not prove MO phenotype is specific to targeted transcript inhibition; most applicable to genes with anatomically restricted gene expression patterns
	Concentration-dependent induction of phenotype	Consistent with MO action at the level of inhibition of translation	Does not prove targeting of correct gene; effects due to non-specificity or chemical impurities may also be concentration-dependent
Demonstrating the capability of a MO to target mRNA in vivo	Quantifying protein product of targeted mRNA using a specific antibody	Indicates that MO definitely inhibits translation of target mRNA	Does not prove MO phenotype is restricted to targeted transcript inhibition
	Splice-site directed MO with analysis of altered splicing	Indicates that MO definitely inhibits processing of target mRNA	Does not prove MO phenotype is specific to targeted transcript inhibition; not applicable to MOs targeted to translation start site
	Stabilization of targeted transcripts detected by *in situ* hybridization	Suggests that MO binds to target mRNA	May be a gene-specific phenomenon; does not prove MO phenotype specific to targeted transcript
	Inhibition of translation from an injected mRNA containing the MO binding site	Indicates that MO has capacity to inhibit translation of mRNA	Does not indicate that the phenotype is due to specific targeting of endogenous RNA

(continued)

Table 19.2
(continued)

Aim of control	Control	Advantages	Disadvantages
Proving the phenotype is specific to MO knockdown of targeted mRNA only	Correlation of phenotype and expression pattern → logical biology of the phenotype	Direct correlation of specific expression and specific phenotype gives confidence in analysis	Does not prove MO phenotype is specific to targeted transcript inhibition. Most applicable to genes with anatomically restricted gene expression patterns
	Direct rescue of phenotype by mRNA lacking MO binding sites	Indicates definitively that rescuing mRNA can functionally substitute for loss of targeted mRNA *in vivo*	Can be problematic to interpret if mRNA itself results in strong dominant overexpression phenotype
	Multiple, preferably non-overlapping, MOs producing the same phenotype	Increases the likelihood that a phenotype is specific to the gene to which MOs are being designed	In a series of MOs, individual MOs that result in an atypical phenotype for the series complicate rather than clarify the interpretation
	Observation of normal development in multiple unaffected tissues	Improves confidence in specificity	Does not prove MO phenotype is specific to targeted transcript inhibition
	Correlation of mor-phant phenotype with hypomorphic and/or null mutant phenotypes	The "gold-standard" specificity control	Only applicable when a char-acterized mutant exists; Not applicable to "discovery" reverse genetic studies with MOs

MO, antisense morpholino oligomer.
*Not recommended as a sufficient single control.

Acknowledgments

The authors thank Pierre Smith, Department of Surgery, Royal Melbourne Hospital for photography and Andrew Badrock from the Heath laboratory for appearing in **Fig. 19.1** and setting up the materials for the other figures.

References

1. Kimmel, C. B., Ballard, W. W., Kimmel, S. R., (1995) Stages of embryonic development of the zebrafish. *Dev Dyn* 203, 253–310.

2. Gilmour, D. T., Jessen, J. R., Lin, S. (2002) in (Nusslein-Volhard C., D. R., ed.) *Zebrafish: Practical Approach*, pp. 121–143, Oxford University Press, New York.

3. Driever, W., Stemple, D., Schier, A., (1994) Zebrafish: genetic tools for studying vertebrate development. *Trends Genet* 10, 152–159.

4. Lane, C. D. (1983) The fate of genes, messengers, and proteins introduced into Xenopus oocytes. *Curr Top Dev Biol* 18, 89–116.

5. Krieg, P. A., Melton, D. A. (1984) Functional messenger RNAs are produced by SP6 in vitro transcription of cloned cDNAs. *Nucleic Acids Res* 12, 7057–7070.

6. Durbin, L., Brennan, C., Shiomi, K., (1998) Eph signaling is required for segmentation and differentiation of the somites. *Genes Dev* 12, 3096–3109.

7. Ando, H., Furuta, T., Tsien, R. Y., (2001) Photo-mediated gene activation using caged RNA/DNA in zebrafish embryos. *Nat Genet* 28, 317–325.

8. Higashijima, S., Okamoto, H., Ueno, N., (1997) High-frequency generation of transgenic zebrafish which reliably express GFP in whole muscles or the whole body by using promoters of zebrafish origin. *Dev Biol* 192, 289–299.

9. Bayer, T. A., Campos-Ortega, J. A. (1992) A transgene containing lacZ is expressed in primary sensory neurons in zebrafish. *Development* 115, 421–426.

10. Stuart, G. W., Vielkind, J. R., McMurray, J. V., (1990) Stable lines of transgenic zebrafish exhibit reproducible patterns of transgene expression. *Development* 109, 577–584.

11. Joore, J. (1999) Promoter analysis in zebrafish embryos. *Methods Mol Biol* 127, 155–166.

12. Liu, W. Y., Wang, Y., Sun, Y. H., (2005) Efficient RNA interference in zebrafish embryos using siRNA synthesized with SP6 RNA polymerase. *Dev Growth Differ* 47, 323–331.

13. Dodd, A., Chambers, S. P., Love, D. R. (2004) Short interfering RNA-mediated gene targeting in the zebrafish. *FEBS Lett* 561, 89–93.

14. Zhao, Z., Cao, Y., Li, M., (2001) Double-stranded RNA injection produces non-specific defects in zebrafish. *Dev Biol* 229, 215–223.

15. Acosta, J., Carpio, Y., Borroto, I., (2005) Myostatin gene silenced by RNAi show a zebrafish giant phenotype. *J Biotechnol* 119, 324–331.

16. Wang, N., Sun, Y. H., Liu, J., et al. (2007) Knock down of gfp and no tail expression in zebrafish embryo by in vivo-transcribed short hairpin RNA with T7 plasmid system. *J Biomed Sci*. epub. Jul 12.

17. Oates, A. C., Bruce, A. E., Ho, R. K. (2000) Too much interference: injection of double-stranded RNA has nonspecific effects in the zebrafish embryo. *Dev Biol* 224, 20–28.

18. Nasevicius, A., Ekker, S. C. (2000) Effective targeted gene 'knockdown' in zebrafish. *Nat Genet* 26, 216–220.

19. Nasevicius, A., Ekker, S. C. (2001) The zebrafish as a novel system for functional genomics and therapeutic development applications. *Curr Opin Mol Ther* 3, 224–228.

20. Summerton, J., Weller, D. (1997) Morpholino antisense oligomers: design, preparation, and properties. *Antisense Nucleic Acid Drug Dev* 7, 187–195.

21. Summerton, J. (1999) Morpholino antisense oligomers: the case for an RNase H-independent structural type. *Biochim Biophys Acta* 1489, 141–158.

22. Heasman, J. (2002) Morpholino oligos: making sense of antisense? *Dev Biol* 243, 209–214.

23. Phillips, B. T., Bolding, K., Riley, B. B. (2001) Zebrafish fgf3 and fgf8 encode redundant functions required for otic placode induction. *Dev Biol* 235, 351–365.

24. Segawa, H., Miyashita, T., Hirate, Y., (2001) Functional repression of Islet-2 by disruption of complex with Ldb impairs peripheral axonal outgrowth in embryonic zebrafish. *Neuron* 30, 423–436.

25. Muller, F., Lakatos, L., Dantonel, J., (2001) TBP is not universally required for zygotic RNA polymerase II transcription in zebrafish. *Curr Biol* 11, 282–287.

26. Shepherd, I. T., Beattie, C. E., Raible, D. W. (2001) Functional analysis of zebrafish GDNF. *Dev Biol* 231, 420–435.

27. Yang, Z., Liu, N., Lin, S. (2001) A zebrafish forebrain-specific zinc finger gene can

induce ectopic dlx2 and dlx6 expression. *Dev Biol* 231, 138–148.

28. Bauer, H., Lele, Z., Rauch, G. J., (2001) The type I serine/threonine kinase receptor Alk8/Lost-a-fin is required for Bmp2b/7 signal transduction during dorsoventral patterning of the zebrafish embryo. *Development* 128, 849–858.

29. Ross, J. J., Shimmi, O., Vilmos, P., (2001) Twisted gastrulation is a conserved extracellular BMP antagonist. *Nature* 410, 479–483.

30. Nasevicius, A., Larson, J., Ekker, S. C. (2000) Distinct requirements for zebrafish angiogenesis revealed by a VEGF-A morphant. *Yeast* 17, 294–301.

31. (2001) Special morpholino edition. *Genesis* 30 (3).

32. Corey, D. R., Abrams, J. M. (2001) Morpholino antisense oligonucleotides: tools for investigating vertebrate development. *Genome Biol* 2, REVIEWS1015.

33. Tang, X., Maegawa, S., Weinberg, E. S., et al. (2007) Regulating Gene Expression in Zebrafish Embryos Using Light-Activated, Negatively Charged Peptide Nucleic Acids. *J Am Chem Soc.* epub. Aug 21.

34. Urtishak, K. A., Choob, M., Tian, X., (2003) Targeted gene knockdown in zebrafish using negatively charged peptide nucleic acid mimics. *Dev Dyn* 228, 405–413.

35. Westerfield, M. (2000) The zebrafish book. A guide for the laboratory use of zebrafish (Danio rerio). University of Oregon Press, Eugene.

36. Amoyel, M., Cheng, Y. C., Jiang, Y. J., (2005) Wnt1 regulates neurogenesis and mediates lateral inhibition of boundary cell specification in the zebrafish hindbrain. *Development* 132, 775–785.

37. Lekven, A. C., Buckles, G. R., Kostakis, N., (2003) Wnt1 and wnt10b function redundantly at the zebrafish midbrain-hindbrain boundary. *Dev Biol* 254, 172–187.

38. Wakahara, T., Kusu, N., Yamauchi, H., (2007) Fibin, a novel secreted lateral plate mesoderm signal, is essential for pectoral fin bud initiation in zebrafish. *Dev Biol* 303, 527–535.

39. Ober, E. A., Verkade, H., Field, H. A., (2006) Mesodermal Wnt2b signalling positively regulates liver specification. *Nature* 442, 688–691.

40. Shimizu, T., Bae, Y. K., Muraoka, O., (2005) Interaction of Wnt and caudal-related genes in zebrafish posterior body formation. *Dev Biol* 279, 125–141.

41. Matsui, T., Raya, A., Kawakami, Y., (2005) Noncanonical Wnt signaling regulates midline convergence of organ primordia during zebrafish development. *Genes Dev* 19, 164–175.

42. Kim, H. J., Schleiffarth, J. R., Jessurun, J., (2005) Wnt5 signaling in vertebrate pancreas development. *BMC Biol* 3, 23.

43. Lele, Z., Bakkers, J., Hammerschmidt, M. (2001) Morpholino phenocopies of the swirl, snailhouse, somitabun, minifin, silberblick, and pipetail mutations. *Genesis* 30, 190–194.

44. Lekven, A. C., Thorpe, C. J., Waxman, J. S., (2001) Zebrafish wnt8 encodes two wnt8 proteins on a bicistronic transcript and is required for mesoderm and neurectoderm patterning. *Dev Cell* 1, 103–114.

45. Ramel, M. C., Buckles, G. R., Baker, K. D., (2005) WNT8 and BMP2B co-regulate non-axial mesoderm patterning during zebrafish gastrulation. *Dev Biol* 287, 237–248.

46. Kim, S. H., Shin, J., Park, H. C., (2002) Specification of an anterior neuroectoderm patterning by Frizzled8a-mediated Wnt8b signalling during late gastrulation in zebrafish. *Development* 129, 4443–4455.

47. Riley, B. B., Chiang, M. Y., Storch, E. M., (2004) Rhombomere boundaries are Wnt signaling centers that regulate metameric patterning in the zebrafish hindbrain. *Dev Dyn* 231, 278–291.

48. Muyskens, J. B., Kimmel, C. B. (2007) Tbx16 cooperates with Wnt11 in assembling the zebrafish organizer. *Mech Dev* 124, 35–42.

49. Fong, S. H., Emelyanov, A., Teh, C., (2005) Wnt signalling mediated by Tbx2b regulates cell migration during formation of the neural plate. *Development* 132, 3587–3596.

50. Wada, H., Tanaka, H., Nakayama, S., (2006) Frizzled3a and Celsr2 function in the neuroepithelium to regulate migration of facial motor neurons in the developing zebrafish hindbrain. *Development* 133, 4749–4759.

51. Cavodeassi, F., Carreira-Barbosa, F., Young, R. M., (2005) Early stages of zebrafish eye formation require the coordinated activity of Wnt11, Fz5, and the Wnt/beta-catenin pathway. *Neuron* 47, 43–56.

52. Lyman Gingerich, J., Westfall, T. A., Slusarski, D. C., (2005) hecate, a zebrafish maternal effect gene, affects dorsal organizer

induction and intracellular calcium transient frequency. *Dev Biol* 286, 427–439.

53. Bellipanni, G., Varga, M., Maegawa, S., (2006) Essential and opposing roles of zebrafish beta-catenins in the formation of dorsal axial structures and neurectoderm. *Development* 133, 1299–1309.

54. Nadauld, L. D., Sandoval, I. T., Chidester, S., (2004) Adenomatous polyposis coli control of retinoic acid biosynthesis is critical for zebrafish intestinal development and differentiation. *J Biol Chem* 279, 51581–51589.

55. Shelton, D. N., Sandoval, I. T., Eisinger, A., (2006) Up-regulation of CYP26A1 in adenomatous polyposis coli-deficient vertebrates via a WNT-dependent mechanism: implications for intestinal cell differentiation and colon tumor development. *Cancer Res* 66, 7571–7577.

56. Meier, N., Krpic, S., Rodriguez, P., (2006) Novel binding partners of Ldb1 are required for haematopoietic development. *Development* 133, 4913–4923.

57. Nyholm, M. K., Wu, S. F., Dorsky, R. I., (2007) The zebrafish zic2a-zic5 gene pair acts downstream of canonical Wnt signaling to control cell proliferation in the developing tectum. *Development* 134, 735–746.

58. Dorsky, R. I., Itoh, M., Moon, R. T., (2003) Two tcf3 genes cooperate to pattern the zebrafish brain. *Development* 130, 1937–1947.

59. Caneparo, L., Huang, Y. L., Staudt, N., (2007) Dickkopf-1 regulates gastrulation movements by coordinated modulation of Wnt/beta catenin and Wnt/PCP activities, through interaction with the Dally-like homolog Knypek. *Genes Dev* 21, 465–480.

60. Ishitani, T., Matsumoto, K., Chitnis, A. B., (2005) Nrarp functions to modulate neural-crest-cell differentiation by regulating LEF1 protein stability. *Nat Cell Biol* 7, 1106–1112.

61. Angers, S., Thorpe, C. J., Biechele, T. L., (2006) The KLHL12-Cullin-3 ubiquitin ligase negatively regulates the Wnt-beta-catenin pathway by targeting Dishevelled for degradation. *Nat Cell Biol* 8, 348–357.

62. Schwarz-Romond, T., Asbrand, C., Bakkers, J., (2002) The ankyrin repeat protein Diversin recruits Casein kinase Iepsilon to the beta-catenin degradation complex and acts in both canonical Wnt and Wnt/JNK signaling. *Genes Dev* 16, 2073–2084.

63. Seiliez, I., Thisse, B., Thisse, C. (2006) FoxA3 and goosecoid promote anterior neural fate through inhibition of Wnt8a activity before the onset of gastrulation. *Dev Biol* 290, 152–163.

64. Waxman, J. S., Hocking, A. M., Stoick, C. L.., (2004) Zebrafish Dapper1 and Dapper2 play distinct roles in Wnt-mediated developmental processes. *Development* 131, 5909–5921.

65. Zhang, L., Zhou, H., Su, Y.., (2004) Zebrafish Dpr2 inhibits mesoderm induction by promoting degradation of nodal receptors. *Science* 306, 114–117.

66. Witzel, S., Zimyanin, V., Carreira-Barbosa, F.., (2006) Wnt11 controls cell contact persistence by local accumulation of Frizzled 7 at the plasma membrane. *J Cell Biol* 175, 791–802.

67. Zhu, S., Liu, L., Korzh, V.., (2006) RhoA acts downstream of Wnt5 and Wnt11 to regulate convergence and extension movements by involving effectors Rho kinase and Diaphanous: use of zebrafish as an in vivo model for GTPase signaling. *Cell Signal* 18, 359–372.

68. Formstone, C. J., Mason, I. (2005) Combinatorial activity of Flamingo proteins directs convergence and extension within the early zebrafish embryo via the planar cell polarity pathway. *Dev Biol* 282, 320–335.

69. Park, M., Moon, R. T. (2002) The planar cell-polarity gene stbm regulates cell behaviour and cell fate in vertebrate embryos. *Nat Cell Biol* 4, 20–25.

70. Carreira-Barbosa, F., Concha, M. L., Takeuchi, M.., (2003) Prickle 1 regulates cell movements during gastrulation and neuronal migration in zebrafish. *Development* 130, 4037–4046.

71. Wada, H., Iwasaki, M., Sato, T.., (2005) Dual roles of zygotic and maternal Scribble1 in neural migration and convergent extension movements in zebrafish embryos. *Development* 132, 2273–2285.

72. Sprague, J., Bayraktaroglu, L., Clements, D..,(2006) The Zebrafish Infromation network (ZFIN): the zebrafish model organism database. *Nucleic Acids Res* 34, D581–D585.

73. Rembold, M., Lahiri, K., Foulkes, N. S., (2006) Transgenesis in fish: efficient selection of transgenic fish by co-injection with a fluorescent reporter construct. *Nat Protoc* 1, 1133–1139.

74. Ogino, H., McConnell, W. B., Grainger, R. M. (2006) High-throughput transgenesis in

Xenopus using I-SceI meganuclease. *Nat Protoc* 1, 1703–1710.

75. Thermes, V., Grabher, C., Ristoratore, F., (2002) I-SceI meganuclease mediates highly efficient transgenesis in fish. *Mech Dev* 118, 91–98.

76. Hermanson, S., Davidson, A. E., Sivasubbu, S., (2004) Sleeping Beauty transposon for efficient gene delivery. *Methods Cell Biol* 77, 349–362.

77. Kawakami, K. (2005) Transposon tools and methods in zebrafish. *Dev Dyn* 234, 244–254.

78. Fisher, S., Grice, E. A., Vinton, R. M., (2006) Evaluating the biological relevance of putative enhancers using Tol2 transposon-mediated transgenesis in zebrafish. *Nat Protoc* 1, 1297–1305.

79. Turner, D. L., Weintraub, H. (1994) Expression of achaete-scute homolog 3 in Xenopus embryos converts ectodermal cells to a neural fate. *Genes Dev* 8, 1434–1447.

80. Pelegri, F. (2003) Maternal factors in zebrafish development. *Dev Dyn* 228, 535–554.

81. Fernandez, J., Valladares, M., Fuentes, R., (2006) Reorganization of cytoplasm in the zebrafish oocyte and egg during early steps of ooplasmic segregation. *Dev Dyn* 235, 656–671.

82. Robu, M. E., Larson, J. D., Nasevicius, A., (2007) p53 activation by knockdown technologies. *PLoS Genet* 3, e78.

83. Kang, J. S., Oohashi, T., Kawakami, Y., (2004) Characterization of dermacan, a novel zebrafish lectican gene, expressed in dermal bones. *Mech Dev* 121, 301–312.

84. Oates, A. C., Ho, R. K. (2002) Hairy/E(spl)-related (Her) genes are central components of the segmentation oscillator and display redundancy with the Delta/Notch signaling pathway in the formation of anterior segmental boundaries in the zebrafish. *Development* 129, 2929–2946.

85. Szeto, D. P., Griffin, K. J., Kimelman, D. (2002) HrT is required for cardiovascular development in zebrafish. *Development* 129, 5093–5101.

86. Draper, B. W., Morcos, P. A., Kimmel, C. B. (2001) Inhibition of zebrafish fgf8 pre-mRNA splicing with morpholino oligos: a quantifiable method for gene knockdown. *Genesis* 30, 154–156.

87. Yan, Y. L., Miller, C. T., Nissen, R. M., (2002) A zebrafish sox9 gene required for cartilage morphogenesis. *Development* 129, 5065–5079.

88. Parker, L. H., Schmidt, M., Jin, S. W., (2004) The endothelial-cell-derived secreted factor Egfl7 regulates vascular tube formation. *Nature* 428, 754–758.

89. Chocron, S., Verhoeven, M. C., Rentzsch, F., (2007) Zebrafish Bmp4 regulates left-right asymmetry at two distinct developmental time points. *Dev Biol* 305, 577–588.

90. Knight, R. D., Nair, S., Nelson, S. S., (2003) lockjaw encodes a zebrafish tfap2a required for early neural crest development. *Development* 130, 5755–5768.

91. Busch-Nentwich, E., Sollner, C., Roehl, H., (2004) The deafness gene dfna5 is crucial for ugdh expression and HA production in the developing ear in zebrafish. *Development* 131, 943–951.

92. Weidinger, G., Stebler, J., Slanchev, K., (2003) dead end, a novel vertebrate germ plasm component, is required for zebrafish primordial germ cell migration and survival. *Curr Biol* 13, 1429–1434.

93. Elsalini, O. A., von Gartzen, J., Cramer, M., (2003) Zebrafish hhex, nk2.1a, and pax2.1 regulate thyroid growth and differentiation downstream of Nodal-dependent transcription factors. *Dev Biol* 263, 67–80.

94. Jopling, C., den Hertog, J. (2005) Fyn/Yes and non-canonical Wnt signalling converge on RhoA in vertebrate gastrulation cell movements. *EMBO Rep* 6, 426–431.

95. Shestopalov, I. A., Sinha, S., Chen, J. K. (2007) Light-controlled gene silencing in zebrafish embryos. *Nat Chem Biol* 3, 650–651.

96. Schneider, S., Steinbeisser, H., Warga, R. M., (1996) Beta-catenin translocation into nuclei demarcates the dorsalizing centers in frog and fish embryos. *Mech Dev* 57, 191–198.

97. Stachel, S. E., Grunwald, D. J., Myers, P. Z. (1993) Lithium perturbation and goosecoid expression identify a dorsal specification pathway in the pregastrula zebrafish. *Development* 117, 1261–1274.

98. van de Water, S., van de Wetering, M., Joore, J., (2001) Ectopic Wnt signal determines the eyeless phenotype of zebrafish masterblind mutant. *Development* 128, 3877–3888.

99. Nojima, H., Shimizu, T., Kim, C. H., (2004) Genetic evidence for involvement of maternally derived Wnt canonical signaling in dorsal determination in zebrafish. *Mech Dev* 121, 371–386.

100. Heisenberg, C. P., Tada, M., Rauch, G. J., (2000) Silberblick/Wnt11 mediates convergent

extension movements during zebrafish gastrulation. *Nature* 405, 76–81.

101. Pai, L. M., Orsulic, S., Bejsovec, A., (1997) Negative regulation of Armadillo, a Wingless effector in Drosophila. *Development* 124, 2255–2266.

102. Marlow, F., Topczewski, J., Sepich, D., (2002) Zebrafish Rho kinase 2 acts downstream of Wnt11 to mediate cell polarity and effective convergence and extension movements. *Curr Biol* 12, 876–884.

103. Nasevicius, A., Hyatt, T., Kim, H., (1998) Evidence for a frizzled-mediated wnt pathway required for zebrafish dorsal mesoderm formation. *Development* 125, 4283–4292.

104. Axelrod, J. D., Miller, J. R., Shulman, J. M., (1998) Differential recruitment of Dishevelled provides signaling specificity in the planar cell polarity and Wingless signaling pathways. *Genes Dev* 12, 2610–2622.

105. Boutros, M., Mihaly, J., Bouwmeester, T., (2000) Signaling specificity by Frizzled receptors in Drosophila. *Science* 288, 1825–1828.

106. Penton, A., Wodarz, A., Nusse, R. (2002) A mutational analysis of dishevelled in Drosophila defines novel domains in the dishevelled protein as well as novel suppressing alleles of axin. *Genetics* 161, 747–762.

107. Tada, M., Smith, J. C. (2000) Xwnt11 is a target of Xenopus Brachyury: regulation of gastrulation movements via Dishevelled, but not through the canonical Wnt pathway. *Development* 127, 2227–2238.

108. Pelegri, F., Maischein, H. M. (1998) Function of zebrafish beta-catenin and TCF-3 in dorsoventral patterning. *Mech Dev* 77, 63–74.

109. Ciruna, B., Jenny, A., Lee, D., (2006) Planar cell polarity signalling couples cell division and morphogenesis during neurulation. *Nature* 439, 220–224.

110. Jenny, A., Darken, R. S., Wilson, P. A., (2003) Prickle and Strabismus form a functional complex to generate a correct axis during planar cell polarity signaling. *Embo J* 22, 4409–4420.

111. Halloran, M. C., Sato-Maeda, M., Warren, J. T., (2000) Laser-induced gene expression in specific cells of transgenic zebrafish. *Development* 127, 1953–1960.

112. Ulrich, F., Krieg, M., Schotz, E. M., (2005) Wnt11 functions in gastrulation by controlling cell cohesion through Rab5c and E-cadherin. *Dev Cell* 9, 555–564.

113. Lewis, J. L., Bonner, J., Modrell, M., (2004) Reiterated Wnt signaling during zebrafish neural crest development. *Development* 131, 1299–1308.

114. Stoick-Cooper, C. L., Weidinger, G., Riehle, K. J., (2007) Distinct Wnt signaling pathways have opposing roles in appendage regeneration. *Development* 134, 479–489.

115. Ueno, S., Weidinger, G., Osugi, T., (2007) Biphasic role for Wnt/beta-catenin signaling in cardiac specification in zebrafish and embryonic stem cells. *Proc Natl Acad Sci USA* 104, 9685–9690.

116. Weidinger, G., Thorpe, C. J., Wuennenberg-Stapleton, K., (2005) The Sp1-related transcription factors sp5 and sp5-like act downstream of Wnt/beta-catenin signaling in mesoderm and neuroectoderm patterning. *Curr Biol* 15, 489–500.

117. Dorsky, R. I., Sheldahl, L. C., Moon, R. T. (2002) A transgenic Lef1/beta-catenin-dependent reporter is expressed in spatially restricted domains throughout zebrafish development. *Dev Biol* 241, 229–237.

118. Eisen, J. S., Smith, J. C. (2008) Controlling morpholino experiments: don't stop making antisense. *Development* 135, 1735–1743.

Chapter 20

Neural Patterning and CNS Functions of Wnt in Zebrafish

Richard I. Dorsky

Abstract

Wnt signaling plays several important roles in the development of the zebrafish central nervous system (CNS). This chapter outlines both the known and postulated roles of Wnts from the earliest step of neural plate induction to relatively late events such as axon pathfinding and synaptogenesis. The common tools useful for examining Wnt function and nervous system development in zebrafish are first reviewed. Examples are then provided for specific phenotypes resulting from gain and loss of Wnt activity at multiple developmental stages. Finally, specific assays and reagents that can be used to investigate the function of novel Wnt pathway components in CNS development are listed.

Key words: Wnt, zebrafish, patterning, neurogenesis, CNS.

1. Introduction

Since the initial characterization of Wnts in vertebrate organisms, it has been apparent that they play important roles in the development of the central nervous system (CNS). In the first report of a vertebrate Wnt homolog, expression of the mouse *Wnt1* gene was observed in the dorsal midline of the CNS *(1)*. Shortly thereafter, *Wnt1* function was found to be required for development of the midbrain and cerebellum *(2)*. Subsequently, many Wnt family members have been localized to the developing vertebrate CNS, and it has become clear that these molecules perform a complex array of functions throughout embryogenesis.

Zebrafish have been particularly useful in uncovering the roles of Wnt signaling during CNS development for several reasons. First, the organization of the zebrafish CNS is highly conserved with other vertebrates at the anatomical and molecular

Elizabeth Vincan (ed.), *Wnt Signaling, Volume II: Pathway Models, vol. 469*
© 2008 Humana Press, a part of Springer Science + Business Media, New York, NY
Book doi: 10.1007/978-1-60327-469-2

level, and serves as a good model for mammals in all regions except the cortical telencephalon. Second, the number of molecular tools available for gene identification and manipulation make a wide variety of assays possible. Third, the rapid development of transparent embryos allows for detailed observation of the CNS in vivo using transgenic reporters to mark cells.

Because Wnt signaling is critical for so many aspects of embryogenesis, it is important to consider each phase of CNS development separately with respect to Wnt function. Therefore this chapter has been broken into sections detailing the major distinct roles reported for Wnts, from maternal gene function through neuronal differentiation. While not comprehensive, it provides an overview of the different and sometimes contradictory ways that one signaling pathway can regulate multiple cellular and molecular processes.

The vertebrate CNS develops through several steps that are conserved in all organisms, and each step requires Wnt signals in a distinct way. *(1)* The neural plate is induced during gastrulation by the division of the ectoderm layer into neural and non-neural territories. *(2)* The unpatterned neural plate acquires anterior/posterior positional information from adjacent tissues, which are then refined by signals within the CNS. *(3)* The neural plate undergoes morphological change to a neural tube through a process of convergent extension. *(4)* Distinct morphological territories with unique gene expression arise and boundaries are created between them. *(5)* The CNS acquires dorsal/ventral positional information from adjacent tissues. *(6)* Neural progenitors begin to exit the cell cycle and express neuron-specific genes. *(7)* Neurons differentiate, adopting unique phenotypes with respect to axon guidance, neurotransmitter expression, and synapse formation.

2. Tools for Analysis

There are several important tools available for investigating Wnt function in zebrafish CNS development. While none is specific to zebrafish, the combination of approaches in a single vertebrate system is unique.

1. Mutants: Several large-scale ENU mutagenesis screens were carried out in the 1990s, followed by smaller screens from individual laboratories in recent years *(3)*. After initial characterization of the mutant phenotypes, many of the genes have been identified by positional cloning. The ability to examine hypomorphic and null alleles for many Wnt pathway components has been extremely valuable in assaying Wnt

functions during development. Mutant alleles have been generated by other means, including gamma-ray mutagenesis *(4)* and recently by retroviral and transposon insertions *(5–7)*. Thanks to the generosity of the zebrafish research community, many of these mutant lines have been made widely available to researchers, allowing them to investigate gene function in their tissue of interest.

2. Morpholinos: Because there is no existing technology to perform targeted gene knockouts in zebrafish and RNAi methods have not yet been successfully adapted to the system, morpholino oligonucleotides are commonly used to knock down gene function *(8)*. Morpholinos are designed to block either mRNA translation or splicing, and when injected into one-cell embryos can inhibit maternal and zygotic gene function, respectively. Although this technique is not as effective as genetic mutations for studying gene function at later developmental stages, most of the events reviewed in this chapter can be examined using morpholinos.

3. mRNA injections: The most widely used technique to overexpress genes in zebrafish is injection of synthetic mRNA at the one-cell stage. These injections lead to widespread gene expression throughout the first 24 hours of development, which can be traced, by co-injection or gene fusion with a marker protein such as GFP. As with morpholinos, most of the events reviewed in this chapter can be examined during this time frame. Unlike in *Xenopus*, it is difficult to target mRNA injections to specific regions of the embryo, mostly because the fate map is not established until later in development *(9, 10)*. Nevertheless this technique is a rapid way to test whether a gene is sufficient to regulate a specific developmental event, and to inhibit protein function using dominant-negative constructs.

4. Functional transgenes: The ability to easily create transgenic zebrafish has also led to innovative methods to investigate gene function *(11)*. Expression constructs under the control of tissue-specific and inducible promoters allow misexpression of wild-type and mutant genes at any time or place of interest. Coupled with fluorescent protein tags, researchers can visualize misexpressing cells *in vivo* and in fixed tissue and carefully analyze resulting phenotypes.

5. Reporter transgenes: As in other organisms, it is possible to fuse upstream gene regulatory regions to reporter molecules such as GFP, and create transgenic zebrafish expressing these constructs. This technology allows researchers to visualize gene activity *in vivo*, and specific gene expression patterns can be used to label discrete cell populations in the CNS. Originally this technique was extremely inefficient, with a

Table 20.1
List of mRNA probes described in this chapter

mRNA probe	Structures labeled	Reference
bozozok	Dorsal organizer	*20, 21*
squint	Dorsal organizer	*18, 23*
chordin	Dorsal organizer	*23*
goosecoid	Dorsal organizer	*23*
soxB1 genes	Neural plate, hypothalamus	*14, 24*
pax6	Dorsal diencephalon	*27*
pax2.1	Midbrain, midbrain-hindbrain boundary	*27, 44*
gbx1	Hindbrain	*27*
nkx2.2a	Hypothalamus	*28*
hgg1	Prechordal plate	*41*
dlx3	Neural plate boundary	*41*
fgf8	Midbrain-hindbrain boundary	*44*
lunatic fringe	Rhombomere boundaries	*47*
foxb1.2	Rhombomere boundaries	*47*
nkx6.1	Ventral spinal progenitors	*50*
iro3	Intermediate spinal progenitors	*51*
pax3	Dorsal spinal progenitors	*52*
isl1	Spinal motoneurons	*53*
en1b	Spinal interneurons	*54*
chx10	Spinal interneurons	*55*
p27, p57	Postmitotic neurons	*58*
cycD1	Mitotic neural progenitors	*59*
neurogenin1	Subclasses of progenitors and neurons	*65*
zash1a	Subclasses of progenitors and neurons	*65*
neuroD	Subclasses of neurons	*65*
vglut2	Glutamatergic neurons	*67*
glyt2	Glycinergic neurons	*67*
gad65/67	GABAergic neurons	*67*

high degree of mosaicism in transient transgenics and a low rate of stable transgenesis. However recent advances in transgene technology, including retroviral and transposon vectors *(5–7)* have greatly increased the transgenesis rate and also permit reporter analysis in transients.

6. *In situ* hybridization: While this is a standard technique for investigating gene function in many developmental systems, it is particularly useful in zebrafish because of embryonic transparency. All cell types of the developing embryo can be visualized throughout the first several days of embryogenesis, and the protocol is relatively simple, requiring a similar level of effort to immunohistochemistry *(12, 13)*. The vast majority of gene expression patterns described in the literature have been examined using this method, and most phenotypic assays for gene function are also performed by *in situ* analysis. **Table 20.1** contains references for all of the mRNA probes listed in this chapter.

7. Biochemistry: Standard biochemical techniques are relatively straightforward in zebrafish, and it is easy to collect material due to the large clutch sizes. In many cases reagents such as antibodies have been developed specifically for zebrafish because versions developed for other species do not work. Recent powerful methods for studying gene function, such as chromatin immunoprecipitation (ChIP) *(14)* and microarrays *(15)*, have been exploited successfully in the zebrafish system to identify Wnt targets.

3. Assays for Wnt Function in the Zebrafish CNS

3.1. CNS Induction via Dorsal Organizer Formation

1. Manipulation of pathway function: Because the dorsal organizer acts as a source of BMP antagonists required for neural induction, establishment of this signaling center is critical for formation of the neural plate. As in other vertebrates, β-catenin accumulates in nuclei on the presumptive dorsal side of the zebrafish embryo prior to the initiation of zygotic transcription *(16)*. This β-catenin activity regulates the expression of organizer-specific genes that then affect neural plate induction (**Fig. 20.1A**). To study maternal signaling components, mRNA overexpression of wild-type and dominant-negative constructs is used in assays for effects on organizer size. In addition, genetic mutants for *β-catenin* and *wnt5a* produce embryos with altered dorsal/ventral patterning, but maternal function must be removed for these phenotypes to appear *(17–19)*. Transgenic methods of gene

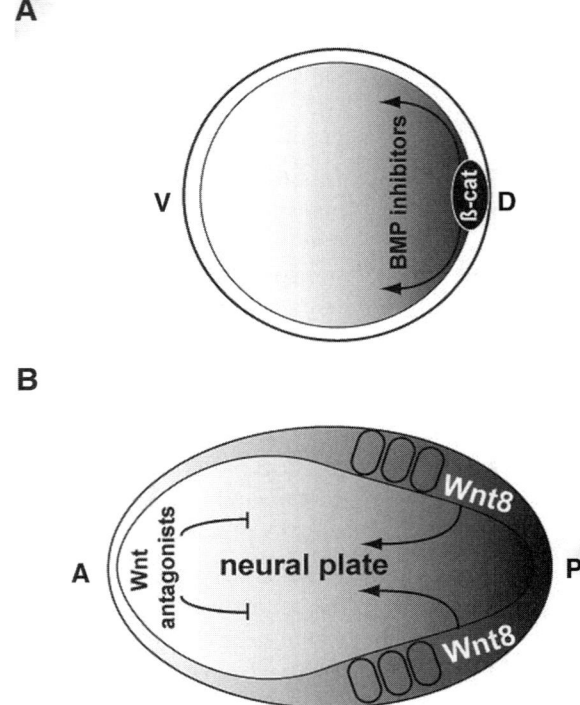

Fig. 20.1. Maternal and early zygotic Wnt function. (**A**) At late blastula stages, β-catenin accumulates in nuclei on the presuptive dorsal (D) side of zebrafish embryos. Transcriptional activity downstream of β-catenin regulates the expression of organizer-specific genes, which in turn lead to the secretion of BMP inhibitors. These inhibitors antagonize BMPs secreted from the ventral (V) side of the embryo and allow neural plate formation from dorsal ectoderm. (**B**) During late gastrulation, Wnt signals from the posterior-ventral mesoderm (P), including Wnt8, induce ventral mesoderm markers as well as posterior markers in the neural plate. Wnt antagonists expressed in the anterior (A) embryo counteract the posterior Wnt signal and allow anterior brain structures to be formed.

misexpression are not effective at these stages because they cannot be activated until the onset of zygotic transcription.

2. Phenotypic assays: There are several key markers useful for assaying dorsal organizer formation and neural plate induction. First, the *bozozok* gene is expressed in dorsal blastoderm and yolk syncytial layer, and has been shown to be a direct target of the Wnt pathway *(20, 21)*. In addition, the Wnt transcriptional reporter TOPdGFP is expressed in a similar pattern *(22)*. Another dorsal-specific gene, *squint*, is dependent on Wnt signaling *(18, 23)* but has not been shown to be a direct target. Expansion or inhibition of any of these markers at late blastula or early gastrula stages is

an indication of gain or loss of Wnt signaling, respectively. Later dorsal mesoderm markers such as *chordin* and *goosecoid* are also used to assay for changes in dorsal/ventral patterning *(23)*, but must be interpreted more carefully. Because these markers are subsequently inhibited by zygotic Wnt signaling, they must be assayed during early gastrulation to provide information about the maternal Wnt pathway. Finally, pan-neural plate markers such as *soxB1* genes can be used to assay for neural induction as a readout of organizer activity *(24)*.

3.2. Anterior/posterior Patterning of the CNS by Mesodermal Wnts

1. Manipulation of pathway function: Zygotic Wnt function, primarily mediated by *wnt8*, acts to restrict the size of the dorsal organizer, and thus inhibit neural induction *(25, 26)*. Consistent with this model, a reporter for Wnt activity is expressed in the ventrolateral mesoderm during late gastrulation *(22)*. Secondarily, *wnt8* in the ventrolateral mesoderm acts directly on the neural plate as a posteriorizing signal *(25, 26)*. This Wnt signal is antagonized by a variety of factors in the anterior embryo, which act at multiple levels of the pathway. During somitogenesis, a Wnt reporter is thus expressed in a posterior to anterior gradient *(22)*. Loss-of-function (mutant and morphant) *wnt8* embryos exhibit expansion of anterior brain structures, while mutants and morphants for *tcf3* and *axin1* exhibit posteriorized brains *(27–31)*. More generally, any down-regulation of Wnt activity would be expected to anteriorize the CNS and up-regulation of the pathway would tend to posteriorize the CNS.

2. Phenotypic assays: Most of the assays for anterior/posterior brain patterning involve in situ detection of regionally expressed markers. Specific markers for forebrain (*pax6*), midbrain (*pax2.1*), and hindbrain (*gbx1*) can be distinguished by the end of gastrulation *(27)*. For example, *pax2*, which marks presumptive midbrain can be expanded anteriorly by moderate activation of Wnt signaling *(29)*. However, this same marker is completely eliminated if the pathway is highly activated, as a hindbrain marker (*gbx1*) spreads to the anterior limit of the CNS *(27)*. It is therefore critical to examine multiple regional markers when characterizing brain phenotypes. This approach can also help distinguish anterior/posterior patterning defects from other effects on regional patterning. For example, markers for dorsal diencephalon (*pax6*) can be used concordantly with markers for ventral diencephalon (*nkx2.2a*) to analyze the status of this brain region *(27, 28)*.

3.3. Regulation of CNS Morphogenesis by Non-canonical Wnt Signaling

1. Manipulation of pathway function: A separate role for non-canonical Wnt signaling has been described in the process of convergence and extension movements, which are critical for transforming the two-dimensional neural plate into the three-dimensional neural keel (**Fig. 20.2**). Mutants and morphants for *knypek* (glypican) *trilobite* (van gogh-like), *prickle*, *dishevelled*, *wnt5b*, and *wnt11* have defects in neural plate convergence and extension *(32–37)*. Secondarily, a mutation in *trilobite* has been shown to affect the polarity of progenitor cells during neural keel intercalation independently from the process of convergence and extension *(38)*. Finally, *trilobite* and *prickle* mutants also lack stereotypical neuronal cell migration behaviors in the hindbrain *(39)*. While less is known about the intracellular signaling cascade of the non-canonical pathway, Rho kinase, and G-protein activity appear to function downstream *(40)* and

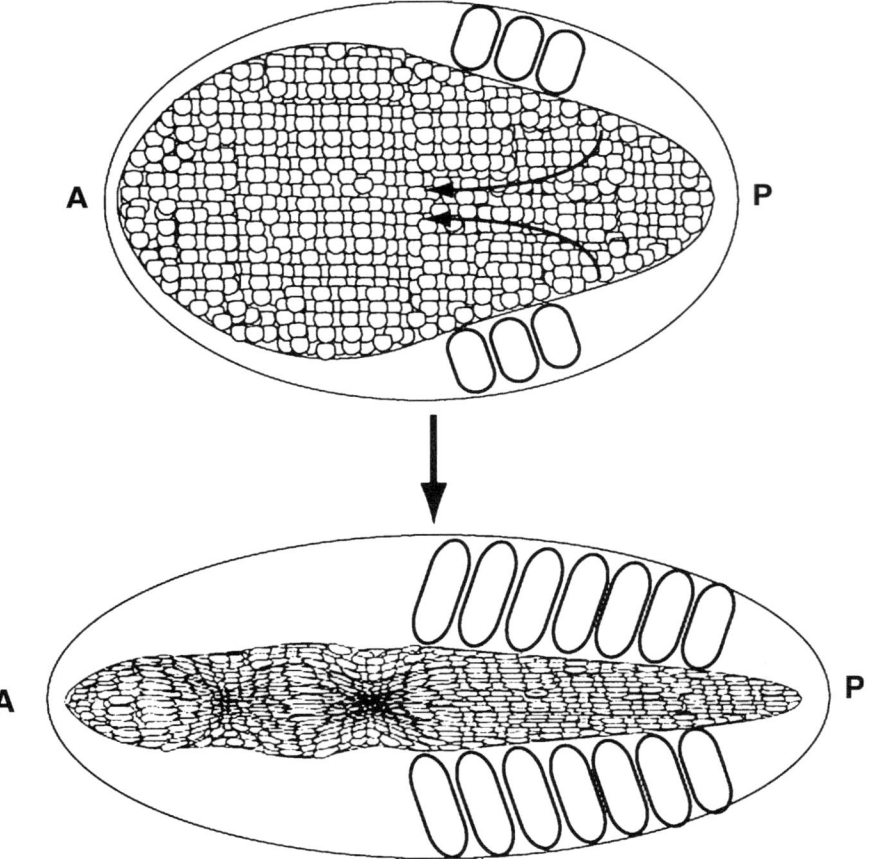

Fig. 20.2. Non-canonical Wnt function during early development. Non-canonical Wnt signaling promotes cell shape changes and motility in the developing neural plate (upper panel). Over time, these behaviors result in convergence and extension of neural plate cells to form the neural keel (lower panel).

their modulation can perturb morphogenetic movements in a similar manner.

2. Phenotypic assays: Because there are few molecular assays for the output of non-canonical Wnt signaling in zebrafish, most phenotypic analysis has been performed using markers for cell morphology and motility. Genetic interactions with known non-canonical Wnt mutants can be used to place new signaling components into the pathway. Measurements of cell polarization relative to the embryonic axis can be used to quantify the extent of convergence *(41)*. Gene expression patterns can help assess the amount of extension, specifically by staining for *hgg1*, which marks the prechordal plate, and *dlx3* which marks the anterior edge of the neural plate *(41)*. Migration of facial motoneurons can most easily be visualized with a transgenic line that expresses GFP under the control of the *isl1* promoter *(42)*.

3.4. Morphological Boundary Formation in the Brain

1. Manipulation of pathway function: Wnt signaling has been demonstrated to regulate the establishment of morphological boundaries in at least two regions of the brain: the midbrain-hindbrain junction and the hindbrain (**Fig. 20.3**). In the first case, signaling downstream of Wnt1, Wnt3a, and Wnt10b is redundantly required to maintain gene expression and morphology of the midbrain/hindbrain boundary (MHB) *(43, 44)*. In the hindbrain, Wnts are expressed at the developing rhombomere boundaries and are required for their establishment *(45, 46)*, as well as neurogenesis in the adjacent rhombomeres. In both cases these boundaries serve as important signaling centers and lack neuronal differentiation during early development.

2. Phenotypic assays: A variety of markers exist for visualizing the developing MHB and rhombomere boundaries in zebrafish. Several genes expressed in the MHB are known to be both required for the development of this structure and targets of

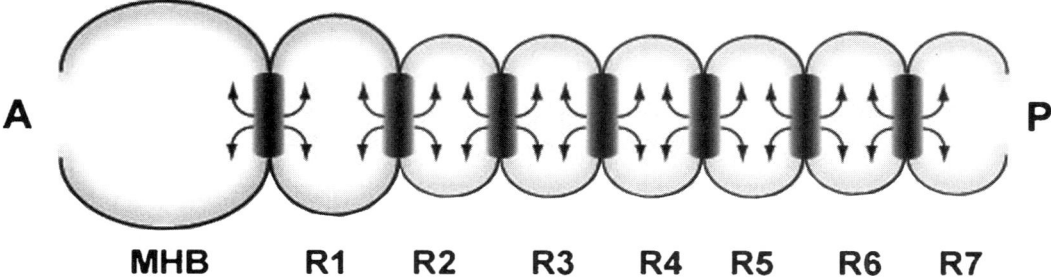

Fig. 20.3. Wnt function in brain segmentation. Wnts are expressed at the developing boundaries between the midbrain and hindbrain (MB-R1) and between hindbrain rhombomeres (R1-7). These Wnts act to promote morphological boundary formation as well as regulating neurogenesis in adjacent tissues.

Wnt signaling, such as *pax2* and *fgf8* *(44)*. In the hindbrain, genes in the Notch pathway such as *lunatic fringe* mark the developing rhombomere boundaries *(47)* as does the transcription factor *foxb1.2* *(45)*. All of these structures can also be visualized morphologically in live embryos or in more detail using cell labeling techniques like phalloidin staining, which marks actin microfilaments. Rhombomere structure can be indirectly assayed by examining the pattern of developing neurons using specific mRNA probes or transgenic lines.

3.5. Dorsal CNS Patterning

1. Manipulation of pathway function: While there is currently little data from zebrafish to support a role for Wnt signaling in dorsal CNS patterning, it is likely that mechanisms similar to those described in amniotes exist in the fish. In chick and mouse, Wnts expressed in the roof plate act as mitogens and induce specific cell fates in the underlying progenitor cells *(48)*. Wnt ligands are expressed in the dorsal roof plate of the zebrafish CNS *(49)*, indicating that they are poised to signal to the developing dorsal neural tube. Knockdown of these Wnts or transgenic manipulation of the pathway would thus be expected to affect dorsal neural tube proliferation and patterning.

2. Phenotypic assays: Genes encoding transcription factors such as *nkx6.1*, *iro3*, and *pax3* are expressed in unique populations of neural tube progenitors, often in stereotyped locations, and there is significant homology between these genes throughout the spinal cord of zebrafish and amniotes *(50–52)*. *In situ* hybridization for these markers could be used to assay for dorsal/ventral patterning defects following Wnt pathway modulation. A dorsal or ventral shift in a particular marker, coupled with a reciprocal shift in a neighboring marker, is a clear indication of a patterning defect. Markers for postmitotic neuronal populations such as *isl1*, *en1b*, and *chx10* can also be used in the same way, to measure the ultimate effect of progenitor mis-specification. Again, many of these genes are expressed in homologous populations of spinal neurons in zebrafish and amniotes *(53–55)*.

3.6. Regulation of Proliferation and Cell Cycle Exit

1. Manipulation of pathway function: Canonical Wnt signaling has been clearly demonstrated to positively regulate mitotic activity in several vertebrate systems (**Fig. 20.4B**). Activation of the pathway induces expression of genes required for cell cycle progression, such as *cyclinD* family members *(56)*. In zebrafish an indirect mechanism has been described, in which Wnt activity regulates *zic2* expression in the dorsal midbrain, which, in turn, regulates proliferation *(57)*. Manipulation of pathway function using mutants, morpholinos, and transgenic lines, could be used to further determine whether the rate

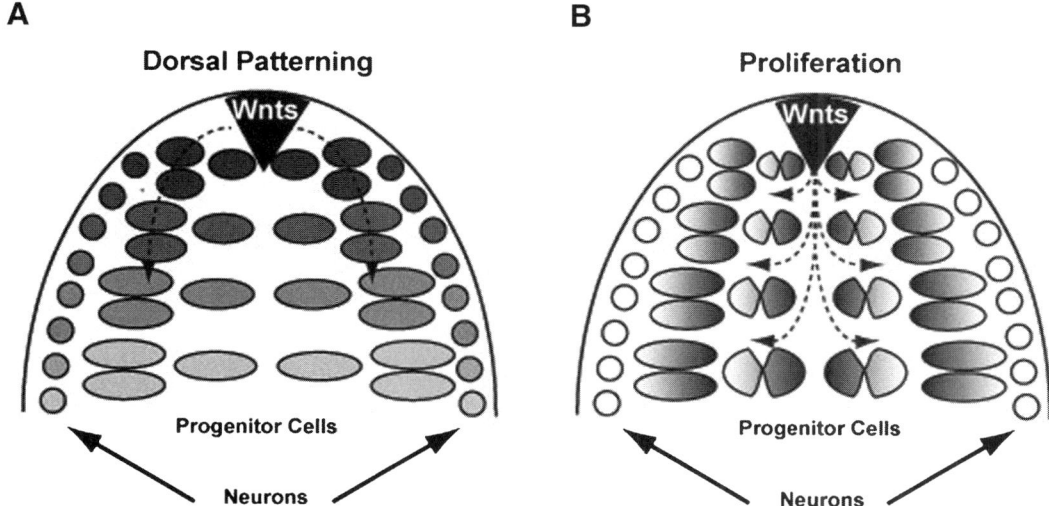

Fig. 20.4. Wnt functions in neural tube patterning and proliferation. (**A**) Wnt molecules secreted from the roof plate of the neural tube act to specify the identity of progenitor cells. In the absence of Wnt signals, dorsal fates are lost and more ventral fates are expanded. While direct evidence for this pathway has not yet been described in zebrafish, all of the relevant Wnt pathway molecules and patterning markers are expressed similarly to other vertebrates. (**B**) Wnts also act in the neural tube to regulate progenitor proliferation. Cells receiving a Wnt signal upregulate cell cycle genes and undergo additional rounds of cell division. Indirect evidence for this pathway has been observed in the zebrafish midbrain, but not yet in other CNS tissues.

of proliferation or cell cycle exit are affected within other specific neuronal populations.

2. Phenotypic assays: Standard techniques for measuring cell proliferation are also commonly used in zebrafish. Cell cycle progression can be measured by BrDU labeling and phospho-histone H3 or PCNA immunohistochemistry. Unless clearly ectopic expression of these markers is detected, it is more reliable to assess the labeling index, which is a measure of the percentage of proliferating cells relative to overall cell number. Cell cycle exit can be detected by in situ hybridization for CDK inhibitor genes such as *p27* or *p57 (58)*. In addition, particular cell cycle genes that are known targets of Wnt signaling, such as *cycD1 (59)*, can be examined. In theory, clonal analysis could also be performed using lineage tracing to determine whether clone sizes are expanded or reduced compared to controls. Tissue size alone is not a reliable way to measure effects on proliferation, because it can also be dependent on cell death or changes in fate specification.

3.7. Neuronal Differentiation

1. Manipulation of pathway function: Recent work indicates that canonical Wnt signaling acts to regulate several steps in the process of neurogenesis within specific populations (**Fig. 20.5**). Because Wnt activity is present broadly in the

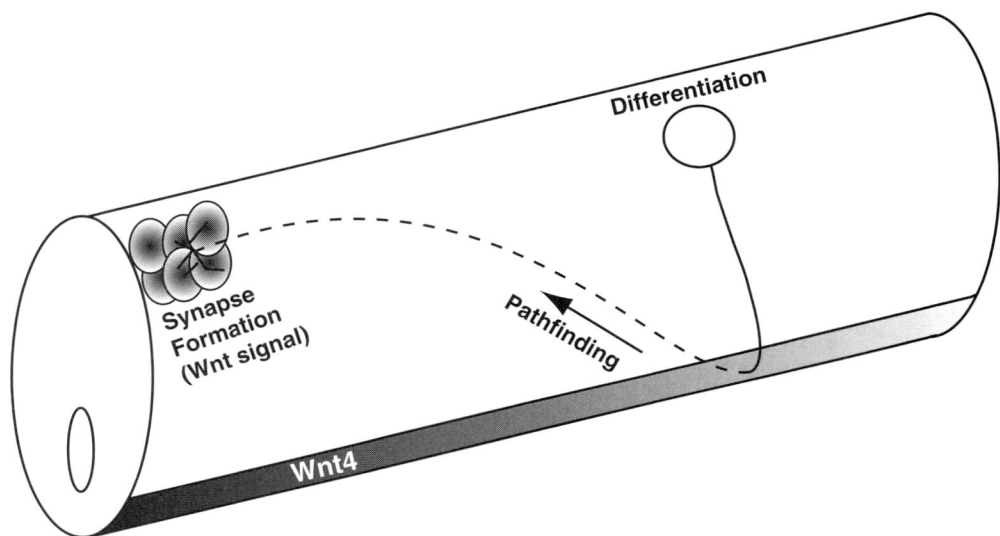

Fig. 20.5. Wnt functions in neuronal development. Wnt signals have been implicated in multiple steps of neuronal differentiation, including neurogenesis, pathfinding, and synapse formation. A prototypical spinal interneuron is shown to depict all of these events. In zebrafish, Wnts regulate the timing of neurogenesis by affecting expression of genes required for this process. Graded Wnt4 activity in the spinal cord has been implicated in axon pathfinding in mammals, and Wnt4 is expressed in a similar pattern in zebrafish. Finally, Wnt signals from post-synaptic cells can regulate synapse formation in pre-synaptic axons in other organisms. Tools for examining axon pathfinding and synaptogenesis are available to determine whether these events are also regulated in the zebrafish CNS.

developing CNS, it may be more appropriate to consider this pathway as a "switch" that regulates the timing of neurogenesis rather than as an inducer of specific phenotypes. For example, Wnt signaling through Lef1 regulates the timing of neurogenesis in the posterior hypothalamus through the direct activation of *sox3 (14)*. Wnt pathway components continue to be expressed in differentiated neuronal populations and thus may regulate postmitotic events such as neurotransmitter expression, axon pathfinding, and synaptogenesis. Evidence exists from other species for Wnt function in the latter two events *(60–64)*, but it is not clear whether canonical or non-canonical pathways are required.

2. Phenotypic assays: Markers for various steps of neuronal differentiation are plentiful. For the process of neurogenesis, proneural bHLH genes such as *ngn1*, *zash1a*, and *neuroD* mark specific populations of cells as they exit the cell cycle *(65)*. Subsequently, specific sets of transcription factors, such as Lim homeodomain proteins *(66)*, are expressed in each subclass of postmitotic neurons. Neurotransmitter expression can be detected using specific antibodies or mRNA probes for the synthetic enzymes, such as *vglut2*, *glyt2*, and *gad65/67 (67)*. For axon pathfinding, immunostaining for

acetylated tubulin labels multiple cell types *(68)*, while more specific antibodies are available to label individual populations. In addition, many transgenic GFP reporter lines labeling subsets of neurons have already been produced *(54, 55)*, and more lines are constantly being made. Transgenes and antibodies can also be used to label synaptic proteins and examine the process of synaptogenesis.

References

1. Wilkinson, D. G., Bailes, J. A., McMahon, A. P. (1987) Expression of the proto-oncogene int-1 is restricted to specific neural cells in the developing mouse embryo. *Cell* 50, 79–88.

2. McMahon, A. P., Bradley, A. (1990) The Wnt-1 (int-1) proto-oncogene is required for development of a large region of the mouse brain. *Cell* 62, 1073–1085.

3. Amsterdam, A., Hopkins, N. (2006) Mutagenesis strategies in zebrafish for identifying genes involved in development and disease. *Trends Genet* 22, 473–478.

4. Fritz, A., Rozowski, M., Walker, C., et al. (1996) Identification of selected gamma-ray induced deficiencies in zebrafish using multiplex polymerase chain reaction. *Genetics* 144, 1735–1745.

5. Amsterdam, A. (2003) Insertional mutagenesis in zebrafish. *Dev Dyn* 228, 523–534.

6. Clark, K. J., Geurts, A. M., Bell, J. B., et al. (2004) Transposon vectors for gene-trap insertional mutagenesis in vertebrates. *Genesis* 39, 225–233.

7. Kawakami, K. (2005) Transposon tools and methods in zebrafish. *Dev Dyn* 234, 244–254.

8. Nasevicius, A., Ekker, S. C. (2000) Effective targeted gene 'knockdown' in zebrafish. *Nat Genet* 26, 216–220.

9. Helde, K. A., Wilson, E. T., Cretekos, C. J., et al. (1994) Contribution of early cells to the fate map of the zebrafish gastrula. *Science* 265, 517–520.

10. Woo, K., Fraser, S. E. (1995) Order and coherence in the fate map of the zebrafish nervous system. *Development* 121, 2595–2609.

11. Amsterdam, A., Becker, T. S. (2005) Transgenes as screening tools to probe and manipulate the zebrafish genome. *Dev Dyn* 234, 255–268.

12. Jowett, T. (2001) Double in situ hybridization techniques in zebrafish. *Methods* 23, 345–358.

13. Oxtoby, E., Jowett, T. (1993) Cloning of the zebrafish krox-20 gene (krx-20) and its expression during hindbrain development. *Nucleic Acids Res* 21, 1087–1095.

14. Lee, J. E., Wu, S. F., Goering, L. M., et al. (2006) Canonical Wnt signaling through Lef1 is required for hypothalamic neurogenesis. *Development* 133, 4451–4461.

15. Thorpe, C. J., Weidinger, G., Moon, R. T. (2005) Wnt/beta-catenin regulation of the Sp1-related transcription factor sp5l promotes tail development in zebrafish. *Development* 132, 1763–1772.

16. Schneider, S., Steinbeisser, H., Warga, R. M., et al. (1996) Beta-catenin translocation into nuclei demarcates the dorsalizing centers in frog and fish embryos. *Mech Dev* 57, 191–198.

17. Bellipanni, G., Varga, M., Maegawa, S., et al. (2006) Essential and opposing roles of zebrafish beta-catenins in the formation of dorsal axial structures and neurectoderm. *Development* 133, 1299–1309.

18. Kelly, C., Chin, A. J., Leatherman, J. L., et al. (2000) Maternally controlled (beta)-catenin-mediated signaling is required for organizer formation in the zebrafish. *Development* 127, 3899–3911.

19. Westfall, T. A., Brimeyer, R., Twedt, J., et al. (2003) Wnt-5/pipetail functions in vertebrate axis formation as a negative regulator of Wnt/beta-catenin activity. *J Cell Biol* 162, 889–898.

20. Leung, T., Soll, I., Arnold, S. J., et al. (2003) Direct binding of Lef1 to sites in the boz promoter may mediate pre-midblastula-transition activation of boz expression. *Dev Dyn* 228, 424–432.

21. Ryu, S. L., Fujii, R., Yamanaka, Y., et al. (2001) Regulation of dharma/bozozok by the Wnt pathway. *Dev Biol* 231, 397–409.

22. Dorsky, R. I., Sheldahl, L. C., Moon, R. T. (2002) A transgenic Lef1/beta-catenin-dependent

reporter is expressed in spatially restricted domains throughout zebrafish development. *Dev Biol* 241, 229–237.

23. Shimizu, T., Yamanaka, Y., Ryu, S. L., et al. (2000) Cooperative roles of Bozozok/Dharma and Nodal-related proteins in the formation of the dorsal organizer in zebrafish. *Mech Dev* 91, 293–303.

24. Okuda, Y., Yoda, H., Uchikawa, M., et al. (2006) Comparative genomic and expression analysis of group B1 sox genes in zebrafish indicates their diversification during vertebrate evolution. *Dev Dyn* 235, 811–825.

25. Lekven, A. C., Thorpe, C. J., Waxman, J. S., et al. (2001) Zebrafish wnt8 encodes two wnt8 proteins on a bicistronic transcript and is required for mesoderm and neurectoderm patterning. *Dev Cell* 1, 103–114.

26. Erter, C. E., Wilm, T. P., Basler, N., et al. (2001) Wnt8 is required in lateral mesendodermal precursors for neural posteriorization in vivo. *Development* 128, 3571–3583.

27. Dorsky, R. I., Itoh, M., Moon, R. T., et al. (2003) Two tcf3 genes cooperate to pattern the zebrafish brain. *Development* 130, 1937–1947.

28. Kapsimali, M., Caneparo, L., Houart, C., et al. (2004) Inhibition of Wnt/Axin/beta-catenin pathway activity promotes ventral CNS midline tissue to adopt hypothalamic rather than floorplate identity. *Development* 131, 5923–5933.

29. Kim, C. H., Oda, T., Itoh, M., et al. (2000) Repressor activity of Headless/Tcf3 is essential for vertebrate head formation. *Nature* 407, 913–916.

30. van de Water, S., van de Wetering, M., Joore, J., et al. (2001) Ectopic Wnt signal determines the eyeless phenotype of zebrafish masterblind mutant. *Development* 128, 3877–3888.

31. Heisenberg, C. P., Houart, C., Take-Uchi, M., et al. (2001) A mutation in the Gsk3-binding domain of zebrafish Masterblind/Axin1 leads to a fate transformation of telencephalon and eyes to diencephalon. *Genes Dev* 15, 1427–1434.

32. Kilian, B., Mansukoski, H., Barbosa, F. C., et al. (2003) The role of Ppt/Wnt5 in regulating cell shape and movement during zebrafish gastrulation. *Mech Dev* 120, 467–476.

33. Heisenberg, C. P., Tada, M., Rauch, G. J., et al. (2000) Silberblick/Wnt11 mediates convergent extension movements during zebrafish gastrulation. *Nature* 405, 76–81.

34. Jessen, J. R., Topczewski, J., Bingham, S., et al. (2002) Zebrafish trilobite identifies new roles for Strabismus in gastrulation and neuronal movements. *Nat Cell Biol* 4, 610–615.

35. Park, M., Moon, R. T. (2002) The planar cell-polarity gene stbm regulates cell behaviour and cell fate in vertebrate embryos. *Nat Cell Biol* 4, 20–25.

36. Topczewski, J., Sepich, D. S., Myers, D. C., et al. (2001) The zebrafish glypican knypek controls cell polarity during gastrulation movements of convergent extension. *Dev Cell* 1, 251–264.

37. Veeman, M. T., Slusarski, D. C., Kaykas, A., et al. (2003) Zebrafish prickle, a modulator of noncanonical Wnt/Fz signaling, regulates gastrulation movements. *Curr Biol* 13, 680–685.

38. Ciruna, B., Jenny, A., Lee, D., et al. (2006) Planar cell polarity signalling couples cell division and morphogenesis during neurulation. *Nature* 439, 220–224.

39. Bingham, S., Higashijima, S., Okamoto, H., et al. (2002) The Zebrafish trilobite gene is essential for tangential migration of branchiomotor neurons. *Dev Biol* 242, 149–160.

40. Lin, F., Sepich, D. S., Chen, S., et al. (2005) Essential roles of G{alpha}12/13 signaling in distinct cell behaviors driving zebrafish convergence and extension gastrulation movements. *J Cell Biol* 169, 777–787.

41. Sepich, D. S., Solnica-Krezel, L. (2005) Analysis of cell movements in zebrafish embryos: morphometrics and measuring movement of labeled cell populations in vivo. *Methods Mol Biol* 294, 211–233.

42. Higashijima, S., Hotta, Y., Okamoto, H. (2000) Visualization of cranial motor neurons in live transgenic zebrafish expressing green fluorescent protein under the control of the islet-1 promoter/enhancer. *J Neurosci* 20, 206–218.

43. Lekven, A. C., Buckles, G. R., Kostakis, N., et al. (2003) Wnt1 and wnt10b function redundantly at the zebrafish midbrain-hindbrain boundary. *Dev Biol* 254, 172–187.

44. Buckles, G. R., Thorpe, C. J., Ramel, M. C., et al. (2004) Combinatorial Wnt control of zebrafish midbrain-hindbrain boundary formation. *Mech Dev* 121, 437–447.

45. Amoyel, M., Cheng, Y. C., Jiang, Y. J., et al. (2005) Wnt1 regulates neurogenesis and mediates lateral inhibition of boundary cell specification in the zebrafish hindbrain. *Development* 132, 775–785.

46. Riley, B. B., Chiang, M. Y., Storch, E. M., et al. (2004) Rhombomere boundaries are Wnt signaling centers that regulate metameric patterning in the zebrafish hindbrain. *Dev Dyn* 231, 278–291.

47. Prince, V. E., Holley, S. A., Bally-Cuif, L., et al. (2001) Zebrafish lunatic fringe demarcates segmental boundaries. *Mech Dev* 105, 175–180.

48. Chizhikov, V. V., Millen, K. J. (2005) Roof plate-dependent patterning of the vertebrate dorsal central nervous system. *Dev Biol* 277, 287–295.

49. Krauss, S., Korzh, V., Fjose, A., et al. (1992) Expression of four zebrafish wnt-related genes during embryogenesis. *Development* 116, 249–259.

50. Cheesman, S. E., Layden, M. J., Von Ohlen, T., et al. (2004) Zebrafish and fly Nkx6 proteins have similar CNS expression patterns and regulate motoneuron formation. *Development* 131, 5221–5232.

51. Lewis, K. E., Bates, J., Eisen, J. S. (2005) Regulation of iro3 expression in the zebrafish spinal cord. *Dev Dyn* 232, 140–148.

52. Seo, H. C., Saetre, B. O., Havik, B., et al. (1998) The zebrafish Pax3 and Pax7 homologues are highly conserved, encode multiple isoforms and show dynamic segment-like expression in the developing brain. *Mech Dev* 70, 49–63.

53. Appel, B., Korzh, V., Glasgow, E., et al. (1995) Motoneuron fate specification revealed by patterned LIM homeobox gene expression in embryonic zebrafish. *Development* 121, 4117–4125.

54. Higashijima, S., Masino, M. A., Mandel, G., et al. (2004) Engrailed-1 expression marks a primitive class of inhibitory spinal interneuron. *J Neurosci* 24, 5827–5839.

55. Kimura, Y., Okamura, Y., Higashijima, S. (2006) alx, a zebrafish homolog of Chx10, marks ipsilateral descending excitatory interneurons that participate in the regulation of spinal locomotor circuits. *J Neurosci* 26, 5684–5697.

56. Megason, S. G., McMahon, A. P. (2002) A mitogen gradient of dorsal midline Wnts organizes growth in the CNS. *Development* 129, 2087–2098.

57. Nyholm, M. K., Wu, S. F., Dorsky, R. I., et al. (2007) The zebrafish zic2a-zic5 gene pair acts downstream of canonical Wnt signaling to control cell proliferation in the developing tectum. *Development* 134, 735–746.

58. Park, H. C., Boyce, J., Shin, J., et al. (2005) Oligodendrocyte specification in zebrafish requires notch-regulated cyclin-dependent kinase inhibitor function. *J Neurosci* 25, 6836–6844.

59. Yarden, A., Salomon, D., Geiger, B. (1995) Zebrafish cyclin D1 is differentially expressed during early embryogenesis. *Biochim Biophys Acta* 1264, 257–260.

60. Liu, Y., Shi, J., Lu, C. C., et al. (2005) Ryk-mediated Wnt repulsion regulates posterior-directed growth of corticospinal tract. *Nat Neurosci* 8, 1151–9.

61. Lu, W., Yamamoto, V., Ortega, B., et al. (2004) Mammalian Ryk is a Wnt coreceptor required for stimulation of neurite outgrowth. *Cell* 119, 97–108.

62. Lyuksyutova, A. I., Lu, C. C., Milanesio, N., et al. (2003) Anterior-posterior guidance of commissural axons by Wnt-frizzled signaling. *Science* 302, 1984–1988.

63. Krylova, O., Herreros, J., Cleverley, K. E., et al. (2002) WNT-3, expressed by motoneurons, regulates terminal arborization of neurotrophin-3-responsive spinal sensory neurons. *Neuron* 35, 1043–1056.

64. Lucas, F. R., Salinas, P. C. (1997) WNT-7a induces axonal remodeling and increases synapsin I levels in cerebellar neurons. *Dev Biol* 192, 31–44.

65. Mueller, T., Wullimann, M. F. (2003) Anatomy of neurogenesis in the early zebrafish brain. *Brain Res Dev Brain Res* 140, 137–155.

66. Dawid, I. B., Chitnis, A. B. (2001) Lim homeobox genes and the CNS: a close relationship. *Neuron* 30, 301–303.

67. Higashijima, S., Schaefer, M., Fetcho, J. R. (2004) Neurotransmitter properties of spinal interneurons in embryonic and larval zebrafish. *J Comp Neurol* 480, 19–37.

68. Wilson, S. W., Ross, L. S., Parrett, T., et al. (1990) The development of a simple scaffold of axon tracts in the brain of the embryonic zebrafish, Brachydanio rerio. *Development* 108, 121–145.

Part VIII

Xenopus

Chapter 21

Studying Wnt Signaling in *Xenopus*

Stefan Hoppler

Abstract

Xenopus is an established and powerful model system for the study of Wnt signaling in vertebrates. Above all, the relatively large size of the embryos enables microinjection experiments, which have led to key discoveries not only about the functional role of Wnt signaling in vertebrate embryos, but also about the molecular mechanisms of Wnt signaling in vertebrate cells.

A major advantage of the Xenopus model is the ability to obtain large numbers of embryos, which develop relatively rapidly and which can be studied in natural separation from sentient adult parental animals. In order to obtain *Xenopus* embryos, ovulation in females is induced with a simple hormone injection, the eggs collected and fertilized with sperm from males.

The *Xenopus* model system has been further strengthened by recent advances such as morpholino technology and efficient transgenic methods, as well as the development of *Xenopus tropicalis* as a diploid genetic model system with a shorter generation time and a genome similar to higher vertebrates.

Key words: *Xenopus*, embryo, egg, sperm, gonadotropin hormone.

1. Introduction

Xenopus has been instrumental as a model system for almost 20 years for key discoveries about Wnt signaling in vertebrates. The dramatic axis duplication phenotype in response to ectopic Wnt ligand expression was the first demonstration of Wnt signaling function in vertebrate embryonic development *(1)*. Dorsal axis development served as a powerful assay to confirm and further study the canonical Wnt signal transduction mechanisms in vertebrates (e.g., refs. *2–10*, *see also* **Chapter 29**), but also for studying the requirement for maternal Wnt signaling components for embryonic development (e.g., refs. *11–13*, *see also* **Chapters 26** and **29**). *Xenopus* embryos were also the eminent model system

---

*All Chapter cross-references in this chapter refer to this volume, unless otherwise specified.

Elizabeth Vincan (ed.), *Wnt Signaling, Volume II: Pathway Models, vol. 469*
© 2008 Humana Press, a part of Springer Science+Business Media, New York, NY
Book doi: 10.1007/978-1-60327-469-2

for the discovery and functional investigation of extracellular signaling components (such as sFrp, e.g., refs. *14, 15*; dickkopf, ref. *16*; and WIF, ref. *17)* and for uncovering mechanisms of Wnt signaling at the membrane (Frizzled receptors: e.g., refs. *18, 19*; *see* **Chapter 28**; LRP6: e.g., refs. *20–23). Xenopus* has also been one of the foremost vertebrate models for unraveling the many functions of β-catenin (refs. *11* and *24*; *see also* **Chapter 23**) and the nuclear mechanisms of Wnt/β-catenin signaling; for instance the *Xenopus* experimental model was prominently involved in establishing the link between β-catenin and Tcf/LEF *(25, 26)* and revealed some of the first direct Wnt/β-catenin target genes in vertebrates (siamois, ref. *27*; Xnr3, ref. *28*; twin, ref. *29*; fibronectin, ref. *30*; engrailed-2, ref. *31*; Xnr5 and 6, ref. *32*; and slug, ref. *33). Many discoveries about non-canonical Wnt signaling in vertebrate development had their origins in *Xenopus* embryos (refs. *34–40*; *see* **Chapters 30, 31**; and Volume 1 **Chapters 10, 13**). Dishevelled functions at a pivotal position at the top of several Wnt signal transduction pathways *(41). Xenopus* experiments helped establish functional domains within dishevelled that are important for differentiating between different Wnt signal transduction pathways *(42–44).*

Xenopus has also contributed to our understanding of the functional role of Wnt signaling in vertebrate embryonic development, not just in the well-known dorsal axis induction (*see* earlier and **Chapters 26** and **29**), but also, for instance, in regulating mesoderm induction *(45, 46)*, mesoderm patterning (e.g., ref. *47)*, morphogenetic movements during gastrulation (e.g., refs. *35, 37,* and *39*; *see* **Chapters 30** and **31**), anteroposterior patterning of the neural plate (e.g., refs. *48, 49)*, during neural crest development (e.g., refs. *50, 33, 51)*, midbrain patterning (e.g., *52)*, and later during organogenesis in eye development *(53, 54)*, heart development *(55, 56, 40)*, and kidney development *(57, 58).*

The success of *Xenopus* as a model system is by no small measure founded on the relative ease with which large numbers of oocytes, eggs or embryos are obtained. Oocytes can be harvested from adult females, manipulated and matured in culture for studying maternal expression and function of Wnt signaling components (**Chapter 26**). By treating adult females with a relatively simple course of hormone injections, oocyte maturation is initiated and egg laying is encouraged (*see* **Subsection 3.1**). These eggs can then be collected to harvest egg extract to study the biochemistry of Wnt signaling *(9)* or as a reagent for the *Xenopus* transgenic protocol (**Chapter 27**). Most applications however rely on fertilizing these eggs with sperm from males, either by in vitro fertilization (*see* **Subsections 3.2** and **3.3**), natural matings (*see* **Note 9**) or by sperm nuclear transplantation (transgenic method, **Chapter 27**) in order to initiate and study embryonic development.

Many approaches have been developed and applied for studying Wnt signaling in *Xenopus*, which benefit from the size, accessibility and large number of *Xenopus* eggs and embryos. Several efficient methods have been developed to study gene expression in *Xenopus* embryos (**Chapter 22**) and oocytes (**Chapter 26**) using antibodies and nucleotide probes. These methods are not only useful for detecting the expression of genes encoding Wnt signaling components, but also to monitor the regulation of downstream genes by Wnt signaling. Additionally, efficient methods have been developed in *Xenopus* embryos to detect the activity of the Wnt/β-catenin pathway by either monitoring nuclear β-catenin *(59, 60)* (see **Chapter 23** in **Volume 2**) or using sophisticated reporter gene constructs in transgenic *Xenopus* embryos (refs. *61* and *62; see* **Chapter 24** in **Volume 2**).

The particular strength of *Xenopus* is, however, as possibly the foremost vertebrate experimental model for testing hypotheses about Wnt signaling in embryonic development with gain- and loss-of-function experiments. Gain-of-function experiments in *Xenopus* embryos are often accomplished by overexpressing the studied gene product, either by injecting mRNA into blastula stage embryos, to study gene function during early embryogenesis (e.g., Wnt-1 *(1), see* **Chapter 25**) or from inducible transgenes in transgenic embryos, to study gene function at later stages of development (e.g., ref. *63, see* **Chapter 27**). Loss-of-function experiments are achieved by overexpressing dominant-negative molecules (i.e., mutated molecules, which are expected to interfere with the function of endogenous gene products, e.g., dominant-negative Wnt8 *(47), see* **Chapter 25**). While studies in *Xenopus* embryos using dominant-negative molecules are successful and informative, they may lack specificity in that they potentially interfere with several related gene products. Morpholino antisense oligonucleotide technology is capable of inhibiting the expression of a particular gene (e.g., ref. *64, see also* **Chapter 25**) and is particularly successful in *Xenopus* because morpholinos can be easily injected into the relatively large blastomeres of *Xenopus* embryos.

The relatively large blastomeres and the availability of detailed fate maps for individual blastomeres (e.g., refs. *65* and *66, see also* www.xenbase.org/anatomy/static/fate-tissue.jsp) allows targeted injections of mRNA and Morpholinos to direct and restrict the experimental manipulation to particular tissues or a particular organ of interest. The large size of *Xenopus* embryos also permits dissection of embryonic explants to study development of isolated organs (e.g., heart development in Doral Marginal Zone (DMZ) explants *(56)*), ectopic development (e.g., ectopic heart development induced by Wnt antagonists in Ventral Marginal Zone [VMZ] explants *(56)*) and differentiation in pluripotent stem-cell-like animal cap explants (e.g., to study synergy between Wnt11 and activin signaling in promoting heart differentiation *(40))*. The large *Xenopus* blastomeres also allow injection of DNA constructs, not just for overexpression experiments (**Chapter 25**), but also for investigating mechanisms of gene regulation through cis-regulatory elements with reporter gene constructs (e.g., siamois *(27)*; Xnr3 *(28)*; twin *(29)*). However, possibly more reliable results can be obtained by integration of such promoter constructs in transgenic embryos (ref. *67*; cf. transgenesis method in **Chapter 27**).

The approaches described so far use gain-of-function (e.g., mRNA injection) and loss-of-function (e.g., morpholino injection) experiments to test particular hypotheses and involve analysis of gene expression of specific genes. These same techniques have also been very successfully used in *Xenopus* for screens to identify novel genes with particular expression patterns or with particular functions, or to detect novel expression patterns and novel functions of already known genes. For instance, gene expression screening using whole-mount RNA *in situ* hybridization (**Chapter 22**) has been successfully used in *Xenopus* to identify so-called synexpression groups (e.g., refs. *68* and *69, see also* http://www.dkfz.de/en/mol_embryology/axeldb.html); expression screens have been used to screen libraries of mRNAs injected into *Xenopus* embryos (**Chapter 25**) to identify Xwnt-8 *(70)*; Dickkopf *(16)*; β-catenin *(49)*; Wise *(71)*; Casein Kinase Iδ, Frizzled 10B and Strabismus *(72)*; and Strabismus, Wnt-11, and FrzA *(73)* by examining for different embryonic phenotypes. Recently, this approach has been extended to loss-of-function screens using libraries of morpholino antisense oligonucleotides injected into *Xenopus* embryos (**Chapter 25**) to identify functional requirement for instance for Flamingo; Frizzled genes 2, 6, 7, 8; and Wnt genes 5b, 8, and 11 for different aspects of development *(74, 75)*.

Other successful approaches for studying Wnt signaling in *Xenopus* used chemical or pharmacological reagents; for instance GSK3 inhibitors like Lithium (e.g., ref. *76*) and more recently BIO *(77)*, which thereby activate Wnt/β-catenin signaling; the Casein Kinase inhibitor CK1-7 *(78)*, which is reported to inhibit Wnt/β-catenin signaling; the JNK inhibitor SP6000125 *(40)*, which inhibits non-canonical Wnt/JNK signaling; and a GSK3-independent Wnt/β-catenin signaling agonist *(79)*. Furthermore, *Xenopus* extracts have been successfully used to reconstitute and biochemically study the Wnt/β-catenin signal transduction pathway *(9)*.

Xenopus research has recently been further invigorated by the use of the diploid species *Xenopus tropicalis*, which has a shorter generation time, is suitable for genetic screens *(80–82)* and benefits from a soon to be completely sequenced genome that shows remarkable similarities with other tetrapod higher vertebrates (recent review in ref. *83)*.

2. Materials

2.1. Animals

At least one *Xenopus* male and several *Xenopus* females of either the species *Xenopus laevis* or the species *Xenopus tropicalis* are required for even the most basic *Xenopus* experiment. Husbandry and maintenance of a *Xenopus* colony for research purposes is governed by local and national regulation in different countries and goes beyond the scope of this article, but consult for instance *(84, 80)*.

2.2. Tools and Plastic Consumables

1. Disposable syringe (1 mL).
2. Syringe needle (27G × 1/2", 0.4 × 12 mm; *see* **Note 1**).
3. Dissection kit with Scalpel and Forceps.
4. 50-mm Petri dishes.
5. 90-mm Petri dishes.
6. Disposable plastic Pasteur pipettes.

2.3. Hormone

1. Pregnant Mare Serum Gonadotropin (PMSG, sold as Foligon in the UK).
2. Human Chorionic Gonadotropin (hCG, sold as Chorulon in the UK; *see* **Note 2**).

2.4. Amphibian Anesthetic

1. Tricaine (Ethyl 3-Aminobenzoate Methanesulfonate, also known as MS222; *see* **Note 3**).

2.5. Solutions

1. Marc's Modified Ringers (MMR) solution (10x stock solution): 20 mM $CaCl_2$, 50 mM HEPES, 20 mM KCl, 10 mM $MgCl_2$, 1 M NaCl, adjust pH to 7.5 with NaOH, autoclave and store at 4°C for up to several months (*see* **Note 4**).
2. Marc's Modified Ringers (MMR) solution (1x working solution). Dilute 10x stock solution ten fold with distilled ddH_2O and store at room temperature for up to several weeks (*see* **Note 4**).
3. Marc's Modified Ringers (MMR) solution (0.1x working solution). Dilute 10x stock solution 100-fold with distilled ddH_2O and store at room temperature for up to several weeks (*see* **Note 4**).
4. Testis Buffer: 1x MMR, 20% (v/v) heat inactivated sheep serum, gentamicin (50 µg/mL). Store in aliquots of 10 mL at –20°C for up to several months (*see* **Note 4**).
5. Cysteine solution: 2% (w/v) cysteine, 0.1x MMR (see above), adjust pH to 8.2 with NaOH pellets. The cysteine solution keeps for only a few hours, we make up fresh solution every 4 hours (*see* **Note 5**).

3. Methods

3.1. Induction of Ovulation in Xenopus Females

Hormone injections into *Xenopus* females stimulate oocyte maturation and egg laying. Most laboratories use a protocol with two hormone injections (the first is usually called priming, the second induction or boosting, *see* **Note 2**). Because the efficiency of egg laying and of subsequent fertilization is variable, several females should be induced, even if only few embryos are required. Different numbers of females are used depending on the particular application (*see* **Note 6**).

3.1.1. Priming (See Note 2)

Prime *Xenopus* females three to seven days (*Xenopus laevis*) or 1 to 2 days (*Xenopus tropicalis*) before planned egg collection.

1. Restrain female by covering head with the palm of your hand and holding the strong hind legs with your fingers in a fixed position on either side of the body. Ensure firm grip to avoid escape reaction and injury during injection (*see* **Note 7**).

2. Inject *Xenopus* female subcutaneously into the back of the lower abdomen or more precisely into the dorsal lymph sac with 150 units (*Xenopus laevis*) or 20 units (*Xenopus tropicalis*) of PMSG. Ensure that the needle is at all times well clear of the spinal cord during injection to avoid serious injury even if the animal suddenly starts to struggle.

Keep primed females separate from the general *Xenopus* colony. Successful priming causes the cloaca to turn a reddish color after several days (*Xenopus laevis*) or already after several hours (*Xenopus tropicalis*).

3.1.2. Induction (Boosting)

Induce *Xenopus* females approximately 12 hours (*Xenopus laevis*) or approximately 4 hours (*X. tropicalis*) before egg collection (*see* **Notes 7** and **8**).

1. Inject *Xenopus* female as above with 750–1000 units (*Xenopus laevis*) or 100 units (*Xenopus tropicalis*) of hCG.

Keep induced females separate from the general colony until egg laying in a quiet darkened place (*see* **Note 9**).

3.2. Isolation of Testes from Xenopus Males

1. Euthanase a *Xenopus* male by terminal anesthesia by placing in at least 500 mL 0.3% Tricaine for at least 20 minutes (*see* **Note 3**). Ensure animal cannot escape by placing a secure lid on beaker or other suitable container.

2. At the same time thaw an aliquot of testis buffer and prepare dissection instruments.

3. Verify death by confirming absence of gag reflex upon inserting little finger into the male's mouth and absence of leg withdrawal reflex upon pinching the webbing between the digits of its hind limbs (*see* **Note 10**).

4. With your dissection tools, remove both testes from the abdomen (they are usually white in color and associated with the yellow fat bodies), remove fat body and blood tissue from the testes, wash them once in MMR and place into thawed testes buffer.

5. Store isolated testes at 4°C. Sperm in the testes keep viable under these conditions for at least 2 days, but if not punctured sometimes up to 1 week (*Xenopus laevis*), but only for several hours (for *Xenopus tropicalis*; *see* **Note 9**).

3.3. Egg Collection and Fertilization

3.3.1. Egg Collection (See Note 9)

Primed and induced *Xenopus* females should start egg laying independently.

1. Once independent egg laying has started, pick up and restrain individual female by covering head with the palm of your hand and holding the strong hind legs with your fingers in a fixed position on either side of the body (*see* **Note 7**).

2. Massage the abdomen particularly on the sides (lateral) and the belly (ventral) and collect eggs emerging from the cloaca in a 9-cm Petri dish with 1x MMR.

Ensure handling of the animal is done confidently and with a firm grip to avoid escape reaction. Egg collection from each female can be repeated several times (usually approximately six times) after an interval of at least 30 minutes.

3.3.2. In Vitro Fertilization (See Notes 5 and 9)

1. Carefully remove excess liquid from between the collected unfertilized eggs with a plastic Pasteur pipette.

2. Dissect a small piece from the previously isolated testis (approximately 1/10 of one testis) with sterilized forceps and a disposable scalpel; crush isolated testis tissue slightly without tearing it apart.

3. Grab the testis tissue with a pair of forceps and drag it over the unfertilized eggs, directly contacting as many eggs as possible.

4. Dissociate the isolated tissue completely with forceps and scalpel in a drop of 1xMMR and distribute sperm suspension with a Pasteur pipette over the eggs. Leave eggs for 3–5 minutes.

5. Cover eggs with 0.1x MMR, leave undisturbed for 20–30 minutes.

Approximately after 20 minutes fertilized eggs will rotate (darker animal pole will point to the top) and approximately after 90 minutes the first cleavage (cell division) of the embryos can be observed. In a good fertilization, 80–95% of eggs will be fertilized (*see* **Notes 5** and **9**).

3.4. Removal of Jelly Coat

1. Carefully remove excess liquid from between the fertilized eggs or embryos with a plastic Pasteur pipette.

2. Add cysteine solution and gently swirl the dish constantly for about three minutes. Replace cysteine solution if desired after a minute or two. The embryos should detach from the bottom of the dish. Be careful not to cause lysis of the embryos by excessive agitation.

Removal of the jelly coat can be monitored by observing tighter packing of embryos (before removal of jelly coat the embryos will not touch each other because of the mostly invisible jelly coat around them, after removal of the jelly coat the embryos will pack close to each other). After removal of the jelly coat, take extra care in all the subsequent steps to prevent embryos from coming in contact with the liquid–air interface, because the tensions cause the embryos to rupture.

3. Rinse embryos repeatedly with 0.1x MMR. Remove unfertilized eggs, debris from ruptured embryos and leftover sperm tissue etc.

Embryos are now ready for injection experiments (**Chapter 25**) or for being left to develop to the appropriate stage of development for analysis of gene expression (**Chapter 22**; *see* **Notes 5 and 11**).

4. Notes

1. We use 27G (0.4-mm) size syringe needles for both *Xenopus laevis* and *Xenopus tropicalis*. However, 25G (0.5-mm) needles are suitable for *Xenopus laevis* and 30G (0.3-mm) needles are optimal for *Xenopus tropicalis*.

2. Some *Xenopus* laboratories make do without PMSG, either by priming with a low dose of hCG, or by not priming at all (i.e., inducing egg laying with only one round of hormone injection equivalent to the procedure described in **Subsection 3.1.2**).

3. Benzocaine is used by some *Xenopus* laboratories instead of Tricaine. Benzocaine is used at a concentration of 0.05% when enthanizing *Xenopus* by terminal anesthesia.

4. Modified Barth's Solution (MBS) is used in many *Xenopus* laboratories instead of MMR. We generally find the two solutions interchangeable for the applications described here. However, we find that the sperm quality is better and lasts for longer if the testes buffer is prepared with MBS (i.e., 1x MBS, 20% (v/v) heat inactivated sheep serum, gentamicin (50 µg/mL), store in aliquots of 10 mL at –20°C for up to several months). Modified Barth's Solution (88 mM NaCl,

1 mM KCl, 0.7 mM CaCl$_2$, 1 mM MgSO$_4$, 5 mM HEPES (pH 7.8), 2.5 mM NaHCO$_3$) is prepared from two stock solutions: 0.1M CaCl$_2$ (autoclaved and stored in aliquots at –20°C) and 10x MBS Salts (888 mM NaCl, 10 mM MgSO$_4$, 50 mM HEPES [pH 7.8], 25 mM NaHCO3, adjust final pH to 7.8 with NaOH, autoclave and store at 4°C for up to several months) by mixing 10% of final volume of 10xMBS salts with 0.7% of final volume of 0.1M CaCl$_2$ and adjust the volume with distilled H$_2$O.

5. For transgenic method, a 2% cysteine solution is prepared with 1x MMR instead of 0.1x MMR (*see* **Chapter 27**). The eggs are not fertilized as in the protocol described in this chapter, instead the jelly coat is removed from unfertilized eggs essentially as described in **Subsection 3.4**, but the unfertilized eggs (not embryos) are treated with this 2% cysteine solution in 1xMMR and the de-jellied unfertilized eggs are rinsed in 1x MMR (instead of 0.1x MMR).

6. The number of females used for egg collection depends on the particular application. We normally induce three to four females for analysis of gene expression in normal embryos (**Chapter 22**) and for mRNA and morpholino injection experiments (**Chapter 25**) and share the embryos between two researchers.

7. *Xenopus* adults have been handled with bare hands (well rinsed to ensure no traces of soap or sweat) for many years. Modern Health and Safety directives, however, insist on researchers using gloves to protect human health. However, normal laboratory gloves containing latex are unsuitable because they often contain substances, which harm the *Xenopus'* skin. We use Nitrile gloves successfully for handling live *Xenopus* adults.

8. Induction of *Xenopus laevis* females is usually done in the evening (e.g., 10 p.m.) before egg collection the next morning; while *Xenopus tropicalis* are usually induced in the early morning before egg collection later that morning. In our laboratory, however, we keep *Xenopus laevis* females after induction (i.e., hCG injection) in a temperature-controlled environment at 16°C, which delays egg laying and allows us to induce them up to 18 hours prior to egg collection (e.g., 5 to 6 p.m. on the day before egg collection starting at around 9 to 10 a.m. in the morning), which incidentally also happens to be more compatible with a social life.

9. An alternative approach to obtaining fertilized embryos is setting up a natural mating. Place one or two primed and induced *Xenopus* females together in a container with one *Xenopus* male (which can also be primed and induced in the

same way as the females) just after induction. In a process called amplexus, the male should grab one of the females and fertilize the eggs that the female lays naturally. Natural matings are convenient for obtaining embryos of a range of different embryonic stages, for instance for analysis of gene expression in normal embryos (**Chapter 22**). It has also been reported that *in vitro* fertilization in *Xenopus tropicalis* is very susceptible to infection of the isolated testis (much more so than in *Xenopus laevis*). Natural matings are often the most reliable method for obtaining *Xenopus tropicalis* embryos.

10. In order to ensure death and prevent any possible suffering or pain, destroy the central nervous system of the adult *Xenopus* male by a procedure called double pithing. Use a secateur (garden instrument normally used to prune roses etc.), place flat secateur blade into the mouth of anesthetized male *Xenopus*, place blunt thick secateur blade above the brain, crush to destroy brain. Use a straight teasing needle to destroy forebrain and then insert teasing needle into the neural canal to destroy the spinal cord, which will cause involuntary muscular movements in the hind legs.

11. If embryos are cultured for more than a day, add an appropriate antibiotic (e.g., gentamicin 50–100 µg/mL), remove dead or dying embryos regularly and replace solution daily. Embryos can be cultured at a range of temperatures to slow down or speed up development (*Xenopus laevis*: 13–23°C; *Xenopus tropicalis*: 20–28°C; however, during cleavage stages, gastrulation and early stages of neurulation keep *X. laevis* between 14 and 21°C and *X. tropicalis* between 22 and 25°C).

Acknowledgments

Thanks to Yvonne Turnbull for technical assistance and to the CSHL practical course on Cell and Developmental Biology of *Xenopus* for originally introducing me and many of my colleagues to *Xenopus* protocols and experimental approaches.

References

1. McMahon, A. P., Moon, R. T. (1989) Ectopic expression of the proto-oncogene int-1 in Xenopus embryos leads to duplication of the embryonic axis. *Cell* 58, 1075–1084.

2. Dominguez, I., Itoh, K., Sokol, S. Y. (1995) Role of glycogen synthase kinase 3 beta as a negative regulator of dorsoventral axis formation in Xenopus embryos. *Proc Natl Acad Sci* USA 92, 8498–8502.

3. Vleminckx, K., Wong, E., Guger, K., et al. (1997) Adenomatous polyposis coli tumor suppressor protein has signaling activity in Xenopus laevis embryos resulting in the induction of an ectopic dorsoanterior axis. *J Cell Biol* 136, 411–420.

4. Zeng, L., Fagotto, F., Zhang, T., et al. (1997) The mouse Fused locus encodes Axin, an inhibitor of the Wnt signaling pathway that regulates embryonic axis formation. *Cell* 90, 181–192.

5. Itoh, K., Krupnik, V. E., Sokol, S. Y. (1998) Axis determination in Xenopus involves biochemical interactions of axin, glycogen synthase kinase 3 and beta-catenin. *Curr Biol* 8, 591–594.

6. Marikawa, Y., Elinson, R. P. (1998) beta-TrCP is a negative regulator of Wnt/beta-catenin signaling pathway and dorsal axis formation in Xenopus embryos. *Mech Dev* 77, 75–80.

7. Yost, C., Farr, G. H., 3rd, Pierce, S. B., et al. (1998) GBP, an inhibitor of GSK-3, is implicated in Xenopus development and oncogenesis. *Cell* 93, 1031–1041.

8. Hedgepeth, C. M., Deardorff, M. A., Klein, P. S. (1999) Xenopus axin interacts with glycogen synthase kinase-3 beta and is expressed in the anterior midbrain. *Mech Dev* 80, 147–151.

9. Salic, A., Lee, E., Mayer, L., et al. (2000) Control of beta-catenin stability: reconstitution of the cytoplasmic steps of the wnt pathway in Xenopus egg extracts. *Mol Cell* 5, 523–532.

10. Ossipova, O., Bardeesy, N., DePinho, R. A., et al. (2003) LKB1 (XEEK1) regulates Wnt signaling in vertebrate development. *Nat Cell Biol* 5, 889–894.

11. Heasman, J., Crawford, A., Goldstone, K., et al. (1994) Overexpression of cadherins and underexpression of beta-catenin inhibit dorsal mesoderm induction in early Xenopus embryos. *Cell* 79, 791–803.

12. Houston, D. W., Kofron, M., Resnik, E., et al. (2002) Repression of organizer genes in dorsal and ventral Xenopus cells mediated by maternal XTcf3. *Development* 129, 4015–4025.

13. Tao, Q., Yokota, C., Puck, H., et al. (2005) Maternal wnt11 activates the canonical wnt signaling pathway required for axis formation in Xenopus embryos. *Cell* 120, 857–871.

14. Leyns, L., Bouwmeester, T., Kim, S. H., et al. (1997) Frzb-1 is a secreted antagonist of Wnt signaling expressed in the Spemann organizer. *Cell* 88, 747–756.

15. Wang, S., Krinks, M., Lin, K., et al. (1997) Frzb, a secreted protein expressed in the Spemann organizer, binds and inhibits Wnt-8. *Cell* 88, 757–766.

16. Glinka, A., Wu, W., Delius, H., et al. (1998) Dickkopf-1 is a member of a new family of secreted proteins and functions in head induction. *Nature* 391, 357–362.

17. Hsieh, J. C., Kodjabachian, L., Rebbert, M. L., et al. (1999) A new secreted protein that binds to Wnt proteins and inhibits their activities. *Nature* 398, 431–436.

18. Yang-Snyder, J., Miller, J. R., Brown, J. D., et al. (1996) A frizzled homolog functions in a vertebrate Wnt signaling pathway. *Curr Biol* 6, 1302–1306.

19. Medina, A., Reintsch, W., Steinbeisser, H. (2000) Xenopus frizzled 7 can act in canonical and non-canonical Wnt signaling pathways: implications on early patterning and morphogenesis. *Mech Dev* 92, 227–237.

20. Tamai, K., Semenov, M., Kato, Y., et al. (2000) LDL-receptor-related proteins in Wnt signal transduction. *Nature* 407, 530–535.

21. Mao, B., Wu, W., Li, Y., et al. (2001) LDL-receptor-related protein 6 is a receptor for Dickkopf proteins. Nature 411, 321–325.

22. Davidson, G., Wu, W., Shen, J., et al. (2005) Casein kinase 1 gamma couples Wnt receptor activation to cytoplasmic signal transduction. *Nature* 438, 867–872.

23. Zeng, X., Tamai, K., Doble, B., et al. (2005) A dual-kinase mechanism for Wnt co-receptor phosphorylation and activation. *Nature* 438, 873–877.

24. Funayama, N., Fagotto, F., McCrea, P., et al. (1995) Embryonic axis induction by the armadillo repeat domain of beta-catenin: evidence for intracellular signaling. *J Cell Biol* 128, 959–968.

25. Behrens, J., von Kries, J. P., Kuhl, M., et al. (1996) Functional interaction of beta-catenin with the transcription factor LEF-1. *Nature* 382, 638–642.

26. Molenaar, M., van de Wetering, M., Oosterwegel, M., et al. (1996) XTcf-3 transcription factor mediates beta-catenin-induced axis formation in Xenopus embryos. *Cell* 86, 391–9.

27. Brannon, M., Gomperts, M., Sumoy, L., et al. (1997) A beta-catenin/XTcf-3 complex binds to the siamois promoter to regulate dorsal axis specification in Xenopus. *Genes Dev* 11, 2359–2370.

28. McKendry, R., Hsu, S. C., Harland, R. M., et al. (1997) LEF-1/TCF proteins mediate

wnt-inducible transcription from the Xenopus nodal-related 3 promoter. *Dev Biol* 192, 420–431.

29. Laurent, M. N., Blitz, I. L., Hashimoto, C., et al. (1997) The Xenopus homeobox gene twin mediates Wnt induction of goosecoid in establishment of Spemann's organizer. *Development* 124, 4905-4916.

30. Gradl, D., Kuhl, M., Wedlich, D. (1999) The Wnt/Wg signal transducer beta-catenin controls fibronectin expression. *Mol Cell Biol* 19, 5576–5587.

31. McGrew, L. L., Takemaru, K., Bates, R., Moon, R. T. (1999) Direct regulation of the Xenopus engrailed-2 promoter by the Wnt signaling pathway, and a molecular screen for Wnt-responsive genes, confirm a role for Wnt signaling during neural patterning in Xenopus. *Mech Dev* 87, 21–32.

32. Yang, J., Tan, C., Darken, R. S., et al. (2002) Beta-catenin/Tcf-regulated transcription prior to the midblastula transition. *Development* 129, 5743–5752.

33. Vallin, J., Thuret, R., Giacomello, E. F., et al. (2001) Cloning and characterization of three Xenopus slug promoters reveal direct regulation by Lef/beta-catenin signaling. *J Biol Chem* 276, 30350–30358.

34. Du, S. J., Purcell, S. M., Christian, J. L., et al. (1995) Identification of distinct classes and functional domains of Wnts through expression of wild-type and chimeric proteins in Xenopus embryos. *Mol Cell Biol* 15, 2625–2634.

35. Deardorff, M. A., Tan, C., Conrad, L. J., et al. (1998) Frizzled-8 is expressed in the Spemann organizer and plays a role in early morphogenesis. *Development* 125, 2687–2700.

36. Sheldahl, L. C., Park, M., Malbon, C. C., et al. (1999) Protein kinase C is differentially stimulated by Wnt and Frizzled homologs in a G-protein-dependent manner. *Curr Biol* 9, 695–698.

37. Djiane, A., Riou, J., Umbhauer, M., et al. (2000) Role of frizzled 7 in the regulation of convergent extension movements during gastrulation in Xenopus laevis. *Development* 127, 3091–3100.

38. Kuhl, M., Sheldahl, L. C., Malbon, C. C., et al. (2000) Ca(2+)/calmodulin-dependent protein kinase II is stimulated by Wnt and Frizzled homologs and promotes ventral cell fates in Xenopus. *J Biol Chem* 275, 12701–12711.

39. Tada, M., Smith, J. C. (2000) Xwnt11 is a target of Xenopus Brachyury: regulation of gastrulation movements via Dishevelled, but not through the canonical Wnt pathway. *Development* 127, 2227–2238.

40. Pandur, P., Lasche, M., Eisenberg, L. M., et al. (2002) Wnt-11 activation of a non-canonical Wnt signaling pathway is required for cardiogenesis. *Nature* 418, 636–641.

41. Sheldahl, L. C., Slusarski, D. C., Pandur, P., et al. (2003) Dishevelled activates Ca2+ flux, PKC, and CamKII in vertebrate embryos. *J Cell Biol* 161, 769–777.

42. Sokol, S. Y. (1996) Analysis of Dishevelled signaling pathways during Xenopus development. *Curr Biol* 6, 1456–1467.

43. Rothbacher, U., Laurent, M. N., Deardorff, M. A., et al. (2000) Dishevelled phosphorylation, subcellular localization and multimerization regulate its role in early embryogenesis. *Embo J* 19, 1010–1022.

44. Wallingford, J. B., Rowning, B. A., Vogeli, K. M., et al. (2000) Dishevelled controls cell polarity during Xenopus gastrulation. *Nature* 405, 81–85.

45. Schohl, A., Fagotto, F. (2003) A role for maternal beta-catenin in early mesoderm induction in Xenopus. *Embo J* 22, 3303–3313.

46. Liu, F., van den Broek, O., Destree, O., et al. (2005) Distinct roles for Xenopus Tcf/Lef genes in mediating specific responses to Wnt/beta-catenin signaling in mesoderm development. *Development* 132, 5375–5385.

47. Hoppler, S., Brown, J. D., Moon, R. T. (1996) Expression of a dominant-negative Wnt blocks induction of MyoD in Xenopus embryos. *Genes Dev* 10, 2805–2817.

48. McGrew, L. L., Hoppler, S., Moon, R. T. (1997) Wnt and FGF pathways cooperatively pattern anteroposterior neural ectoderm in Xenopus. *Mech Dev* 69, 105–114.

49. Domingos, P. M., Itasaki, N., Jones, C. M., et al. (2001) The Wnt/beta-catenin pathway posteriorizes neural tissue in Xenopus by an indirect mechanism requiring FGF signaling. *Dev Biol* 239, 148–160.

50. LaBonne, C., Bronner-Fraser, M. (1998) Neural crest induction in Xenopus: evidence for a two-signal model. *Development* 125, 2403–2414.

51. De Calisto, J., Araya, C., Marchant, L., et al. (2005) Essential role of non-canonical Wnt signaling in neural crest migration. *Development* 132, 2587–2597.

52. Kunz, M., Herrmann, M., Wedlich, D., et al. (2004) Autoregulation of canonical Wnt signaling controls midbrain development. *Dev Biol* 273, 390–401.

53. Rasmussen, J. T., Deardorff, M. A., Tan, C., et al. (2001) Regulation of eye development by frizzled signaling in Xenopus. *Proc Natl Acad Sci* USA 98, 3861–3866.

54. Maurus, D., Heligon, C., Burger-Schwarzler, A., et al. (2005) Noncanonical Wnt-4 signaling and EAF2 are required for eye development in Xenopus laevis. *Embo J* 24, 1181–1191.

55. Marvin, M. J., Di Rocco, G., Gardiner, A., et al. (2001) Inhibition of Wnt activity induces heart formation from posterior mesoderm. *Genes Dev* 15, 316–327.

56. Schneider, V. A., Mercola, M. (2001) Wnt antagonism initiates cardiogenesis in Xenopus laevis. *Genes Dev* 15, 304–315.

57. Saulnier, D. M., Ghanbari, H., Brandli, A. W. (2002) Essential function of Wnt-4 for tubulogenesis in the Xenopus pronephric kidney. *Dev Biol* 248, 13–28.

58. Lyons, J. P., Mueller, U. W., Ji, H., et al. (2004) Wnt-4 activates the canonical beta-catenin-mediated Wnt pathway and binds Frizzled-6 CRD: functional implications of Wnt/beta-catenin activity in kidney epithelial cells. Exp *Cell Res* 298, 369–387.

59. Schneider, S., Steinbeisser, H., Warga, R. M., et al. (1996) Beta-catenin translocation into nuclei demarcates the dorsalizing centers in frog and fish embryos. *Mech Dev* 57, 191–198.

60. Schohl, A., Fagotto, F. (2002) Beta-catenin, MAPK and Smad signaling during early Xenopus development. *Development* 129, 37–52.

61. Geng, X., Xiao, L., Lin, G. F., et al. (2003) Lef/Tcf-dependent Wnt/beta-catenin signaling during Xenopus axis specification. *FEBS Lett* 547, 1–6.

62. Denayer, T., Van Roy, F., Vleminckx, K. (2006) In vivo tracing of canonical Wnt signaling in Xenopus tadpoles by means of an inducible transgenic reporter tool. *FEBS Lett* 580, 393–398.

63. Wheeler, G. N., Hamilton, F. S., Hoppler, S. (2000) Inducible gene expression in transgenic Xenopus embryos. *Curr Biol* 10, 849–52.

64. Heasman, J., Kofron, M., Wylie, C. (2000) Beta-catenin signaling activity dissected in the early Xenopus embryo: a novel antisense approach. *Dev Biol* 222, 124–134.

65. Moody, S. A. (1987) Fates of the blastomeres of the 16-cell stage Xenopus embryo. *Dev Biol* 119, 560–578.

66. Moody, S. A. (1987) Fates of the blastomeres of the 32-cell-stage Xenopus embryo. *Dev Biol* 122, 300–319.

67. Polli, M., Amaya, E. (2002) A study of mesoderm patterning through the analysis of the regulation of Xmyf-5 expression. *Development* 129, 2917–2927.

68. Gawantka, V., Pollet, N., Delius, H., et al. (1998) Gene expression screening in Xenopus identifies molecular pathways, predicts gene function and provides a global view of embryonic patterning. *Mech Dev* 77, 95–141.

69. Pollet, N., Muncke, N., Verbeek, B., et al. (2005) An atlas of differential gene expression during early Xenopus embryogenesis. *Mech Dev* 122, 365–439.

70. Smith, W. C., Harland, R. M. (1991) Injected Xwnt-8 RNA acts early in Xenopus embryos to promote formation of a vegetal dorsalizing center. *Cell* 67, 753–65.

71. Itasaki, N., Jones, C. M., Mercurio, S., et al. (2003) Wise, a context-dependent activator and inhibitor of Wnt signaling. *Development* 130, 4295–4305.

72. Voigt, J., Chen, J. A., Gilchrist, M., et al. (2005) Expression cloning screening of a unique and full-length set of cDNA clones is an efficient method for identifying genes involved in Xenopus neurogenesis. *Mech Dev* 122, 289–306.

73. Chen, J. A., Voigt, J., Gilchrist, M., et al. (2005) Identification of novel genes affecting mesoderm formation and morphogenesis through an enhanced large scale functional screen in Xenopus. *Mech Dev* 122, 307–331.

74. Kenwrick, S., Amaya, E., Papalopulu, N. (2004) Pilot morpholino screen in Xenopus tropicalis identifies a novel gene involved in head development. *Dev Dyn* 229, 289–299.

75. Rana, A. A., Collart, C., Gilchrist, M. J., et al. (2006) Defining synphenotype groups in Xenopus tropicalis by use of antisense morpholino oligonucleotides. *PLoS Genet* 2, e193.

76. Klein, P. S., Melton, D. A. (1996) A molecular mechanism for the effect of lithium on development. *Proc Natl Acad Sci U S A* 93, 8455–9.

77. Meijer, L., Skaltsounis, A. L., Magiatis, P., et al. (2003) GSK-3-selective inhibitors derived from Tyrian purple indirubins. *Chem Biol* 10, 1255–1266.

78. Lee, E., Salic, A., Kirschner, M. W. (2001) Physiological regulation of [beta]-catenin stability by Tcf3 and CK1epsilon. *J Cell Biol* 154, 983–993.

79. Liu, J., Wu, X., Mitchell, B., et al. (2005) A small-molecule agonist of the Wnt signaling pathway. *Angew Chem Int Ed Engl* 44, 1987–1990.

80. Grammer, T. C., Khokha, M. K., Lane, M. A., et al. (2005) Identification of mutants in inbred Xenopus tropicalis. *Mech Dev* 122, 263–272.

81. Noramly, S., Zimmerman, L., Cox, A., et al. (2005) A gynogenetic screen to isolate naturally occurring recessive mutations in Xenopus tropicalis. *Mech Dev* 122, 273–287.

82. Goda, T., Abu-Daya, A., Carruthers, S., et al. (2006) Genetic screens for mutations affecting development of Xenopus tropicalis. *PLoS Genet* 2, e91.

83. Amaya, E. (2005) Xenomics. *Genome Res* 15, 1683–1691.

84. Major, N., Wassersug, R. J. (1998) Survey of Current Techniques in the Care and Maintenance of the African Clawed Frog (Xenopus laevis) *Contemp Top Lab Anim Sci* 37, 57–60.

Section A

Methods for Studying Wnt Signaling in *Xenopus* Embryos

Chapter 22

Analysis of Gene Expression in *Xenopus* Embryos

Danielle L. Lavery and Stefan Hoppler

Abstract

Determining the expression pattern of a gene of interest is critical to understanding when, where, and how it may function during development. This chapter describes methods for determining the localization and expression levels for both mRNA and protein. Some of these methods can be described as quantitative or semi-quantitative while others are considered more qualitative in nature. To determine the spatial localization of mRNA expression, RNA *in situ* hybridization can be used on both whole-mount or sectioned embryos. Northern blot and qPCR are more quantitative methods for analyzing mRNA expression levels. For determining protein localization, antibody staining either by immunocytochemistry or immunofluorescence can be performed on both whole-mount or sectioned embryos; whereas Western blot is the method generally used for quantifying protein expression levels.

Key words: *Xenopus*, RNA, histology, antibody, Northern blot, Western blot, *in situ*, immunocytochemistry.

1. Introduction

Gene expression patterns can be analyzed in terms of both mRNA and protein localization. These in turn can be further analyzed in terms of their spatial and temporal patterns as well as their quantitative levels. Tissue-based qualitative methods generally give spatial and temporal information about gene expression; while gel-based methods are generally used for analyzing temporal expression on a more quantitative level. Much can be inferred from a gene's expression pattern in terms of its potential function. The timing and location of expression can suggest hypotheses as to what role a gene may play during normal development. Protein localization within the cell can also help determine what other

Elizabeth Vincan (ed.), *Wnt Signaling, Volume II: Pathway Models, vol. 469*
© 2008 Humana Press, a part of Springer Science + Business Media, New York, NY
Book doi: 10.1007/978-1-60327-469-2

proteins it may interact with and through what mechanisms it may be signaling. This is particularly true for β-catenin, which plays two very distinct but necessary roles in different compartments of the cell. It acts as a transcription factor when it translocates into the nucleus upon activation of the canonical Wnt signaling pathway (*see* **Chapter 8, Volume 1**) and is also an essential part of the cadherin complex at the cell membrane necessary for cell-cell adhesion. Analysis of gene expression is also important for interpreting the result of functional experiments with Wnt signaling components (*see* **Chapter 25, Volume 2**), either by studying the regulation of downstream target genes or by studying the resulting embryonic phenotype with tissue-specific marker genes.

RNA *in situ* hybridization is a powerful tool utilized for both spatial and semi-quantitative localization of mRNA during *Xenopus* development. This technique was first developed for *Xenopus* embryos by Richard Harland *(1)*. Embryos or tissue sections can be analyzed at different stages of development to give the overall temporal expression pattern of the gene of interest. This technique is based upon complementary binding of a digoxigenin- or fluorescein-labeled antisense RNA probe to the sense strand of the endogenous mRNA expressed from the gene of interest. Analysis of RNA expression on histological sections reveals better information about gene expression in internal tissues, such as the developing organs. Quantitative PCR (qPCR) is a quantitative method for looking at mRNA expression levels, which can be used to measure expression during different stages of development, different tissues as well as changes in gene expression in response to different experimental conditions. Expression levels determined by qPCR are generally described as relative expression levels but can also be expressed as copy number if quantified standards are used to generate a standard curve. Northern blot can be used for semi-quantitative analysis of mRNA expression levels. It is particularly powerful at detecting different size transcripts or splice products of a gene. It can be used on either whole embryos or microdissected tissues and at different stages of development.

RNase protection assays are another option for looking at mRNA expression levels, which are not further covered in this chapter. It is based on binding of a labeled (dig or radio) antisense probe to your gene of interest which then protects this target mRNA from subsequent digestion with a nuclease specific for single-stranded RNA. It can be used to give information about exon/intron boundaries, identification of translation initiation and termination sequences and also detection of tissue-specific alternative splicing. In addition, it can be used to analyze expression levels of multiple transcripts at the same time provided that the probes designed against each gene are of different lengths. *See* Ma et al. *(2)* or the Ambion Web site (www.ambion.com) for more details and protocols. Microarray analysis is also not covered in this chapter, but have been developed for both *Xenopus tropicalis* and *Xenopus laevis* and can be used to detect changes in expression levels of genes that are arrayed on chips in response to experimental conditions (ref. *3*, see also Affymetrix [www.affymetrix.com], Agilent [www.chem.agilent.com], and Operon [www.operon.com]).

Western blotting is a semi-quantitative method that can be used to determine relative protein levels over different developmental stages or to detect changes in protein levels following experimental treatments. Tissue-specific Western blots on dissected organs and tissues of embryos can be used to determine spatial expression pattern of a protein. Cell fractionation techniques allow separate analysis of cytoplasmic versus nuclear localization of a protein, such as β-catenin, which allows conclusions to be drawn about the activity of Wnt signaling *(4)*. Antibodies can also be used as reagents to detect the localization of Wnt signaling components and other proteins in *Xenopus* embryonic tissue. Considerably different protocols are used to detect protein localization in whole-mount fixed embryos as opposed to histology sections and depending on the detection method associated with the secondary antibody (fluorochrome conjugated versus enzyme conjugated). While some of these protocols share similar steps, the researcher will follow a particular protocol that suits the antibody tools available.

2. Materials

2.1. Whole-mount RNA In Situ Hybridization

2.1.1. Solutions
(See **Note 1**)

1. DEPC-treated ddH$_2$O. To prepare, add DEPC (Sigma) to ddH$_2$O at a concentration of 1 mL/L, place at 37°C overnight and autoclave. Store at room temperature.

2. MEM Salts (10X stock): 1 M MOPS (Sigma), 20 mM EGTA (Sigma), 10 mM MgSO$_4$ (Sigma), adjust pH to 7.4 with NaOH, treat with DEPC (1 mL/L), place at 37°C overnight, autoclave, and store-wrapped in aluminum foil at 4°C.

3. MEMFA: 1X MEM salts, 4% (w/v) formaldehyde (Sigma) in DEPC-treated ddH$_2$O. Always made up fresh when needed.

4. DEPC-treated phosphate-buffered saline (PBS): Dissolve five tablets of PBS (Sigma) in 1 L ddH2O, treat with 1 mL/L DEPC by incubating overnight at 37°C and autoclaving. (*See* ref. 5 for alternatives.)

5. PBST: 0.1% (v/v) Tween-20 in DEPC-treated PBS.

6. Bleaching solution: 5% (v/v) formamide, 0.5X SSC, 1% (v/v) H2O2 in DEPC- treated ddH2O.

7. Triethanol-amine (Sigma): 0.1 M made up in DEPC-treated ddH2O.

8. Acetic anhydride (Sigma): 2.5 µL/mL made up freshly in 0.1 M triethanol-amine.

9. 20X SSC: 0.15 M NaCl, 0.015 M sodium citrate, add 1 mL/L DEPC, adjust the pH to 7.0 with NaOH, incubate overnight at 37°C and autoclave.

10. Hybridization buffer: 50% formamide, 5X SSC, 1 mg/mL Torula RNA, 100 µg/mL heparin, 1X Denharts (2% BSA [bovine serum albumin, Sigma-Fraction V]), 2% PVP-20, 2% Ficoll 400), 0.1% (v/v) Tween-20, 5 mM EDTA in DEPC-treated ddH$_2$O. Store at –20°C.

11. 5X Maleic Acid Buffer (MAB): 500 mM maleic acid (Sigma), 750 mM NaCl (Biochema), adjust pH to 7.5 with NaOH, autoclave and store at room temperature for several months. Dilute to 1X before use.

12. Blocking reagent stock solution: 10% blocking Reagent (Roche) made up in 1X MAB, heat until it is in solution, autoclave, and store in aliquots at –20°C.

13. Sheep serum: heat inactivate at 65°C for 30 minutes and store in aliquots at –20°C.

14. Blocking Solution: 2% Blocking Reagent, 20% (v/v) heat inactivated sheep serum in 1X MAB. Prepare from stock solutions before use.

15. Alkaline Phosphate buffer (AP Buffer): 0.1 M Tris-HCl, pH 9.0, 50 mM MgCl$_2$, 0.1 M NaCl, 0.1% (v/v) Tween-20 in DEPC-treated ddH$_2$O.

16. Staining solution: 50 µL NBT and 35 µL BCIP (both from Roche) in 10 mL AP buffer. Prepare fresh prior to use.

2.1.2. Embryos

Embryos to be analyzed by whole-mount RNA *in situ* hybridization should first be fixed in MEMFA for 1 to 2 hours with gentle agitation at room temperature in 8-mL glass vials. Following fixation, wash the embryos two to three times in 100% methanol with gentle agitation (to prevent the embryos from sticking to one another) for 2 hours total at room temperature or at 4°C overnight. Store the embryos at 4°C until ready to use (up to 6 months; *see* **Note 2**).

2.1.3. Digoxygenin- and Fluorescein-labeled RNA Probes

We use the Megascript High Yield Transcription Kit (SP6, T7 or T3 kits) from Ambion for generation of digoxygenin and/or fluorescein-labeled RNA probes. Add digoxygenin-11-UTP or fluorescein-12-UTP (Roche, depending on what you are using to label your RNA) at an appropriate ratio to normal UTP (1:2 or 1:3) in your transcription reaction mix (*see* **Note 3**). These probes can be used for analysis of both whole embryos and tissues. As templates for *in vitro* transcription, suitable plasmid vectors are required containing the cDNA or partial cDNA of the gene of interest, suitable restriction sites and prokaryotic RNA phage promoter sequences. In order to synthesize anti-sense probes, the plasmid DNA is linearized with a restriction enzyme that cuts close to the start site of the gene. The probe is transcribed with the RNA polymerase corresponding to the RNA phage promoter (SP6, T7, or T3) at the 3' end of the coding region of the gene (*see* **Notes 4** and **5**).

2.2. RNA In Situ Hybridization on Histological Sections

2.2.1. Solutions (see **Note 1**)

1. Bleaching solution: 8.3 mL of 30% (v/v) H_2O_2 in 41.7 mL 100% methanol.
2. PBST: 0.1% (v/v) Tween-20 in DEPC-treated PBS.
3. Proteinase K stock solution: 20 mg/mL in H_2O, store at −20°C.
4. Glycine (Sigma).
5. Paraformaldehyde: 4% in DEPC-treated PBS.
6. Gluteraldehyde (Grade I: 25% aqueous solution, Sigma cat No. G-5882).
7. Hybridization mix: 50% formamide, 5X SSC, pH 4.5 (adjust pH of 20x SSC stock solution with HCl), 50 µg/mL yeast or Torula RNA, 1% (w/v) SDS, 50 µg/mL heparin. (Do not autoclave!)
8. Solution I: 50% (v/v) formamide, 5X SSC, pH 4.5, 1% (w/v) SDS. (Do not autoclave!)
9. Solution II: 50% (v/v) formamide, 2x SSC, pH 4.5. (Do not autoclave!)
10. MAB: 100 mM maleic acid, 150 mM NaCl, adjust pH to 7.5 with NaOH pellets.
11. MAB with 2% Blocking reagent (*see* **Note 6**).
12. MABT: 0.1% (v/v) Tween-20 in MAB.
13. AP Buffer: 0.1 M Tris-HCL, pH 9.5, 50 mM $MgCl_2$, 0.1 M NaCl, 0.1% (v/v) Tween-20.
14. AP Buffer with 5% (v/v) Polyvinyl alcohol (PVA [MW 31,000–50,000]; *see* **Note 7**).
15. BM Purple: one part AP buffer with 5% (v/v) PVA, four parts BM Purple.

2.2.2. Paraffin Sections of Xenopus Embryos

Fix embryos of the appropriate stage of development in 4% paraformaldehyde (PFA) (or MEMFA, see **Subsection 2.1.2**) for 30 minutes at room temperature, change PFA twice and keep embryos at 4°C for at least 1 week. Wash fixed embryos three times in PBST, dehydrate them through an ethanol series, wash them once in xylene:paraffin (1:1) for 45 minutes at 70°C and then three times in 100% paraffin for at least 30 minutes each at 70°C. Place embryos in suitable plastic containers; orient them under a dissecting microscope using forceps and leave to solidify overnight. Use a microtome to cut 10-μm sections, place them on 3-aminopropyltriethoxy-silane treated microscope slides and leave to dry overnight.

2.2.3. Digoxygenin-labeled RNA Probes

See Subsection 2.1.3.

2.3. Northern Blot Analysis (See Note 1)

1. Ammonia acetate: 5 M made with DEPC-treated ddH$_2$O.

2. Ethanol: 70% made with DEPC-treated ddH$_2$O.

3. RNA ladder: Invitrogen 0.24–9.5 Kb RNA Ladder

4. Glyoxal: prepared by deionizing with Amberlite (or BioRad AG501-X8) until pH is more than 5.0 and stored as 50-μL aliquots at –20°C.

5. 0.1 M NaPO$_4$: Add 3.9 mL of 1 M NaH$_2$PO$_4$, 6.1 mL of 1 M Na$_2$HPO$_4$, 90 mL of ddH$_2$O, adjust pH to 7, and store at room temperature.

6. Loading Buffer: 50% (v/v) glycerol (Sigma), 10 mM NaPO$_4$ (0.1 M NaPO$_4$ diluted 1:10), 0.25% (w/v) bromophenol blue, 0.25% (w/v) xylene cyanol.

7. Agarose gel(100 mL): 1 g agarose (1% w/v final), 10 mL 0.1 M NaPO$_4$ 90 mL DEPC-treated ddH$_2$O.

8. Running buffer: 10 mM NaPO$_4$, pH 7. Add 3.9 mL 1 M NaH$_2$PO$_4$, 6.1 mL 1 M Na$_2$HPO$_4$ and 990 mL DEPC-treated ddH$_2$O.

9. 20X SSC: 0.15 M NaCl, 0.015 M sodium citrate, DEPC 1 mL/L, incubate overnight at 37°C and autoclave.

10. 10X SSC: Dilute 20X SSC stock solution in DEPC-treated ddH$_2$O.

11. 0.02% (w/v) methylene blue made up in 0.5 M NaOAc, pH 5.2 (DEPC-treated).

12. Northern Starter Kit from Roche according to manufacturer's instructions to synthesize a Digoxigenin-labeled probe.

2.4. Western Blotting and Protein Analysis

1. PBS: dissolve five tablets of PBS (Sigma) in 1 L ddH$_2$O and autoclave.

2. RIPA Buffer (kept on ice): 50 mM Tris-HCl, pH 7.4, 150 mM NaCl, 1% (v/v) nonidet P-40, 0.5% (w/v) deoxycholic acid, 0.1% (w/v) SDS, 1x complete protease inhibitor mixture (Roche; *see* **Note 27**).

3. BCA Protein Assay Kit from Novagen, which contains all required solutions, apart from RIPA Buffer. Prepare BCA working reagent by combining (for each sample) 200 μL

BCA solution (provided in the kit) with 4-µL cupric sulfate (provided in the kit).

4. NuPAGE 10X Sample Reducing Agent (Invitrogen).

5. NuPAGE 4X LDS Sample Buffer (Invitrogen).

6. Pre-cast 4–12% Bis-Tris NuPAGE gel (Invitrogen).

7. Multi-Mark protein size marker ladder (Invitrogen).

8. MES or MOPS SDS Running Buffers, depending on the protein size separation on the gel (Invitogen).

9. PVDF membrane.

10. NuPAGE Transfer buffer (Invitrogen): dilute 50X stock solution to 1X working solution before use, e.g., 50 mL 20X NuPAGE Transfer buffer (Invitrogen), 200 mL 100% methanol, 750 mL ddH$_2$O.

11. Novex or Biorad Transfer apparatuses (see manufacturer instructions or ref. 5 for alternatives).

12. Tris-buffered saline (TBS): 137 mM NaCl, 10 mM Tris-base. Adjust the pH to 7.4, autoclave, and store at room temperature.

13. TBST: 0.1% (w/v) Tween-20 in TBS. Store at 4°C.

14. Blocking Solution: 5% (w/v) powdered milk, 5% (w/v) BSA in TBST.

2.5. Whole-mount Immunoflurescence Staining

2.5.1. Embryos

Fix embryos in Dent's Fixative (20% DMSO, 80% Methanol) overnight at –20°C. The next day replace Dent's Fixative with 100% methanol and store at –20°C until ready to use (*see* **Note 8**).

2.5.2. Solutions

1. PBS: see **Subsection 2.4., item 1**.

2. PBST: 1% (w/v) Tween-20 in PBS.

3. Bleaching solution: 5% (v/v) formamide, 1% (v/v) H2O2, 0.5X SSC in ddH2O.

4. 5% (w/v) BSA in PBS.

5. Blocking Solution: 5% (w/v) powdered milk, 5% (w/v) BSA in TBST.

6. Murray's clear: 1 part benzyl alcohol, 2 parts benzyl benzoate.

2.6. Immunofluorescence Staining

2.6.1. Sections of Xenopus Embryos

Fix embryos Dent's Fixative (20% DMSO, 80% Methanol) overnight at –20°C, the next day replace Dent's Fixative with 100% methanol and store at –20°C until ready to further process. Rehydrate embryos by washing through a decreasing methanol series 5 mL/wash made up in PBST: 75% methanol/25% PBST; 50% methanol/50% PBST; 25% methanol/PBST. Perform each wash for 5 minutes at room temperature with gentle agitation on a rocking platform. Remove the PBS, equilibrate the embryos in OTC for 1 hour at room temperature. Replace the OTC with fresh OTC, transfer the embryos to a small square weigh boat and carefully orient the embryos by gently moving them using a plastic pipette tip (may require more than one adjustment) and then slowly freeze on ice. Once the OTC has solidified to the point where the embryos can no longer move, place them on a level surface at –20°C to complete the freezing process. Trim the block of OTC containing the embedded embryos with a razor blade and attach to a sectioning block in the desired position (depending on what type of sections are required) with fresh OTC, allow to freeze and section with a cryosectioner at –20°C. Place sections onto polylysine coated slides (VWR), allow to dry at room temperature overnight and store at –20°C until ready to use (*see* **Note 9**).

2.6.2. Solutions

1. 5% (w/v) BSA in PBS.
2. PBST: 0.1% (w/v) Triton X-100 (Sigma) in PBS.
3. Blocking solution: 5% (w/v) BSA, 10% (v/v) normal donkey serum (or the serum corresponding to what your secondary antibodies are raised in) in PBS.
4. Vectorshield mounting media (Vector Labs) or any other suitable mounting media.

2.7. Whole-mount Immunocytochemistry Staining

2.7.1. Embryos

Fix embryos in Dent's Fixative (20% DMSO, 80% Methanol) overnight at –20°C, the next day replace Dent's Fixative with 100% methanol and store at –20°C until ready to use (*see* **Note 8**).

2.7.2. Solutions

1. Bleaching solution: 5% (v/v) formamide, 0.5X SSC, 1% (v/v) H_2O_2 in ddH_2O.
2. TBS: Dissolve one packet of TBS powder (Sigma) in 1 L of ddH_2O, which yields 0.05 M Tris buffered saline, pH 8.0.
3. 5% (w/v) BSA in TBS.
4. Blocking solution: 5% (w/v) BSA, 10% (v/v) normal donkey serum (or of whatever species your secondary antibody is raised in) in TBS.
5. Staining solution (always make up fresh, immediately prior to use): 1 DAB Tablet (Sigma), 15 mL TBS, 12 µL H_2O_2 (30% stock solution; *see* **Note 10**).
6. Murray's Clear (optional): 2 volumes of benzyl benzoate (Sigma), 1 volume benzyl alcohol (Sigma).

2.8. Immuncytochemistry Staining

Fix embryos in Dent's Fixative (*20% DMSO, 80% Methanol*) overnight at –20°C, the next day replace Dent's Fixative with

2.8.1. Sections of Xenopus Embryos

100% methanol and store at –20°C until ready to further process. Replace Dent's fixative with 5 mL of 100% methanol and incubate for 5 minutes at room temperature. Replace 100% methanol with fresh 100% methanol and incubate for 5 minutes. Remove Methanol and replace with 100% Xylene and incubate for 30 minutes at room temperature in a laminar flow hood. Repeat the Xylene wash. Remove the Xylene and add a mixture of 50% melted paraffin with 50% Xylene and incubate the embryos for 2 hours at 80°C. Replace Xylene/paraffin mixture with 100% paraffin and incubate for 6 hours to overnight at 80°C. Carefully embed embryos in fresh paraffin in square weigh boats placed on top of a hot plate so that the paraffin remains liquid while you orient the embryos appropriately. With tadpole stage embryos and older, it is easiest to let them fall flat onto the bottom of the weight boat so that they are lying on one of their sides. Once the embryos are in the desired position carefully move them off the hot plate without jarring the embryos and move them onto a cool surface to solidify for several hours. The paraffin blocks containing the embryos can be cut with a razor blade and attached to the sectioning block in an appropriate orientation with fresh paraffin. Section the embryos and place sections onto polylysine-coated slides (VWR) and store at room temperature until ready to use (*see* **Note 11**).

2.8.2. Solutions

1. TBS: see **Subsection 2.7.2., item 2**.

2. 5% (w/v) BSA in TBS.

3. 0.1% (v/v) Triton X-100 in TBS.

4. Blocking solution: see **Subsection 2.7.2., item 4**.

5. Staining Solution: see **Subsection 2.7.2., item 5**.

6. Non-aqueous mounting media such as DPX (Sigma).

3. Methods

3.1. Whole-mount RNA In Situ Hybridization (See Note 12)

1. Rehydrate embryos by washing through a decreasing methanol series made up in DEPC-treated PBST: 75% methanol/25% PBST; 50% methanol/50% PBST; 25% methanol/PBST; 100% PBS without Tween. Perform each wash for 5 minutes at room temperature with gentle agitation on a rocking platform.

2. Remove the last PBS wash, add the bleaching solution and place the embryos under a fluorescent light with gentle agitation. Allow the embryos to bleach for 1 to 2 hours (depending on the level of pigmentation) until the majority of the pigmentation has been removed (*see* **Note 13**).

3. Once the embryos are finished bleaching, wash three times five minutes in DEPC-treated PBS.

4. Wash embryos three times five minutes in 0.1 M triethanolamine; to the second and third washes add 2.5 µL/mL acetic anhydride.

5. Wash the embryos twice 5 minutes with DEPC-treated PBST and carefully transfer the embryos to 2-mL eppendorf tubes with a 3-mL disposable plastic Pasteur pipette with the tip cut off (*see* **Note 14**).

6. Wash the embryos with 0.5 mL of hybridization buffer and leave to stand for a few minutes at room temperature. Once the embryos have all sunk to the bottom of the tube, replace the hybridization buffer with 0.5 mL fresh hybridization buffer and prehybridize for at least 6 hours or overnight at 65°C with gentle agitation on a rocking platform.

7. Replace the hybridization buffer with 0.5 mL of fresh hybridization buffer containing digoxygenin labeled probe (0.1–10 µg/mL) and incubate at 65°C overnight.

8. After hybridization, the hybridization mix containing the digoxygenin labeled RNA probe is taken off and can be saved for reuse (*see* **Note 15**). Wash the embryos sequentially (2 mL of solution per wash) with: 50% formamide/5X SSC for 10 minutes at 65°C, 25% formamide/2X SSC for 10 minutes at 65°C, 12.5% formamide/2X SSC for 10 minutes at 65°C, 2X SSC/0.1% Tween for 10 minutes at 65°C, 0.2X SSC/0.1% Tween for 30 minutes at 65°C.

9. At room temperature with gentle agitation, wash the embryos three times 5 minutes in PBST at room temperature, once in maleic acid buffer (MAB) for 5 minutes, and once in 2% block reagent made up in MAB for 5 minutes.

10. Block the embryos for 4 to 5 hours in blocking solution at room temperature with gentle agitation.

11. Replace the blocking solution with fresh blocking solution containing a 1/2000 dilution of anti-digoxygenin or anti-fluorescein Fab fragment antibody (depending on the type of probe used, *see* **Subsection 2.1.3.**) conjugated to alkaline phosphatase (AP) and incubate overnight at 4°C with gentle agitation on a rocking platform (*see* **Note 16**).

12. Wash the embryos 8–10 times in 0.1% Tween/MAB at room temperature with gentle agitation; each wash is done for approximately 30 minutes.

13. Wash the embryos twice for 10 minutes in AP buffer. Replace the AP buffer with the staining solution (prepare fresh prior to use) and allow the color reaction to develop protected from light, at room temperature, with gentle agitation on a

rocking platform. Highly expressed genes and strong probes generally developed within 1 to 2 hours. For some genes with low expression it is necessary to incubate overnight at 4°C and to continue the staining the next day at room temperature (*see* **Note 17**).

14. Once the staining is satisfactory (and before the background starts to come up too much in for instance control embryos hybridized with the sense control probe), the staining is stopped by washing the embryos three times in PBST and then fixing them for 1 hour in MEMFA at room temperature with gentle agitation. Make sure to stop your control embryos at the same time as your experimental embryos for each probe. Wash the embryos three times in 100% methanol (the first two washes while rocking gently at room temperature for 1 hour to prevent the embryos from sticking to one another) and store at 4°C or –20°C (*see* **Note 18**).

15. For best results wait at least 2 days before taking pictures because letting the embryos sit in methanol helps reduce background staining.

3.2. RNA In Situ Hybridization on Histological Sections (See Note 19)

The following protocol is adapted from ref. *6*.

1. Dewax slides in xylene 3 × 5 minutes at room temperature (RT).

2. Wash slides in methanol 2 × 5 minutes at RT.

3. Incubate slides with gentle shaking in bleaching solution for 3 hours at RT.

4. Rehydrate slides in 75, 50, 25, and 0% MeOH in PBST, 5 minutes, each at RT.

5. Wash slides 3 × 5 minutes in PBST at RT.

6. Treat slides with proteinase K (12.5 µg/mL in PBST) for 15 minutes at RT (1 mL/slide). This solution is freshly made from a 20 mg/mL proteinase K stock solution.

7. Stop reaction by washing with glycine (2 mg/mL in PBST) 2 × 5 minutes (1 mL/slide). Glycine solution is freshly made by dissolving the reagent in PBST.

8. Refix in 4% PFA, 0.2% gluteraldehyde in PBS for 30 minutes at RT (1 mL/slide). This solution is freshly made by adding gluteraldehyde in previously made 4% PFA.

9. Wash 2 × 5 minutes in PBST at RT.

10. Prehybridization: Incubate slides in hybridization mix at 70°C for 2 hours (120 µL/slide). Line the box with dampened tissue. Cut the parafilm larger than the slides as it will shrink at the high temperature. Meanwhile prepare solutions I and II for the next day washes.

11. Hybridization: Replace prehybridization solution with hybridization mix containing probe (0.1–10 µg/mL) and

incubate slides overnight at 70°C (120 µL/slide). Line the box with dampened tissue. Cut the parafilm larger than the slides as it will shrink at the high temperature.

12. Warm solution I to 70°C then wash slides 2 × 30 minutes at 70°C.

13. Warm solution II to 65°C then wash slides 2 × 30 minutes at 65°C.

14. Wash slides in MAB 3 × 5 minutes at RT.

15. Block with 2% blocking reagent in MAB for at least 2 hours (1 mL/slide).

16. Dilute anti-DIG antibody 1:2000 (or anti-fluoresceine) in 2% blocking reagent in MAB. Incubate for 2 hours at room temperature (1 mL/slide). Do not use Parafilm in this step.

17. Wash slides with gentle shaking in MABT 3 × 20 minutes at RT, under gentle rocking.

18. Wash in AP buffer 3 × 5 minutes.

19. Develop color with BM Purple in AP buffer containing 5% PVA (1 part AP buffer + PVA, 4 parts BM Purple) 120 µL/slide. Keep boxes at RT in the dark.

20. If development takes place for more than 3 days, check if slides are not drying out. To prevent drying, put 20 µL of AP buffer in the border of the parafilm every 3 days.

3.3. Quantitative PCR (qPCR)

3.3.1. RNA Extraction

RNA from embryos or dissected tissue can be extracted in a number of different ways, several of these methods involve Trizol. In our lab, we have found that the Qiagen RNeasy RNA extraction kit in combination with the Qiashredder columns (for the homogenization step) is a very simple and reliable method. For each experimental condition or developmental stage a total of 15 embryos are harvested, however, they are divided into three separate extractions (five embryos/tube). This is done to prevent overloading the columns with too much RNA, which would lead to decreased yields. Five embryos are placed into a 1.5-mL Eppendorf tube and 600 µL of RT lysis buffer containing 2-mercaptoethanol is added and the embryos are homogenized by pipeting up and down with a 1 mL pipette tip. The lysis mixture can be frozen down at this point at −80°C. Once the extraction is complete, the three tubes for the same experimental condition are recombined.

3.3.2. cDNA Synthesis

The QantiTect Reverse Transcription kit from Qiagen works well for making cDNA as this kit incorporates a genomic DNA removal step and uses a primer mix containing both random and oligo-dT primers (which will generate cDNA from mRNAs with and without polyA tails). For all reactions the maximum possible amount of RNA was used, this of course depends on the sample with the lowest concentration. In most cases the RNA concentrations are sufficiently high enough to enable the use of 1 μg of total RNA per 20 μL reverse transcription (RT) reaction, the maximum amount recommended for use with this kit. Following the RT reaction, the newly generated cDNA is often diluted 1:10 (or correspondingly less if less than 1 μg of RNA was used to synthesize the cDNA) with RNase-Free ddH$_2$O (*see* **Note 20**).

3.3.3. Primer Optimization

For qPCR, primer pairs are usually designed to amplify a product between 100–300 bp in size but each machine has specific instructions. There are several programs available to aid in the design of primers including vector NTI. It is a good idea to test and optimize all of your primer sets in traditional RT-PCR reactions prior to use in qPCR. It is important to try and use a PCR program and reaction set up that will be similar to what is used on your qPCR machine. It is also best to try and use a PCR enzyme that has similar properties to the enzyme that will be used in the qPCR reaction. In the case described, Qiagen HotstarTaq DNA Polymerase (Qiagen) is used. cDNA generated from wild-type Xenopus embryos of an appropriate developmental stage (a stage when you know or expect your gene of interest to be expressed) is used as template. Test primers using annealing temperatures close to but below their Tm (melting temperature) to prevent nonspecific amplification products. The Tm is usually available from the primer manufacturer, but an easy calculation to estimate the primer Tm is 2°C for every A or T and 4°C for every G or C in the sequence.

A sample PCR reaction is listed below but will have to be adjusted according to the manufacturer's guidelines for the enzyme you are using:

PCR sample reaction:

10 μL of cDNA

10 μL primer mix (2 μM each primer)

5 μL 10X reaction buffer (MgCl$_2$ concentration is 1.5 mM)

1 μL 100 mM dNTPs (Invitrogen)

23.75 μL ddH2O

0.25 μL HotstarTaq DNA Polymerase

50 μL Total volume.

A general PCR program is listed below, but the program will depend on the type of enzyme used and also the Tm of the primer pair of interest:

Program for PCR

94°C for 10 minutes (to activate hotstart enzyme).

40 cycles of: 94°C for 30 seconds (denaturing).

60°C for 30 seconds (annealing)

72°C for 30 seconds (elongation)
72°C for 10 minutes (extension time)
Cool to 4°C
All reactions are run out on a 2% agarose TAE gel to verify that they amplified a single, appropriately sized band (bands can also be cut out and purified to send off for sequencing). Half of the reaction (25 µL) was held back to be used to make a standard curve in subsequent qPCR reactions (see later). The standard was diluted 1:10 and stored at –20°C.

3.3.4. qPCR Reagents and Programs

The following described conditions are for use on an Opticom II machine with Dynamo hotstart SYBR green enzyme mix (Finnzymes). Run all samples in triplicate. Make up a standard curve from your test sample (see earlier). It is a good idea when starting out to use a large standard curve to make sure the expression of your gene of interest comes up within your standard curve. If it does not, your results won't be valid (*see* **Note 21**).

qPCR sample reaction:
5 µL cDNA (diluted appropriately, *see* earlier and **Note 20**)
5 µL Primer mix (2 µM each)
10 µL Dynamo hotstart SYBR green enzyme mix
20 µL Total volume

Sample program for qPCR machine:
94°C for 10 minutes (to activate hotstart enzyme).
40 cycles of: 94°C for 30 seconds (denaturing)
60°C for 30 seconds (annealing)
72°C for 1 minute (elongation)
Plate Read (reading product and primer dimmer)
80°C for 20 seconds (melting any primer dimer)
Plate Read (reading only product not dimer)
Melting curve from 70 to 90°C (reading every 0.3°C, hold 1 second).
Cool to 4°C

3.3.5. qPCR Analysis and Normalization

Normalize your qPCR data from samples showing the relative expression levels of your genes of interest to the relative expression levels of a housekeeping gene such as ODC or Histone H4 (it is a good idea to try more than one housekeeping gene to make sure you get similar results and that your results are not biased by your choice of housekeeping gene). If you are comparing expression of your gene of interest under different experimental conditions, all your samples should be normalized to the control sample so that your control sample value equals 1. At this point, average your values for each triplicate and calculate your standard deviations. Expression levels can then be visualized as bar graphs with error bars (*see* **Note 22**).

3.4. Northern Blot (See Note 1)

RNA used in Northern blots can be extracted as previously described using the RNeasy RNA extraction kit from Qiagen (*see* **Subsection 3.3.1.**). Northern membranes can be prepared using the glyoxal method or formaldehyde gels, the method described here is the glyoxal method.

3.4.1. Preparation of Samples

1. Ethanol precipitate 1–3 µg of each sample RNA (more for lowly expressed genes, less for highly expressed genes) by adding $1/20^{th}$ volume of 5 M ammonia acetate (RNase-free) and 2 volumes of 100% ethanol (ice cold) and placing at −20°C for at least 30 minutes.

2. Pellet the RNA by centrifuging at $17,900 \times g$, at 4°C for 30 minutes. Wash the pellet once with 70% ethanol and respin at $17,900 \times g$ 4°C for 10 minutes. Carefully remove the 70% ethanol and allow the pellets to air-dry for 5–10 minutes.

3. Resuspend the pellets in 2.7 µL DEPC ddH_2O/sample. To each RNA sample or 2.7 µL of RNA ladder add 2.7 µL glyoxal, 8 µL DMSO and 1.5 µL 0.1 M $NaPO_4$. Mix the samples by vortexing and incubate at 50°C for 60 minutes. Place the samples very briefly on ice and add 3 µL of loading buffer (*see* **Note 23**).

3.4.2. Preparation and Running of the Gel

1. Prepare an 1% agarose gel (e.g., a 100 mL gel, *see* **Note 24**): add 1 g (1% w/v) of agarose to 60 mL of DEPC ddH_2O and bring the mixture to a boil in the microwave.

2. Allow the mixture to cool to approximately 60°C and add 10 mL of 0.1 M $NaPO_4$ and 30 mL DEPC ddH_2O and mix well.

3. Pour the gel and allow it to set.

4. Use a recirculating gel tank (*see* **Note 25**). Load the samples into the gel and set the gel to run at 80 mA.

5. Confirm pH of the buffer every 30 minutes with pH paper strips even if using a recirculating tank (*see* **Note 25**).

6. Once the gel has run a sufficient distance from the origin, stop the gel and wash it three times in 10X SSC.

3.4.3. Gel Transfer to a Nylon Membrane

1. Set up the transfer apparatus: Fill a glass dish about halfway with 20X SSC and place a glass plate over the top which is shorter than the dish is long. Cut a piece of 3 mm Whatman paper about 1 cm wider than the gel on either side, wet with 20X SSC and place over the glass plate so that either end lays in the 20X SSC in the dish. Take care to remove any bubbles by rolling either a pen or disposable plastic pipette over

the paper (bubbles with inhibit transfer of the RNA to the membrane).

2. Place the gel on top of the Whatman paper so that it is upside down (it is a good idea to cut off the loading wells at the top of the gel so that the gel lies flatter).

3. Place a piece of Nylon membrane cut to the size of the gel and which has been previously equilibrated in ddH_2O for 10 minutes followed by a quick wash in 20X SSC, on top of the gel, again with care taken to remove any bubbles.

4. Cut the lower left-hand corner from both the gel and the membrane with a razor blade or a scalpel, so that the orientation of the gel/membrane can be determined later.

5. Cut two pieces of Whatman paper slightly larger than the gel and wet with 20X SSC then place on top of the membrane again with care taken to remove any bubbles.

6. On top of this, place a stack of paper towels, cut to approximately the same size as the two pieces of Whatman paper, and weigh down with a glass plate, with a bottle on top for extra weight.

7. Leave the gel to transfer overnight at room temperature by capillary action.

8. In the morning, disassemble the transfer apparatus and allow the membrane to dry.

9. The RNA can be fixed to the membrane by UV-crosslinking at 12,000 J or baking in an oven.

10. Stain the membrane with 0.02% (w/v) methylene blue in 0.5 M NaOAc pH 5.2.

11. Destain in DEPC-treated ddH_2O to visualize the rRNA bands (18S and 28S ribosomal subunits) and the RNA ladder. This will show how well your RNA ran, how equally loaded the samples were and if the RNA was at all degraded (the 18S and 28S ribosomal subunits should appear as two distinct bands with a lighter smear of RNA between them).

12. Allow the membrane to dry at room temperature and then wrap up in plastic wrap or a plastic bag, scan an image and store in a cool dry place until needed.

13. The lane containing the RNA ladder is cut off prior to hybridization so that it can be used later as a size reference.

3.4.4. Northern Hybridization

We use the Northern Starter Kit from Roche, which works well in our hands for Northern hybridizations. It includes materials for generation of digoxigenin-labeled Northern probes. Follow the kit instructions and use your labeled RNA probe at a concentration of 100 ng/mL.

3.4.5. Stripping the Membrane

1. Bring a 0.1% SDS solution to a boil, pour over the membrane and allowed to cool whilst on a rocker. Do this twice.

2. Wash the membrane briefly in 2X SSC and either re-probe or wrapped in a sealed bag to prevent drying out and stored at −20°C (*see* **Note 26**).

3.5. Western Blotting and Protein Analysis

3.5.1. Protein Extraction from Xenopus Embryos

1. Once your embryos have reached the appropriate developmental stage, rinse them twice in PBS at room temperature.

2. After the second rinse, remove as much of the PBS as possible. Add 10 µL of ice cold RIPA buffer (*see* **Note 27**) per embryo and homogenize by pipeting up and down with a 1 mL pipette tip and placed immediately on ice.

3. Clarify the extracts one to two times by centrifuging at 17,900 x*g* for 20 minutes at 4°C and transferring to a new tube each time.

4. Freeze extracts at −20°C for short-term storage or −80°C for longer-term storage.

3.5.2. Quantification of Protein Extracts (BCA Assay; Optional)

If equal numbers of embryos and equal amounts of lysis buffer are used then your extracts should have relatively the same protein concentration. However, if you want a more accurate calculation of your protein concentration, the BCA (bicinchoninic acid) Protein Assay Kit from Novagen works really well. Follow the micro-scale protocol and dilute your protein extracts in the range of 1:20 to 1:100 (depends on expected protein concentration; *see* **Note 28**).

3.5.3. Preparation of Protein Samples and Running of Protein Gels

The amount of protein loaded onto a gel is generally dependent on the sample with the lowest concentration. Samples are set up as follows: equal amounts of protein made up in a total of 14 µL of ddH$_2$O, 2 µL of NuPAGE 10X Sample Reducing Agent (Invitrogen) and 4 µL 4X NuPAGE LDS Sample Buffer (Invitrogen). You may want to omit the reducing agent depending on your antibody and gene of interest.

1. Mix the samples and then place at 95°C for 5 minutes to denature the protein.

2. Place the samples briefly on ice and spin down quickly before loading onto a pre-cast 4–12% Bis-Tris NuPAGE gel (Invitrogen; *see* **Note 29**).

3. Load your samples and 10 µL of a protein ladder such as Multi-Mark (Invitrogen) as a size standard (*see* **Note 29**).

4. Run the gel at 200 V until the dye front has reached the bottom (or longer if you need more separation of your proteins).

3.5.4. Transferring the Protein Gel to a PVDF Membrane

1. Cut a piece of PVDF membrane (always handle with forceps or gloves) to the same size of the gel and equilibrate in 100% methanol for at least 10 minutes.

2. Rinse the membrane briefly in 1X transfer buffer when assembling the transfer apparatus. Cut two pieces of Whatman paper approximately 1 cm larger than the gel on all sides and soak in 1X transfer buffer.

3. Once the gel has run a sufficient distance, remove it from its plastic case with a gel spatula. Carefully cut off the loading wells and the dye front.

4. Assemble the transfer apparatus as follows: Layer the gel between the membrane and one piece of Whatman paper, place the other piece of Whatman paper on the other side of the membrane. Make sure there are no bubbles by rolling over the "sandwich" with a pen or a plastic Pipette. Place the Whatman paper/gel/membrane/Whatman paper sandwich between transfer sponges soaked with transfer buffer and placed into the transfer apparatus but ensure that the membrane ends up closest to the positive electrode and the gel closest to the negative electrode since protein is negatively charged and will move away from the negative and towards the positive charge.

5. Transfer the gel for 1 hour at 200 mA (~30–40V) or longer for very large proteins.

3.5.5. Immunodetection of Western Blot

1. After transferring the membrane, block it in blocking solution for at least 1 hour at room temperature or overnight at 4°C with gentle agitation.

2. Replace the blocking solution with fresh blocking solution containing primary antibody diluted appropriately (*see* **Note 30**).

3. Incubate for at least 1 hour at room temperature or overnight at 4°C with gentle agitation.

4. Wash the membranes in TBST six to eight times, 15 minutes per wash, at room temperature with gentle agitation.

5. Remove the last wash and replace with blocking solution containing the secondary antibody (should detect the species of animal that your primary antibody was raised in) diluted appropriately (*see* **Note 30**) for at least 1 hour at room temperature or overnight at 4°C with gentle agitation.

6. Wash the membranes in TBST six to eight times 15 minutes at room temperature with gentle agitation.

7. Visualize antibody detection with appropriate staining protocol:
   a. If you are using a secondary conjugated to horseradish peroxidase (HRP), use an enhanced chemiluminesence substrate such as the ECL Western Blotting Substrate from Pierce Biosciences. Follow the manufacturer's instructions; place the membrane under clear plastic and expose to Xray film for variable lengths of time to achieve the optimal exposure.

b. An alkaline phosphatase (AP) conjugated secondary antibody can also be used but this will need to be used in combination with a colorimetric substrate such as BCIP and NBT. Follow the manufacturer's instructions and when the appropriate level of staining is reached stop the staining by washing twice 10 minutes in ddH$_2$O.

8. Western blots can be stripped and re-blotted with the Re-Blot Western Blot Recycling Kit (Chemicon) and re-probed as described above (*see* **Note 31**).

3.6. Whole-mount Immunoflurescence Staining

1. Rehydrate embryos by washing through a decreasing methanol series made up in PBST: 75% methanol/25% PBST; 50% methanol/50% PBST; 25% methanol/PBST. Perform each wash for 5 minutes at room temperature with gentle agitation on a rocking platform.

2. Wash the embryos once in PBS for 5 minutes before adding the bleaching solution.

3. Add the bleaching solution and place the embryos under a fluorescent light and allow the embryos to bleach for one to two hours (depending on the level of pigmentation) until the majority of the pigmentation has been removed (*see* **Note 13**).

4. Wash the embryos twice in PBS/5% BSA for 20 minutes each wash, at room temperature on a rocking platform.

5. Block the embryos for at least 1 hour at room temperature or overnight at 4°C in 2 mL blocking solution with gentle agitation.

6. Replace the blocking solution with 1 mL fresh blocking solution containing the primary antibody diluted appropriately (*see* **Notes 30** and **32**).

7. Incubate embryos standing upright in a rack with gentle agitation at 4°C overnight.

8. Wash the embryos three times for 2 hours per wash, in 5 mL PBS at room temperature with gentle agitation.

9. Remove the last wash and add 1 mL of the secondary antibody diluted appropriately (*see* manufacturer's instructions) in blocking solution at 4°C overnight with gentle agitation (*see* **Note 33**).

10. Wash the embryos three times for 2 hours per wash, in 5 mL PBS at room temperature, with gentle agitation (*see* **Note 5**).

11. Dehydrate embryos in an ethanol series (25% ethanol/75% PBS, 50% ethanol/50% PBS, 75% ethanol/25% PBS, 100% ethanol) and store in 100% ethanol at 4°C in the dark until ready to capture images (*see* **Note 33**).

12. (Optional) Embryos can be cleared with Murray's clear, which makes embryos transparent and helps with imaging internal

structures that will require a confocal microscope. (For example, if you are trying to image the pronephric tubules mount the embryos in concave microscope slides containing Murray's clear just prior to imaging. Position the embryos so that the pronephritic tubules are immediately under the cover slip.)

3.7. Immunofluorescence Staining on Xenopus Sections (See Note 34)

1. Remove slides from –20°C and allow to thaw at room temperature for 5–10 minutes.

2. Briefly wash the slides in warm ddH$_2$O (~30°C) to remove any remaining OTC and dry at room temperature.

3. Encircle the sections on the slides in wax to form a liquid tight ring using a Papp pen (Sigma). Allow to dry completely before continuing.

4. Wash sections twice for 10 minutes per wash, in PBS/5% BSA.

5. Permeabilize sections in PBS/0.1% Triton X-100 for 20 minutes, at room temperature.

6. Block the slides for 1 hour at room temperature in 250 µL/slide of blocking solution (depends on the size of your slides).

7. Replace block solution (gently tip the slide on its side onto absorbent paper to allow block solution to run off the slide) with 250 µL/slide of blocking solution with the appropriately diluted primary antibody (*see* **Note 30**) and incubate for either 4 hours at room temperature, or overnight at 4°C (agitation is not necessary as long as the slide is completely wet; *see* **Notes 32** and **35**).

8. Wash slides three times 5 minutes in PBS.

9. Add secondary antibodies diluted appropriately alone or in combination in blocking solution (*see* manufacturer's instructions and **Note 30**), and incubate for 1-2 hours at room temperature, protected from light (*see* **Note 36**).

10. Wash slides three times 5 minutes in PBS.

11. Mount slides by applying several drops of mounting media with or without DAPI or propidium iodide (both optional for nuclear staining) and anti-fade (or your mounting media of choice) to each slide and place a cover slip carefully on top, with care taken to remove any air bubbles. Seal the cover slips with nail varnish and store at 4°C in the dark, until ready to image on a fluorescent microscope. Best if imaged within 2 weeks.

3.8. Whole-mount Immunocytochemistry in Xenopus Embryos

1. Rehydrate embryos by washing through a decreasing methanol series made up in PBST: 75% methanol/25% PBST; 50% methanol/50% PBST; 25% methanol/PBST. Perform each wash for 5 minutes at room temperature with gentle agitation on a rocking platform.

2. Wash the embryos once in PBS for 5 minutes before adding the bleaching solution.

3. Add the bleaching solution and place the embryos under a fluorescent light and allow the embryos to bleach for 1 to 2 hours (depending on the level of pigmentation) until the majority of the pigmentation has been removed (*see* **Note 13**).

4. Wash the embryos twice in TBS/5% BSA for 20 minutes each wash, at room temperature on a rocking platform.

5. Block the embryos for at least 1 hour at room temperature or overnight at 4°C in 2 mL blocking solution with gentle agitation.

6. Replace the blocking solution with 1 mL fresh blocking solution containing appropriately diluted primary antibody (*see* manufacturer's instructions and **Note 30**). Incubate embryos standing upright in a rack with gentle agitation at 4°C overnight (*see* **Note 32**).

7. Wash the embryos three times for 2 hours per wash, in 5 mL TBS at room temperature with gentle agitation.

8. Remove the last wash and add 1 mL of the secondary antibody conjugated to HRP diluted appropriately (*see* manufacturer's instructions and **Note 30**) in blocking solution. Incubate at 4°C overnight with gentle agitation.

9. Wash the embryos three times for 2 hours per wash, in 5 mL TBS at room temperature, with gentle agitation.

10. Incubate embryos in 1 mL of DAB staining solution and incubate until satisfactory staining is observed (5–60 minutes). Stop the color reaction by replacing the DAB solution with 100% methanol. Stained embryos can then be photographed in methanol.

11. (optional) Embryos can be cleared in Murray's Clear which makes the embryos transparent (*see* **Note 37**).

3.9. Immuncyto-chemistry Staining on Xenopus Sections (See Note 34)

1. Dewax slides by washing in xylene three times for 5 minutes at room temperature in a laminar-flow hood.

2. Wash in 100% methanol twice for 5 minutes and allow to dry.

3. Encircle the sections on the slides in wax to form a liquid tight ring using a Papp pen (Sigma). Allow to dry completely before continuing on with the next step.

4. Rehydrate sections by a series of decreasing methanol washes: made up in TBS: 75% methanol/25% TBS; 50% methanol/50% TBS; 25% methanol/TBST; 100% TBS, 5 minutes per wash.

5. Wash sections twice for 10 minutes per wash, in TBS/5% BSA.

6. Permeabilize sections by incubation in TBS/0.1% Triton X-100 for 20 minutes, at room temperature.

7. Block the slides for 1 hour at room temperature in approximately 250 µL/slide of blocking solution (volume depends on the size of your slides).

8. Replace block solution (gently tip the slide on its side onto absorbent paper to allow block solution to run off the slide) with approximately 250 µL/slide (volume depends on the size of your slides) of fresh blocking solution containing appropriately diluted primary antibody (*see* manufacturer's instructions and **Notes 30, 32,** and **35**).

9. Incubate slides for either 4 hours at room temperature, or overnight at 4°C (agitation is not necessary as long as the slide is completely wet).

10. Wash slides three times for 5 minutes in TBS.

11. Add secondary antibodies diluted appropriately alone or in combination in blocking solution (*see* manufacturer's instructions and **Note 30**), and incubate for 1 to 2 hours at room temperature.

12. Wash slides three times for 5 minutes in TBS.

13. Incubate embryos in 1 mL of DAB staining solution and incubate until satisfactory staining is observed (5–60 minutes). Stop the color reaction by placing slides in 100% methanol.

14. Mount slides in a non-aqueous mounting media.

4. Notes

1. Prepare all solutions to be used for Northern Analysis or either RNA hybridization protocol in autoclaved milliQ H_2O, use brand new autoclaved Duran Bottles and sterile falcon tubes, or alternatively pre-treat solutions with 0.1% or 1 mL/L of diethylpyrocarbonate (DEPC), incubate overnight at 37°C and autoclave to prevent any RNase activity.

2. Prior to fixing, remove the vitelline membrane manually with steel forceps from embryos stage 22 and older to prevent curling. It is recommended to pierce very carefully embryos of stages 10–20 prior to fixation, with a glass needle in an area of the embryo that is not important for staining of the gene of interest (if you know) to prevent trapping of the probe and to reduce background staining.

3. If you are doing double *in situs*, label the probe corresponding to the more highly expressed gene with fluorescein and the probe corresponding to the more weakly expressed gene with digoxygenin (*see* **Notes 16–18**).

4. Sense probes can be used as a control. For generation of sense probes, plasmid DNA is linearized with a restriction enzyme that cuts near the termination codon, of the gene at the 3′ end. The sense control probe is transcribed with the RNA polymerase corresponding to the 5′ end of the coding region of the gene. The sense control probe will not be able to bind to the mRNA transcript of your target gene and is used at the same concentration as your anti-sense probe to gauge the level of background staining.

5. LNA probes are short DNA oligonucleotides with increased affinity to RNA due to the addition of about 30% Locked Nucleic Acid (LNA). LNA is a high-affinity RNA analog in which the furanose ring of the ribose sugar is chemically locked in an RNA-mimicking conformation by an O2′, C4′ methylene bridge *(7)*. This modification results in an increased affinity for complementary DNA and RNA molecules *(7)*. The company Exiqon manufactures and distributes them in Europe and they also offer a custom design service (*see* www.Exiqon.com). LNA probes have been previously shown to work well for detecting microRNAs in mouse, *Xenopus*, and zebrafish embryos *(7)* and we have had some success using them for detection of full-length RNAs in Xenopus embryos. We used them according to the whole-mount hybridization protocol at a concentration of 10 nM diluted in hybridization buffer at a hybridization temperature of 50°C (or at least 20°C below their Tm).

6. Prepare MAB + Blocking reagent in a 50 mL falcon tube, add blocking reagent to previously made MAB, heat at 70°C until Block reagent dissolves. Store MAB + Block at –20°C.

7. Prepare AP Buffer + PVA in a 50 mL falcon tube, add PVA to previously made AP Buffer, heat at 70°C to dissolve PVA. After a while, if PVA does not dissolve completely do not worry, the undissolved stuff will go to the bottom of the tube, use the jelly like solution at the top. Keep AP Buffer + PVA at RT.

8. Prior to fixing, remove the vitelline membrane manually with steel forceps from embryos stage 22 and older to prevent curling. Perform all steps in 8-mL glass vials.

9. Paraffin sections will not work well for immunoflurescence because the embryos are heated at relatively high temperatures (up to 80°C) during the embedding protocol, which results in high auto-fluorescence background.

10. Add 3 mg/mL NiSO$_4$ (Sigma) to staining solution if you want a purple color rather than reddish-brown.

11. Embryos can also be embedded in OTC and cryosectioned as described in **Subsection 2.6.**, but if so, follow **steps 1 and 2** of the protocol described in **Subsection 3.7.** before continuing with **step 5** of the protocol described in **Subsection 3.9.**

12. Prior to the pre-hybridization step all washes and treatments are done in 5-mL volumes in 8-mL glass vials. From the pre-hybridization step onward all washes and treatments are done in 2-mL volumes in 2-mL eppendorf tubes unless otherwise stated.

13. It is important not to over bleach the embryos because this makes them very fragile and causes them to fall apart during subsequent steps.

14. Be sure to use approximately the same number of embryos for your control samples as for your experimental samples to make sure that the ratio of probe to embryo is consistent across all your samples so that any quantitative differences in staining can be attributed to your experimental conditions.

15. Probes can be reused up to 12 times depending on the probe by storing at –20°C.

16. If doing double *in situs*, incubate with the antibody against the probe to be developed first.

17. If doing double *in situs*, set up first staining by incubating in AP buffer containing just BCIP (yields a turquoise blue color on its own) and incubate at RT for color development. When satisfied with staining, wash three times in PBS/Tween-20. At this point, you can take photographs in order to compare after the second staining and locate regions of co-expression.

18. If doing double *in situs*, fix in MEMFA for 1 hour. At this point transfer the embryos to fresh Eppendorf tubes then wash twice in methanol for 30 minutes at RT. Wash in MAB for 5 minutes and then twice in MAB/Tween (MAB + 0.1% Tween 20) for 5 minutes. Reblock in MAB-block solution for 1 hour or longer and then replace with fresh block solution containing the second antibody. Repeat **steps 11 and 12**. Replace AP buffer with AP buffer containing 35 µL NBT/10 mL or alternatively BM-Purple (both of these staining solutions yield a purple color, which can be distinguished from the turquoise stain used to detect the first probe (*see* **Note 17**). After staining fix in MEMFA for 1 hour, wash three times in 100% methanol and store in 100% methanol at –20°C.

19. Use 40–50 mL coplin jars for most steps, but perform **steps 6–8, 10, 11, 15, 16, and 19** in plastic airtight boxes with

"racks" for holding the slides flat and off the base of the box (tip box inserts work well as they have holes for allowing humidity from the wet paper towels lining the bottom of the boxes to permeate). When using coplin jars, incubate with gentle shaking on a rocking table. When using boxes, pipette the solution directly onto the slide. When setting prehybridization and hybridization, cover with parafilm to prevent evaporation.

20. If your gene of interest is expressed at very low levels it is best to use the cDNA undiluted in your PCR reaction. The RT reactions can be scaled up linearly depending on the amount of cDNA required to assay a set of genes.

21. It is a good idea to run one of your housekeeping genes such as ODC or histone H4 first in order to check and make sure your cDNA is working well.

22. Note that there will be no error bars on your control samples. Though most people state the expression of their gene of interest in terms of relative expression, it is possible to calculate copy number. You must calculate the amplicon size for each gene product and determine the concentration and dilution of your standards used in your standard curve. You can then calculate the copy number of your gene of interest and in respect to copy number of your housekeeping gene.

23. Do not leave your samples on ice too long or it may cause the amberlite to fall out of solution.

24. Adjust the volumes given here accordingly if preparing a larger or smaller gel.

25. If you are not using a recirculating gel tank, the running buffer (which is composed of 10 mM $NaPO_4$) will need to be changed every 30 minutes because there is no buffering capacity in 10 mM $NaPO_4$ and a pH gradient will be established over time. This is to be avoided since glyoxal dissociates from RNA at pH higher than 8.0. Only the first batch of buffer needs to be made up in DEPC water, once the samples have run into the gel, regular milliQ ddH_2O will suffice.

26. Only membranes that have not dried out can be reprobed. Drying of the membrane results in increased background.

27. Instead of RIPA buffer, you can use any lysis buffer of your choice, e.g. NuPAGE native lysis buffer (Invitrogen) containing 1% digitonin also works well.

28. Make sure you make up your standard curve and make your dilutions in RIPA buffer or whatever lysis buffer you used on your samples.

29. The type and percentage of gel to be used depends on the size and characteristics of your protein of interest. The type

of running buffer you choose to use will depend on what type of gel you are running. Invitrogen makes a number of pre-made buffers including MES and MOPS SDS Running Buffers but you can also make these up yourself. Use a protein ladder that has fragments in the size range of your protein of interest (if known).

30. If using a commercially available antibody, use according to manufacturer's instructions or try 0.1–10 µg/mL. The correct dilution of antibody, particularly if using a non-commercial antibody may require some optimization, before analyzing protein localization in precious experimental samples.

31. Always blot with the antibody that gives you the weakest signal first since repeated stripping and re-blotting can result in loss of some of your protein on your membrane and weaker signals.

32. It is also a good idea to include a set of controls. Incubate control embryos without primary antibody or with isotype control to match the primary antibody (i.e., normal rabbit IgG for a rabbit polyclonal primary antibody) to see what your background staining is.

33. For **steps 9–11** above protect your embryos from light by wrapping vials in aluminium foil or placing in a light protective box because the flurophores conjugated to your secondary antibody are light-sensitive.

34. Use 40–50 mL coplin jars for most steps, but perform **steps 6, 7 and 9** in plastic boxes with "racks" (*see* **Note 19**).

35. If staining for two different proteins, mix both primary antibodies together before adding to slide. Use isotype controls and no primary antibody controls for each of your primary antibodies to determine background staining of your secondary antibodies.

36. For **steps 9–11** above protect your slides from light by placing in a light protective box because the flurophores conjugated to your secondary antibody are light sensitive.

37. Photograph immediately after placing in Murray's clear since prolonged exposure to Murray's clear will cause the embryos to turn a brown color.

38. If staining for two different proteins, mix both primary antibodies together before adding to slide. Use isotype controls and no primary antibody controls for each of your primary antibodies to determine background staining of your secondary antibodies.

References

1. Harland, R. M. (1991) In situ hybridization: An improved whole-mount method for Xenopus embryos in (Kay, B. K., Peng, H. B., eds.) *Xenopus laevis: Practical Uses in Cell and Molecular Biology*, vol. 36, pp. 679–695. Acadmic Press, San Diego.

2. Ma, Y. J., Dissen, G. A., Rage, F., et al. (1996) RNase Protection Assay. *Methods* 10, 273–278.

3. Urban, A. E., Zhou, X., Ungos, J. M., et al. (2006). FGF is essential for both condensation and mesenchymal-epithelial transition stages of pronephric kidney tubule development. *Dev Biol* 297, 103–117.

4. Liao, G., Tao, Q., Kofron, M., et al. (2006). Jun NH2-terminal kinase (JNK) prevents nuclear beta-catenin accumulation and regulates axis formation in Xenopus embryos. *Proc Natl Acad Sci USA* 103, 16313–16318.

5. Sambrook, J., Russell, D. (2001) *Molecular Cloning: A Laboratory Manual.* 3rd edit, Cold Spring Harbor Laboratory Press, Cold Spring Harbor.

6. Ciau-Uitz, A., Walmsley, M., Patient, R. (2000) Distinct origins of adult and embryonic blood in Xenopus. *Cell* 102, 787–796.

7. Kloosterman, W. P., Wienholds, E., de Bruijn, E., et al. (2006). In situ detection of miRNAs in animal embryos using LNA-modified oligonucleotide probes. *Nat Methods* 3, 27–29.

Chapter 23

Detection of Nuclear β-catenin in *Xenopus* Embryos

François Fagotto and Carolyn M. Brown

Abstract

Immunodetection of β-catenin accumulation in the nucleus is the most direct and reliable method to determine the intensity and the spatial/temporal patterns of Wnt-dependent signaling activity. Due to the large size of the *Xenopus* embryo, staining must be done on sections. We present here a simple protocol to prepare cryosections and produce high-quality images of the early embryo using immunofluorescence. We also provide comments on various conceptual and technical issues from fixation to image collection, which may assist in optimizing immunodetection in embryos and tissues beyond the specific scope of β-catenin localization.

Key words: β-catenin, Wnt, signal transduction, *Xenopus*, embryonic development, embryonic patterning, immunofluorescence.

1. Introduction

1.1. Nuclear β-catenin as Ultimate Readout for Wnt Signaling

Nuclear localization of β-catenin can be considered the hallmark of Wnt signaling. In non-stimulated cells, β-catenin is primarily localized to the plasma membrane due to its interaction with cadherins. Very low levels of β-catenin are also detectable in the cytoplasm. The punctuate pattern observed probably corresponds to the transient pool captured by the Axin-based complex and ultimately targeted for fast degradation (unpublished results). Wnt stimulation induces the slow accumulation of a soluble β-catenin pool, which appears to enter the nucleus and interact directly with TCF/Lef-1 transcription factors and components of the transcription machinery.

Elizabeth Vincan (ed.), *Wnt Signaling, Volume II: Pathway Models, vol. 469*
© 2008 Humana Press, a part of Springer Science + Business Media, New York, NY
Book doi: 10.1007/978-1-60327-469-2

Detection of nuclear β-catenin is presently the most reliable indicator of Wnt-induced β-catenin signaling as it corresponds to the most downstream step and thus reflects the immediate state of the pathway. Reporter constructs for TCF-dependent activity are detectable only above a certain threshold achieved over time (several hours) and can consequently obscure the dynamics of the signal. For instance, we have observed appearance and disappearance of the signal within less than 1 hour during early embryonic development (1), a time frame below the sensitivity of gene reporter systems.

1.2. Detection of Nuclear β-catenin in Various Systems

Despite the necessity of monitoring nuclear β-catenin, this remains a difficult task in most model systems. Nuclear β-catenin is often prominent in immunocytochemical stainings of cancer biopsies, but this accumulation may be transient (2, 3) and a direct correlation with Wnt activity is difficult to establish in these samples. In cell culture, immunofluorescence only detects nuclear β-catenin reliably under artificial conditions such as LiCl or proteasome treatment (4), or in cells showing anomalously elevated levels of β-catenin such as SW480 cells (5).

Biochemical fractionation is frequently used as an alternative to immunolocalization to document nuclear accumulation of β-catenin (6). However, this method has serious caveats. First, soluble nuclear proteins readily leak during cell fractionation. Additionally, "nuclear" fractions are heavily contaminated with plasma membranes as well as other dense insoluble materials, many of which are associated with the cytoskeleton, which happens to be enriched with several components of the Wnt pathway (unpublished, see also ref. 7).

In *Xenopus* embryos, β-catenin can be readily detected in the dorsal side of the early blastula (1, 8, 9). This corresponds to the well-characterized maternal signal necessary for establishment of the primary dorso-ventral axis (10–12) via activation of direct target genes such as Siamois, Twin and Nodal-related 3 (13–17). By stage 9, maternal β-catenin activity spreads all around the equator (see **Fig. 23.1A**), where it participates in primary mesoderm induction (9). Later, waves of zygotic Wnts stimulate β-catenin accumulation in various parts of the embryo. The earliest zygotic signal appears in the ventro-lateral side of the early gastrula (1), where Wnt8 regulates further patterning of the mesoderm (18–22). It is followed by a strong activation at the dorsal blastopore lip (1), probably related to the activity of Wnt3a (23) linked to anterior-posterior patterning of the neural plate (24, 25).

Accumulation of nuclear β-catenin has also been detected in sea urchin (26), ascidian (27), mouse (28), chicken (29), and zebrafish (8, 30) embryos. Notably, the β-catenin homolog in Drosophila, armadillo, does not seem to accumulate, but rather

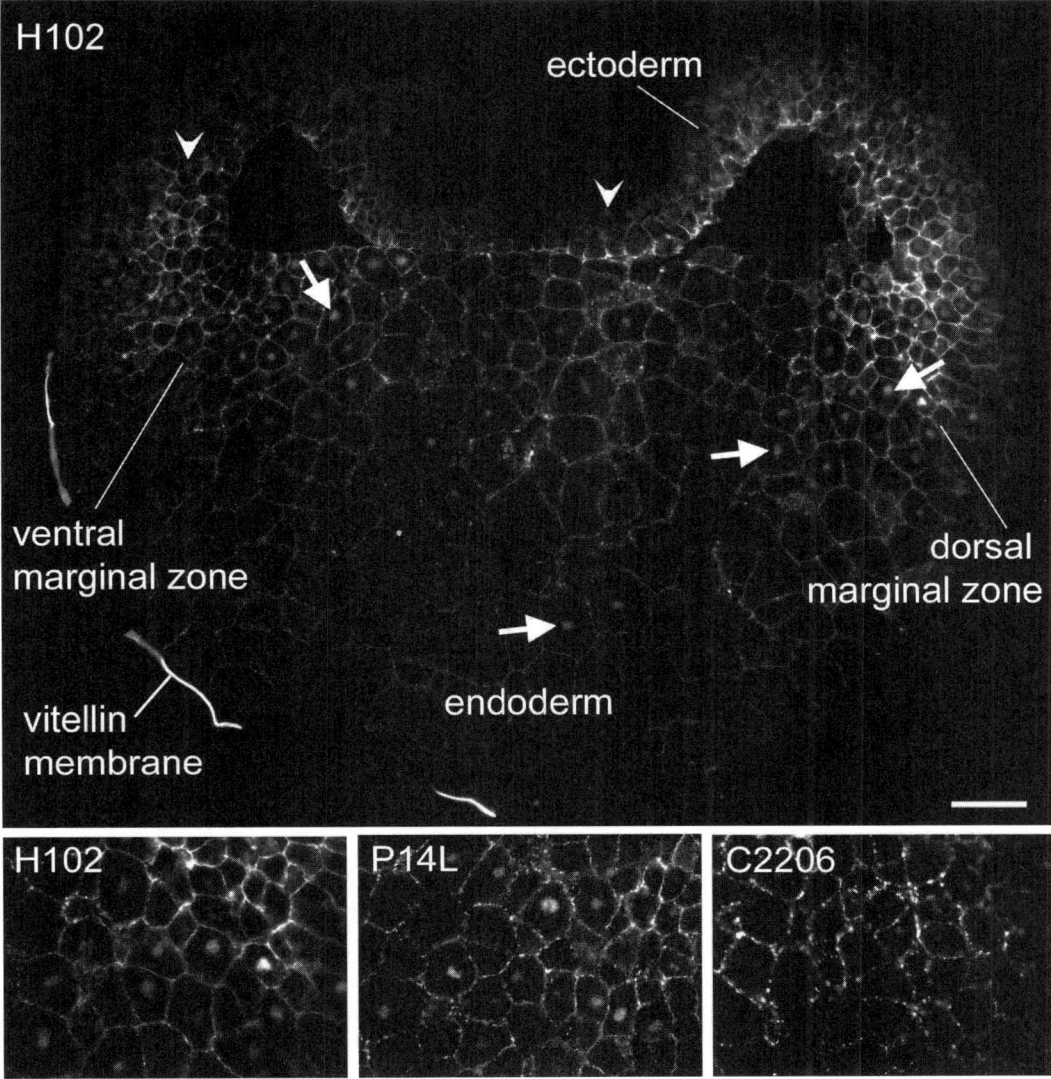

Fig. 23.1. Examples of β-catenin detection using various antibodies. (**A**) Image of a whole sagittal section of a stage 9 blastula immunolabelled with the H102 antibody and goat Alexa488 anti-rabbit antibody. The image was produced by stitching nine images of contiguous fields collected using a Leica inverted microscope equipped with a 20x oil immersion objective and a narrow band fluorescein filter set. Note the broad field of positive nuclei spanning the whole marginal zone and a large part of the endoderm (arrows). Nuclei of ectodermal cells show only a very low signal (arrowheads). Membrane staining is observed for all cells. Bar: 10 μm. (**B–D**) Details of the dorsal marginal zone of stage 9 blastula labeled with three different anti-β-catenin antibodies. P14L and H102 give a very similar staining, while C2206 reacts well with membrane-bound β-catenin but only very weakly with nuclear β-catenin. For comparison of more antibodies, see **Table 23.1.**

equilibrates between the nucleus and the cytoplasm *(31–33)*. This variation in the degree of nuclear β-catenin accumulation among models may be due to both biological and technical parameters (see later).

1.3. Conceptual and Practical Issues in Detecting β-catenin

β-catenin can diffuse freely through the nuclear pore *(34)*. Consequently, levels of nuclear β-catenin probably result from a dynamic equilibrium with cytoplasmic β-catenin. This equilibrium can be influenced by variable retention of β-catenin by interacting partners in the nucleus, in the cytoplasm and at the plasma membrane *(33–36)*. β-catenin distribution can be modulated experimentally by increasing the number of binding sites in a particular compartment either by overexpressing β-catenin binding proteins (e.g., cadherins [37]) or by altering the localization of endogenous β-catenin binding proteins. The latter situation can be created effectively by inhibiting CRM-1-mediated nuclear export for long periods (hours). This leads to nuclear accumulation of APC and Axin *(6, 38–41)* along with many other components. The abundance of endogenous nuclear and cytoplasmic binding sites for β-catenin remains unknown, and is likely to vary between cell types and organisms.

Variable retention of β-catenin after fixation both within cells and among model systems must be considered. We suspect that the high enrichment of nuclear versus cytoplasmic β-catenin in many embryos may be due to a relatively better fixation of nuclear β-catenin and a loss of the more diffusible cytosolic pool during washing. In the *Drosophila* embryo, cytoplasmic armadillo may be more efficiently fixed, while, on the contrary, in vertebrate cell lines, β-catenin may leak more readily from both the cytoplasm and the nucleus. Possible causes for such variations may include differences in the relative levels of β-catenin, density and/or fixability of the cytoplasm and the nucleoplasm. Obviously, detecting and analyzing cytosolic β-catenin by immunofluorescence is far from trivial: in the future, efforts should be made to better quantify and minimize the problem of variable β-catenin retention.

In addition to the fixation problems, one has to consider the complexity resulting from the multiple co-existing pools of β-catenin. It is often impossible to discriminate by immunofluorescence between the soluble cytosolic pool of β-catenin and that bound to cellular components. Early *Xenopus* embryos are a case in point. They contain a huge cytoplasmic pool of small vesicles rich in maternal cadherin and associated β-catenin (unpublished). These small vesicles will provide the new plasma membrane during cleavage. This vesicular β-catenin is virtually indistinguishable from "true" cytosolic β-catenin at the resolution of fluorescence microscopy. Given this difficulty, it follows that the small difference in "cytoplasmic" β-catenin levels that has been detected between the dorsal and ventral side of early cleaving embryos *(42, 43)* should be confirmed by estimating the β-catenin/cadherin ratio in these two regions.

Finally, many antibodies show some degree of nonspecific staining in the nucleus. This is in particular the case for the anti-dephospho-β-catenin antibody 8E7, at least in mammalian culture cells (unpublished; Fornerod, personal communication). Proper controls, including depletion of the antigen (e.g., using anti-β-catenin morpholino antisense oligonucleotides *(9)* or Axin over-expression) are thus crucial.

1.4. Immunostaining in Xenopus

Although widely used, whole-mount staining techniques are plagued by problems due to poor penetrability, especially for large yolk-rich embryos such as *Xenopus*. In Dent's-fixed embryos, antibody diffusion is slow, necessitating overnight incubation. In PFA-fixed embryos, acceptable staining of the center is virtually unachievable. Consequently, the signal in the yolk-poor and superficial tissues is systematically overrepresented. This led to the initial incorrect assumption that nuclear β-catenin was restricted to the superficial cells in the early blastula *(8)*. We now know that the activated regions span the whole embryo *(1)* and **Fig. 23.1A**. Even staining of half embryos shows very poor signal in the yolky vegetal/endodermal cells (e.g., Smad2 staining, c.f. refs. *44* and *1*). Note that this problem is not restricted to protein immunostaining, but also to *in situ* hybridization (e.g., Wnt8 localization, c.f. refs. *18* and *45*). Difficulties associated with embryo penetrability are not restricted to *Xenopus* and sections have recently been used to improve staining in Zebrafish embryos *(30)* and Drosophila ovaries (Dansereau, personal communication).

Sectioning is the only reliable approach for immunostaning in *Xenopus*. Two methods are presently available: sectioning of polyacrylamide-embedded embryos *(46)* and cryosectioning *(1, 47, 48)*. Our laboratory routinely uses cryosectioning. The protocol presented below can yield perfect sections for all stages from uncleaved eggs to late tadpoles, with optimal antigen preservation. Another advantage is its speed, as serial sections through whole embryos can be prepared in less than 20 minutes.

1.5. Important Parameters to be Considered for Preparation and Staining of Embryos

1.5.1. Fixation

Xenopus embryos are large and yolky. This makes good preservation of structure and antigenicity a challenge, especially for the large vegetal and later endodermal cells. Although good results are generally obtained with 4% paraformaldehyde (PFA) fixation, some antibodies only work with lower PFA concentrations of 2% or even 1%. Acceptable preservation was obtained with 2% PFA. When using 1%, however, it must be kept in mind that the center of the embryo will be incompletely fixed, likely because PFA becomes limiting due to the large amount of yolk proteins. Methanol-based fixations are inadequate, because soluble proteins, including nuclear β-catenin, are lost and many antigens, including membrane proteins, can be partially washed away.

1.5.2. Permeabilization

Membrane permeabilization is required for efficient antibody infiltration (see later). We have obtained comparable results with both Dent's solution (methanol/DMSO *(49)*) and Triton X-100

(0.1%), although the former may be slightly better, possibly by enhancing fixation and thus improving antigen retention. Furthermore, embryos can be stored at –20°C in Dent's for long periods.

1.5.3. Infiltration

Historically, the extreme heterogeneity of *Xenopus* eggs due to the dense protein crystals of the yolk has been the main obstacle for cryosectioning. By infiltrating the embryonic cytoplasm with a thick medium similar in consistency to the yolk, the embryo is rendered more homogeneous (*see* **Note 1**). We use cold-water fish gelatin because it is liquid at room temperature and can be used at high concentrations. The concentration of gelatin required depends on the fixation: for embryos fixed in Dent's solution *(48)*, 15% gelatin is sufficient, while for the harder PFA-fixed embryos, 25% gelatin is required. Due to the high viscosity of gelatin, it is essential to perform infiltration progressively, starting with a low concentration (15%).

1.5.4. Cryopreservation

Although not as essential as for electron microscopy, cryopreservation is useful in optical microscopy as it minimizes cellular damage caused by water crystals. For this purpose, 15% sucrose is included in the infiltration medium.

1.5.5. Autofluorescence

The strong autofluorescence of the yolk is perhaps the greatest challenge encountered when using immunodetection on *Xenopus* embryos. Autofluorescence of *Xenopus* yolk spans virtually all wavelengths. One solution is to quench the autofluorescence and shift it to higher wavelengths using Eriochrome Black *(50)*. When using the appropriate Eriochrome staining intensity, set empirically for each experiment, autofluorescence in the fluorescein and Cy3 color range can be completely abolished. The strong fluorescence obtained in the deep red after quenching can then be used as counterstaining.

1.5.6. Double-staining

Quenching autofluorescence with Eriochrome Black allows observation of specific fluorescence in the green-orange range with very low background. Double-staining requires the use of narrow band filters for two reasons: *(1)* orange fluorescence (Cy3, Alexa546) must be separated from the red fluorescence of Eriochrome Black staining yolk; and *(2)* the spectra of the green fluorescent dyes (fluorescein and derivatives such as DTAF or Alexa488) and of orange fluorescent dyes (Cy3, Alexa546) are partially overlapping. One must thus ensure that there will be no fluorescence bleed-through.

1.5.7. Choice of Antibody

We have screened several commercial antibodies, and compared the signal with the original staining obtained with the P14L antibody (refs. *1* and *8*; **Fig. 23.1** and **Table 23.1**). At least two

Table 23.1
Comparative reactivity of various anti-β-catenin antibodies on *Xenopus* embryo cryosections

Antibody name	Species	Antigen/ epitope	Source	Optimal Fixation (% PFA)	Conc./ Dilut.	Signal		Notes
						Nuclear	Plasma membrane	
P14L	Rabbit	C-term peptide	Dr. Peter Hausen	2–4%	?	+++	+++	Limited supply
H102	Rabbit	aa680-781	Santa-Cruz	2–4%	2 µg/mL	++	++	
6F9	Mouse	?	Sigma	N/A	N/A	–	–	Strong nuclear non-specific signal
C2206	Rabbit	peptide aa786-781	Sigma		1:500 (whole serum)	+/–	+++	
CAT-5H10	Mouse	C-term 100 aa	Zymed	2%	1 µg/mL	+	+++	
8E7	Mouse	aa33-44	Upstate	2–4%	4µg/mL	+/–	+/–	Inconsistent signal
7D11	Mouse	N-term (exon 2)	Upstate		N/A	–	–	

commercial antibodies gave an excellent staining, virtually identical to P14L. However, other antibodies stain plasma membranes only, or do not react at all in *Xenopus*.

2. Materials

2.1. Fixation and Embedding

1. PFA fixative: 4% paraformaldehyde, 100 mM NaCl, 100 mM Hepes-NaOH, pH 7.4. To insure reproducible conditions, we use paraformaldehyde rather than formalin solution. To prepare paraformaldehyde solution, dissolve 4 g PFA powder in 100 mL of 100 mM NaCl and 100 mM Hepes, pH 7.4 by heating at 60°C. The pH is checked and adjusted to pH 7.4 with NaOH. The fixative is then stored as frozen aliquots (–20°C).
2. Dent's solution: 20% (v/v) DMSO, 80% (v/v) methanol, store at –20°C. (Flammable!)
3. Fish gelatin: 45% solution "Hipure liquid gelatin" from Norland Products Inc., Cranbury, NJ (*see* **Note 2**).
4. Infiltration solutions: 15 and 25% gelatin, 15% sucrose in H_2O (45% gelatin stock diluted 1/3 and 1/1.8, respectively).
5. Embedding solution: 20% gelatin, 15% sucrose in H_2O.
6. Glass vials (2 mL).
7. Rotator (slow speed less than 10 revolution/minute).

2.2. Freezing and Sectioning

1. Motorized cryotome required.
2. Real histology knife. If sharp, knives yield much better sections compared to disposable blades.
3. Plastic or rubber histology molds, dimensions: at least 8 mm deep and 10 × 10 mm wide.
4. Embedding medium: Tissue-Tek® O.C.T. (Sakura Finetek) or HistoPrep (Fisher).
5. Razor blades.
6. Pretreated or pre-coated glass slides (*see* **Notes 3** and **4**).
7. Hotplate (37°C).

2.3. Immunofluorescence

1. Phosphate Saline Buffer (PBS): 145 mM NaCl, 2 mM KCl, 10 mM Na_2HPO_4, 10 mM KH_2PO_4, filtered.
2. Blocking buffer: 5% (w/v) nonfat milk powder in PBS, dissolve with medium heat, filter.
3. Washing buffer: 1% (w/v) nonfat milk powder in PBS.

4. Primary antibodies. *See* **Table 23.1**.

5. Secondary antibodies: Alexa488 and Alexa546 coupled goat anti-rabbit and anti-mouse antibodies (Molecular Probes, Invitrogen).

6. DAPI or Hoechst 0.4 μg/mL in PBS, from a 2 mg/mL stock in H_2O.

7. 0.2 % Eriochrome Black (Aldrich) in PBS, filtered.

8. Mounting medium: SlowFade medium, Molecular Probes.

9. Fan.

10. Humid chamber (*see* **Note 5**).

11. Glass cover slips No. 1.

12. Nail polish (transparent, non-fluorescent).

3. Methods

3.1. Fixation and Embedding

1. Fix embryos 45–60 minutes in 4% PFA (*see* **Note 6**).

2. Remove PFA fixative, add quickly cold (–20°C) Dent's solution and keep at –20°C overnight (*see* **Notes 7** and **8**).

3. Wash embryos with 100 mM NaCl, 100 mM Tris-HCl, pH 7.3 (30–60 minutes; *see* **Note 9**).

4. Infiltrate sequentially with 15% fish gelatin followed by 25% fish gelatin, each step for 16–24 hours, with continuous gentle mixing (e.g., rotator).

5. Store up to two weeks at 4°C in gelatin.

3.2. Freezing and Sectioning (See Notes 10 and 11)

3.2.1. Cryotome Settings

1. Remove anti-rolling plastic plate from knife holder (*see* **Note 12**).

2. Set the cryotome at –20 to –25°C (*see* **Note 13**).

3. The knife angle should be 4°.

3.2.2. Sample Orientation, Embedding, and Preparation for Sectioning

1. Samples should be frozen and sectioned the same day (*see* **Note 14**).

2. Samples (up to three to four embryos/block) are gently transferred from their last infiltration (25% gelatin) to the mold filled with 20% gelatin using a plastic pipette (*see* **Note 15**) with an opening wider than 1.5 mm diameter.

3. Orient the embryos considering that the bottom face will correspond to the sectioning plane (*see* **Note 16**).

4. Freeze on dry ice (*see* **Note 17**). Let it harden for 15 minutes.

5. Put mold with samples in the cryotome to warm up slightly until soft (15–20 minutes, up to about –15 to –25°C) before removing from the mold.

6. Trim coarse square blocks (about 10-mm sides) with razor blade.

7. Glue the block to the holder using a large drop of "embedding medium" (*see* **Notes 18** and **19**).

8. Fix holder to cryotome.

9. Re-trim to the final size and shape of a ~7 × 7 mm square, making sure that the sides have clean sharp edges (*see* **Notes 20** and **21**).

10. Orient the block precisely. Unlike in standard sectioning methods, the block's edges must not be oriented parallel to the edge of the knife as sections will stick, but at 45° (**Fig. 23.2**).

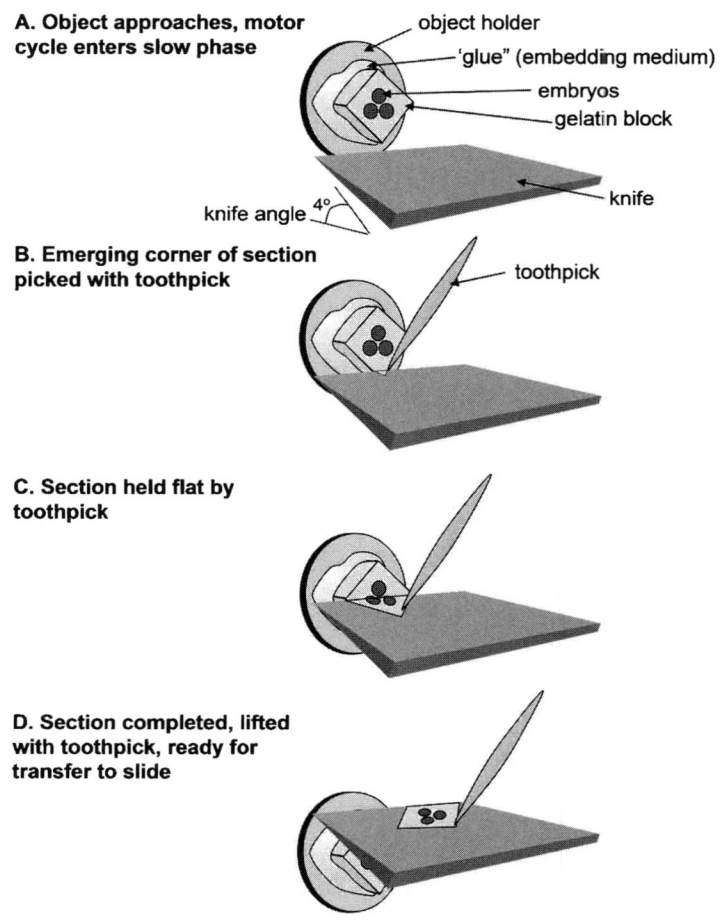

Fig. 23.2. Collection of gelatin-embedded crysections.

Additionally, the surface of the block must be as parallel as possible to the edge of the blade, as it is desirable to obtain complete sections from the beginning.

3.2.3. Sectioning

1. Section thickness: 10 µm (*see* **Note 22**).
2. Slow phase of motor's cycle: low-intermediate speed, starting when block ~1 cm above knife (**Fig. 23.2A**), ending just below the knife (**Fig. 23.2D**).

3.2.4. Collecting Sections

1. Grab with a toothpick the first corner of the section as it comes off the blade, to prevent curling of the section (**Fig. 23.2B**). Do not touch the embryos (*see* **Note 12**).
2. Keep holding the tip of the section as it slides along the knife (**Fig. 23.2C**). Ideally, the complete section will lay flat on the knife, sticking only weakly to the blade with its upper edge.
3. Bring the slide close to the knife (*see* **Note 23**).
4. Pull the section gently off the knife (**Fig. 23.2D**) and place carefully on pretreated glass slides (*see* **Note 24**). The move should be firm, as hesitation can cause shearing or tearing, and quick, to prevent curling.
5. Dry sections for 30 minutes on a warm plate (37°C).
6. Store sections at –80°C. Avoid repeated freeze–thawing.

3.3. Immunofluorescence

1. Thaw slides and dry extensively (1 hour) with a fan.
2. Dip for 1 minute in acetone, let dry (1 to 2 minutes).
3. Lay the slides flat, sections up, slightly heightened, in a wet chamber.
4. Re-hydrate by carefully covering the sections with PBS (2 minutes; *see* **Note 25**).
5. Incubate with blocking buffer for 60 minutes.
6. Incubate for 2 hours with the primary antibody in blocking buffer (*see* **Note 26**).
7. Rinse three times in 1% milk/PBS (5–10 minutes each wash).
8. Incubate for 1 hour at room temperature with conjugated secondary antibody in blocking buffer.
9. Rinse once in 1% milk/PBS, twice in PBS.
10. Incubate for 10 minutes with DAPI or Hoechst (nuclear staining, optional).
11. Rinse twice in PBS.
12. Quench autofluorescence of the yolk with 0.2% Eriochrome Black for 5–10 minutes (monitor color of embryos, *see* **Note 27**).

13. Rinse twice in PBS.

14. Mount in a glycerin-PBS medium containing an antifading reagent (Slow Fade; *see* **Note 28**).

15. Seal coverslips with nail polish.

16. Keep mounted sections in the dark at 4°C for up to a few weeks, or at − 80°C for long-term storage.

3.4. Microscopy

We routinely observe embryos using oil or water immersion, 20x or 25x objectives. Because an embryo section is much larger than the field of one of these objectives, we use an x-y motorized stage that allows automated collection of multiple images, which are then stitched (*see* **Note 29**). Several image processing software include a stitching function, we use either Metamorph (Perkin Elmer) or AnalySIS (Olympus-SIS).

Data can be obtained from up to four color channels: blue for the nuclei (DAPI; *see* **Note 30**), green and orange for two antibodies, and red for the yolk (Eriochrome Black). It is advisable to couple the weakest antibody to a green fluorescent secondary antibody, rather than orange, since, provided a good Eriochrome Black staining, the background is lower and the sensitivity higher.

Our microscopes are equipped with standard DAPI and FITC (narrow band) filter sets. For the orange channel, we use a narrow band Cy3 orange filter set. Regular Rhodamine/TRITC filter sets cannot be used, because they will pick equally the orange antibody signal and the red Eriochrome staining of the yolk. On the other hand, we also often use a FITC filter set with a broad emission filter, which provides in a single image both green antibody staining and red yolk counterstain.

4. Notes

1. Classical cryosectioning embedding media such as O.T.C. do not penetrate tissues.

2. The same gelatin is sold by Sigma/Fluka under the name "cold water fish gelatin," but we had problems with these preparations, which although theoretically identical to Norland's, appeared too liquid, at least in some batches.

3. Glass slides: We use SuperFrost Plus (Fisher Scientific Co., Springfield, NY, USA), but home-made gelatin or polylysine-coated slides work very well.

4. Do not keep Superfrost slide stocks longer than 6 months, they seem to loose their adhesiveness over time.

5. We use plastic boxes with lid. We cover the bottom with wet paper towels, and lay the slides flat on two cylindrical supports

(e.g., plastic 5- to 10-mL pipettes, but we use custom-made metal rods), to avoid direct contact with the humid bottom of the box and facilitate handling of the slides.

6. For some antibodies, PFA must be decreased to 2% or even 1%, *see* comment in **Subsection 1.5.1**.

7. Dent's solution permeabilizes the sample by dissolving membranes.

8. Samples in Dent's solution can be stored for at least a couple of weeks at −20°C.

9. Methanol must be washed out before infiltration with gelatin, which may otherwise precipitate at contact with the solvent. Amines such as Tris are used to quench any un-reacted formaldehyde and thus prevent fixation of gelatin at the surface of the embryo.

10. General remarks on cryosectioning:

For successful cryosectioning, it is important that everything is set at the optimal temperature.

Be aware that temperature equilibration of large objects (holder, knife, …) takes time and should be done well ahead. Thus, it is best to permanently keep the cryotome at operating temperature, as it can take several hours to equilibrate. Insert knife into its holder well before sectioning (at least 30 minutes). Consider that any manipulation will cause slight temperature changes. In particular, temperature variations affect the gelatin block, causing it to shrink or expand. For instance, both trimming and fixing the sample to the holder will warm the gelatin and cause temporary expansion. Always keep this in mind and wait for temperature to re-equilibrate.

11. Common problems encountered with cryosectioning and suggested solutions are listed in **Table 23.2**.

12. The "anti-roll" plate commonly used to collect O.C.T.-embedded sections cannot be used with gelatin samples, because sections stick to it and are damaged. Brushes also tend to damage sections.

13. The optimal temperature depends on the gelatin batch and on the fixation. The color of frozen fish gelatin varies with temperature and is a good indication of its consistence: ideally it should be light grey with a slight hint of yellow. Darker yellowish gelatin is close to melting. White gelatin (low temperature) is too hard for sectioning.

14. Samples must be frozen on the day of sectioning. Frozen blocks dry fast, even at −80°C. Best cut within 4 to 5 hours.

15. The embryos will sink to the bottom of the mold. Embryos float in 25% gelatin.

Table 23.2
Troubleshooting strategies

Problem	Possible diagnostic	Solutions
Sections are sticky, of a rubbery consistence	Blocks have been frozen too early and have partially dried up	Freeze samples just before sectioning. Thaw and re-embed (!not advised! repeated thawing–freezing likely to affect structure)
Sections melting	Temperature too high	Wait for temp. equilibration; lower cryotome temperature
Sections are not homogenous and/or brittle	Temperature too low	Increase temperature
	Inhomogeneous temperature	Make sure all parts (cryotome chamber, knife, object, …) have equilibrated
	Sectioning speed	Slow down sectioning speed
	Bad knife's blade	Look at the edge of the blade under stereomicroscope for blunt blade and/or notches, re-sharpen
	Fixation too strong	Check PFA solution; shorten fixation
	Incomplete gelatin infiltration	Check gelatin conc. 15% gelatin step required
Sections are of irregular thickness	Something is improperly tightened	Check knife, knife holder, object holder, object
	Sectioning speed	Slow down speed
	Bad knife's blade	See above
Sections do not come off the knife	Irregular edges of the gelatin block	Straighten the edges by trimming with a razorblade
	Sections may be slightly melting during sectioning	Decrease cryotome temperature
Sections curl during transfer and land folded on the slide	Ambient conditions (humidity, static electricity); area of section too small	Use larger blocks, as weight of section balances upward curling. To save curling sections: maintain on knife for few seconds to reduce tension before transferring to slide; stretch slightly section by pulling gently with the toothpick as it is being cut
Sections bend down ward during transfer, causing folds and wrinkles to form on the slide.	Section too large, and therefore too heavy	Decrease size of block

16. Embryos should be placed close together (maximal total surface 0.3 × 0.3 mm), but should not touch each other.

17. Because embryos tend to roll back to yolk facing down, one should monitor freezing under a stereomicroscope and readjust orientation until gelatin solidifies. Alternatively, the mold filled with gelatin can be pre-cooled on ice (gelatin will become more viscous) to facilitate orientation.

18. Add only as much embedding medium as required for tight adherence to the holder. The gelatin block should stick out of the embedding medium by at least 3 mm. Avoid, however, too tall gelatin blocks (>10 mm), as they will be too flexible and the quality of the sections will be affected.

19. Gelatin will temporarily warm up, wait until embedding medium has completely solidified (white!) before continuing (*see* **Note 10** and **Table 23.2**).

20. The size of the surface of the block is important for sectioning and its size should be adjusted according to conditions (*see* **Table 23.2**).

21. Clean edges are crucial for easy retrieval of the sections.

22. We routinely prepare 10-μm sections. Resolution is satisfactory, antibodies are able to penetrate the sample completely, and most 20–40x objectives can focus adequately. If desired, this method allows thinner sections, down to 5 μm. Sectioning more than 20 μm causes too much distortion of the block and sections.

23. Slides are kept outside of the cryotome chamber, as a temperature difference is essential for adherence.

24. We usually collect serial sections distributed over 6–10 slides (generally 8) as explained in **Fig. 23.3**. This sequence of collection provides sections from the entire span of the embryo on a single slide. It also allows staining of consecutive sections, for example with two antibodies raised in the same species.

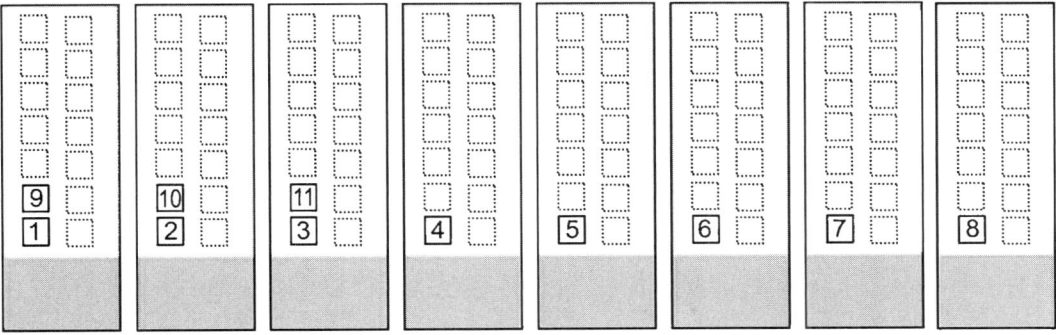

Fig. 23.3. Diagram of collection of consecutive serial cryosections.

25. From this point on, the sections should always be covered with liquid. Even a short time dry may cause precipitates. For this reason, buffers or antibodies are replaced rapidly, slide by slide: the liquid is discarded by draining, the slide is laid flat again and covered with new buffer.

26. Incubation length may vary according to the antibody (e.g., for detection of tags, such as Myc using the 9E10 antibody, a 1-hour incubation is sufficient). Avoid overnight incubations, as they may cause increased background.

27. The embryos on the sections should become blueish (5–10 minutes). Grey appearance corresponds to incomplete quenching: autofluorescence may still be visible in the green and orange channels, and/or partial quenching will cause a strong background signal in the orange channel. Dark blue indicates that counterstaining is too strong, which may lead to quenching of the specific antibody signal.

28. Frozen sections are very fragile. Avoid air bubbles, and do not press the cover slip too hard, it will smash them.

29. Care should be given to align the fluorescent lamp (Hg or equivalent) to produce a homogenous field of light. This can be best checked by testing stitching at small scale (3 × 3 images). Readjustment: test stitching should be repeated until satisfactory homogeneity is obtained. In addition to x-y alignment of the bulb and x-y-z alignment of the mirror, the position of the condenser is also crucial.

30. DAPI filter cubes cause strong UV-blue excitation. Avoid long exposures of samples to these wavelengths, since they will photobleach fluorescein and derivatives.

Acknowledgements

The work of FF laboratory is supported by funds NCIC #17162 and NSERC # 261679. We would like to acknowledge the invaluable contribution of Mrs. Anne Schohl to the establishment of the original protocol for β-catenin nuclear localization.

References

1. Schohl, A., Fagotto, F. (2002) β-catenin, MAPK and Smad signaling during early *Xenopus* development. *Development* 129, 37–52.

2. Jung, A., Schrauder, M., Oswald, U., et al. (2001) The invasion front of human colorectal adenocarcinomas shows co-localization of nuclear β-catenin, cyclin D1, and

p16INK4A and is a region of low proliferation. *Am J Pathol* 159, 1613–1617.

3. Brabletz, T., Jung, A., Reu, S., et al. (2001) Variable beta-catenin expression in colorectal cancers indicates tumor progression driven by the tumor environment. *Proc Natl Acad Sci USA* 98, 10356–10361.

4. Staal, F. J. T., van Noort, M., Strous, G. J., et al. (2002) Wnt signals are transmitted through N-terminally dephosphorylated β-catenin. *EMBO Reports* 3, 63–68.

5. Hendriksen, J., Fagotto, F., van der Velde, H., et al. (2005) RanBP3 enhances nuclear export of active β-catenin independently of CRM1. *J Cell Biol* 171, 785–797.

6. Henderson, B. R. (2000) Nuclear-cytoplasmic shuttling of APC regulates β-catenin subcellular localization and turnover. *Nature Cell Biol* 2, 653–660.

7. Say, Y. H., Hooper, N. M. (2007) Contamination of nuclear fractions with plasma membrane lipid rafts. *Proteomics* 7, 1059–1064.

8. Schneider, S., Steinbeisser, H., Warga, R. M., et al. (1996) β-catenin translocation into nuclei demarcates the dorsalizing centers in frog and fish embryos. *Mech Dev* 57, 191–198.

9. Schohl, A., Fagotto, F. (2003) A role for β-catenin in mesoderm induction in *Xenopus*. *EMBO J* 22, 3303–3313.

10. Heasman, J., Crawford, A., Goldstone, K., et al. (1994) Overexpression of cadherins and underexpression of β-catenin inhibit dorsal mesoderm induction in early *Xenopus* embryos. *Cell* 79, 791–803.

11. Tao, Q., Yokota, C., Puck, H., et al. (2005) Maternal Wnt11 activates the canonical Wnt signaling pathway required for axis formation in *Xenopus* embryos. *Cell* 120, 857–871.

12. Heasman, J. (2006) Patterning the early *Xenopus* embryo. Development *133*, 1205–1217.

13. Carnac, G., Kodjabachian, L., Gurdon, J.B., et al. (1996) The homeobox gene Siamois is a target of the Wnt dorsalisation pathway and triggers organiser activity in the absence of mesoderm. *Development* 122, 3055–3065.

14. Brannon, M., Kimelman, D. (1996) Activation of Siamois by the Wnt pathway. *Dev Biol* 180, 344–347.

15. Brannon, M., Gomperts, M., Sumoy, L., et al. (1997) A β-catenin/XTcf-3 complex binds to the siamois promoter to regulate dorsal axis specification in *Xenopus*. *Genes Dev* 11, 2359–2370.

16. Laurent, M.N., Blitz, I. L., Hashimoto, C., et al. (1997) The *Xenopus* homeobox gene twin mediates Wnt induction of goosecoid in establishment of Spemann's organizer. *Development* 124, 4905–4916.

17. McKendry, R., Hsu, S. C., Harland, R. M., et al. (1997) LEF-1/TCF proteins mediate wnt-inducible transcription from the *Xenopus* nodal-related 3 promoter. *Dev Biol* 192, 420–431.

18. Christian, J. L., Moon, R. T. (1993) Interactions between Xwnt-8 and Spemann organizer signaling pathways generate dorsoventral pattern in the embryonic mesoderm of *Xenopus*. *Genes Dev* 7, 13–28.

19. Hoppler, S., Brown, J. D., Moon, R. T. (1996) Expression of a dominant-negative Wnt blocks induction of MyoD in *Xenopus* embryos. *Genes Dev* 10, 2805–2817.

20. Yasuo, H., Lemaire, P. (2001) Role of Goosecoid, Xnot and Wnt antagonists in the maintenance of the notochord genetic programme in *Xenopus* gastrulae. *Development* 128, 3783–3793.

21. Shi, D. L., Bourdelas, A., Umbhauer, M., et al. (2002) Zygotic Wnt/β-catenin signaling preferentially regulates the expression of Myf5 gene in the mesoderm of *Xenopus*. *Dev Biol* 245, 124–135.

22. Reintsch, W. E., Habring-Mueller, A., Wang, R. W., et al. (2005) β-catenin controls cell sorting at the notochord-somite boundary independently of cadherin-mediated adhesion. *J. Cell Biol.* 170, 675–686.

23. Wolda, S. L., Moody, C. J., Moon, R. T. (1993) Overlapping expression of Xwnt-3A and Xwnt-1 in neural tissue of *Xenopus* laevis embryos. *Dev Biol* 155, 46–57.

24. Saint-Jeannet, J. P., He, X., Varmus, H. E., et al. (1997) Regulation of dorsal fate in the neuraxis by Wnt-1 and Wnt-3a. *Proc Natl Acad Sci USA* 94, 13713–13718.

25. Kazanskaya, O., Glinka, A., Niehrs, C. (2000) The role of *Xenopus* dickkopf1 in prechordal plate specification and neural patterning. *Development* 127, 4981–4992.

26. Logan, C. Y., Miller, J. R., Ferkowicz, M. J., et al. (1999) Nuclear β-catenin is required to specify vegetal cell fates in the sea urchin embryo. *Development* 126, 345–357.

27. Imai, K., Takada, N., Satoh, N., et al. (2000) β-catenin mediates the specification of endoderm cells in ascidian embryos. *Development* 127, 3009–3020.

28. Mohamed, O. A., Clarke, H. J., Dufort, D. (2004) β-catenin signaling marks the prospective site of primitive streak formation in the mouse embryo. *Dev Dyn* 231, 416–424.

29. Roeser, T., Kessel, M. (1999) Nuclear β-catenin and the development of bilateral symmetry in normal and LiCl-exposed chick embryos. *Development* 126, 2955–2965.

30. Momoi, A., Yoda, H., Steinbeisser, H., et al. (2003) Analysis of Wnt8 for neural posteriorizing factor by identifying Frizzled 8c and Frizzled 9 as functional receptors for Wnt8. *Mech Dev* 120, 477–489.

31. Peifer, M., Wieschaus, E. (1990) The segment polarity gene armadillo encodes a functionally modular protein that is the Drosophila homolog of human plakoglobin. *Cell* 63, 1167–1176.

32. Peifer, M., Sweeton, D., Casey, M., et al. (1994) wingless signal and Zeste-white 3 kinase trigger opposing changes in the intracellular distribution of armadillo. *Development* 120, 369–380.

33. Tolwinski, N. S., Wieschaus, E. (2001) Armadillo nuclear import is regulated by cytoplasmic anchor Axin and nuclear anchor dTCF/Pan. *Development* 128, 2107–2117.

34. Wiechens, N., Fagotto, F. (2001) Crm1- and Ran-independent nuclear export of beta-catenin. *Curr Biol* 11, 18–27.

35. Tolwinski, N. S., Wehrli, M., Rives, A., et al. (2003) Wg/Wnt signal can be transmitted through arrow/LRP5,6 and Axin independently of Zw3/Gsk3beta activity. *Dev Cell* 4, 407–418.

36. Krieghoff, E., Behrens, J., Mayr, B. (2006) Nucleo-cytoplasmic distribution of Beta-catenin is regulated by retention. *J Cell Sci* 119, 1453–1463.

37. Fagotto, F., Funayama, N., Gluck, U., et al. (1996) Binding to cadherins antagonizes the signaling activity of beta-catenin during axis formation in *Xenopus*. *J Cell Biol* 132, 1105–1114.

38. Rosin-Abersfeld, R., Townsley, F., Bienz, M. (2000) The APC tumor suppressor has a nuclear export function. *Nature* 406, 1009–1012.

39. Neufeld, K. L., Zhang, F., Cullen, B. R., et al. (2000) APC-mediated downregulation of beta-catenin activity involves nuclear sequestration and nuclear export. *EMBO Reports* 1, 519–523.

40. Wiechens, N., Heinle, K., Englmeier, L., et al. (2004) Nucleo-cytoplasmic Shuttling of Axin, a Negative Regulator of the Wnt-β-Catenin Pathway. *J Biol Chem* 279, 5263–5267.

41. Cong, F., Varmus, H. (2004) Nuclear-cytoplasmic shuttling of Axin regulates subcellular localization of β-catenin. *Proc Natl Acad Sci USA* 101, 2882–2887.

42. Larabell, C. A., Torres, M., Rowning, B. A., et al. (1997) Establishment of the dorsoventral axis in *Xenopus* embryos is presaged by early asymmetries in β-catenin that are modulated by the Wnt signaling pathway. *J Cell Biol* 136, 1123–1136.

43. Rowning, B. A., Wells, J., Wu, M., et al. (1997) Microtubule-mediated transport of organelles and localization of β-catenin to the future dorsal side of *Xenopus* eggs. *Proc Natl Acad Sci USA* 94, 1224–1229.

44. Faure, S., Lee, M. A., Keller, T., et al. (2000) Endogenous patterns of TGFβ superfamily signaling during early *Xenopus* development. *Development* 127, 2917–2931.

45. Lemaire, P., Gurdon, J. B. (1994) A role for cytoplasmic determinants in mesoderm patterning: cell-autonomous activation of the goosecoid and Xwnt-8 genes along the dorsoventral axis of early *Xenopus* embryos. *Development* 120, 1191–1199.

46. Hausen, P., Dreyer, C. (1981) The use of polyacrylamide as an embedding medium for immunohistochemical studies of embryonic tissues. *Stain Technol* 56, 287–293.

47. Fagotto, F. (1999) The Wnt pathway in *Xenopus* development in (Guan, J. -L., ed.) *Signaling through Cell Adhesion*, pp. 303–356, Boca Raton, CRC Press.

48. Fagotto, F., Gumbiner, B. M. (1994) β-catenin localization during *Xenopus* embryogenesis: accumulation at tissue and somite boundaries. *Development* 120, 3667–3679.

49. Dent, J. A., Cary, R. B., Bachant, J. B., et al. (1992) Host cell factors controlling vimentin organization in the *Xenopus* oocyte. *J Cell Biol* 119, 855–866.

50. Torpey, N. P., Heasman, J., Wylie, C. C. (1992) Distinct distribution of vimentin and cytokeratin in *Xenopus* oocytes and early embryos. *J Cell Sci* 101, 151–160.

Chapter 24

Transgenic Reporter Tools Tracing Endogenous Canonical Wnt Signaling in *Xenopus*

Tinneke Denayer, Hong Thi Tran, and Kris Vleminckx

Abstract

Activation of the canonical Wnt pathway leads to the transcriptional activation of a particular subset of downstream Wnt target genes. To track this localized cellular output in a living organism, reporter constructs can be designed containing multimerized consensus lymphoid enhancer binding factor (LEF)-1/T cell factor (TCF) transcription factor binding sites, generally referred to as TCF optimal promoter (TOP) sites. In *Xenopus*, several Wnt-responsive reporter systems have been designed containing a number of these TOP sites that, in combination with a minimal promoter, drive the expression of a reporter gene. Following transgenic integration in *Xenopus* embryos, a Wnt reporter tool reveals the spatiotemporal delineation of endogenous Wnt pathway activities throughout development. Assumed to be a general readout of the Wnt pathway, such reporters can assist in elucidating unknown functional implications in developing *Xenopus* embryos.

Key words: canonical Wnt signaling, *Xenopus*, reporter, transgenesis, β-gal, GFP, fluorescent imaging, histological sectioning.

1. Introduction

The canonical Wnt pathway is recurrently activated during embryonic development and in specific cell populations throughout adult life. Following Wnt pathway activation, cytoplasmic β-catenin accumulates, translocates to the nucleus and subsequently associates with lymphoid enhancer binding factor (LEF)-1/T cell factor (TCF) transcription factors, resulting in the establishment of a heterodimeric transcriptional complex *(1, 2)*. In a given cellular context, a particular subset of targets is activated. The spatio-temporal demarcation of a signaling event within a developing

Elizabeth Vincan (ed.), *Wnt Signaling, Volume II: Pathway Models, vol. 469*
© 2008 Humana Press, a part of Springer Science+Business Media, New York, NY
Book doi: 10.1007/978-1-60327-469-2

organism often provides clues to its exact biological role. However, the presence of Wnts, Frizzled, or LRP-5/-6 receptors does not guarantee activation of the pathway because many negative regulators are known to intersect at different subcellular levels *(1)*. Moreover, the Wnt pathway is bifurcated downstream of Dishevelled into a β-catenin dependent and β-catenin independent pathway, depending on the cellular context. Also, it was reported that β-catenin might be activated independently of a Wnt ligand *(3, 4)*. Hence, a reporter able to trace the molecular output of the Wnt pathway in specific cell populations within a developing embryo is an important tool to gain knowledge about the potential involvement of this pathway in tissue formation, organogenesis and homeostasis. Important to note, this reporter reflects sensu strictu the downstream transcriptional activation of Wnt targets, induced by binding of the β-catenin-LEF-1/TCF transcriptional complex onto the corresponding binding sites, which is generally believed to be the key event for canonical Wnt signaling. However, canonical Wnt signaling in which nuclear factors other than LEF-1/TCF are involved, such as Pitx2, retinoic acid receptor and several Sox members, will not be traced by a LEF-1/TCF-dependent Wnt-responsive reporter system *(5–7)*. Also, other molecules that recognize LEF-1/TCF binding sites independent of canonical Wnt signaling activity cannot be excluded and may influence the outcome of the reporter. Endogenous promoters of Wnt target genes generally get input from different signaling pathways, implying that the signal strength detected with the Wnt reporter is not informative about the strength of the Wnt signal in reality. Consequently, a weak Wnt reporter signal could be a reflection of the strong activation of a specific endogenous promoter. Nevertheless, keeping in mind these potential limitations, Wnt reporters can generally be assumed as a helpful and mostly faithful tool to elucidate the molecular spatiotemporal mechanisms of gene regulation of the Wnt targets and their concordant functional implications *(8)*.

In the β-catenin-LEF-1/TCF transcriptional complex established upon Wnt pathway activation, β-catenin essentially provides the transactivation domain, while LEF-1/TCF supplies the DNA-binding domain recognizing specific consensus binding sites. The latter specific promoter sequence has been narrowed down to an AGATCAAAGGG motif *(9)*, which is generally referred as a TCF optimal promoter (TOP) site. Multiple copies of this consensus binding motif fused to a minimal promoter and cloned upstream of a reporter gene constitute the basic outline of a classical Wnt reporter. First- and best-known is the TOPFLASH construct, in which three TOP sites upstream of a minimal c-fos promoter are driving luciferase gene expression *(9)*. So far, several Wnt reporters have demonstrated their effectiveness in registering Wnt signaling activity not only *in vitro* but as well in different species including zebrafish *(10)* and mice *(11–15)*. Also in *Xenopus*, the successful employment of similar Wnt-responsive reporter tools in transgenic embryos have been yet described, driving the expression of β-galactosidase (β-gal), green fluorescent protein (GFP), DsRed or a destabilized GFP variant (refs. *16* and *17*, Tran et al., unpublished results). Due to distinct characteristics linked with a specific chosen reporter gene, it is possible for the analysis of endogenous Wnt signaling to span early to late development. Serving as a negative control, mutant binding motifs or Far From Optimal (FOP) sites are used (= AGGCCAAAGGG; = FOP sites; ref. *9*).

This chapter specifically covers the design of transgenic Wnt reporter systems and their subsequent spatiotemporal and dynamic analysis in *Xenopus* embryos. Protocols for testing the reporter efficacy, β-gal staining, fluorescent *in vivo* imaging, and histological analysis on sections are included. Details on the transgenesis procedure in *Xenopus* have not been incorporated, but the used procedure is essentially as described previously (*see* **Note 1**).

2. Materials

All solutions should be made up in sterile deionized water.

2.1. Testing Functionality of Wnt Reporters in Xenopus Embryos

2.1.1. β-gal or Luciferase Quantification Assay on DNA-Injected Embryos

1. General microinjection equipment, mainly including a stereomicroscope (e.g., Stemi 2000, Zeiss) supplied with an ocular micrometer for calibration of the injection volume, a cold-light source (e.g., KL750, Schott), a 3-D micromanipulator (e.g., Narishige M40B) and a micro-injector (e.g., PLI-100, Harvard Bioscience).

2. Microinjection needles, produced from borosilicate glass capillaries (e.g., with 1.0 and 0.58 mm as outer and inner diameter, respectively; Sutter Instruments Co.) using a micropipette puller (e.g., Model P-97, Sutter Instruments Co.).

3. Injection dish: a 35- or 60-mm Petri dish onto which a mesh (e.g., 700-µm mesh; Spectrum Laboratories) is fixed with a few drops of chloroform or superglue. The grids correspond with the size of an embryo.

4. Adult female *Xenopus laevis*, induced with 700 units human chorionic gonadotropin (HCG) the evening prior to injection. Freshly dissected or stored testis (at 4°C in 100% fetal calf serum up to a couple of days). Details on the fertilization procedure are described elsewhere *(18)*.

5. 1x Marc's Modified Ringers (MMR): 0.1 M NaCl, 1.8 mM KCl, 2 mM CaCl$_2$, 1 mM MgCl$_2$, 5 mM Hepes, pH 7.6.

6. 2% cysteine solution in water, adjusted to pH 8.0 with NaOH (800 µL of 10 M NaOH in a total of 50 mL) to dejelly the fertilized embryos.

7. Ficoll (Amersham Pharmacia).

8. Lysis buffer: reporter assay lysis buffer (Roche Applied Science) or lysis solution supplied by Galacto-Star kit (Tropix; Applied Biosystems).

9. Luciferase substrate: 40 mM Tricine, 2.14 mM (MgCO$_3$)$_4$Mg(OH)$_2$, 5.34 mM MgSO$_4$, 66.6 mM DTT, 0.2 mM EDTA, 521 µM coenzyme A, 734 µM ATP, 940 µM luciferin.

10. β-gal substrate: 50x concentrate Galacto-Star Substrate diluted with supplied Reaction Buffer Diluent (Tropix; Applied Biosystems)

11. 96-well black microplates and TopSeal sealing film (Packard).

12. Luminometer (e.g., Topcount NXT, Packard)

2.1.2. LiCl Treatment of Transgenic Embryos

1. Rotator.

2. Lithium chloride (LiCl).

3. 0.1x MMR (*see* **Subheading 2.1.1., item 5**).

2.2. Whole-mount Spatiotemporal Analysis of Transgenic Wnt Reporter Gene Expression

2.2.1. Whole-mount β-gal Staining

1. Pair of sharp tweezers (Sigma) to remove the vitteline membrane.

2. 5-mL screw-cap glass vials (Wheaton).

3. Rotator.

4. MEMFA: 0.1 M MOPS, pH 7.4, 2 mM EGTA, 1 mM MgSO4, 3.7% (v/v) formaldehyde. A 10x stock solution without formaldehyde can be made. Note that this solution) turns yellow if autoclaved or aged.

5. PTW: PBS (Cambrex) supplemented with 0.1% (v/v) Tween-20 (Sigma). Make a 10x stock of PBS in water and autoclave.

6. X-gal staining solution: PTW (pH 7.3) containing 5 mM K$_3$Fe(CN)$_6$, 5 mM K$_4$Fe(CN)$_6$, 2 mM MgCl$_2$ and 1 mg/mL X-gal (5-bromo-4-chloro-3-indolyl-β-D-galactosidase, Sigma), stored at 4°C in the dark. X-gal can be prepared as a 20 mg/mL stock in N,N,-dimethylformamide (DMF, Sigma) and stored at –20°C in the dark. K$_3$Fe(CN)$_6$ and K$_4$Fe(CN)$_6$ can be prepared as 0.5M (100x) stocks and stored at 4°C in the dark.

7. Ethanol.

8. Mayor's bleaching solution: 1% (v/v) H2O2, 5% (v/v) formamide, 0.5x SSC (freshly made). Make sure that H2O2 and formamide are diluted into the SSC and not reversely, because an explosive mixture can be formed between undiluted H2O2 and formamide.

9. BB/BA (or Murray's) clearing solution: 2:1 volumes benzyl benzoate/benzyl alcohol. Caution is needed because of toxicity and irritation of these reagents. BB/BA dissolves polystyrene, use of glass receptacles or plates is preferred.

2.2.2. Fluorescent In Vivo Imaging

1. 0.02% (w/v) MS222 (3-aminobenzoic acid ethyl ester or tricaine; Sigma) in 0.1x MMR. Make a 100x stock in water and store as aliquots at –20°C.

2. 0.1x MMR (*see* **Subheading 2.1.1., item 5**).

3. Fluorescence stereomicroscope (e.g., Leica MZ FLIII) and/ or confocal microscope (e.g., Leica SP5 AOBS) equipped with a digital color camera and corresponding software.

2.3. Histological Analysis on Sections

2.3.1. Plastic Sections

1. MEMFA (*see* **Subheading 2.2.1., item 4**).

2. PBS.

3. 5-mL screw-cap glass vials (Wheaton).

4. JB-4 embedding kit (Polysciences 00226; stored in a cool dark place), containing JB-4 Solution A (Monomer), JB-4 Solution B (Accelerator) and JB-4 Catalyst (Benzoyl Peroxide, Plasticized). JB-4 is a glycol methacrylate based polymer of which the polymerization is exothermic. According to the manufacturer's instructions, 100 mL of catalyzed JB-4 Infiltration Solution is prepared by mixing 100 mL Solution A with 1.25 g of Catalyst while stirring until dissolved (~10–20 minutes). Measurement of the catalyst is critical, as it will control the rate of polymerization of the plastic and the exothermic reaction. This infiltration solution can be stored for up to two weeks in the dark at 4°C. JB-4 Embedding Solution is prepared by thoroughly mixing 25 mL infiltration solution and 1 mL JB-4 solution B. Use the JB-4 embedding kit under a fume hood with appropriate gloves and avoid skin or eye contact.

5. Plastic block holders (Polysciences 15899). These are important to exclude oxygen from the molds during the polymerization process. Alternatively, the molds can be covered with an airtight film. If anaerobic conditions are not maintained, the JB-4 may polymerize incompletely or not at all.

6. Polyethylene molding cup trays (Polysciences 17177A).

7. Ethanol.

8. Microtome (for dry sectioning) and appropriate glass, Ralph, or tungsten carbide knife.

9. Forceps.

10. Room temperature water bath, with a few drops of NH_4OH to aid flattening of the sections.

11. Pre-cleaned and ready-to-use glass microscope slides and cover slips (Menzel-Gläser).

12. Hot plate.

13. Aquatex mounting medium (Merck).

2.3.2. Paraffin Sections

1. 4% (w/v) paraformaldehyde (PFA) in PBS. Heat at 60°C and neutralize the cloudy solution with few drops of 10 M NaOH until clear while stirring on a magnetic stirrer. A 20% (w/v) PFA stock solution can be made and stored at –20°C.

2. 5-mL screw-cap glass vials (Wheaton).

3. Ethanol.

4. Histoclear (VWR International), being a less toxic xylene substitute.

5. Paraplast (Sigma).

6. Incubation oven at 60°C with rocking platform, to melt the paraplast and to keep glass and wide-bore plastic pipettes warm for orienting and handling of the embryos.

7. Embedding cassettes (Klinipath).

8. Hot plate.

9. Microtome (e.g., HM360, Microm).

10. Pre-cleaned glass microscope slides and coverslips (Menzel-Gläser).

11. DPX mounting medium (BDH Laboratory Supplies).

2.3.3. Cryosections

1. 4% PFA/PBS (*see* **Subheading 2.3.2., item 1**).

2. 5-mL screw-cap glass vials (Wheaton).

3. 30% (w/v) sucrose in PBS.

4. Tissue-Tek O.C.T compound (Bayer).

5. Hand-made molds of aluminum foil or commercially available molds (Bayer).

6. Cryostat (e.g., HM500 OM, Microm).

7. SuperFrost Plus microscope slides and coverslips (Menzel-Gläser).

8. DAPI-containing Vectashield mounting medium (Vector Laboratories). DAPI stains the cell nuclei resulting in a blue fluorescent signal.

9. Fluorescence microscope (e.g., Olympus BX61) equipped with a digital camera and corresponding software.

3. Methods

3.1. Design of Transgenic Xenopus Wnt-responsive Reporter Tools

A classical Wnt-responsive reporter construct comprises several TOP sites that in combination with a minimal promoter are driving the expression of a reporter gene. This general layout is attuned to the specific aims for a given experimental setup, e.g., the developmental stage of analysis, type of analysis, etc. The number of TOP/FOP sites, the minimal promoter, the choice of the reporter gene, the inducibility of the system and the use of insulators are discussed later.

3.1.1. TOP Versus FOP Sites

1. TOP site is an artificial AGATCAAAGGG consensus LEF-1/TCF binding motif *(9)*. The number of TOP sites in Wnt reporters is generally varying between three and eight. A high number of TOP sites might increase the signal to noise ratio. Nonetheless, the number of TOP sites in reporters is often higher than in native Wnt target gene enhancers, many of which enclose only one or two LEF-1/TCF binding sites *(8)*. This can probably be explained by cooperative signaling inputs that are present in the latter case.

2. Along with and based on the design of a TOP-based Wnt reporter, a negative control is constructed by interchanging all the TOP sites for mutant binding motifs, or FOP sites (AGGCCAAAGGG; ref. *9)*. Theoretically, this control should be inactive since LEF-1/TCF binding is no longer ensuing.

3.1.2. Minimal Promoter

Various different minimal TATA-containing promoters have been cloned into Wnt reporter transgenes designed for mice, zebrafish or *Xenopus (8)*. Most commonly used promoters include a minimal *c-fos* promoter *(10, 11, 16)*, a minimal *siamois* promoter *(13, 15)* a minimal thymidine kinase (TK) promoter *(12, 17)* or a minimal TATA box (Tran et al., unpublished results). Noteworthy, the minimal *c-fos* promoter contains a cAMP response element *(19)*, and hence may report not only Wnt signaling but extraneous events as well, making it less suitable for specific detection. Both the *siamois* and *c-fos* gene are also considered as transcriptional Wnt targets. Even though their corresponding promoters do not include any of the known TCF binding sites, it is conceivable that they might be somehow competent to respond to LEF-1/TCF activation *(8)*. In conclusion, the use of a minimal *TK* promoter or minimal TATA box should be favored instead.

3.1.3. Choice of the Reporter Gene

The nature of the reporter gene is dependent on the developmental stage and kind of analysis. Due to distinct features of a specific chosen reporter, it is possible to span early to late development. Below, most widely used reporter genes are discussed.

The luciferase gene is not included, because it cannot be utilized as reporter for *in situ* spatial delineation. Conversely, luciferase is preferred for quantitative analysis on embryo lysates, being rapid and highly sensitive, e.g., testing functionality following reporter gene DNA injection (*see* **Subheading 3.2.1.**).

1. The *E.Coli lacZ* gene encoding β-galactosidase (β-gal) is a classical histochemical reporter gene that can be detected using a variety of substrates, of which X-gal is most commonly used. The use of β-gal reporter is of particular interest for early stage embryos, showing quite some autofluorescence due to high yolk protein concentrations. Hence, at these stages the use of a fluorescent protein as *in vivo* reporter is less likely. In addition, GFP is also too stable to register the fast successive dynamic Wnt patterns that are representative for early development. The disadvantage of β-gal is that its reporter activity is detected by X-gal staining or whole mount in situ hybridization for β-gal, consequently making the analysis only semi-dynamically due to the retrospective context. We also found that X-gal staining is rather "patchy" in early embryos.

2. The GFP isolated from the jellyfish *Aequoria victoria* has been extensively used as a reporter owing to its potential for the direct real-time monitoring of gene expression *in vivo* (*20*). Moreover, *Xenopus* embryos are considerably transparent at later stages favoring the use of fluorescent proteins in general. However, during early *Xenopus* stages no advantage can be taken of the *in vivo* reporting capacity of GFP due to the high degree of autofluorescence. Accordingly, these expression patterns need to be determined by means of a more sensitive *in situ* hybridization (*34*). The GFP protein is in addition very stable with a half-life of ~26 hours in mammalian cells (*21*), and several days in *Xenopus* embryos (own unpublished results). Enhanced GFP (eGFP) is a widely used variant that has an even higher sensitivity and stability. The high stability vastly hampers the study of dynamic expression patterns, like Wnt signaling processes. As such, GFP fluorescence originating from early reporter activation continues for several days and obscures the interpretation of fluorescent patterns at later developmental stages. Noteworthy, too high and continuously expressed levels of GFP might cause toxicity in embryos (*22*). Hence, it is preferable to design an inducible system that can give exact information by means of GFP at a specific developmental stage when it is triggered to become active (*see* **Subheading 3.1.4.**), or to use a less stable GFP variant like destabilized GFP.

3. A destabilized form of GFP (dGFP) can be employed as an alternative for GFP (ref. *23*; *see* **Note** 2), featuring a higher

turnover rate and a much lower half-life of only several hours in *Xenopus*, which in comparison with the fast successive dynamic Wnt patterns might still be relatively long. This GFP variant does not accumulate as much as normal GFP and hence generates a weaker signal, but generally seems to be strong enough to detect via life imaging.

4. DsRed (DsRed-express) is an engineered mutant of red fluorescent protein from the *Discosoma sp.* reef coral and can also be used as *in vivo* reporter. Similar to GFP, it is also a very stable protein (ref. *24*; *see* **Note 3**). A destabilized version, being DsRed-Express-DR, exists as well.

3.1.4. Inducible Versus Non-inducible Systems

Although fluorescent *in vivo* markers like GFP or DsRed are widely used as reporter genes, their major disadvantage is the high protein stability. Consequently, fluorescence originating from early reporter activation continues for several days and obscures the interpretation of fluorescent patterns at later developmental stages. In this context, an inducible system could be preferred, which avoids capture of pre-occurring signals by remaining dormant until triggered to become active. Hence, such a reporter will be able to provide exact information on spatial expression at the specific developmental stage of induced activation without interference with earlier upcoming signals. An example of this is the TOPTK-iGFP reporter, which is inducible by the synthetic glucocorticoid, dexamethasone (Dex; ref. *17*). The TOPTK promoter drives the expression of a GAL4VP16GR transactivator, and further downstream, 14 GAL4-binding upstream activating sequences (UAS) elements and an E1b promoter direct expression of eGFP. Fusing the hormone-binding domain of the glucocorticoid receptor (GR) to the GAL4 DNA-binding domain (GAL4) and the transactivation domain of the *Herpes simplex* virus VP16 protein (VP16) renders the activity of GAL4VP16GR dependent on dexamethasone (Dex). The half-life of GAL4 alone or the GAL4VP16 fusion protein has been previously shown to be only ~60 minutes being favorably much less stable than GFP. The GAL4/UAS induction loop also potentiates amplification, and thereby increases the sensitivity of the reporter. Other binary inducible systems have demonstrated their effectiveness in transgenic *Xenopus* embryos, including a system controlled by the progesterone analogue RU-486 binding a mutant progesterone receptor *(25, 26)*, and a doxycycline controlled Tet-on based inducible system *(26)*.

3.1.5. Use of Insulators

In order to improve transgene expression, chromosomal insulators can additionally be integrated to flank the transgene. Acting as a cis-regulatory element, such insulators can overcome chromosomal position effect and putative heterochromatin induced silencing *(27)*. As such, position-independent expression of integrated

Table 24.1
Transgenic Xenopus Wnt reporters

Transgene	TOP sites	Promoter	Reporter gene	Reference
TOPTK-β-gal	3	TK	β-gal	17
p-LEF$_7$-fos-GFP	7	c-fos	GFP	16
TOPTK-iGFP	3	TK	eGFP	17
8TOP-dGFP	8	TATA	dGFP	Tran, unpublished
8TOP-DsRed	8	TATA	DsRed	Tran, unpublished

transgenes can be guaranteed. A functional insulator element in *Xenopus* comprises two copies of the 250 bp core sequence containing hypersensitive site at the 5′ end of the chicken β-globin *(28)*.

3.1.6. Overview of Known Wnt Reporter Tools in Xenopus

An overview of several faithful Wnt reporters in *Xenopus* is given in **Table 24.1**. A selection marker (e.g., a known promoter driving fluorescent gene expression; ref. *17*) can additionally be integrated into the transgene cassettes to screen for the *bona fide* transgenic embryos featuring non-mosaic transgene integration at an early stage.

3.2. Testing Functionality of Wnt Reporters in Xenopus Embryos

Having created a transgenic Wnt reporter, it is important to assess its functionality and efficacy *in vivo*. This can be achieved via *(1)* DNA microinjection into 4-cell stage *Xenopus* embryos. At mid-blastula stage (stage 8.5), embryos become dorsally enriched for β-catenin following local activation of the maternal canonical Wnt pathway. As such, the dorsal-vegetal region displays active Wnt signaling during midblastula and early gastrula stages, characterized by expression of several Wnt target genes like *siamois* and *twin (29)*. Consistently, higher transcriptional activity is to be expected in the dorsal-vegetal versus the ventral-vegetal region of the injected embryos. Since this experiment is performed at early stage, the analysis is limited to a β-gal or luciferase based reporter of which the activity is measured in a quantification assay. Also, TOP versus FOP variants can be assessed in this way. To even more specifically determine the responsiveness of a Wnt reporter, *(2)* a LiCl treatment can be performed on embryos having transgenically integrated the reporter transgene *(16, 17)*. LiCl is a known inhibitor of glycogen synthase kinase-3β, which in absence of a Wnt signal, phosphorylates β-catenin thereby targeting it for degradation *(18)*. Blocking this event results in constitutive Wnt pathway activity throughout the embryo. In parallel, FOP control constructs need to be included in the analysis and are expected to reveal no signal at all. Transgenic embryos are generated as described previously (*see* **Note 1**).

3.2.1. β-gal or Luciferase Quantification Assay on DNA-injected Embryos

1. Collect eggs from female frogs in 1x MMR, fertilize and dejelly as previously described *(18)*.

2. Transfer the embryos to 1x MMR supplemented with 6% Ficoll, and inject at four-cell stage (staged according to Nieuwkoop and Faber *[33]*), with two bilateral injections per embryo. Generally, around 250 pg DNA of the reporter plasmid is injected per blastomere. In case of β-gal reporter analysis, an empty vector containing the luciferase gene can be coinjected to normalize the values, and vice versa. It is advisable to perform at least two independent injection experiments.

3. Grow the embryos in 0.1x MMR till early gastrula stage, and lyse them in ice-cold lysis buffer (100 µL per three embryos) by pippeting in Eppendorf tubes.

4. Keep the tubes on ice for 5 minutes, centrifuge for 15 minutes at maximum speed at 4°C and transfer the supernatants to fresh tubes (about 95 µL if lysed in 100 µL).

5. For each sample, in a black 96-well microplate, mix 95 µL embryo lysate with 40 µL luciferase substrate or 100 µL β-gal substrate. In case of luciferase, proceed immediately by carrying out the measurements with a luminometer. For β-gal activity, incubate the mixture for at least 40 minutes prior to measuring. In case of coinjection with luciferase or β-gal, split the lysate in two and continue as described earlier. All measurements should at least be performed in triplicate.

3.2.2. LiCl Treatment on Transgenic Embryos

1. Collect transgenic and non-transgenic embryos (e.g., early gastrulas)

2. Transfer the embryos to a petri dish containing 0.3 M LiCl in 0.1x MMR, and incubate at 18-20°C for 10 minutes on an agitating rotator to equalize the LiCl over the embryos (*see* **Note 4**).

3. Rinse the embryos extensively in 0.1x MMR, transfer them to a clean Petri dish and let them develop for another 1 to 2 hours at 18°C.

4. Fix the embryos in MEMFA for 1 hour at room temperature or overnight at 4°C. Further process as suited, either by X-gal staining (*see* **Subheading 3.3.1.**) for β-gal detection or whole-mount ISH for any reporter gene used as described elsewhere *(34)*.

3.3. Whole-mount Spatiotemporal Analysis of Transgenic Wnt Reporter Gene Expression

The type of analysis is dependent on the reporter gene used and time point of analysis. Whole mount staining can technically be performed till stage 35, and later on the staining needs to be done on sections. Alternatively, a fluorescent reporter gene can be used that allows *in vivo* real-time analysis during later development.

3.3.1. Whole-mount β-gal Staining

As the use of β-gal reporter activity is generally preferred for early stages, the whole mount procedure is described using X-gal as a substrate (*see* **Note 5**). For later stages (past stage 35), β-gal activity can be determined on cryosections, which is more labor intensive (*see* **Note 6**). Whole mount β-gal stained embryos can afterwards be sectioned for more profound histological analysis (*see* **Subheading 3.4.1.**).

1. Collect the embryos at the appropriate developmental stage, remove the vitelline membrane if needed (advisable for neurula and tailbud) and fix them in MEMFA for 1 hour at room temperature, while vials are rocking end over end (*see* **Notes 7** and **8**). Care needs to be taken that the fixation doesn't last longer or β-gal activity can be impaired (*see* **Note 9**).

2. Rinse the embryos three times for 5 minutes in PTW.

3. Replace the PTW by the X-gal staining solution. Incubate the embryos in the dark at 30°C (or 37°C, resulting in faster staining), observe in between and develop the color reaction until the desired intensity is reached. This generally takes about 15 minutes up to several hours, depending on the strength of the signal, incubation temperature and localization of the signal. It usually does not help to stain the embryos overnight, as this can vastly increase the background. If done, perform the incubation at lower temperature, e.g., room temperature.

4. After staining is completed, rinse the embryos three times for 5 minutes in PTW.

5. Refix the embryos in MEMFA for 15–20 minutes. This will help to stabilize the stain and can be prolonged to 1 hour.

6. Dehydrate the embryos in 100% ethanol, rinse twice for 5 minutes, replace with fresh 100% ethanol and store indefinitely at –20°C (*see* **Note 10**).

7. If desired to bleach the pigment, transfer the embryos into Mayor's bleaching solution overnight on a rotator at room temperature or for 1 hour under UV light.

8. Dehydrate the embryos by rinsing twice in 100% ethanol and store at –20°C.

9. In case of deep inner staining, clear the embryos by transferring them into 2:1 BB/BA. This clearing is reversible and embryos can afterwards be transferred back to 100% ethanol and stored.

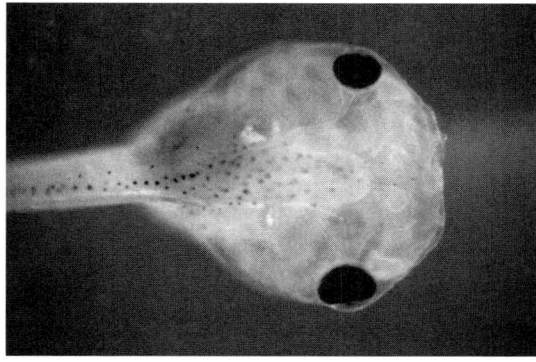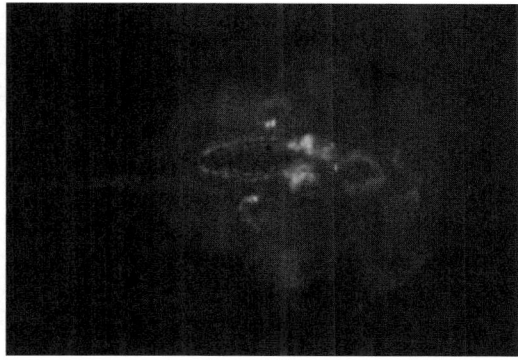

Fig. 24.1. *In vivo* tracing of Wnt signaling activity in transgenic tadpoles. Bright field (left) and fluorescent (right) images of a living *Xenopus tropicalis* tadpole with an integrated 8TOP-dGFP transgene. GFP signals are present in regions in the brain, the inner ear, and the olfactory epithelium.

3.3.2. Fluorescent In Vivo Imaging

For gastrula and early neurula embryos featuring a high degree of yolk proteins and as such a high level of autofluorescence, whole mount *in situ* hybridization for the corresponding mRNA will be required *(34)*. For older stages fluorescent images can be recorded as described later.

1. If the embryos are free-swimming tadpoles, anesthetize them in 0.1x MMR supplemented with MS222, and let them settle for a couple of minutes.

2. Take photographs with the appropriate microscope and process using the corresponding software. An example is given in **Fig. 24.1**.

3.4. Histological Analysis on Sections

For a more profound analysis of reporter gene expression, the embryos can be sectioned. In case of whole-mount stained embryos, the plastic method can be followed that yields superior tissue preservation and cell morphology *(18)*. Plastic sections can afterwards be stained for routine histological or histochemical procedures, while immunohistochemical procedures are not possible. Alternatively, sections can be made following embedding in paraffin giving lower level of morphological detail than plastic sectioning. Embryos expressing fluorescent proteins are preferentially investigated on cryosections. All three protocols are described underneath.

3.4.1. Plastic Sections

1. Fix the embryos in MEMFA for 1 hour at room temperature or overnight at 4°C with gentle rocking.

2. Rinse the embryos three times for 5 minutes in PBS.

3. Prepare some catalyzed JB-4 Infiltration Solution, and meanwhile dehydrate the fixed embryos in a series of 10 minute washes through an increasing gradient of ethanol:

PBS (30:70, 50:50, 70:30, and 95:5) ending with two final washes in 100% ethanol.

4. Substitute the ethanol for JB-4 Infiltration Solution and infiltrate the embryos in three to four changes (1 to 1.5 hours each, the last change can be extended to overnight) at 4°C with rocking.

5. Prepare the embedding solution, fill the molds with this solution (approximately 1.5 to 2 mL per mold), drop the embryo(s) into it and orient them as desired.

6. If the embryos are oriented, add some more embedding solution, place a plastic block holder on top of it and allow polymerization. Room temperature polymerization will be completed in 1 to 2 hours for small blocks, resulting in a color change of the blocks to light yellow or amber. To prevent polymerization from occurring too fast and possible overheating of the tissue (e.g., for larger blocks), the polymerization process can be slowed down by incubating at 4°C and extending the polymerization from several hours to overnight.

7. Remove the block from the mold, install on the microtome, trim and start sectioning at desired thickness. Plastic sectioning allows a cutting thickness of only 1 μm (ideal for studying cell morphology), though to visualize staining, cutting at 10 μm is more appropriate.

8. Capture the sections with forceps, transfer them onto the water bath surface to flatten and collect them on glass microscope slides (*see* **Note 11**).

9. Dry the sections for 1 to 2 minutes on a hot plate. Mount with a few drops of Aquatex, put a cover slip on top and view under a light microscope.

3.4.2. Paraffin Sections

1. Fix the embryos in 4% PFA/PBS for 1 hour at room temperature with rocking.

2. Rinse for three times for 5 minutes in PBS.

3. Dehydrate the embryos by passing through a series of 10-minute washes with increasing gradient of ethanol:PBS (30:70, 50:50, 70:30, and 95:5) ending with two final washes in 100% ethanol. It is important that all water is removed, as it would interfere with paraffin infiltration.

4. Replace 50% of the ethanol with Histoclear. After 10 minutes, substitute all of the solution with Histoclear and leave another 10 minutes. Repeat and leave overnight at room temperature. Melt overnight the required amount of paraplast in a glass beaker.

5. Replace 50% of the Histoclear with molten paraplast and keep in the 60°C heated incubation oven while rocking. After 10

minutes, remove the solution as much as possible, substitute with 100% molten paraplast and keep in the oven for 1 hour. Replace with new molten paraplast twice more and leave the embryos overnight in the final paraplast change.

6. Pour a shallow layer of molten paraplast into an embedding cassette, quickly transfer an embryo from the vial using a heated wide-bore plastic pipette and orientate quickly with a heated glass pipette as desired. This can be done under the microscope and usually in such a way that its head is facing one side of the block. Once the paraplast around the embryo(s) is hardened a little, add more molten paraplast and put aside to set overnight (*see* **Note 12**).

7. Install the cassette on the microtome. Trim the top and bottom surfaces with a scalpel until parallel and making it trapezoidal with the surface closer to the knife being wider, which will prevent curling of the sections.

8. Section at 10 μm or thicker in case of weak staining. Collect the sections from the warm water bath of the microtome onto microscope slides, let them dry and put overnight at 40°C. At this step, sections can be stored indefinitely at room temperature.

9. Deparaffinize the sections by washing twice for 2 minutes in Histoclear (*see* **Note 11**). Cover each section with a few drops of DPX mounting medium and a cover slip, and view under a light microscope.

3.4.3. Cryosections

1. In case of free-swimming tadpoles, anesthetize them in 0.1x MMR containing MS222 prior to fixation.

2. Fix the embryos in 4% PFA/PBS for 1 hour at room temperature with rocking (*see* **Notes 13** and **14**).

3. Wash the embryos for three times for 5 minutes in PBS.

4. Transfer into 30% sucrose to cryoprotect, let them settle down and further embed overnight at 4°C with rocking.

5. Transfer the embryos through a 1:1 mixture of Tissue-Tek OCT:sucrose, and then into appropriate molds filled with Tissue-Tek OCT medium (*see* **Note 14**) while positioning them in the desired orientation (e.g., depending on sagittal versus transversal sections; usually its head facing one side of the block) using a stereomicroscope and forceps. Once in position, gradually freeze the block on a dry ice/ethanol bath or liquid nitrogen layer. Blocks can be stored in a –70°C freezer prior to cutting.

6. Cool the cryostat (generally around –20 or –30°C), then place your block into it and allow equilibration to the cutting temperature.

7. Carefully mount the cryoblock onto the frozen platform of a cryostat holder with a small amount of fresh Tissue-Tek OCT. Trim the block, make cryosections of desired thickness (generally 10–20 µm; if cellular staining needs to be retrieved, cut at 10 µm thickness or less) and collect them on SuperFrost Plus microscope slides. Sections can be stored in a box with silica gel at –70°C until required (*see* **Note 13**).

8. Dry the slides at room temperature for 2 hours, rehydrate in PBS and mount with a few drops of DAPI-containing Vectashield and a cover slip. View directly with a fluorescence microscope and process the images with the corresponding software. The mounted slides are stored in the dark at 4°C.

4. Notes

1. Transgenic F_0 embryos of *Xenopus laevis* are raised essentially according to the restriction enzyme mediated integration (REMI) method of Amaya and Kroll *(30)*. For *Xenopus tropicalis*, the method described by Hirsch et al. *(31)* is followed. To evaluate transgene integration (sites and copy number), genomic DNA can be extracted from individual tadpoles, e.g., from the tail, for analysis by Southern blotting and PCR *(32)*. Transgenic reporter lines can be raised. This is preferentially done using *Xenopus tropicalis* being diploid and having a much shorter generation time compared with the pseudo-tetraploid *Xenopus laevis*. A transgenic *Xenopus tropicalis* F_0 embryo can be raised in 3–5 months till sexually mature (protocol as described by Grainger lab at http://faculty.virginia.edu/tropicalis/husbandry/TroptadcareNew. htm), and is ready to be used for matings. By preference, lines are selected with a strong and reproducible expression pattern, a single integration and a limited number of transgene copies.

2. The most commonly used dGFP is d2eGFP being a destabilized variant of eGFP. A portion of the mouse ornithine decarboxylase gene containing a PEST sequence was fused to eGFP, rendering the protein for degradation and resulting in rapid turnover *(23)*. Other dGFP variants can be found on www.clontech.com.

3. DsRed-Express is a rapidly maturing variant of *Discosoma* sp. red fluorescent protein (DsRed). It contains nine amino acid substitutions, which enhance its solubility, reduce its green emission, and accelerate its maturation *(24)*. Other DsRed variants exist, though are less favorable to be used as a reporter. Other reef coral fluorescent proteins include

cyan, yellow, and far-red fluorescent proteins (www.clon-tech.com). Of note, far-red fluorescent HcRed in our hands is much weaker and less suitable as a reporter.

4. If a hormone-inducible Wnt reporter is used (e.g., TOPTK-iGFP), a combinatorial hormone/LiCl treatment will be needed. In case of TOPTK-iGFP, embryos were pre-treated with 10 µM Dex for 2 hours to functionally induce the systems prior to LiCl treatment. Following re-exposure to Dex for another 3 hours, embryos were analyzed showing ubiquitous eGFP expression only if Dex was added *(17)*.

5. If dealt with dynamic signaling patterns (e.g., fast successive early Wnt signaling events during midblastula-gastrula stages), this staining might be shortcoming and be only a delayed reflection. To appropriately accommodate this, the more sensitive ISH technique can also be used to detect β-gal mRNA as described elsewhere *(33)*.

6. For later stages, β-gal activity can also be determined on cryosections (*see* **Subheading 3.4.3.**). Sections are fixed in 0.25% glutaraldehyde/PBS for 15 minutes, washed for 5 minutes in PBS and stained overnight at 30–37°C with X-gal staining solution in a humid chamber. Wash three times for 5 minutes in PBS, counterstain if desired (*see* **Note** 11) and mount.

7. Glass vials need to be filled with solution almost till the edge, leaving only a small air bubble that during rotation will keep the inner solution in motion. To prevent lysis of the embryos, always leave them in suspension and avoid contact with the air when removing solution and refilling. Early stage embryos are particularly fragile for this. As a good average, about 20–25 embryos can be put together in one 5-mL screw cap glass vial. Larger number of embryos will tend to clump.

8. Alternatively, embryos can be fixed for 1 hour on ice in freshly made PBS containing 2% formaldehyde, 0.2% gluteraldehyde, 0.02% NP-40, and 0.01% sodium deoxycholate *(18)*.

9. If the X-gal staining cannot be performed immediately, fix the embryos only for 30 minutes, wash and store them in 1x PBS at 4°C. Later on, refix them in MEMFA for 30 minutes prior to the staining.

10. If required, a subsequent whole mount ISH, e.g., for a specific Wnt target gene or a marker gene, can now be performed on the embryos as described elsewhere *(33)*.

11. For sections made following whole mount staining, counterstaining is usually not required. Nonetheless if wanted,

a hematoxylin/eosin counterstain can easily be performed. Sections are stained with hematoxylin (Hematoxylin's Mayer; Sigma) and eosin (0.5%, Eosin Y, Sigma) for 30 seconds each, with in between and after rinsing with water. Alternatively, counterstaining with 1% (w/v) Toluidine Blue (Sigma) can be performed for 30–60 seconds at 50–60°C. For plastic sections, counterstain before drying and processing as described. For paraffin sections, hydrate the deparaffinized sections by successive 2-minute washes with 100% ethanol, 95% ethanol and water prior to counterstain. Afterwards, dehydrate in successive washes with water, 95% ethanol, 100% ethanol, and Histoclear prior to mounting.

12. To better control the paraplast solidification, the molds can be placed onto a heated plate and moved aside after orientation of the embryos is completed. If needed, the paraplast can also be remelted.

13. When studying fluorescent proteins it is advisable to proceed with every step of the protocol as fast as possible so as to not lose any signal.

14. If the analyzed embryos are early gastrulas or younger, an alternative protocol is preferred using fish gelatin as cryoprotectant. Fix the embryos in Dent's fixative (80% methanol/20% DMSO) and rinse three times for 5 minutes in PBS. Embed in a mixture of 15% cold-water fish gelatin (Sigma) and 15% sucrose at room temperature for 24 hours. Similarly, transfer the embryos to a mold, orient, freeze on a dry ice surface and section.

Acknowledgments

TD is a postdoctoral fellow of the Research Foundation—Flanders (FWO). Research is supported by the Belgian Foundation against Cancer and the Research Foundation—Flanders (FWO).

References

1. Logan, C. Y., Nusse, R. (2004) The Wnt signaling pathway in development and disease. *Annu Rev Cell Dev Biol* 20, 781–810.

2. Hoppler, S., Kavanagh, C. L. (2007) Wnt signaling: variety at the core. *J Cell Sci* 120, 385–393.

3. Xu, Q., Wang, Y., Dabdoub, A., et al. (2004) Vascular development in the retina and inner ear: control by Norrin and Frizzled-4, a high-affinity ligand-receptor pair. *Cell* 116, 883–895.

4. Kazanskaya, O., Glinka, A., del Barco Barrantes, I., et al. (2004) R-Spondin2 is a

secreted activator of Wnt/beta-catenin signaling and is required for *Xenopus* myogenesis. *Dev Cell* 7, 525–534.

5. Kioussi, C., Briata, P., Baek, S. H., et al. (2002) Identification of a Wnt/Dvl/beta-Catenin --> Pitx2 pathway mediating cell-type-specific proliferation during development. *Cell* 111, 673–685.

6. Easwaran, V., Pishvaian, M., Salimuddin, et al. (1999) Cross-regulation of beta-catenin-LEF/TCF and retinoid signaling pathways. *Curr Biol* 9, 1415–1418.

7. Zorn, A. M., Barish, G. D., Williams, B. O., et al. (1999) Regulation of Wnt signaling by Sox proteins: XSox17 alpha/beta and XSox3 physically interact with beta-catenin. *Mol Cell* 4, 487–498.

8. Barolo, S. (2006) Transgenic Wnt/TCF pathway reporters: all you need is Lef? *Oncogene* 25, 7505–7511.

9. Korinek, V., Barker, N., Morin, P. J., et al. (1997) Constitutive transcriptional activation by a beta-catenin-Tcf complex in APC-/- colon carcinoma. *Science* 275, 1784–1787.

10. Dorsky, R. I., Sheldahl, L. C., Moon, R. T. (2002) A transgenic Lef1/beta-catenin-dependent reporter is expressed in spatially restricted domains throughout zebrafish development. *Dev Biol* 241, 229–237.

11. DasGupta, R., Fuchs, E. (1999) Multiple roles for activated LEF/TCF transcription complexes during hair follicle development and differentiation. *Development* 126, 4557–4568.

12. Staal, F. J., Meeldijk, J., Moerer, P., et al. (2001) Wnt signaling is required for thymocyte development and activates Tcf-1 mediated transcription. *Eur J Immunol* 31, 285–293.

13. Maretto, S., Cordenonsi, M., Dupont, S., et al. (2003) Mapping Wnt/beta-catenin signaling during mouse development and in colorectal tumors. *Proc Natl Acad Sci USA* 100, 3299–3304.

14. Mohamed, O. A., Clarke, H. J., Dufort, D. (2004) Beta-catenin signaling marks the prospective site of primitive streak formation in the mouse embryo. *Dev Dyn* 231, 416–424.

15. Nakaya, M. A., Biris, K., Tsukiyama, T., et al. (2005) Wnt3a links left-right determination with segmentation and anteroposterior axis elongation. *Development* 132, 5425–5436.

16. Geng, X., Xiao, L., Lin, G. F., et al. (2003) Lef/Tcf-dependent Wnt/beta-catenin signaling during *Xenopus* axis specification. *FEBS Lett* 547, 1–6.

17. Denayer, T., Van Roy, F., Vleminckx, K. (2006) *In vivo* tracing of canonical Wnt signaling in *Xenopus* tadpoles by means of an inducible transgenic reporter tool. *FEBS Lett* 580, 393–398.

18. Sive, H. L., Grainger, R. M., Harland, R. M. (eds.) (2000) *Early Development of Xenopus laevis, A Laboratory Manual.* Cold Spring Harbor Laboratory Press, NY.

19. Staal, F. J., Burgering, B. M., van de Wetering, M., et al. (1999) Tcf-1-mediated transcription in T lymphocytes: differential role for glycogen synthase kinase-3 in fibroblasts and T cells. *Int Immunol* 11, 317–323.

20. Chalfie, M., Tu, Y., Euskirchen, G., et al. (1994) Green fluorescent protein as a marker for gene expression. *Science* 263, 802–805.

21. Corish, P., Tyler-Smith, C. (1999) Attenuation of green fluorescent protein half-life in mammalian cells. *Protein Eng* 12, 1035–1040.

22. Huang, W. Y., Aramburu, J., Douglas, P. S., et al. (2000) Transgenic expression of green fluorescence protein can cause dilated cardiomyopathy. *Nat Med* 6, 482–483.

23. Li, X., Zhao, X., Fang, Y., et al. (1998) Generation of destabilized green fluorescent protein as a transcription reporter. *J Biol Chem* 273, 34970–34975.

24. Bevis, B. J., Glick, B. S. (2002) Rapidly maturing variants of the Discosoma red fluorescent protein (DsRed) *Nat Biotechnol* 20, 83–87.

25. Chae, J., Zimmerman, L. B., Grainger, R. M. (2002) Inducible control of tissue-specific transgene expression in *Xenopus tropicalis* transgenic lines. *Mech Dev* 117, 235–241.

26. Das, B., Brown, D. D. (2004) Controlling transgene expression to study *Xenopus laevis* metamorphosis. *Proc Natl Acad Sci USA* 101, 4839–4842.

27. Recillas-Targa, F., Pikaart, M. J., Burgess-Beusse, B., et al. (2002) Position-effect protection and enhancer blocking by the chicken beta-globin insulator are separable activities. *Proc Natl Acad Sci USA* 99, 6883–6888.

28. Sekkali, B., Tran, H. T., Crabbe, E., et al. (2008) Chicken beta-globin insulator overcomes variegation of transgenes in *Xenopus* embryos. *Faseb J* 22, 2534–2540.

29. De Robertis, E. M., Larrain, J., Oelgeschlager, M., et al. (2000) The establishment of Spemann's organizer and patterning of the vertebrate embryo. *Nat Rev Genet* 1, 171–181.

30. Amaya, E., Kroll, K. L. (1999) A method for generating transgenic frog embryos in (Sharpe, P., Mason, I., eds.) *Molecular Embryology: Methods and Protocols*, Humana, Totowa, NJ, pp. 393–414.

31. Hirsch, N., Zimmerman, L. B., Gray, J., et al. (2002) *Xenopus tropicalis* transgenic lines and their use in the study of embryonic induction. *Dev Dyn* 225, 522–535.

32. Jansen, E. J., Holling, T. M., van Herp, F., et al. (2002) Transgene-driven protein expression specific to the intermediate pituitary melanotrope cells of *Xenopus laevis*. *FEBS Lett* 516, 201–207.

33. Nieuwkoop, P. D., Faber, J. (eds.) (1994) *Normal table of Xenopus laevis (Daudin): a systematical and chronological survey of development from the fertilized egg till the end of metamorphosis.* Garland Science, NY.

34. Broadbent, J., Read, E. M. (1999) Wholemount in situ hybridization of *Xenopus* and zebrafish embryos in (Guille, M., ed.) *Molecular Methods in Developmental Biology, Xenopus and Zebrafish*, Humana, Totowa, NJ, pp. 57–67.

Chapter 25

Gain-of-Function and Loss-of-Function Strategies in *Xenopus*

Danielle L. Lavery and Stefan Hoppler

Abstract

Xenopus embryos are particularly suited for functional experiments to investigate vertebrate embryonic development. Due to the large size of embryos and their development outside of the mother organism, they are very accessible, easy to manipulate, and allow for immediate observation of developmental phenotypes. Powerful methods have been established for both gain- and loss-of-function strategies, which build on these inherent advantages. This chapter describes injection methods used to overexpress gene products and inhibit gene expression as well as pharmacological approaches to manipulate Wnt signaling in *Xenopus* embryos.

Key words: *Xenopus*, microinjection, Morpholino, small molecule inhibitors, BIO.

1. Introduction

1.1. Gain-of-function Strategies in Xenopus

Experimental overexpression or ectopic expression of Wnt signaling components provides information about the potential function of the endogenous gene; it shows what the gene product is capable of and whether it is sufficient in a given context to influence Wnt signaling pathways or embryonic development. Such overexpression experiments can be efficiently achieved in *Xenopus* embryos by mRNA injections and by injection of DNA constructs containing strong promoters driving the expression of the desired gene product. Wnt signaling components act as activators (agonists) or as regulators (antagonists) of Wnt signaling pathways. Depending on the gene and its normal function in the pathway, the result of gain-of-function experiments can therefore activate (e.g., Wnt *(1)* or β-catenin *(2)*) or inhibit Wnt signaling (e.g., axin *(3)*).

Elizabeth Vincan (ed.), *Wnt Signaling, Volume II: Pathway Models, vol. 469*
© 2008 Humana Press, a part of Springer Science+Business Media, New York, NY
Book doi: 10.1007/978-1-60327-469-2

1.1.1. Gain-of-function Studies with Injected mRNA

Capped mRNAs can be synthesized (*see* **Subsection 2.2.1**) and injected into *Xenopus* embryos (*see* **Subsection 3.2**). These capped mRNAs will be translated into protein in the developing embryo and provide insight into how this protein may function by observing the resulting developmental phenotype.

These injections can be targeted to specific blastomeres that will give rise to specific tissues of the embryo (refs. *4–6*, see fate map on Xenbase at www.xenbase.org/anatomy/static/fate-tissue. jsp). Often tracer molecules such as dextran dyes (*see* **Subsection 2.2.4** and ref. *7*) or mRNA encoding GFP *(8)* or β-galactosidase *(9)* are co-injected to monitor correct tissue targeting of the mRNA of interest.

Overexpression of Wnt signaling components has lead to discoveries involving the early developmental roles of these signaling molecules. For example, mRNA injection of Wnt ligands and β-catenin first lead to the discovery that this pathway functions in early axis determination and have lead to the axis duplication assay by which Wnt ligands can be tested for their ability to activate the canonical Wnt signaling pathway *(1,2*; *see* **Note 1**; *see* also **Chapter 29** in **Volume 2**). Overexpression of the molecule GBP, a negative regulator of GSK3 can also lead to activation of canonical Wnt signaling *(10)*. However, gain-of-function experiments can also lead to inhibition of the Wnt signaling pathway by overexpression of negative regulators such as axin or GSK3 *(3)*, groucho *(11)*, or members of the secreted frizzled related protein (sFRP) family *(12)*.

There have been several mutant proteins created for Wnt signaling components, which are constitutively active, including a point mutated β-catenin (ref. *32*; *see also* **Table 25.1**) that is insensitive to phosphorylation and subsequent degradation via the ubiquitylation pathway, as well as, a constitutively active LRP receptor, which is activated even in the absence of a Wnt ligand (ref. *13*; *see also* **Table 25.1**). There are also several different dishevelled mutants that selectively activate either the canonical or non-canonical Wnt signaling pathways (**Table 25.1**; *see also* refs. *14* and *15)*.

1.1.2. Gain-of-function Studies with Injected DNA

mRNA injection is useful for studying very early events. However, some Wnt signaling components such as Wnt8 are not active until after mid-blastula transition, after the start of zygotic transcription (e.g., ref. *16)*. In order to perform appropriate overexpression experiments of these types of molecules, researchers often inject plasmid DNA (e.g., CSKA-Xwnt8 *(17)*; *see* **Subsections 2.2.2** and **3.2**). In the developing embryo, generally no transcription occurs until after MBT (there are a few rare exceptions [i.e., Xnr5], *see* ref. *18)* therefore, no mRNA will be made and proteins produced at these early stages are translated from existing mRNA of maternal origin. When exogenous mRNA is injected it

Table 25.1
Mutated Wnt signaling components used.

Wild-type protein	Native function	Mutated construct	Suggested mechanism of mutated construct	Reference
β-catenin	Activated by binding of a Wnt ligand to Frz and LRP co-receptors, whereby it translocates to the nucleus and relieves repression of LEF/TCF transcription factors by the co-repressors groucho/legless and activates transcription of Wnt responsive target genes.	β-catenin point mutated	Constitutively activates canonical Wnt signaling because phosphorylation sites that normally led to its degradation are mutated; this results in a stable and nuclear β-catenin.	32
Wnt	Secreted glycoprotein which binds to the frizzled family of proteins andLRP5/6 co-receptors (in the case of canonical signaling) to activate the Wnt signaling cascade (both canonical and non-canonical)	dnWnt	Dominant-negative inhibitor of both canonical Wnt signaling (see dnWnt8 (*16*)) and also non-canonical Wnt signaling (see dnWnt11 (*33*)).	*16, 33*
Frizzled	Receptor for the Wnt family of secreted glyco-proteins. Functions in both canonical and non-canonical Wnt signaling.	Frz-CRD	Acts as a dominant negative inhibitor of Wnt signal-ing by binding Wnt ligands and preventing them from binding to endogenous frizzled receptors.	*19*
Glycogen synthase kinase-3	Part of the destruction complex, phosphorylates β-catenin, targeting it for ubiquitination/degra-dation.	dnGSK3 (kinase dead GSK3)	Leads to enrichment of β-catenin levels and subse-quent activation of canonical Wnt signaling by inhibiting endogenous GSK-3 activity.	*32*

(continued)

Table 25.1
(continued)

Wild-type protein	Native function	Mutated construct	Suggested mechanism of mutated construct	Reference
LRP6	Acts as a co-receptor for the frizzled family of proteins in binding Wnt ligands. Required only for canonical and not for non-canonical Wnt signaling.	LRP6ΔC	Lacks the C-terminal intracellular domain of LRP6; acts as a dominant inhibitor of Canonical Wnt signaling (presumably through sequestering Wnt proteins).	34
		LRP6m5	Alanine has been substituted in the invariant S/T residue in all five PPP(S/T)P motifs, which are required for LRP6 mediated activation of canonical Wnt signaling. Acts as a dominant inhibitor of canonical Wnt signaling.	13
		LRP6ΔN	Lacks the N-terminal extracellular domain of LRP6 and leads to constitutive activation of canonical Wnt signaling.	13
LEF/TCF	Functions as a transcriptional acitvator when β-catenin associates with it on the promoters of Wnt responsive genes.	NTCF3	Contains just the N-terminal β-catenin binding domain and Inhibits canonical Wnt signaling by binding to β-catenin and preventing it from binding to endogenous LEF/TCF transcription factors on the promoters of Wnt-responsive genes.	31, 35
		ΔNTCF3; ΔNLEF1	These molecules can still bind to the promoter DNA of Wnt responsive genes but lack the N-terminal β-catenin binding domain and therefore inhibit canonical Wnt signaling by constitutively repressing expression of Wnt responsive genes.	31, 35

Groucho	Functions as a co-repressor when complexed with LEF/TCF transcription factors on the promoters of Wnt target genes.	dnGroucho	Activates canonical Wnt signaling by inhibiting endogenous groucho function.	11
Dishevelled	It is downstream of Wnt binding to Frz and is required for both canonical and non-canonical Wnt signaling. Its exact role in these signaling pathways is net yet well understood.	Xdd1 = XdshD4	Dominant negative dishevelled, this molecules lacks the internal PDZ domain inhibits both canonical and non-canonical Wnt signaling.	14
		XdshΔDIX	The DIX domain is required for activation of the canonical Wnt signaling pathway, overexpression of this dishevelled mutant lacking the DIX domain leads to selective activation of non-canonical Wnt signaling.	15
		XdshDIX	The DIX domain of dishevelled binds β-catenin and if overexpressed sequesters it away from the nucleus thereby inhibiting canonical Wnt signaling.	15
		XdshΔPDZ	The PDZ domain is required for activation of the non-canonical Wnt signaling pathway. Overexpression of this dishevelled mutant lacking the PDZ mutation leads to selective activation of canonical Wnt signaling and inhibition of non-canonical Wnt signaling.	15

can be immediately translated into protein since transcription is not needed. However, injecting plasmid DNA, which will first require transcription of the mRNA of interest before it can be made into a protein, delays protein synthesis until after mid-blastula when zygotic transcription begins. As with mRNA injection, these can be targeted to specific blastomeres and co-injected with tracer molecules. However, there is a limit as to the amount of DNA that can be injected because too much ectopic DNA will cause apoptosis. We describe in this chapter injection procedures, however, these same constructs can also be used to create transgenic embryos (c.f., **Chapter 27** in **Volume 2**).

1.2. Loss-of-function Strategies in Xenopus

Experimentally inhibiting protein function or endogenous gene expression of a Wnt signaling component, informs about the normal need or requirement for that molecule in normal embryonic development or in normal Wnt signal transduction. Because Wnt signaling components include both activators and regulators (inhibitors) of signaling pathways, loss-of-function experiments may, depending on the molecule, result not only in insufficient but also excessive Wnt signaling. For instance, inhibition of GSK3, which is a negative regulator of canonical Wnt signaling, results in up-regulation of this pathway *(3)*.

1.2.1. Loss-of-function Studies Using Dominant-negative Constructs

The use of dominant-negative molecules to inhibit endogenous gene function has been particularly successful in *Xenopus*, because of the ease with which these reagents can be expressed in *Xenopus* embryos. There have been many dominant-negative constructs designed for different components of the Wnt signaling pathway, which result in inhibition of endogenous gene function (*see* **Table 25.1**). Several constructs which are mutated versions of Wnt activators such as dnWnt ligands, LRP6ΔC, LRP6m5, Frz-CRD, ΔNTCF3, and NTCF3 result in inhibition of the canonical Wnt signaling pathway; whereas overexpression of mutant constructs for negative regulators of canonical Wnt signaling such as dnGroucho, dnGSK3 (i.e., kinase dead GSK3) all result in up-regulation of the canonical Wnt signaling pathway. The dnWnt and Frz-CRD constructs are capable of inhibiting both the canonical and non-canonical Wnt-signaling pathways. There is one dishevelled mutant construct, XdshΔPDZ, which is selectively activating the canonical Wnt signaling pathway while at the same time, inhibiting non-canonical Wnt signaling *(15)* and another Xdd1, which inhibits both the canonical and non-canonical Wnt signaling pathways at the same time *(14)*. It is important to keep in mind that the inhibitory effect of many of these constructs, including dnWnt ligands *(16)* or Frz-CRD constructs, are likely not to be specific for the product of only one gene and instead function as inhibitors of Wnt signaling in general *(19)*. All of these constructs can be delivered as mRNA (c.f., **Subsection 1.1.1**), plasmid DNA (**Subsection 1.1.2**) or from inducible transgenes in transgenic *Xenopus* embryos (**Chapter 27** in **Volume 2**). The method and timing of delivery of dominant-negative molecules will affect the phenotypic outcome. Thus, the embryonic stage and the tissue of interest will determine your choice of appropriate method for delivery of a dominant-negative molecule.

1.2.2. Loss-of-function Studies Using Morpholino (MO) Oligonucleotides

It is currently not possible to create complete targeted knockout *Xenopus* embryos as in mouse; and while overexpression of dominant-negative constructs can be effective at inhibiting wild-type function, they can also be relatively non-specific as in the case of dnWnt ligands or Frz-CRD constructs. Instead, a gene knockdown strategy has been devised involving anti-sense morpholino (MO) oligonucleotides (ref. *20*; *see* **Subsection 2.2.3**), which can be directly injected into early *Xenopus* embryos (*see* **Subsection 3.2**). MOs are modified oligonucleotides composed of nucleotides attached to morpholine rings and connected by non-ionic phosphorodiamidate intersubunit linkages. They are generally between 18 and 23 nucleotides in length, are not readily degraded in the embryo, have very low toxicity associated with them in comparison to regular anti-sense oligonucleotides and work by a steric block mechanism whereby they bind to their complementary sense sequence in the endogenous transcript. The company Genetools, LLC manufactures and supplies MO and also offer a custom design service, whereby if you provide them with the cDNA sequence of your target gene, they will design specific MOs for your gene of interest (*see* their Web site for more information visit www.gene-tools.com). MOs can be designed to bind just before or to the start translation site, in order to inhibit translation initiation of your mRNA of choice. MOs can also be designed to bind across intron/exon boundaries, in order to interfere with pre-mRNA splicing and thereby inhibiting proper splicing and gene expression of your gene of interest. However, for the design of splice MOs, you must know the genomic sequence. Splice MOs do have an advantage over MOs designed to inhibit translation initiation because RT-PCR can be used to monitor effective knockdown of the gene of interest since the targeted mRNA will not be able to be processed correctly. With the translation initiation MOs, effective knock down must be monitored with an antibody detecting endogenous protein expression or by *in vitro* translation assays to test for inhibition of protein synthesis. Genetools also sell a standard control MO, as well as, custom control MOs, which may be designed as inverted to your MO sequence or as complimentary to your MO sequence. Using two or more non-overlapping MOs targeting the same gene product is a further useful control to monitor development of similar phenotypes and rule out off-target effects of individual MOs.

1.3. Pharmacological Inhibitors of Wnt Signaling Components

Lithium has long been known to be an inhibitor of GSK3, a negative regulator of canonical Wnt signaling, and has been used for many years to activate canonical Wnt signaling *(21)*. However, lithium also has several side effects and can interfere with other signaling pathways *(22)*. Recently a small molecule, cell-permeable BIO (6-bromoindirubin-3'-oxime [Calbiochem]) has been shown to inhibit GSK3 very specifically *(23, 24)*. Treating embryos with BIO results in stabilized β-catenin levels and activated canonical Wnt signaling. The small molecule MeBIO (1-methyl-6-bromoindirubin-3'-oxime [Calbiochem]) has essentially the same structure as BIO, but has a methyl substitution at the N-1 position, which makes it unable to inhibit GSK3 activity *(23)*. Therefore, MeBIO is an excellent control molecule.

Another molecule was discovered in small molecule screens, which leads to activation of Wnt signaling, called Wnt Agonist (ref. *25; see also* Calbiochem). It activates Wnt signaling through a yet undefined mechanism but is not believed to act via inhibition of GSK-3.

CK1-7 [N-(2-aminoethyl)-5-chloro-isoquinoline-8-sulfonamide (Alexis Biochemicals)] is a specific small molecule inhibitor of casein kinase 1 (CK1), a positive regulator of canonical Wnt signaling *(26)*. Treatment with CK1-7 results in inhibition of canonical Wnt signaling *(27, 28)*.

Several Wnt ligands, which are capable of activating the non-canonical Wnt signaling pathway, for example Wnt4 *(29)* and Wnt11 *(30)*, have been shown to do so by activating JNK signaling. Therefore, the JNK inhibitor SP6000125 [Anthra(1,9-cd)pyrazol-6(2H)-one; 1,9-Pyrasoloanthrone (US Biological)] has been utilized to inhibit the non-canonical Wnt signaling pathway *(29)*.

Xenopus experiments involving pharmacological small molecule reagents (**Subsection 2.3**) are similar to RNA, DNA, or MO injection experiments described in **Subsection 3.2** but usually do not involve injection of the reagent (*see* **Subsection 3.3**)

2. Materials

2.1. Solutions

1. Marc's modified ringers (MMR) solution (10X stock solution): 20 mM $CaCl_2$, 50 mM HEPES, 20 mM KCl, 10 mM $MgCl_2$, 1 M of NaCl, adjust pH to 7.5 with NaOH, autoclave, and store at 4°C for up to several months.

2. MMR solution (0.1X working solution): dilute 10X stock solution 100 fold with distilled ddH_2O and store at room temperature for up to several weeks.

3. 3% Ficoll solution: add 3% (w/v) of Ficoll to 0.1X MMR, stir until dissolved, keep covered and store at 4°C for up to 2 weeks.

4. 3-(*N*-MO)-propanesulfonic acid (MOPs) solution (10X stock solution): 200 mM MOPs, 50 mM sodium acetate, 10 mM EDTA adjust pH to 7.0. Treat with DEPC 1 mL/L place at 37°C overnight, autoclave, wrap in foil to protect from light and store at 4°C for up to 6 months.

5. 1X MOPS buffer: dilute 10X MOPs stock 1:10 with DEPC treated MilliQ ddH$_2$O, made up fresh when needed.

2.2. Reagents

2.2.1. Synthesis of Capped RNA for mRNA Injections

The mMessage mMachine Kit (SP6, T7 or T3) from Ambion is very good for generation of capped mRNA for injection into *Xenopus* embryos. Linearize the plasmid DNA encoding the gene of interest with a restriction enzyme that cuts just 3′ of both the termination codon and the PolyA sequence. The mRNA is transcribed with the RNA polymerase corresponding to the RNA phage promoter (SP6, T7 or T3) 5′ from the start ATG at the beginning of the coding region of the gene of interest. The resulting RNA concentration can be measured using a spectrophotometer. The capped RNA can be further analyzed for possible degradation and to ensure that the correct size template is generated, by running 1 μL on a 1% agarose MOPS gel. Prepare your samples as follows: 1 μL of each capped RNA mixed with 9 μL of the RNA loading buffer supplied in the mMessage mMachine kit, and then heated at 95°C for 3 minutes before running on a 1% agarose MOPS gel containing ethidium bromide. RNA bands can be visualized using a UV light box and documented photographically.

2.2.2. DNA Constructs for Injections

DNA constructs to be used for overexpressing Wnt signaling components or expressing dominant-negative molecules should contain a promoter that is active in *Xenopus* embryos to drive transcription and optimal translation initiation and polyadenylation sites to ensure subsequent optimal protein production. Previously successfully used DNA constructs include pCSKA *(17)* and CS2+ (http://sitemaker.umich.edu/dlturner.vectors/home) and other CMV promoter containing plasmids. Such DNA plasmids are injected as circular plasmids (no need to linearise them, unlike for the transgenic procedure, *see* **Chapter 27** in **Volume 2**).Generally, it is best to keep the amount of DNA injected under 100 pg/blastomere to prevent artifacts, which are usually detectable as apoptosis of groups of cells just before initiation of gastrulation. Injection of the same amount of a DNA construct lacking a cDNA insert is a good control to monitor such artifacts.

2.2.3. MO Oligonucleotides for Microinjection

Design and purchase MOs as described in **Subsection 1.2.2** Lyophilized MO should be resuspended according to manufacturer's instructions (*see* **Note 2**). Dilute MO and control MO to desired concentration (usually between 1 and 60 ng/μL). Follow the methods for RNA injection as described in **Subsection 3.2**, but ensure you heat the morpholino at 65°C for 3 minutes and place briefly on ice before loading the needle (*see* **Note 3**).

2.2.4. Tracer Dye

Molecular Probes supplies several Dextran dyes that are conjugated to a number of different fluorophores (*see* Molecular Probes section of the Invitrogen Web site; *see also* ref. *7*). Dextrans are hydrophilic polysaccharides, which have high molecular weight, good water solubility, low toxicity, and are relatively inert and resistant to cleavage by most endogenous cellular glycosidases and so are relatively stable when injected into embryos. These are generally used at concentrations similar to what one would use for morpholino injections, (i.e., around 10–20 ng/10 nL injection) and can be detected using a fluorescent microscope with appropriate filters.

2.3. Small Molecule Inhibitors

The concentrations listed below are what we have found to work for our applications; these will have to be optimized for each application (*see* **Subsection 3.3.**).

1. LiCl (use at different concentrations starting around 0.3 M for 5–30 minutes) Use NaCl as a control, however, we have noticed that NaCl at the same concentration is often causing more unspecific damage to embryo development than LiCl *(31)*.

2. BIO (6-bromoindirubin-3′-oxime [Calbiochem]) and for a control MeBIO (1-methyl-6-bromoindirubin-3′-oxime [Calbiochem]). Use at a concentration of 10 μM.

3. Wnt agonist (2-amino-4-(3,4-(methylenedioxy) benzyl-amino]-6-(3-methoxyphenyl)pyrimidine) (Calbiochem). Use at a concentration of 5–10 μM.

4. CK1-7   (N-(2-aminoethyl)-5-chloro-isoquinoline-8-sulfonamide) (Alexis Biochemicals) Use at a concentration of 2 mM.

5. JNK inhibitor SP6000125 (Anthra(1,9-cd)pyrazol-6 (2H)-one; 1,9-Pyrasoloanthrone (US Biological)) use at a concentration of 12 μM.

2.4. Injection Needles and Equipment

1. Sutter Instrument Co. P30 vertical needle puller with a P30T30 filament or equivalent.

2. Harvard Medical Instruments PPL-100 microinjector.

3. Micromanipulator.

4. Light microscope.

5. Drummund glass microcaps (size: 30 µL).

6. Injection tray of choice.

3. Methods

3.1. Calibration and Needle Loading for mRNA, DNA or Morpholino Injection

It is important to use appropriately sized needles when injecting mRNA, plasmids or MOs into *Xenopus* blastomeres to ensure survival of the manipulated embryos. Needles that work best are very sharp, but relatively short to prevent breakage of the needle. There are several different types of needle pullers available; some are vertical while others are horizontal. The conditions described here are for a Sutter Instrument Co. P30 vertical needle puller with a P30T30 filament. Drummund glass microcaps (size: 30 µL) are very good needles as these glass capillaries are very thin and when pulled to an appropriate size, leave only a very small hole when injected into *Xenopus* blastomeres. The settings which we have been found to work well for making RNA needles is to set the filament to heat: 750 and pull: 850 pull distance 5mm. This however is just a guideline and needs to be retested and adjusted every time a filament is replaced. Therefore you may find different settings work better for you.

1. Insert the pulled needle into the micromanipulator attached to the light microscope and watch as you carefully clip off the very tip of the needle with fine forceps.

2. Calibrate the needles by calculating the number of injections required to eject 1 µL (i.e., 1000 nL) of water (ddH$_2$O) with the injection time set to about 1 second (i.e., 990 ms) on the Harvard Medical Instruments PPL-100 microinjector; for an appropriately sized needle it will take between 8 and 16 injections (i.e., about 8000–16000 ms) to expel completely the 1 µL (= 1000 ng) of water. Discard needles that take considerably longer or considerably shorter to expel the water. Adjust the injection time for your individual needle to result in single injections of 10 nL. For example, this calculates an injection time of 80 ms for an injection volume of 10 nL/injection with a needle that took 8 injections to expel the 1 µL water with (i.e., with this exemplary needle 1000 nL are injected in 8,000 ms, therefore 1/100 of this volume, i.e., 10 nL are injected in 8,000/100 ms, i.e., 80 ms). We do all of our injections in a 10 nL injection volume, which is approximately 1% of the total volume of a *Xenopus* oocyte, which is approximately 1 µL. Generally, we find needles with injection times in the range of 80–160 ms work very well with *Xenopus laevis* embryos.

3. Load your needle by pipeting 2–6 μL of a previously synthesized, quantified and appropriately diluted capped RNA, DNA or MO onto a piece of paraffin placed on the lower left corner of the injection tray filled with 3% ficoll/ 0.1X MMR. Lower the needle into the droplet and fill the needle by pressing the fill button on the Harvard PLI-100 microinjector. Lower the needle into the ficoll solution and push the injection button to make sure the RNA, DNA or MO are expelling correctly from the needle (this can be seen since the RNA, plasmid and MOs are made up in ddH$_2$O and are a less dense solution than the ficoll).

3.2. mRNA, MO or DNA Injections

1. Fertilize *Xenopus* embryos and de-jelly as previously described (**Chapter 21** in **Volume 2**).

2. Allow the embryos to develop to the appropriate stage (*see* **Note 1**).

3. Once the desired stage is reached, sort the embryos under the light microscope and place the regularly cleaving embryos into a Petri dish containing 3% ficoll/0.1X MMR (*see* **Note 4**).

4. Transfer the embryos to be injected into an injection tray, filled with 3% ficoll solution, and carefully place the embryos into the grooves of the tray (*see* **Note 5**). Orient the embryos by moving the tray and manipulating the needle with care taken so as not to break the needle. Once the embryo is in the desired position, drive the needle into the embryo until the tip of the needle penetrates the cell, and press the foot pedal or the inject button to deliver the RNA, DNA, or MO. Repeat this procedure until all the embryos in the tray are injected.

5. Transfer the injected embryos to a labeled Petri dish containing fresh 3% ficoll/0.1X MMR and place into an incubator at an appropriate temperature (between 14 and 24°C) overnight (*see* **Note 6**).

6. In the morning remove any dead or dying embryos and transfer the healthy embryos to a fresh Petri dish containing 0.1X MMR/100 μg/mL gentamicin and leave to develop to the desired stage and harvest appropriately depending on the intended analytic procedure (*see* **Chapter 22** in **Volume 2**).

3.3. Treatment with Small Molecule Pharmacological Reagents

Treatment of *Xenopus* embryos with small molecule pharmacological reagents is similar to that described previously for injection experiments,

1. Fertilize *Xenopus* embryos and de-jelly as previously described (**Chapter 21** in **Volume 2**).

2. Allow the embryos to develop to the desired stage (*see* **Note 7**).

3. Transfer the embryos into a new Petri dish containing the appropriate concentration of your small molecule of choice diluted in 0.1X MMR and treat for the appropriate time (*see* **Note 8**).

4. To stop treatment either fix the embryos or wash them by transferring them at least 3 x 5 minutes to new Petri dishes containing fresh 0.1X MMR without the small molecule and allow them to develop to desired stage.

4. Notes

1. If testing a Wnt ligand for its ability to activate canonical Wnt signaling, this will involve injection into the marginal zone of one ventral blastomere of a four-cell embryo (*see* also **Chapter 29** in **Volume 2**). The amount of mRNA you will need to inject to induce an ectopic axis will depend on the potency of your Wnt ligand. Generally it is a good idea to try several concentrations ranging between 5 pg and 200 pg.

2. MOs are delivered in a powder form and should be resupended in nuclease-free ddH_2O but not DEPC treated ddH_2O. Genetools have stated that the DEPC can cause degradation of MOs.

3. Genetools recommends that diluted MOs be heated for 3 minutes at 65°C and placed briefly on ice just prior to loading the needle, to prohibit secondary structure formation.

4. The reason for placing embryos that are to be injected into ficoll solution is to equilibrate them since this is the solution in which they will be injected. Ficoll is a sugar solution, which aids in the healing of the embryos after they have been injected. Some people use 6% ficoll solution to perform their injections, but if this is done it is important to remove the embryos from the ficoll solution before they reach gastrulation because this higher concentration of ficoll has been shown to adversely affect the gastrulation process. We have found that 3% ficoll is sufficient for the healing process without side effects on gastrulation.

5. Different labs use different kinds of injection trays; we prefer rectangular dishes, which have vertical grooves so that the embryos can be lined up for injection. The high wall of the groove is furthest away from the needle tip so that the wall of this groove can be used to orient the embryo. The low wall of the groove is closest to the needle to prevent breakage of the

needle. Other research laboratories glue an appropriately sized mesh to the bottom of a 3.5-cm Petri dish to immobilize the embryos during injections, while others carefully hold individual embryos with dissection forceps during injections.

6. Some research groups argue that MOs are most effective if the embryos are allowed to develop at the highest temperature possible (22–24°C for *Xenopus laevis*).

7. It is necessary to remove the vitelline membrane from embryos that are to be treated with CK1-7, as it does not appear to permeate the membrane. Use very fine steel forceps or a very gentle proteinase K treatment (5 μg/mL for a few minutes; be careful too long and it will damage the embryos). It is not necessary to remove the vitelline membrane from embryos to be treated with BIO, MeBIO, Wnt agonist, or JNK inhibitor SP6000125.

8. Make sure you have control embryos, in the case of BIO the obvious control is MeBIO, but if treating with CK1-7 or other small molecules, add the same amount of the appropriate solvent (solution which your small molecule is dissolved in) to a dish of control embryos.

References

1. Sokol, S., Christian, J. L., Moon, R. T., et al. (1991) Injected Wnt RNA induces a complete body axis in *Xenopus* embryos. *Cell* 67, 741–752.

2. Funayama, N., Fagotto, F., McCrea, P., et al. (1995) Embryonic axis induction by the armadillo repeat domain of beta-catenin: evidence for intracellular signaling. *J Cell Biol* 128, 959–968.

3. Itoh, K., Krupnik, V. E., & Sokol, S. Y. (1998) Axis determination in *Xenopus* involves biochemical interactions of axin, glycogen synthase kinase 3 and beta-catenin. *Curr Biol* 8, 591–594.

4. Moody, S. A. (1987) Fates of the blastomeres of the 32-cell-stage *Xenopus* embryo. *Dev Biol* 122, 300–319.

5. Moody, S. A. (1987) Fates of the blastomeres of the 16-cell stage *Xenopus* embryo. *Dev Biol* 119, 560–578.

6. Bauer, D. V., Huang, S., & Moody, S. A. (1994) The cleavage stage origin of Spemann's Organizer: analysis of the movements of blastomere clones before and during gastrulation in *Xenopus*. *Development* 120, 1179–1189.

7. Urban, A. E., Zhou, X., Ungos, J. M., et al. (2006) FGF is essential for both condensation and mesenchymal-epithelial transition stages of pronephric kidney tubule development. *Dev Biol* 297, 103–117.

8. Borchers, A., David, R., & Wedlich, D. (2001) *Xenopus* cadherin-11 restrains cranial neural crest migration and influences neural crest specification. *Development* 128, 3049–3060.

9. Witta, S. E., Sato, S. M. (1997) XIPOU 2 is a potential regulator of Spemann's Organizer. *Development* 124, 1179–1189.

10. Yost, C., Farr, G. H., 3rd, Pierce, S. B., et al. (1998) GBP, an inhibitor of GSK-3, is implicated in *Xenopus* development and oncogenesis. *Cell* 93, 1031–1041.

11. Roose, J., Molenaar, M., Peterson, J., et al. (1998) The *Xenopus* Wnt effector XTcf-3 interacts with Groucho-related transcriptional repressors. *Nature* 395, 608–612.

12. Wang, S., Krinks, M., Lin, K., et al. (1997) Frzb, a secreted protein expressed in the Spemann organizer, binds and inhibits Wnt-8. *Cell* 88, 757–766.

13. Tamai, K., Zeng, X., Liu, C., et al. (2004) A mechanism for Wnt coreceptor activation. *Mol Cell* 13, 149–156.

14. Sokol, S. Y. (1996) Analysis of Dishevelled signalling pathways during *Xenopus* development. *Curr Biol* 6, 1456–1467.

15. Rothbacher, U., Laurent, M. N., Deardorff, M. A., et al. (2000) Dishevelled phosphorylation, subcellular localization and multimerization regulate its role in early embryogenesis. *Embo J* 19, 1010–1022.

16. Hoppler, S., Brown, J. D., & Moon, R. T. (1996) Expression of a dominant-negative Wnt blocks induction of MyoD in *Xenopus* embryos. *Genes Dev* 10, 2805–2817.

17. Christian, J. L. & Moon, R. T. (1993) Interactions between Xwnt-8 and Spemann organizer signaling pathways generate dorsoventral pattern in the embryonic mesoderm of *Xenopus*. *Genes Dev* 7, 13–28.

18. Yang, J., Tan, C., Darken, R. S., et al. (2002) Beta-catenin/Tcf-regulated transcription prior to the midblastula transition. *Development* 129, 5743–5752.

19. Hsieh, J. C., Rattner, A., Smallwood, P. M., et al. (1999) Biochemical characterization of Wnt-frizzled interactions using a soluble, biologically active vertebrate Wnt protein. *Proc Natl Acad Sci USA* 96, 3546–3551.

20. Heasman, J., Kofron, M., & Wylie, C. (2000) Beta-catenin signaling activity dissected in the early *Xenopus* embryo: a novel antisense approach. *Dev Biol* 222, 124–134.

21. Klein, P. S. & Melton, D. A. (1996) A molecular mechanism for the effect of lithium on development. *Proc Natl Acad Sci USA* 93, 8455–9.

22. Phiel, C. J. & Klein, P. S. (2001) Molecular targets of lithium action. *Annu Rev Pharmacol Toxicol* 41, 789–813.

23. Meijer, L., Skaltsounis, A. L., Magiatis, P., et al. (2003) GSK-3-selective inhibitors derived from Tyrian purple indirubins. *Chem Biol* 10, 1255–1266.

24. Sato, N., Meijer, L., Skaltsounis, L., et al. (2004) Maintenance of pluripotency in human and mouse embryonic stem cells through activation of Wnt signaling by a pharmacological GSK-3-specific inhibitor. *Nat Med* 10, 55–63.

25. Liu, J., Wu, X., Mitchell, B., et al. (2005) A small-molecule agonist of the Wnt signaling pathway. *Angew Chem Int Ed Engl* 44, 1987–1990.

26. Gao, Z. H., Seeling, J. M., Hill, V., et al. (2002) Casein kinase I phosphorylates and destabilizes the beta-catenin degradation complex. *Proc Natl Acad Sci USA* 99, 1182–1187.

27. Lee, E., Salic, A., & Kirschner, M. W. (2001) Physiological regulation of [beta]-catenin stability by Tcf3 and CK1epsilon. *J Cell Biol* 154, 983–993.

28. Braun, M. M., Etheridge, A., Bernard, A., et al. (2003) Wnt signaling is required at distinct stages of development for the induction of the posterior forebrain. *Development* 130, 5579–5587.

29. Maurus, D., Heligon, C., Burger-Schwarzler, A., et al. (2005) Noncanonical Wnt-4 signaling and EAF2 are required for eye development in *Xenopus* laevis. *Embo J* 24, 1181–1191.

30. Pandur, P., Lasche, M., Eisenberg, L. M., et al. (2002) Wnt-11 activation of a non-canonical Wnt signalling pathway is required for cardiogenesis. *Nature* 418, 636–641.

31. Hamilton, F. S., Wheeler, G. N., & Hoppler, S. (2001) Difference in XTcf-3 dependency accounts for change in response to beta-catenin-mediated Wnt signalling in *Xenopus* blastula. *Development* 128, 2063–2073.

32. Yost, C., Torres, M., Miller, J. R., et al. (1996) The axis-inducing activity, stability, and subcellular distribution of beta-catenin is regulated in *Xenopus* embryos by glycogen synthase kinase 3. *Genes Dev* 10, 1443–1454.

33. Tada, M. & Smith, J. C. (2000) Xwnt11 is a target of *Xenopus* Brachyury: regulation of gastrulation movements via Dishevelled, but not through the canonical Wnt pathway. *Development* 127, 2227–2238.

34. Tamai, K., Semenov, M., Kato, Y., et al. (2000) LDL-receptor-related proteins in Wnt signal transduction. *Nature* 407, 530–535.

35. Molenaar, M., van de Wetering, M., Oosterwegel, M., et al. (1996) XTcf-3 transcription factor mediates beta-catenin-induced axis formation in *Xenopus* embryos. *Cell* 86, 391–399.

Chapter 26

How The Mother Can Help: Studying Maternal Wnt Signaling by Anti-sense-mediated Depletion of Maternal mRNAs and the Host Transfer Technique

Adnan Mir and Janet Heasman

Abstract

Early development in *Xenopus laevis* is controlled by maternal gene products synthesized during oogenesis. The dorsal/ventral and anterior/posterior axes are established as a result of canonical Wnt signaling activity. The functions of maternal genes in embryonic development are most effectively studied by introducing anti-sense, oligos complementary to their mRNAs into oocytes and culturing the oocytes long enough to allow for the breakdown of the target RNAs and the turnover of existing cognate proteins before fertilization. This method has been used to establish the role of Wnt signaling in *Xenopus* axis formation. Here we describe the methodology for targeting of maternal mRNAs and for successful fertilization of mRNA-depleted oocytes.

Key words: Host transfer, maternal mRNA depletion, oocytes, anti-sense, oligo.

1. Introduction

Many new technologies have emerged recently to study gene function in *Xenopus*. The most common involves the use of anti-sense morpholino oligonucleotides (oligos) injected into fertilized eggs to block the translation or processing of target mRNAs *(1)*. This is especially useful for genes that are transcribed zygotically, and not supplied maternally. Zygotic transcription begins at the 4000-cell stage (mid blastula, stage 8) and, since in this case there is no stored maternal protein, oligos can block the translation

Elizabeth Vincan (ed.), *Wnt Signaling, Volume II: Pathway Models, vol. 469*
© 2008 Humana Press, a part of Springer Science+Business Media, New York, NY
Book doi: 10.1007/978-1-60327-469-2

of nascent transcripts, resulting in effective loss of function. However, maternal mRNAs, which control the earliest stages of development, are synthesized during oogenesis and translated before mid-blastula transition, rendering injection of morpholino anti-sense oligos into fertilized eggs less effective.

Depletion of maternal RNAs from oocytes through an RNase H-mediated mechanism is possible by injection of anti-sense DNA oligos. However, although full-grown oocytes can be matured with progesterone, and can be activated after maturation, they are very difficult to fertilize *in vitro*. Instead, we have optimized the host transfer technique for oocyte fertilization whereby manipulated oocytes are implanted back into the body cavity and pass through the oviducts of the host female to be fertilized by standard *Xenopus in vitro* fertilization methods (**Fig. 26.1**). Though the technique can be used for oocytes injected with any reagents, we will concentrate here on the use of anti-sense DNA oligos to deplete mRNAs through an RNase H-mediated mechanism. We describe the collection of ovary, oligo selection and testing, oocyte injection, host transfer, and fertilization. Although a challenging technique, this method has proved invaluable in functional studies of the roles of maternal transcription factors *(2–4)*, signaling pathway components *(5–7)*, cytoskeletal elements and adhesion molecules *(8, 9)*. In particular, with regards to the maternal Wnt signaling pathway in *Xenopus*, this method has been used to study the roles of many components including Wnt11, LRP6, FRL1, exostosin1, axin, X frizzled 7, GBP, β catenin, pygopus, and XTcf1, 3 and 4 *(2, 3, 6, 7, 10–14)*.

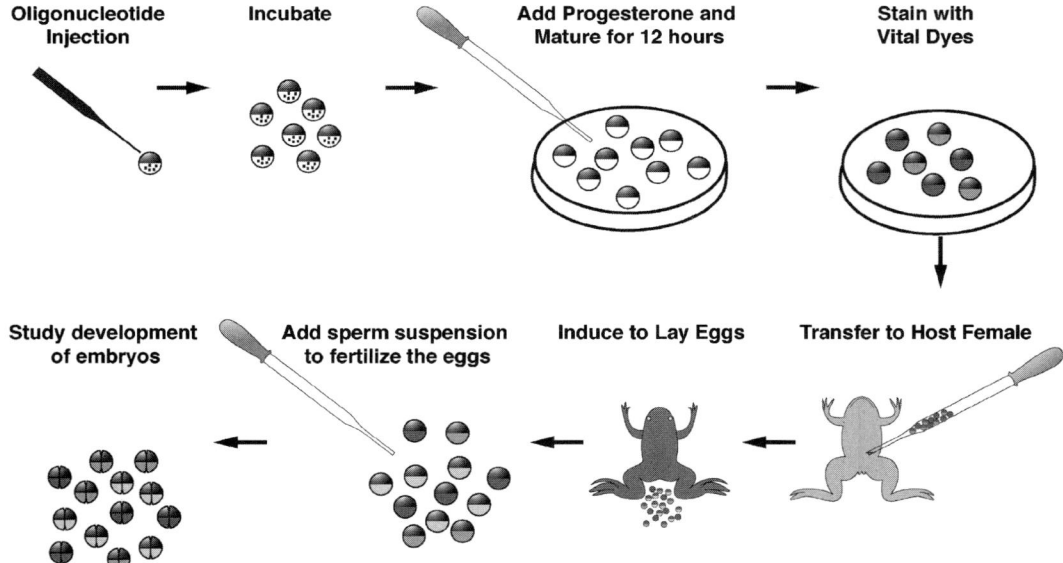

Fig. 26.1. An overview of the host transfer process. This procedure is described in detail in the text.

2. Materials

2.1. Collection of Oocytes

1. Tricaine (MS-222; Aquatic Eco-Systems), stored in 2 g aliquots at −20°C and dissolved in 1 L water for use.

2. Oocyte Culture Media (OCM), made fresh and stored at 14–16°C for up to one week. For 800 mL: 320 mL sterile water, 480 mL Liebovitz medium (L-15) with glutamine (Sigma), 0.32 g bovine serum albumen (BSA; Sigma), 4 mL penicillin/streptomycin (Fisher); Adjust pH to 7.6–7.8 with 1 M NaOH.

3. 4-0 silk suture, black braided c-17, with 12 mm 3/8 circle needle (Surgical Specialties).

2.2. Host Transfer

1. Human Chorionic Gonadotropin (Chorulon; Intervet)

2. Progesterone (Sigma), stored at 1 mM in absolute ethanol at −20°C.

3. Vital dyes. Add the indicated amount of dye to 50 mL H_2O and shake vigorously for 30 minutes (2 hours for Nile Blue). Spin in clinical centrifuge at maximum speed for 20 minutes, and make 1 mL aliquots of supernatant for use. Store at −20°C.

   1) Neutral Red (Sigma), 0.125 g/50 mL.

   2) Bismark Brown (Sigma). Use 0.5 g/50 mL.

   3) Nile Blue (Sigma). 0.05 g/50 mL.

2.3. Egg Collection and Fertilization

1. MMR. For 10X stock, 1L: 58.44 g NaCl, 1.5 g KCl, 2.94 g $CaCl_2$, 2.03 g $MgCl_2$, 35.7 g HEPES; Adjust pH to 7.6 with 5M NaOH, sterile filter, and store at 4°C.

2. High salt solution. For 10X stock, 4 L: 280 g NaCl, 7.2 g KCl, 14.1 g $CaCl_2$, 9.7 g $MgCl_2$, 17.12 g HEPES; Adjust to pH 7.6 with NaOH.

3. Dejellying solution: 2% cysteine (Sigma) in 0.1X MMR, pH 7.6–7.8.

3. Methods

3.1. Collection of Oocytes

1. Frog surgeries are performed with aseptic technique. Surgical scissors, needle holders, and forceps are sterilized using a glass bead sterilizer at 250°C for 5 minutes. The surgical bench is disinfected with 70% ethanol.

2. Ovary donor females are submerged in 0.2% Tricaine just until the frog is unresponsive to the swallowing reflex, evoked by gently stroking the animal's throat. This usually takes 10–20 minutes depending on the frog.

3. A small (<1.0 cm) incision is placed in the skin on one side of the lower abdomen using a razor blade. The incision is widened using sterile curved scissors until it is <1.0 cm in width. A similarly sized incision is then made through the underlying abdominal wall fascia and muscle, using curved scissors and being careful to lift the tissue away from the underlying organs to avoid nicking them and causing extra damage and/or infection. Using forceps, a small amount of ovary is pulled out from the abdominal cavity. It is cut with scissors, and placed in OCM for evaluation under a dissecting microscope.

4. The quality of ovary is subjectively evaluated (*see* **Note 1**). If an ovary is rejected, the female is sewn up and allowed to recover in water. The body wall is sutured first, being sure to include both muscle and overlying fascia in each stitch to ensure strength of the suture. The overlying skin is sutured separately.

5. If an ovary is accepted, enough is removed for the size of the intended experiment (*see* **Note 2**). If a female is completely depleted of ovary, she is sacrificed. Otherwise, the animal should be sewn up and allowed to recover and rest for at least three months before being reused. During this time, the scar should recover strength, and the suture should dissolve (*see* **Note 3**). If a frog is reused for ovary, the next surgery should be distant from previous ones to avoid scar tissue and adhesions. Ovary is cut into small-sized pieces and stored in OCM with 6–12 pieces per dish.

6. Oocytes are manually defolliculated in OCM (*see* **Note 4**). Full-grown, stage VI oocytes should be chosen, though there can be some variation in size. The presence of an equatorial band (less pigmented ring around the equator) is often indicative of suitable oocytes, though not all ovaries will have clear bands. Marked and damaged oocytes are discarded. Oocytes and media should be treated with aseptic technique. Pipettes and forceps should be sterilized prior to use.

7. Oocytes are tested for maturation before a full experiment is started. Several defolliculated oocytes, including the range of sizes to be used, are incubated for 5–8 hours at room temperature or 10–12 hours at 18°C in 2 µM progesterone. Maturation is assessed by germinal vesicle breakdown (GVBD), marked by the appearance of a white spot at the animal pole of the oocyte.

8. Ovary is stored at 14–16°C, and should be used within 1 or 2 days after it is removed from the animal.

3.2. Oligo Selection and Testing

1. 6–10 anti-sense oligos are selected by hand, scanning through the mRNA sequence of interest. 18-mers are designed as reverse complements of stretches of the 5' end of the coding sequence, with care to avoid poly-N stretches of 3 or more, and to include roughly equal A/T:G/C content. If no suitable regions are present at the 5' end of the sequence, stretches are chosen from within the rest of the sequence.

2. Potential oligo sequences are BLASTed back against EST databases to check that they are unique sequences.

3. Initially, unmodified DNA oligos are screened for effective depletion of the target mRNA. Two doses (usually 5 and 10 ng) of each oligo, reconstituted in water at 1–5 ng/nL, are injected into oocytes. Oligos are microfuged for 10 mins at maximum speed to remove particles that might block the injection needle. Injections and incubations are carried out in OCM. For mRNAs localized to a particular region of the oocyte, injections are aimed directly at that region. For ubiquitous mRNAs, injections are made at the equator. This is important, as the site of oligo injection can affect depletion, as well as phenotype.

4. Oocytes are cultured at 18°C for at least 24 hours to allow for RNase-H mediated mRNA degradation, and then frozen for RT-PCR. Oocytes may be matured in 2 μM progesterone 10–12 hours before freezing (*see* **Note 5**).

5. Real-time RT-PCR is used to measure the depletion of the target mRNA relative to uninjected controls (*see* **Note 6**). Two examples of successful oligo depletions are shown in **Figs**. **26.2** and **26.3**.

6. A new synthesis of the most effective oligo is then generated, modified with phosphorothioate bonds between the last three nucleotides on each end of the sequence, and tested again as before. This modification stabilizes the oligo, and often increases the level of depletion (*see* **Note 7**). A satisfactory depletion is typically below 20% of controls. If none of the oligos are effective, more are designed and tested.

3.3. Oligo Injection and Oocyte Culture

1. Anti-sense oligos are injected in OCM, being careful to discard abnormal or damaged oocytes. Typically, oocytes are injected with a volume of around 10 nL. If the oocytes are sticking together, they are gently pipetted up and down to separate them.

2. Oocytes are cultured in 60 x 15 mm Petri dishes in OCM at 18°C. Up to 150 oocytes are kept in each dish. The length

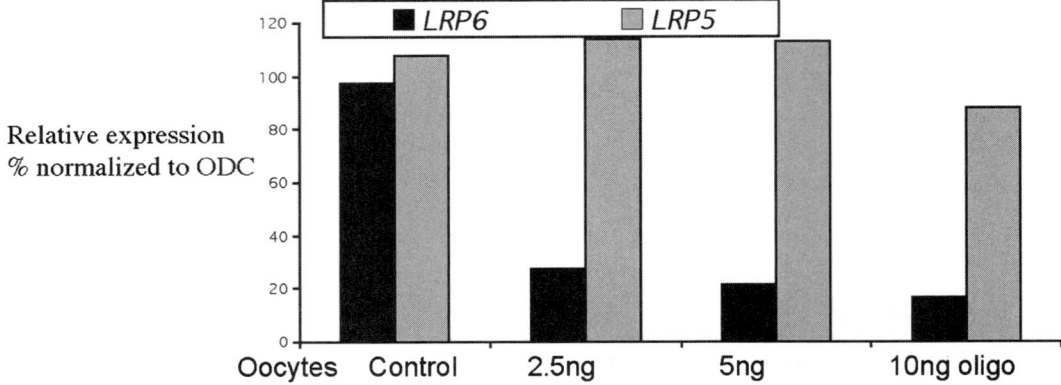

Fig. 26.2. Anti-sense oligo depletion of a maternal mRNA. An anti-sense oligo directed against LDL receptor related protein 6 (LRP6) depletes LRP6 mRNA and not LRP5 in a dose–responsive fashion.

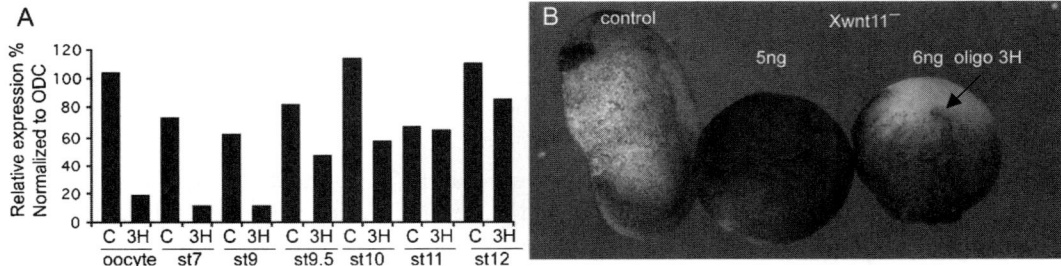

Fig. 26.3. The phentoype caused by depleting maternal Wnt11 mRNA. (**A**) An anti-sense oligo directed against maternal Wnt11 depletes Wnt11 mRNA and not zygotic Wnt11. C = control; 3H = 5ng of oligo 3H injected. (**B**) The phenotype of maternal Wnt11 depleted embryos. Figure is reproduced with permission from ref. 7.

of culture should be at least 48 hours for phosphorothioate-modified oligos, or 24 hours for unmodified oligos, to allow for oligo breakdown. However, extra time in culture to allow for protein turnover may enhance the phenotype (*see* **Note 8**).

3. Oocytes are transferred to fresh medium each day, and sick or dead oocytes are discarded.

3.4. Host Transfer

1. 12–14 hours before the host transfer surgery, host females are injected with 1,000 units of HCG, and kept at 18°C. In general, if one host female is to be used, three frogs are injected. This is to ensure that at least one female is laying eggs. Once the oocytes have been matured, the experiment cannot be interrupted. Thus, there must be a laying host or the experiment is lost.

2. 8–10 hours before the surgery, the oocytes are matured by adding progesterone to the culture medium to a final concentration of 2 μM (*see* **Note 9**).

3. After they have been allowed to mature, the oocytes are examined for GVBD. Unmatured oocytes should not be discarded, as they may still have time to mature before fertilization.

4. Oocytes are frozen from each group to assess the level of depletion by real-time RT-PCR. If an antibody is available, oocytes are frozen for Western blot analysis of depletion.

5. The oocytes are dyed by adding vital dye to the OCM into the dish containing the oocytes in the amounts shown in **Table 26.1**. Oocytes are dyed for 10 minutes at room temperature with gentle rocking. Typically, there are four colors available to differentiate four experimental groups: brown, red, blue, and purple. In the event that another color is needed, green is used. This can be difficult to tell apart from brown or blue, however, and its use is avoided when possible. **Table 26.1** lists the appropriate quantities of each solution for each color (*see* **Note 10**). The oocytes are then transferred to a 90-mm dish containing OCM without vital dye to rinse away excess vital dye. They remain in this dish until they are implanted in the host female.

6. A host female that has just begun laying eggs is selected (*see* **Note 11**) while the oocytes are being colored, and anesthetized in 0.2% Tricaine until the frog is just unresponsive.

7. A small incision is made in the skin on one side of the midline on the lower abdomen of the frog. Another incision is made through the body wall, again raised away from the underlying organs with forceps to avoid damaging them. The body wall incision should be just large enough for the tip of the glass pipette.

8. One edge of the body wall is grasped with forceps and held up, away from the abdomen. It is critical that this edge be held up until the first stitch is placed to avoid the loss of experimental oocytes.

9. Using a glass pipette, the oocytes are deposited inside the body cavity, taking care to transfer in as little media as possible. They join the host frog's own ovulated oocytes, and lie in the space between the ovary and the body wall. (NB the aim is not to put the oocytes in the ovary itself, but into the coelomic cavity). To avoid crowding the oocytes, they are spread throughout the area around the incision by holding the pipette at an angle and turning it around underneath the incision as the oocytes are expelled (*see* **Note 12**).

Table 26.1
Amounts of vital dyes used to produce different colors

Color	Treatment
Red	100 µL red for 10 minutes
Blue	100 µL blue for 10 minutes
Brown	75 µL brown for 10 minutes
Purple	100 µL red + 100 µL blue for 10 minutes
Green	75 µL brown + 75 µL blue for 10 minutes, then 75 µL blue alone for 5 minutes
Orange	75 µL brown + 75 µL red for 10 minutes, then 75 µL red alone for 5 minutes

10. The body wall and skin are sutured as during the ovarectomy, and the frog is allowed to recover from anesthesia in water (*see* **Note 13**).

3.5. Egg Collection and Fertilization

Over the subsequent few hours, the colored experimental oocytes will be carried to and down the oviducts with the host female's own eggs. Depending on preference, the further applications of the experiment, and the quality of the host female, there are two different ways to collect and fertilize eggs from the host female. The first is the standard technique of actively squeezing eggs out of the host, and the second is to allow the host to release the eggs into a high salt solution. Using the former method, multiple fertilizations are done over the course of 2 or more hours, while with the latter, the eggs are collected and fertilized all at once.

3.5.1. Standard Fertilization

1. The host animal is allowed to recover in water, and is monitored closely for release of colored eggs.

2. When the first colored egg appears, the frog is gently squeezed into a Petri dish. If no colored eggs are released within three hours after the surgery, the frog is squeezed anyway. It is important to avoid letting the female destroy the eggs as they exit through the cloaca. This can often be achieved by gently holding the female with her legs pulled back without squeezing until she begins to release the eggs on her own, and then applying gentle pulses of pressure to aid their release.

3. The eggs are fertilized with a small amount of a dense suspension of minced testis in 1X MMR. The testis should be

tested before it is used on transfer embryos to be sure that it fertilizes non-experimental eggs normally

4. After 5 minutes, the eggs are flooded with 0.1X MMR and allowed to develop as usual. After at least 1 hour, the eggs are dejellied with 2% cysteine and sorted from the host embryos.

3.5.2. High Salt Fertilization

1. The host animal is transferred to 2 L 1X high salt.

2. The frog is periodically monitored for the release of colored eggs. However, unlike water, eggs released into high salt are fertilizable for up to several hours, as the jelly coat remains dehydrated in hypertonic conditions.

3. 1.5–2 hours after the first colored egg is released, the female is gently squeezed into the high salt solution to release the remaining colored eggs (*see* **Note 14**). If few or no colored eggs have been released by 5 hours after the surgery, the frog should be squeezed.

4. The high salt eggs are collected in a Petri dish, and the colored eggs are sorted away from the host eggs into a new dish using a pipette (*see* **Note 15**). Alternatively, if there is a small number of eggs, or just a small number of uncolored, host eggs, they may all be fertilized together and sorted after fertilization.

5. Excess high salt solution is removed from the eggs by decanting and using a pipette, being careful not to lose any colored eggs, as the high salt prevents them from sticking to the dish.

6. A large piece of testis is minced directly in the dish with the eggs, and spread around dish using a plastic transfer pipette.

7. After 5 minutes, the eggs are flooded with 0.1X MMR.

8. The embryos are allowed to develop as usual (*see* **Note 16**).

4. Notes

1. There is no way to predict accurately the success of any particular ovary in a host transfer experiment. The most important aspect to examine is the ease of defolliculation without damaging the oocytes. Other qualities that may be useful in assessing the quality of ovary are even pigmentation (good), rigidity of oocytes (good if not too floppy), density of full-grown oocytes on the ovary (few full grown oocytes in a young ovary may not provide enough for the experiment, although they often fertilize well), and the presence of abnormal light or dark spots in the pigment (bad). None of these are completely predictive, however, and even the

most seemingly perfect oocytes may fail to fertilize, cleave, or develop normally. Several attempts are often necessary to achieve a successful experiment. Animal husbandry and overall colony health has proven to be quite important. Sick or dead frogs should be isolated immediately, and frogs should be kept at 15–18°C at all times.

2. The best transfer experiments may result in a high rate of fertilization (70–80%). Most often, however, not all of the oocytes defolliculated are returned, and not all of those returned develop normally. Thus, it is important to include large numbers of oocytes in each experimental group, keeping in mind that a maximum of about 800–1000 oocytes can be transferred into a host female. Typically, 50–200 oocytes are included for each experimental group, depending on the downstream applications, number of different treatments or doses, etc. The amount of ovary to be removed is approximated based on the density of full-grown oocytes, with a tendency to overestimate the amount that will be needed.

3. Silk suture does not need to be removed from the animal after it heals, and so closure of a wound is done by separately sewing up the body wall and skin. Despite the resistance of silk to reabsorption in mammals, it does dissolve in frogs.

4. As enzymatically defolliculated oocytes are not fertilizable, oocytes must be manually defolliculated. This technique is particularly laborious, and requires a great deal of practice to become proficient. The ovary is anchored with one pair of forceps, while another pair of finer forceps is used to peel the target oocyte out of the ovary, being careful not to mark the cell. It is often helpful to use a pair of forceps that are curved gently so that they only meet at the tips.

5. Maturing oocytes with progesterone prior to freezing for oligo-dT primed real-time RT-PCR is necessary if the mRNA of interest is not polyadenylated until maturation.

6. Northern blots are effective for assessing depletion, but standard block PCR is not appropriate, as it is not sufficiently sensitive. Typically, PCR primers are 3'-biased for oligo-dT primed cDNA. However, if there is a stable break-down product, it may be detected by such primers. If no suitable oligos are found, it is worth designing more 5'-biased primers to see if the message is actually being affected.

7. Phosphorothioate modification usually increases the effectiveness of an oligo. Unmodified oligos, however, can be used in host transfer experiments and can give good results. If no oligos result in good depletions of the target mRNA, combinations of partially effective oligos can be used and may increase effectiveness. As with any anti-sense techniques, rescue experiment are helpful in assessing the specificity of the oligos used.

8. The minimum culture period after oligo injection is 24 hours, to allow for oligo break down. Performing a host transfer after overnight incubation with a modified oligo may result in toxicity, if it is not completely broken down. Most depletion experiments, however, benefit from longer culture periods, as the stability of the protein is variable. Incubations of up to 4 days are sometimes necessary to achieve the most severe phenotypes. The existence of an antibody to the protein of interest can be exceptionally useful in optimizing the dose and culture length.

9. The amount of time between addition of progesterone to the oocytes and the host transfer surgery should not exceed 12–14 hours, as oocytes start to become unhealthy after they peak in maturation. Though oocytes can be matured and frogs injected with HCG at the same time, 10–12 hours before the surgery, it is better to inject the frogs 2 to 3 hours before maturing the oocytes to ensure that they are laying eggs by the time of the oocyte transfer procedure.

10. It has been observed that at low temperatures (14–16°C), brown dye can be toxic to embryos. If embryos are to be cooled overnight, the amount of brown dye should be reduced to 50 µL/10 mL of media. Also, there is some variability between different batches of dyes. If one batch seems to wash out or fade, it is acceptable to increase the amount of dye used as necessary. Blue and red dyes are non-toxic, and can be used at double the recommended concentrations. They do, however, autofluoresce and red or blue dyed embryos cannot be used for immunostaining using fluorescently labeled secondary antibodies.

11. Frogs that have fewer than three scars are chosen for injection. Very small females or very large females are avoided, as are thin or unhealthy ones. If the time from progesterone addition to the oocytes exceeds 12 hours and no females have begun laying yet, a female that looks like it will likely begin to lay (with an engorged cloaca) can be chosen as a host. Females that are laying eggs with thick jelly, or laying poor-quality eggs, should not be used.

12. Occasionally, a host female will fail to return any colored eggs. This is usually due to overcrowding of the oviducts, which become clogged with transferred oocytes. Unfortunately, there is little that can be done in this situation. Spreading the oocytes around during the surgery reduces the chances that they will clog the oviducts.

13. Anesthetized females should be monitored until they have recovered. Females that are slow to recover should be raised to the surface of the water occasionally to prevent them from drowning.

14. Though the period between the surgery and fertilization can exceed 5 hours, this should be avoided. If the female does not lay eggs 4 hours after the operation, she should be gently squeezed.

15. The pipette, glass or plastic, is first coated with OCM. The BSA prevents high salt eggs from sticking to the inside of the pipette.

16. Low yield of normally cleaving embryos is often encountered with a host transfer experiment. The most likely cause is poor ovary quality, and unfortunately, this cannot be predicted. Other possible causes are a poor host (this can be ruled out if the host embryos cleave normally), improper treatment of oocytes during culture (cells should be fed daily and kept sterile), bad high salt solution or leaving eggs in high salt too long, or bad testis. A good experiment sometimes takes several attempts, and one should be prepared to try the experiment several times to achieve success.

References

1. Heasman, J. (2002) Morpholino oligos; making sense of antisense *Dev Biol* 243, 209–14.

2. Houston, D. W., and Wylie, C. (2005) Maternal *Xenopus* Zic2 negatively regulates Nodal-related gene expression during anteroposterior patterning *Development* 132, 4845–55.

3. Standley, H. J., Destree, O., Kofron, M., Wylie, C., and Heasman, J. (2006) Maternal XTcf1 and XTcf4 have distinct roles in regulating Wnt target genes. *Dev Biol* 289, 318–28.

4. Zhang, J., Houston, D. W., King, M. L., Payne, C., Wylie, C., and Heasman, J. (1998) The role of maternal VegT in establishing the primary germ layers in *Xenopus* embryos *Cell* 94, 515–24.

5. Birsoy, B., Kofron, M., Schaible, K., Wylie, C., and Heasman, J. (2006) Vg 1 is an essential signaling molecule in *Xenopus* development. *Development* 133, 15–20.

6. Kofron, M., Klein, P., Zhang, F., Houston, D. W., Schaible, K., Wylie, C., and Heasman, J. (2001) The role of maternal axin in patterning the *Xenopus* embryo *Dev Biol* 237, 183–201.

7. Tao, Q., Yokota, C., Puck, H., Kofron, M., Birsoy, B., Yan, D., Asashima, M., Wylie, C., Lin, X., and Heasman, J. (2005) Maternal Wnt11 activates the canonical Wnt signaling pathway required for axis formation in *Xenopus* embryos *Cell* 120, 857–71.

8. Lloyd, B., Tao, Q., Lang, S., and Wylie, C. (2005) Lysophosphatidic acid signaling controls cortical actin assembly and cytoarchitecture in *Xenopus* embryos. *Development* 132, 805–16.

9. Tao, Q., Lloyd, B., Lang, S., Houston, D., Zorn, A., and Wylie, C. (2005) A novel G protein-coupled receptor, related to GPR4, is required for assembly of the cortical actin skeleton in early *Xenopus* embryos. *Development* 132, 2825–36.

10. Belenkaya, T. Y., Han, C., Standley, H. J., Lin, X., Houston, D. W., Heasman, J., and Lin, X. (2002) Pygopus encodes a nuclear protein essential for wingless/Wnt signaling *Development* 129, 4089–101.

11. Heasman, J., Crawford, A., Goldstone, K., Garner-Hamrick, P., Gumbiner, B., McCrea, P., Kintner, C., Noro, C. Y., and Wylie, C. (1994) Overexpression of cadherins, and underexpression of β catenin inhibit dorsal mesoderm induction in early *Xenopus* embryos. *Cell* 79, 791–803.

12. Kofron, M., Birsoy, B., Houston, D., Tao, Q., Wylie, C., and Heasman, J. (2007)

Wnt11/{beta}-catenin signaling in both oocytes and early embryos acts through LRP6-mediated regulation of axin *Development* 134, 503–13.

13. Sumanas, S., Strege, P., Heasman, J., and Ekker, S. C. (2000) The putative Wnt receptor *Xenopus* frizzled-7 functions upstream of beta- catenin in vertebrate dorsoventral mesoderm patterning *Development* 127, 1981–90.

14. Yost, C., Farr, G. H., 3rd, Pierce, S. B., Ferkey, D. M., Chen, M. M., and Kimelman, D. (1998) GBP, an inhibitor of GSK-3, is implicated in *Xenopus* development and oncogenesis *Cell* 93, 1031–41.

Chapter 27

Inducible Gene Expression in Transient Transgenic *Xenopus* Embryos

Grant N. Wheeler, Danielle L. Lavery, and Stefan Hoppler

Abstract

Xenopus laevis has for many years been successfully used to study Wnt signaling during early development. However, because loss of function and gain of function experiments generally involve injecting RNA, DNA, or morpholinos into early embryos (1- to 32-cell), major phenotypes are often observed before the embryo has reached later stages of development. The combined use of transgenics and a heat shock inducible system has overcome these problems and enables investigations of Wnt signaling at later stages of *Xenopus* embryonic development, including organogenesis.

Key words: *Xenopus*, transgenics, egg extract, sperm nuclei preparation, nuclear transplantation, heat shock promoter.

1. Introduction

Wnt signaling is important in multiple events during development *(1,2)*. In early *Xenopus* embryogenesis it is crucial in initial axis formation *(3)*. **Chapters 25** and **26** in **Volume 2** describe how Wnt signaling is studied during these early developmental events. However, whole-mount *in situ* analysis with Wnt signaling components in *Xenopus*, plus results from other model systems such as *Drosophila*, zebrafish, and mouse has suggested Wnt signaling is also important in many later developmental events such as mesoderm induction and patterning, gastrulation, neural crest induction and organogenesis of the heart, kidneys and many

Elizabeth Vincan (ed.), *Wnt Signaling, Volume II: Pathway Models, vol. 469*
© 2008 Humana Press, a part of Springer Science+Business Media, New York, NY
Book doi: 10.1007/978-1-60327-469-2

other tissues/organs *(4, 5)*. These later events are hard to study because manipulating Wnt signaling early causes phenotypes, which mask any later effects Wnt signaling may have on development. In this chapter we describe a method using inducible gene expression, which enables study of these later events *(6)*. In this system the gene of interest is placed downstream of a heat shock protein promoter (*Xhsp70 [7]*). This means that the gene of interest will only be expressed when the embryo is exposed to a higher temperature, which activates the heat shock promoter. While others report some success with using simple DNA injection of heat shock promoter constructs into the cytoplasm of blastula stage embryos *(8)*, we find that we obtain much more reproducible results and higher inducibility when our heat shock promoter DNA constructs are stably integrated into the genomic DNA of the chromosomes with *Xenopus* transgenic techniques *(9–12)*. We have used this method to show Wnt signaling is important for head formation, ventral mesoderm patterning and Frizzled involvement in convergent/extension movements *(6)*. Transiently over expressing the Wnt signaling pathway protein β-catenin using the heat shock protocol on pre-gastrula embryos shows its importance in patterning the mesoderm *(13)*.

The heat shock inducible system, as well as being useful to study Wnt signaling, is also applicable for study of any other gene or protein during later stages of development (**Table 27.1** contains a comprehensive list of heat shock constructs so far published). Using the heat shock system, studies have been carried out looking at regeneration of tail and limbs and at metamorphosis *(14–16)*.

2. Materials

In the following subsections we will describe solutions and materials that can be prepared well in advance followed by those solutions that need to be prepared fresh on the day of use for that particular method.

2.1. Sperm Nuclei Preparation

The following chemicals and solutions can be prepared in advance for the Sperm Nuclei preparation (**Subsection 3.1**):

1. 5 M NaCl.

2. 1 M HEPES, pH8.55, store at room temperature.

3. 1.5 M sucrose (filter sterilized), store in aliquots of 10 mL at –20°C for 6 months up to 1 year.

4. 10 mM spermidine (filter sterilized), store in aliquots of 7.5 mL at –20°C for 6 months up to one year.

Table 27.1
Heat shock-inducible *Xenopus* transgene constructs

Name	Protein under control of HS promoter	Marker promoter + protein	Reference
pHS	(Empty vector)		6
pHSGFP	GFP	HS + GFP	6
pHSXwnt8HSG	XWnt8	HS + GFP	6
pHSXfz7HSG	Xfz7	HS + GFP	6
pHSXfz7CRDHSG	Xfz7CRD	HS + GFP	Wheeler and Hoppler unpublished
pHSβ-catHSG	β-catenin	HS + GFP	13
HSP70-Alk3- gCrys-GFP	Alk3	γ-crystallin + GFP	14
HSP70-eveMsx1-gCrys-GFP	eveMsx1	γ-crystallin + GFP	14
HSP70-deltaNMsx1-gCrys-GFP	deltaNMsx1	γ-crystallin + GFP	14
HSP70-NICD-gCrys-GFP	NICD	γ-crystallin + GFP	14
HSP70-Noggin-gCrys-GFP	Noggin	γ-crystallin + GFP	14
HSP70-tBR-gCrys-GFP	tBR	γ-crystallin + GFP	14
HSP70-tBR-gCrys-RFP	tBR	γ-crystallin + RFP	14
HSP70-NotchICD-gCrys-GFP	NotchICD	γ-crystallin + GFP	14
pCGHSwG	Stromelysin 3	γ-crystallin + GFP	15
pCGHSmG	Stromelysin 3		15
pHSS1	(empty vector)		8
pHsS/EGFP	EGFP	HS + EGFP	8
pHsS1/β-catenin	β-catenin	Coinjected with pHsS/EGFP	8

5. 10 mM spermine (filter sterilized), store in aliquots of 3 mL at –20°C for 6 months up to 1 year.

6. 100 mM DTT (filter sterilized), store at –20°C for 6 months up to 1 year.

7. 0.3 M PMSF (Phenylmethylsulfonyl Fluoride) in ethanol, store in 15 μL aliquots at –20°C for 6 months up to 1 year.

8. 1xMMR (for 1 L 10 x MMR): 200 mL of 5 M NaCl, 10 mL of 2M KCl, 10 mL of 1 M MgCl$_2$, 20 mL of 1 M CaCl$_2$, 50 mL of 1 M HEPES, not sodium salt (make fresh 11.9 g for 50 mL, don't pH, Sigma H-3375), 710 mL of water, pH to 7.5 with NaOH.

9. 0.5 M EDTA, pH 7.7 (filter sterilized).

10. 10% (w/v) BSA (bovine serum albumin) in sterile ddH$_2$O, adjust pH to 7.6 with KOH, store 5 mL aliquots at –20°C.

11. 10 mg/mL digitonin (in DMSO; *see* **Note 1**).

12. Hoechst: store in 10 µL aliquots in the dark.

The following solutions should be prepared immediately before use and placed on ice:

13. 10% BSA solution (0.5 g in 5 mL ddH$_2$O).

14. 2x nuclear preparation buffer (NPB; for 30 mL): Add 14.18 mL of ddH$_2$O, 10 mL of 1.5 M sucrose (filter sterilized, final conc. in 1 × NPB is 250 mM), 900 mL of 1M HEPES, pH 8.55 (so that the pH at final concentration is 7.7, and final conc. in 1 × NPB is 15 mM), 3 mL of 10 mM spermidine (final conc. in 1 × NPB is 0.5 mM), 1.2 mL of 10 mM spermine (final conc. in 1 × NPB is 0.2 mM), 600 mL of 100 mM DTT (final conc. in 1 × NPB is 1 mM), 120 mL of 0.5 M EDTA, pH 8.0 (final conc. in 1 × NPB is 1 mM).

Use this 2x stock solution to make 30 mL of the following 1 × NPB solutions:

15. 1 × NPB + 3% BSA + protease inhibitors (for 10 mL): Add 5 mL of 2 × NPB, 3 mL of 10% BSA, 1.4 mL of 7x complete mini protease inhibitors (Roche), 10 µL of 0.3 M PMSF, 0.6 mL of ddH$_2$O.

16. 1 × NPB + 0.3% BSA (for 5 mL): Add 2.5 mL of 2x NPB, 0.15 mL of 10% BSA, 2.35 mL of ddH2O.

17. Sperm storage buffer (for 1 mL): Add 500 µL of 1x NPB containing, 30 µL of 10% BSA, 300 µL of 100% glycerol, 170 µL of dH$_2$O.

2.2. Egg Extract Preparation

These chemicals and solutions are only required for the Kroll and Amaya transgenics protocol (*see* **Subsection 3.3.1**) but not for the simplified *Xenopus* transgenics protocol (*see* **Subsection 3.3.2**).

The following solutions can be prepared in advance:

1. 2 M KCl, store at room temperature.
2. 1 M MgCl$_2$, store at room temperature.
3. 1 M CaCl$_2$ (filter sterilized), store at room temperature.
4. 1 M HEPES, pH 7.7, store at room temperature.
5. 1.5 M sucrose (filter sterilized, *see* **Subsection 2.1., item 3**).
6. 0.5 M EGTA, pH 7.7 (filter sterilized), store at room temperature.
7. 1 M CaCl$_2$ (filter sterilized), store at room temperature.
8. Energy Mix (make up 2 mL and store in 100 µL aliquots at –20°C for several months.), 150 mM creatine phosphate

(Boehringer Mannheim Biochemicals), 20 mM ATP (Invitrogen), 20 mM $MgCl_2$.

9. 250 mL 20 × Extract Buffer Salt Stock: 2 M KCl, 2 mM CaCl, 20 mM $MgCl_2$, MilliQ ddH_2O. Autoclave and store at 4°C for up to a week.

The following solutions should be prepared immediately before use:

10. 500 mL 1x extract buffer (EB): 1x extract buffer salt stock, 50 mM sucrose, 10 mM HEPES, pH 7.7, ddH_2O. Make fresh just prior to use.

11. 7.5 mL of protease inhibitor mix (dissolve 5 tablets in 7.5 mL of ddH_2O to make a 7 × stock) Roche complete mini, EDTA-free made just prior to use and kept on ice.

12. 50 mL cytostatic factor-extract buffer: 1 × EB salts, 2 mM $MgCl_2$, 10 mM HEPES, pH 7.7, 50 mM sucrose, 5 mM EGTA, 1x protease inhibitor mix, ddH_2O.

2.3. Transgenesis by Sperm Nuclear Transplantation

The following items, chemicals and solutions should be prepared in advance for the Sperm nuclear transplantation (*see* **Subsection 3.3**).

1. 2 M KCl, store at room temperature.

2. 0.5 M EGTA, pH 7.7 (filter sterilized), store at room temperature.

3. 1.5 M sucrose (filter sterilized), store in aliquots of 10 mL at −20°C for 6 months up to 1 year.

4. 10 mM spermidine (filter sterilized.), store in aliquots of 7.5 mL at −20°C for 6 months up to 1 year.

5. 10 mM spermine (filter sterilized), store in aliquots of 3 mL at −20°C for 6 months up to 1 year.

6. Sperm Dilution Buffer (SDB), for 10 mL: 1.67 mL 1.5 M sucrose (final conc. is 250 mM), 0.375 mL 2 M KCl (final conc. is 75 mM), 0.5 mL 10 mM spermidine (final conc. is 0.5 mM), 0.2 mL 10 mM spermine (final conc. is 0.2 mM spermine), 7.255 mL water, pH 7.3–7.5 with NaOH (optional: add 10 mM HEPES, pH7.6, to make it a proper buffer). Store in 1 mL aliquots at −20°C.

7. Agarose-coated dishes: Melt 350 mL 2.5% or 2% agarose in 0.1x MMR and pour into 12 15-cm Petri dishes. Before the agarose sets, place plastic square boats into it so as to cause a square-like slight depression. Once the agarose sets, the plastic square weighing boats should be removed; the dishes should be immersed with 0.1 × MMR, sealed with parafilm and kept at 4°C for up to several weeks.

8. 0.4 × MMR/6% Ficoll: 10 mL of 10x MMR, 15 g Ficoll made up to 250 mL with dH$_2$O and filter sterilized.

9. 0.1 × MMR/6% Ficoll: 2.5 mL of 10x MMR, 15 g Ficoll made up to 250 mL with dH$_2$O and filter sterilized.

10. Needles: The needles used for nuclear transfer are unlike standard needles used for DNA and RNA injection (*see* **Chapter 25** in **Volume 2**) in that a long sloping taper from the wide part of the needle to the tip is needed to control the flow rate through the very large needle tip. The width of the tip is determined by which of the two transgenic methods are being used. For the Kroll/Amaya method (**Subsection 3.3.1**) we use tips of 60–80 µm and for the simplified method (**Subsection 3.3.2**) where the nuclei are not decondensed we use needles with a diameter of 40 µm. Generally, the length of this tapered region in our transplantation needles is about 15 mm, versus 3–5 mm for most DNA or RNA injection needles. 30 µL Drummond micropipettes (Fisher, cat. #: 21-170J) are used to make needles. Needles can be made using several commercial pullers. While settings for every puller will vary, generally, use of a high heat setting and a low pulling force and velocity is desirable. The high heat will melt a greater amount of glass, which the slow pull will draw to a long, tapered needle. For our Sutter Instrument Co. P30 vertical needle puller with a P30T30 filament we use a heat setting of 990 and a pull setting of 190. The dimensions of a good transplantation needle (using a single pull) are shown in **Fig. 27.1**. Clip the needle with forceps to produce a beveled tip of the required diameter using the ocular micrometer of a dissecting microscope or a stage micrometer for measurement. It is essential the tip is the correct width or nuclei passing through will be damaged. When clipping tips, it often helps to use forceps with slightly unmatched tips and to pull outward at a 20- or 30-degree angle from the needle as the forceps contacts the needle.

11. Prepare in advance some 200 µL yellow pipette tips that have had their end cut with a razor blade diagonally approximately 5 mm from the end of the pipette. Onto this end attach a 1 cm length of Tygon tubing. These prepared tips are autoclaved and then stored.

12. Preparation of linearized DNA. Linearize plasmid DNA (5 µg) for transgenesis by restriction enzyme digest and purify using a silica based DNA purification resin such as Geneclean II kit (Q-Bio gene). Make sure the final concentration of eluted linearized DNA in dH$_2$O is 250 ng/µL (*see* **Note 3**).

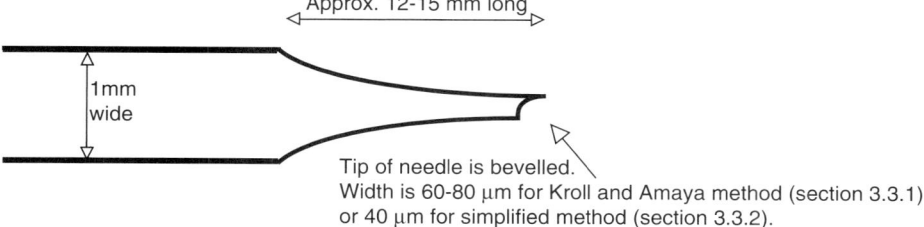

Fig. 27.1. Dimensions of transplantation needle. It is important that the transplantation needle is of the correct size; if the opening at the tip of the needle is too wide, the unfertilized eggs will be unnecessarily injured; if the opening is too narrow, the sperm nuclei will be damaged. Note that the different transgenic protocols described in this chapter (*see* **Subsections 3.3.1** and **3.3.2**) require transplantation needles of different sizes (*see* text for details).

13. Prime and induce four female frogs with PMSG and Chorulon (*see* **Chapter 21**; **Note 4** in **Volume 2**).

3. Methods

3.1. Sperm Nuclei Preparation (See Notes 2 and 5)

Ensure that all necessary solutions (*see* **Subsection 2.1.**) are ready including those needing to be freshly prepared before starting this protocol.

1. Dissect and isolate testes from one or two adult male *Xenopus* (*see* **Chapter 21**). Roll the dissected testes on a bone-dry towel to remove any attached blood vessels. Place testes in ice cold 1x MMR (keep cold by placing on a bed of ice). Using forceps clear the testes of any attached pieces of fat body and debris. Wash the testes twice in $1 \times$ MMR and then wash for 5 minutes in ice cold $1 \times$ NPB in a tilted small Petri dish. Then transfer the testes to a 35-mm Petri dish and macerate them using forceps, until no visible clumps are seen (*see* **Note 6**). Gently resuspend the macerate in 2 mL of 1x NPB by pipeting the solution up and down through a fine polished truncated Pasteur pipette with an opening of approximately 3 mm in diameter (*see* **Note 7**).

2. Filter the macerate through a 30 µm Nylon net filter (Millipore NY3004700) into a 15 round bottom polypropylene tube (Fisher 2059) by washing twice with 3 mL and 5 mL of 1x NPB, respectively.

3. Pellet the sperm by centrifuging the filtrate at $1400 \times g$ (e.g., Labfuge-400R, Heraeus centrifuge 3000 rpm) for 10 minutes at 4°C.

4. Gently resuspend the sperm pellet in 10 mL of 1x NPB (*see* **Note 5**) and centrifuge again at $1400 \times g$ at 4°C for 10 minutes. While this spin is in progress put 1 mL 1x NPB for next step at room temperature.

5. Gently resuspend this pellet in 1 mL of 1x NPB and then add 50 µL of 10 mg/mL digitonin (in DMSO; *see* **Note 1**). Incubate the sperm for 5 minutes with gentle agitation at RT, following which add 9 mL of freshly prepared 1x NPB + 3% BSA solution (including protease inhibitors and PMSF) and then centrifuge at $1400 \times g$ for 10 minutes at 4°C.

6. Resuspend the sperm pellet in 5 mL of freshly prepared 1x NPB + 0.3% BSA solution using a clipped tip to prevent shearing of chromosomal DNA.

7. Centrifuge the sperm solution at $1400 \times g$ for 10 minutes at 4°C.

8. Resuspend the final pellet in 500 µL of sperm storage buffer using a clipped tip and place at 4°C while counting the sperm nuclei.

9. To count the sperm nuclei density, use a hemocytometer (Fisher MNK-420-010N). Dilute 2 µL of sperm nuclei suspension (remembering to use a clipped tip) in 200 µL sperm dilution buffer (SDB) and add 1 µL of a 1:100 dilution of Hoechst stain to visualize the sperm heads under a fluorescence microscope (*see* **Note 8**). A count of 75–125 sperm nuclei/nL was considered a good count (*see* **Note 9**).

10. Split sperm nuclei suspension into aliquots of 10 µL, snap freeze in liquid nitrogen, and store at –80°C (*see* **Note 10**).

3.2. High-Speed Egg Extract Preparation

This protocol only needs to be done if the Kroll and Amaya method (**Subsection 3.3.1**) is to be used to generate transgenic embryos. If the simplified method (**Subsection 3.3.2**) is to be used then preparation of high-speed egg extract is not necessary.

All solutions (*see* **Subsection 2.2.**) should be prepared before beginning the extract preparation since the procedure should be carried through all steps promptly once it is initiated; optimally, the high speed spin should begin within 45–60 minutes of dejellying the eggs.

1. The rotors and centrifuges we use for this preparation are Beckman rotor JS13.1 with Beckman centrifuges Avanti J-20 and Beckman rotor TLA100.3 with Beckman Ultima TLX. We have indicated the approximate xg forces so that equivalent rotors can be used if necessary.

2. Induce 12 adult *Xenopus* females as described (*see* **Chapter 21** in **Volume 2**). Keep the females in individual buckets in 1xMMR solution (*see* **Note 11**).

3. Collect eggs. Gently expel eggs manually from each frog into a large beaker of 1×MMR. Collect unbroken eggs

with even pigmentation. Good eggs can also be collected from the 1x MMR in the frog buckets. Total volume of eggs should be 100 mL or greater before dejellying (if necessary, keep eggs from one collection at 14°C for up to an hour before collecting another batch of eggs from the same females).

4. Remove as much liquid as possible from the eggs. Dejelly the eggs in 300 mL of 2% cysteine (Sigma C-7755) made up in 1x MMR, pH adjusted to 8.0 with NaOH within 1 hour of use (see **Note 12**).

5. Wash eggs in 1x MMR and finally in 1x extract buffer (XB with HEPES/sucrose). We use about 35 mL for each wash and do four washes in total.

6. Wash eggs in CSF-XB with 1x protease inhibitors. We do two 25 mL washes.

7. Transfer eggs into Beckman ultraclear tubes. For these volumes, we typically use 14 x 95 mm tubes. If multiple tubes will be used, try to transfer an equal volume of eggs per tube. Remove as much CSF-XB as possible.

8. Pack eggs by centrifugation. Spin at 2°C in a Beckman JS13.1 swinging bucket rotor for about 60 seconds at 1000 rpm (~150 g) and then 30 seconds at 2000 rpm (~600xg). Eggs should be packed after this spin but unbroken. Remove the excess CSF-XB and then rebalance the tubes.

9. Crush the eggs by centrifugation. Spin the tubes in rubber adapters for 10 minutes at 10,000 rpm (~16,000xg) at 2°C in a Beckman JS13.1 swinging bucket rotor. After this three layers should be seen: lipid (top), cytoplasm (center), and yolk (bottom). Collect the cytoplasmic layer from each tube with an 18-gauge needle by inserting the needle at the base of the cytoplasmic layer and withdrawing slowly. Transfer cytoplasm to a fresh Beckman tube on ice.

10. Add 7x protease inhibitors (to achieve a final concentration of 1x) to the isolated cytoplasm. Recentrifuge the cytoplasm in Beckman tubes for an additional 10 minutes at 10,000 rpm (16,000xg) to clarify. Collect the clarified cytoplasm as before (see **Note 13**).

11. Add Energy Mix (1/20th volume of the ATP-regenerating system). Transfer the clarified cytoplasm into Beckman ultra clear thick wall polycarbonate tubes. Tubes hold about 3 mL each and should be at least half full.

12. Add $CaCl_2$ to each tube to a final concentration of 0.4 mM (see **Note 14**). Incubate at room temperature for 15 minutes then balance for the high-speed spin.

13. Spin the tubes in a Beckman tabletop TL-100 ultracentrifuge in a TLA100.3 rotor (gold top; fixed angle,) at 70,000 rpm (~265,000 × g) for 1.5 hours at 4°C.

14. Four layers are produced, top: lipid, second: cytosol, third: membranes/mitochondria, bottom: glycogen/ribosomes. Collect the cytosol, which is clear (~1 mL if 2–3 mL was loaded into the tube) by inserting a syringe into the top of the tube through the lipid layer, transfer to fresh TL-100 tubes and spin again at 70,000 rpm, for 20 minutes, at 4°C.

15. Collect the cytosol supernatant and split into 6 µL aliquots in 0.5 mL eppendorf tubes. Snap-freeze the aliquots in liquid nitrogen and store at –80°C until use (*see* **Note 15**).

3.3. Xenopus Transgenics Protocol

Two methods to generate *Xenopus* transgenic embryos are currently being used. The first, which was developed, by Kroll and Amaya *(9, 10)* involves the decondensing of the sperm nuclei before transplantation (i.e. injecting) into the egg which necessitates a large needle tip and can lead to a large number of abnormal embryos due to nuclear damage. The second method is a variation of this procedure in which the decondensing step is left out *(11, 12)*. As a consequence the nuclei are smaller allowing the use of smaller needles and causing less damage to the nuclei. This has the effect of increasing the efficiency of nuclear transfer and more normal embryos but a decrease in the efficiency of transgenesis. Both methods have been used successfully by our research groups and protocols for both are detailed here. Alternative *Xenopus* transgenic methods have recently been developed *(17–19)*, but have not so far been tested with this heat shock inducible gene expression system.

Ensure all required reagents and solutions are ready (*see* **Subsections 2.3**), including linearized DNA (*see* **Note 3**), before starting the procedure.

Then continue with either:

3.3.1. Kroll and Amaya Method (9, 10; See Note 16)

1. Incubate 5 µL of linearized plasmid DNA (150–250 µg/mL) with 10 µL of sperm nuclei (~4 × 10^5 nuclei) in a 1.5 mL microcentrifuge tube, mix gently by pipeting up and down using a white pipetmen tip with a clipped end so as not to damage the sperm nuclei, and incubate for 5 minutes at RT (not on ice!).

2. After 5 minutes, add 0.5 µL of a 1:40 dilution of XbaI, SalI or NotI along with 2 µL 100mM MgCl$_2$ (*see* **Note 17**), 6 µL high-speed egg extract and 20 µL of Sperm Dilution Buffer (SDB). Mix gently using a clipped yellow tip and incubate for 10 minutes at RT (not on ice; *see* **Note 18**).

3.3.2. Simplified Xenopus Transgenics Protocol (11, 12)

Or:

1. Incubate linearized plasmid DNA (200–250 ng/µL) with 4 µL of sperm nuclei (approximately 4×10^5 nuclei) in a 1.5 mL microcentrifuge tube, mix gently using a white pipetmen tip with a cut end so as not to damage the sperm nuclei (*see* **Note 18**), and incubate for 5 minutes at RT (not on ice!).

2. After the 5-minute incubation, add 15 µL of sperm dilution buffer (SDB) to the DNA/SN mix tube. Mix gently using a yellow pipetmen tip with a cut end so as not to damage the sperm nuclei (*see* **Note 18**). Incubate for 15 minutes at RT (not on ice!).

3.3.3. Nuclear Transplantation

3. During the above incubation, collect unfertilized eggs *Xenopus* females (*see* **Chapter 21** in **Volume 2**) in a beaker containing 1x MMR and subsequently dejelly the eggs (as described earlier in **Section 3.2**, **step 4**; or in **Chapter 21** in **Volume 2**). Finally, carefully transfer the eggs with a Pasteur pipette with wide opening into a 15-cm agarose-coated Petri dish bathed in 0.4x MMR/6% Ficoll. Place the injection dish on top of a large 14-cm Petri dish containing ice before and during the injection process to cool the injected eggs to below approximately 18°C.

4. After the incubation, in a new Eppendorf tube gently add 2 µL of the DNA/sperm nuclei reaction to 148 µL of SDB. Mix very gently with a yellow tip capped with Tygon tubing (see Subsection 2.3 item 11) by carefully pipeting up and down for 10 times. Make certain that no air bubbles are introduced (see Note 18). Carefully take up the complete solution into the pipette tip and gently remove the tip from the end of the pipette.

5. Then insert the back end of the transgenic needle (60–80 µm diameter for Kroll and Amaya method **Subsection 3.3.1** and 40 µm for the simplified method **Subsection 3.3.2**) into the Tygon tubing and point the needle tip downward. The liquid gently fills the needle. If the liquid is not flowing in easily, gently push down on the open end of the pipette tip to create a small amount of pressure. Make sure that no air bubbles are present inside the needle. If there are bubbles the needle should be discarded.

6. Once the needle is filled transfer it to the Tygon tubing filled with mineral oil attached to an infusion pump (from Harvard Apparatus). The pump is set to deliver 10 nL/sec or 0.6 µL/minute. Attach the needle to the micromanipulator and confirm whether the solution is flowing freely from the needle (*see* **Note 19**).

7. Transplant sperm nuclei into unfertilized eggs by piercing the plasma membrane of each egg with a single, jabbing motion and drawing the needle out slightly more slowly (*see* **Note**

20). The sperm nuclei have in theory been diluted to a concentration of approximately one nucleus per 10 nL. This means that though some eggs will receive two or more nuclei and some none, about a third of injected embryos will receive a single nucleus.

8. After the injection, incubate the embryos at 16–18°C. Normally dividing embryos at four-cell stage should be transferred to a dish containing 0.1x MMR/6% Ficoll (**Fig 27.2**). This is done using a wide bore glass Pasteur pipette, which has been previously flamed over a Bunsen burner to smooth the glass (as stated earlier, we expect one-third of injected embryos to be injected with one nucleus and so develop normally; *see* **Note 21**). At embryonic stage 7, transfer the embryos to a dish containing 0.1x MMR and 50 µg/mL gentamycin (*see* **Note 22**). We commonly culture transplanted embryos in 35-mm dishes with about 10–30 embryos per dish because culturing embryos at high density can compromise their health (*see* **Note 23**).

3.4. Heat Shock Protocol: Induction of Gene Expression

1. Embryos are allowed to develop at 14–16°C to the desired stage for induction of gene expression. Prior to reaching the desired stage, remove embryos from the experiment that are either deformed (*see* **Note 24**) or express GFP uninduced (observed with a stereomicroscope equipped with UV lamp and GFP filters; *see* **Note 25**).

2. For induction of gene expression, transfer embryos with a disposable plastic Pasteur pipette with a clipped tip between a pre-cooled Petri dish of $0.1 \times$ MMR at 16°C and a prewarmed Petri dish of $0.1 \times$ MMR at 34 or 37°C. Treat embryos between embryonic stages 13 and 18–20 at 34°C for no longer than 15 min, e.g., 3 to 4 times at 34°C for 15 minutes with an interval of 15 minutes at 16°C. Treat embryos older than stages 18–20 at 37°C, e.g., three to four times for 15 minutes at 37°C with an interval of 15 minutes at 16°C (*see* **Notes 26** and **27**).

3. Observe induced expression of the marker gene GFP with a stereomicroscope equipped with UV lamp and GFP filters (e.g., Leica). Maximal expression is usually observed 4–12 hours after induction but GFP is often visible after 1 to 2 hours and usually still visible in live embryos several days later (**Fig 27.3**).

4. When embryos have reached the desired stage of development, fix them with an appropriate fixative (e.g., MEMFA for whole-mount RNA *in situ* hybridization or Dent's fixative for antibody staining, *see* **Chapter 22** in **Volume 2**). If embryos are fixed in MEMFA and subsequently stored in PBS for morphological analysis, GFP can still be observed (*see* **Note 28**).

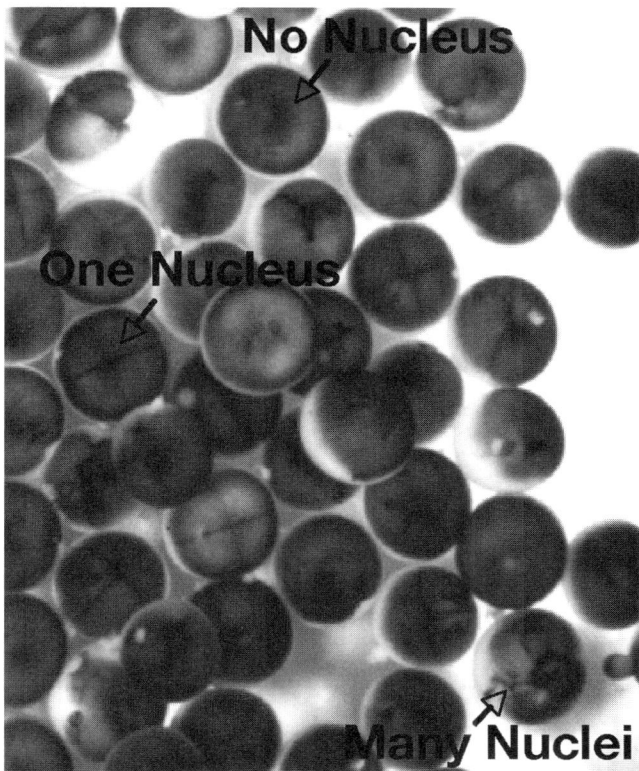

Fig. 27.2. Cleavage patterns of *Xenopus* embryos following nuclear transplantation. Example of a dish of *Xenopus* embryos approximately 2 to 3 hours after nuclear transplantation. During nuclear transplantation, the continuous flow from the needle and the dilution of the sperm nuclei means some eggs have not received a sperm nucleus and are therefore not dividing; some have received one nucleus and have started cleavage and are at the four-cell stage; and some have received more than one sperm nucleus and are showing abnormal cleavage. Only those embryos showing a symmetrical cleavage at the four- to eight-cell stages are picked and allowed to develop.

5. Visualize GFP expression with fluorescence microscope with GFP filter (e.g., a Zeiss Stemi V-11) and take images using an appropriate imaging system (e.g., a RS photometrics Cool SNAP digital camera with Improvision Openlab and Adobe Photoshop software on a Macintosh computer).

4. Notes

1. As an alternative to digitonin, 50 µL of lysolecithin (10 mg/ mL) can be added to the suspension.

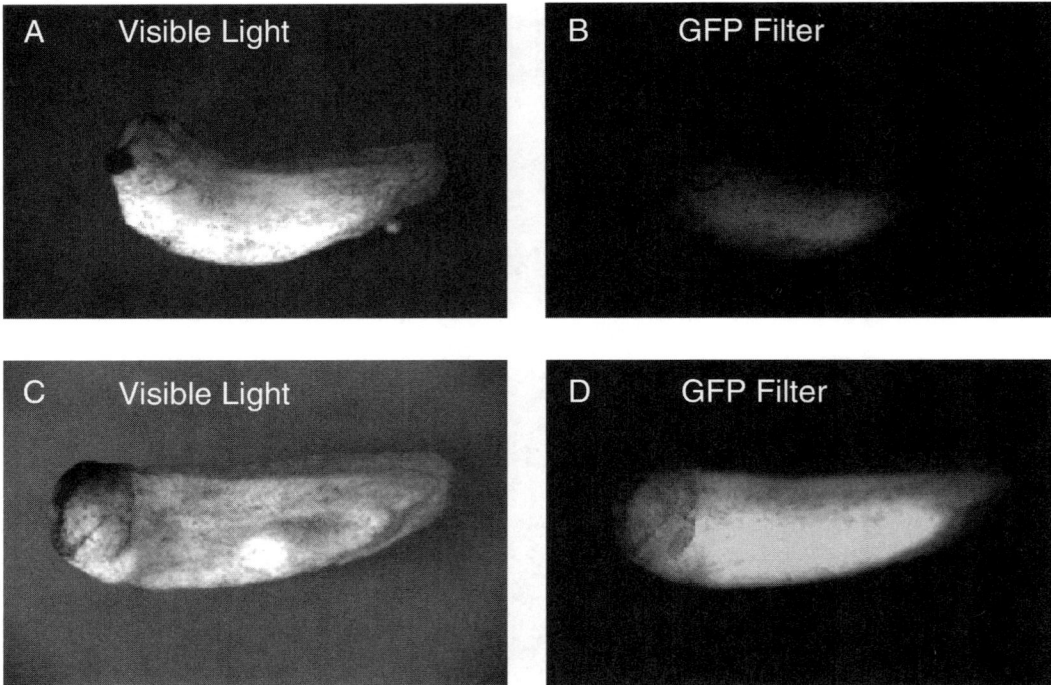

Fig. 27.3. Induced expression of GFP following heat shock: *Xenopus* tadpole embryos that were subjected to the transgenic protocol using the pHSXwnt8HSG DNA including nuclear transplantation are viewed in visible light (**A, C**) and through a GFP filter (**B, D**). Note that the successful transgenic embryos with induced gene expression (**C**) can be easily identified by their expression of GFP (**D**), while non-transgenic embryos (**A** and **B**) are ideal controls, because they have been subjected to the same procedure but fail to express the transgene.

2. The protocol is basically taken from ref. *20*, except that protease inhibitors are omitted from many steps to avoid transfer into the final mixture, which is diluted for egg injections.

3. We generally find that DNA prepared and stored at –20°C 1 or 2 days before use gave better results than DNA stored at –20°C for longer periods.

4. Large numbers of good quality eggs are necessary for this procedure.

5. Sperm nuclei should never be drawn through a small (<100 µm) orifice such as a pipette tip, as chromosomes will be sheared. For resuspensions, use a 5- or 10-mL pipette, or clip the ends of plastic pipetteman tips (P-200 or P-1000) with a clean razor blade to widen the orifice.

6. The thoroughness with which the testis is macerated in this step is a major factor in determining the final yield of nuclei obtained from this preparation.

7. Cut the end of a Pasteur pipette to the desired diameter (~3 mm) using a diamond pen and then hold briefly in a hot Bunsen burner flame to smooth the edges.

8. The sperm nuclei have a characteristic crescent shape when viewed under the fluorescent microscope with a 10x lens.

9. If the concentration of sperm nuclei is found to be substantially less (50 sperm nuclei/nL or less) repellet the sperm nuclei and resuspend in a smaller volume of sperm storage buffer.

10. We have successfully used sperm frozen for up to 3 months. Sperm nuclei can be kept at 4°C for up to 48 hours and used for transgenics. However we have found no significant difference in transgenic efficiency between fresh and frozen sperm.

11. If possible, keep induced females at relatively cold temperatures overnight (14–18°C) and place individually into container with at least 2 L 1 × MMR (because one frog with lysing or activating eggs can compromise the whole extract preparation, we prefer to separate the frogs for the ovulation). The next morning the egg quality from each container is screened before mixing all the eggs and starting the extract preparation. All the eggs released from a frog, which lays mottled, lysing or dying eggs, are left out of the extract preparation.

12. Add a small amount at a time, swirl eggs, and partially replace with fresh cysteine solution several times during dejellying. Remove broken eggs with a pipette during dejellying. Dejellying can be performed separately for different batches of eggs; those that show breakage or egg activation are discarded.

13. If large volumes of darkly pigmented eggs are used, the cytoplasmic layer may be grayish rather than golden at this step. After a second spin to clarify this extract, it should be golden. Expect to get about 0.75–1 mL of unpurified cytoplasm per batch of eggs collected from one frog.

14. This inactivates CSF and pushes the extract into interphase.

15. We typically obtain 1–2 mL of purified high-speed cytosol from preparations of this scale. Sperm nuclei should be incubated in an aliquot of extract and stained with Hoechst as previously described (**Subsection 3.1**, **step 9**) to determine whether extract is effective. If active interphase extract has been prepared, nuclei should swell visibly (thicken and lengthen) within 10 minutes of addition to extract at room temperature.

16. Transgenic experiments should be carried out using two people for a single experiment. This allows one person to harvest and prepare the eggs while the other prepares the sperm nuclei. With the simplified method (**Subsection 3.3.3**) this can be done by one person. However the nuclear transplantation injections should be done by two people to maximize the number of nuclear transplantations from one reaction.

17. Adding restriction enzyme to the transgenic reaction and MgCl$_2$ to aid the enzymatic reaction is optional. Also note that the restriction enzyme added at this stage does not need to be the same as the one used to linearize the transgene construct DNA. We do find that addition of restriction enzyme and MgCl$_2$ to the transgenic reaction increases the rate of transgenesis we observe, but sometimes also increases gastrulation and neurulation defects that are possibly associated with aneuploidy (*see* **Note 24**).

18. When mixing transgenic reactions do not push the tip against the bottom of the tube and do not introduce bubbles as this rough handling will damage chromosomes and shear nuclei.

19. Using the micromanipulator place the tip of the needle close to an egg. The solution in the needle has a different optical density than the ficoll-containing solution in the dish so it is easy to see if the solution is flowing from the tip. Moving the tip slightly will also help to visualize if the solution is flowing.

20. Transplanting Sperm nuclei into unfertilized eggs *(21)*.

    a. Move the needle fairly rapidly from egg to egg, piercing the plasma membrane of each egg with a single sharp motion then drawing the needle out slightly more slowly. Keep the needle inside each egg for approximately 1 second. The angle of the needle should be perpendicular to the membrane surface (rather than glancing) to avoid unnecessary tearing of the plasma membrane. Be certain that you are piercing the plasma membrane; if a slow motion is used, the membrane will deform around the needle instead of being punctured.

    b. The rate of flow should be robust enough that the needle does not flow or clog with cytoplasm during injections and slow enough to be manageable. At the flow and injection rates we generally use, about 10 nL volume is injected in each injection, so a 1:500–1:1000 dilution of the original sperm stock allows approximately one sperm to be injected in that volume.

    c. If the needle becomes clogged, as evidenced by no flow out or visible cell debris in the needle, bring the tip to the air–liquid interface of the solution covering the embryos in the dish. Sometimes the surface tension of the interface removes the cytoplasm plug in the end of the needle. If the needle cannot be unblocked, fill a new one and continue. If a needle tip is too narrow, or if it becomes partially clogged with debris during transplantation, the injected nuclei will be damaged during transplantation and haploid embryos will result. Haploid tadpoles have shortened trunks and tails, are thicker than normal throughout the trunk region with a "pigeon-chested" appearance and often have heads and tails which curl toward the dorsal side; these tadpoles will live for a while but usually become edemic and die around the time of feeding *(21)*.

d. Correct injection technique: A hole about the diameter of the needle tip should be visible on the egg and should remain open for a couple of seconds after injection; the hole will have a clean smooth edge (not jagged) and a small amount of translucent fluid is visible just over the opaque white cytoplasm in the wound.

e. If there is insufficient flow rate, the hole created in the egg by the needle instantly closes after injection and little or no volume is delivered (sometimes a ring of white cytoplasm will be drawn from the hole as the needle is withdrawn).

f. If there is excessive flow rate, the surface of the egg near the injection site may ripple or the site of injection may expand in size significantly.

21. The embryos that have received one nucleus have a regular cleavage pattern. The most obvious criteria we find is that the first and second cleavages are at right angles, even if other aspects may not be entirely normal at this stage. With optimal conditions, one third of embryos will be four-cell embryos about 2 to 3 hours after transplantation with rectangular cleavage, one third polyspermic with irregular and excessive cleavage patterns and one third aspermic with no or only pseudo-cleavage. If conditions are not optimal, less than one-third of embryos are four-cell embryos with rectangular cleavage, either because there are too many aspermic or too many polysperimic embryos. We found that increasing or decreasing, respectively, the amount of sperm nuclei suspension added to the initial incubation with DNA (**Subsection 3.3.1** or **3.3.2, step 1**) is the most efficient way to correct the conditions, but that adjusting the flow rate does not normally help.

22. Keeping embryos in ficoll can adversely affect gastrulation. So removing the ficoll before gastrulation is important.

23. It is important to remove dying embryos promptly because they can also affect the healthy embryos. Because of the large needle tip used for transplantation, embryos may develop large blebs at the site of injection. These blebs occur when cells are forced out of the hole left in the vitelline membrane at the injection site but they generally do not affect development. The blebs usually fall off at the

neurula or tailbud stage, but can be manually removed with dissection tools once the embryos have reached late blastula stages.

24. The manipulation of the sperm nuclei can induce chromosomal damage, which cause developmental abnormalities (mostly gastrulation and neurulation defects) that may obscure phenotypes induced by the overexpression of the transgene or may even mislead the investigation (however, note that the chromosomal damage phenotype will also be present in non-transgenic and non-induced embryos). If chromosomal damage is suspected, ensure that you are very careful and gentle when manipulating sperm nuclei in suspension, particularly if using the Kroll/Amaya method, which involves decondesation of the chromatin. Additionally, reduce the duration of the incubation in SDB (**Subsection 3.3.1** or **3.3.2, step 2**), although this may also reduce the transgenic rate (*see* **Note 26**).

25. The uninduced GFP expression is likely due to enhancer trap integration of the transgene or due to cellular stress in deformed embryos. When embryos are kept at 16°C we do not detect any background activity of the heat-shock promoter.

26. If you observe a low transgenesis rate with your DNA construct, test a previously successfully used construct. One that we use regularly is Pax6::GFP *(22)*. If the transgenesis procedure is working fine, it may be necessary to reprep and re-purify your DNA. Finally, if you are using the simplified protocol (**Subsection 3.3.2**), try the Kroll/Amaya method (**Subsection 3.3.1**). Additionally, extend the duration of the incubation in SDB (**Subsection 3.3.1, step 2**), although this may also induce developmental defects associated with chromosomal damage (*see* **Note 24**).

27. More cycles (up to six times) can be done with no affect on the embryos. Artefacts due to heat treatment were only observed in embryos that were treated during gastrulation stages (st. 10–12) and occasionally at early neurulation stages (st. 13–16). These include incorrect or incomplete blastopore closure, which results in arrest of development at this late gastrulation stage or further development with spina bifida. We also observe a transient ruffling of the ectoderm, which however does not appear to affect later development.

28. Placing the fixed embryos into ethanol destroys the GFP fluorescence. Therefore it is best to photograph GFP positive embryos before processing for whole-mount RNA *in situ* hybridization.

Acknowledgments

We would like to thank Enrique Amaya and Kris Kroll for introducing us to their transgenic protocol and the Cold Spring Harbor "Early Development of *Xenopus laevis*" course for providing an introduction to the *Xenopus* system for all three of us.

References

1. Gradl, D., Kuhl, M., Wedlich, D. (1999) Keeping a close eye on Wnt-1/wg signaling in *Xenopus*. *Mech Dev* 86, 3–15.

2. Hoppler, S., Kavanagh, C. L. (2007) Wnt signalling: variety at the core. *J Cell Sci* 120, 385–393.

3. Tao, Q., Yokota, C., Puck, H., et al. (2005) Maternal wnt11 activates the canonical wnt signaling pathway required for axis formation in *Xenopus* embryos. *Cell* 120, 857–871.

4. Wheeler, G. N., Hoppler, S. (1999) Two novel *Xenopus* frizzled genes expressed in developing heart and brain. *Mech Dev* 86, 203–207.

5. Huelsken, J., Birchmeier, W. (2001) New aspects of Wnt signaling pathways in higher vertebrates. *Curr Opin Genet Dev* 11, 547–553.

6. Wheeler, G. N., Hamilton, F. S., Hoppler, S. (2000). Inducible gene expression in transgenic *Xenopus* embryos. *Curr Biol* 10, 849–852.

7. Bienz, M. (1984) *Xenopus* hsp 70 genes are constitutively expressed in injected oocytes. *Embo J* 3, 2477–2483.

8. Michiue, T., Asashima, M. (2005) Temporal and spatial manipulation of gene expression in *Xenopus* embryos by injection of heat shock promoter-containing plasmids. *Dev Dyn* 232, 369–376.

9. Kroll, K. L., Amaya, E. (1996) Transgenic *Xenopus* embryos from sperm nuclear transplantations reveal FGF signaling requirements during gastrulation. *Development* 122, 3173–3183.

10. Amaya, E., Kroll, K. L. (1999) A method for generating transgenic frog embryos. *Methods Mol Biol* 97, 393–414.

11. Sparrow, D. B., Latinkic, B., Mohun, T. J. (2000) A simplified method of generating transgenic *Xenopus*. *Nucleic Acids Res* 28, E12.

12. Smith, S. J., Fairclough, L., Latinkic, B. V., et al. (2006) *Xenopus* laevis transgenesis by sperm nuclear injection. *Nat Protoc* 1, 2195–2203.

13. Hamilton, F. S., Wheeler, G. N., Hoppler, S. (2001) Difference in XTcf-3 dependency accounts for change in response to beta-catenin-mediated Wnt signalling in *Xenopus* blastula. *Development* 128, 2063–2073.

14. Beck, C. W., Christen, B., Slack, J. M. (2003) Molecular pathways needed for regeneration of spinal cord and muscle in a vertebrate. *Dev Cell* 5, 429–439.

15. Fu, L., Ishizuya-Oka, A., Buchholz, D. R., et al. (2005) A causative role of stromelysin-3 in extracellular matrix remodeling and epithelial apoptosis during intestinal metamorphosis in *Xenopus* laevis. *J Biol Chem* 280, 27856–27865.

16. Beck, C. W., Christen, B., Barker, D., et al. (2006) Temporal requirement for bone morphogenetic proteins in regeneration of the tail and limb of *Xenopus* tadpoles. *Mech Dev* 123, 674–688.

17. Hamlet, M. R., Yergeau, D. A., Kuliyev, E., et al. (2006) Tol2 transposon-mediated transgenesis in *Xenopus* tropicalis. *Genesis* 44, 438–445.

18. Pan, F. C., Chen, Y., Loeber, J., et al. (2006) I-SceI meganuclease-mediated transgenesis in *Xenopus*. *Dev Dyn* 235, 247–252.

19. Sinzelle, L., Vallin, J., Coen, L., et al. (2006) Generation of trangenic *Xenopus* laevis using the Sleeping Beauty transposon system. *Transgenic Res* 15, 751–760.

20. Murray, A. W. (1991) Cell cycle extracts in (Kay, B. K. & Peng, H. B., eds.) *Xenopus laevis: Practical Uses in Cell and Molecular Biology*, vol. 36, pp. 581–604, Academic Press, San Diego.

21. Gurdon, J. B. (1960) Factors responsible for the abnormal development of embryos obtained by nuclear transplantation in *Xenopus* laevis. *J Embryol Exp Morphol* 8, 327–340.

22. Hartley, K. O., Nutt, S. L., Amaya, E. (2002) Targeted gene expression in transgenic *Xenopus* using the binary Gal4-UAS system. *Proc Natl Acad Sci USA* 99, 1377–1382.

Chapter 28

Wnt-Frizzled Interactions in *Xenopus*

Herbert Steinbeisser and Rajeeb K. Swain

Abstract

The Wnt signaling cascades are regulatory modules which are involved in embryonic patterning, cell differentiation, morphogenesis, and diseases *(1, 2)*. The Wnt pathways are activated when secreted Wnt ligands interact with 7-trans-membrane receptors of the Frizzled (Fz) family. Specific readouts are determined by the ligand/receptor combinations and the cellular context. Here we describe two methods for the analysis of Wnt/Frizzled interactions in *Xenopus* embryos. Physical interaction of ligand and receptor are demonstrated by co-immunoprecipitation assays. The activation of Wnt targets in *Xenopus* animal cap tissue provides a versatile test system for activating and inhibitory components of the Wnt/ß-catenin pathway.

Key words: Wnt, Frizzled, co-immunoprecipitation, Wnt/β-catenin target genes, *Xenopus*, animal cap assay.

1. Introduction

In vertebrates, 19 Wnt proteins and 10 Frizzled (Fz) receptors have been described. The Wnts can be classified in two groups: the Wnt-1 group, which is able to transform mouse mammary epithelial cells and induce secondary body axes in *Xenopus* embryos and the non- or weakly transforming Wnt 5a group. Transformation of cells and axis induction are achieved by the activation of the Wnt/β-catenin pathway *(1, 3)*. In this signaling module, β-catenin protein is stabilized in the cytoplasm and moves into the nucleus where it associates with TCF/LEF transcription factors and regulates expression of target genes. In contrast, the

Elizabeth Vincan (ed.), *Wnt Signaling, Volume II: Pathway Models, vol. 469*
© 2008 Humana Press, a part of Springer Science + Business Media, New York, NY
Book doi: 10.1007/978-1-60327-469-2

Wnt-5a-type ligands trigger ß-catenin-independent Wnt pathways such as Wnt/Ca⁺⁺ and planar cell polarity (PCP) pathway *(2, 4)*. The β-catenin independent Wnt cascades modulate rearrangements of the cytoskeleton and regulate morphogenetic processes. These different branches of the Wnt signaling pathway can be activated by over expression of specific Fz receptors or by combinations of Wnts and Fzs. Due to the large number of Wnt and Fz, the question of specificity of Wnt/Fz interactions remains unclear. However, it is well accepted that Fz receptors are promiscuous in two aspects. They are not restricted to the interaction with one specific Wnt ligand or a specific group of Wnts. Wnt/β-catenin dependent and independent pathways can be initiated by the same Fz receptor. The mechanisms by which this pathway selection is achieved are subject of intensive research. The analysis Wnt/Fz interaction is complicated by the fact that numerous secreted proteins interact with ligands and/or receptors and modulate the readout. Proteins such as Cerberus and Dickkopf inhibit Wnt/β-catenin signaling, whereas Norrin and R-Spondin augment it. Secreted Frizzled related proteins (sFRPs) are able to acts as inhibitors or activators *(3)*.

The *Xenopus* embryos provide assay systems to address experimentally the interaction of Wnts, Fzs and their modulators. Specific Wnt-Fz combinations can be expressed in embryonic cells by microinjection of synthetic mRNAs. The physical interaction can be analyzed in co-immunoprecipitation experiments *(5)*. The functional interaction of Wnts and Fzs can be assessed by expression analysis of Wnt/β-catenin targets such as nodal-related-3 (xnr-3) in embryonal ectoderm (animal cap assay; ref. *6*).

2. Methods

2.1. Animals and Embryos

1. Adult *Xenopus laevis* were commercially purchased from *Xenopus* Express (France; www.xenopus.com).

2. Modified Bath's solution (MBS): 88 mM NaCl, 1 mM KCl, 2.4 mM Na_2HCO_3, 0.82 mM $MgSO_4$, 0.33 mM, $Na(NO_3)_2$, 0.41 mM $CaCl_2$, 10 mM HEPES, pH 7.4, 10 μg/mL streptomycin sulfate, 10 μg/mL penicillin, sterile filtrated (can be prepared as 10X solution and stored at 4°C for several months).
   Human chorionic gonadotropin (HCG): 1000 U/mL (Sigma)

3. Cysteine solution: 2% (w/v) cysteine in water, pH adjusted to 8.0 with NaOH (*see* **Note** 1).

2.2. Capped Synthetic mRNA and Microinjection

1. mMessage mMachine kit (Ambion, Austin, TX; www.ambion.com).

2. Stop solution: 3 M NH$_4$Ac, 100 mM EDTA.

3. Phenol/chloroform/isoamylalcohol: 25:24:1 (Sigma).

4. Isopropanol.

5. 75% (v/v) ethanol.

6. Injector: IM300 Microinjector, (Narishige Japan; www.narishige.co.jp/).

7. Glas capillaries, 6.66 µL (Hirschmann; www.hirschmann-laborgeraete.de/).

8. Petridishes coated with 0.5% (w/v) agarose in 1X MBS.

2.3. RNA Extraction

1. Trizol reagent (Invitrogen).

2. High salt precipitation solution: 0.8 M sodium citrate and 1.2 M NaCl.

3. RNAse-free water (not DEPC-treated, available from Fermentas or Ambion).

2.4. RT-PCR and Electrophoretic Separation of DNA Fragments

2.4.1. Reverse Transcription

1. Random 6mer nucleotides: 500 ng/µL (Amersham).

2. dNTPs solution: 10 mM (Fermentas).

3. RNAse Inhibitor: 40 units/µL (Fermentas).

4. 5x reverse transcription (RT) buffer: 200 mM Tris-HCl, pH 8.3, 250 mM KCl, 20 mM MgCl$_2$, 50 mM DTT.

5. M-MLV reverse transcriptase: 200 units/µL (Fermentas).

6. JETquick PCR purification kit (Genomed).

2.4.2. PCR

1. dNTPs solution: 2 mM of each dATP, dCTP, dGTP, and dTTP.

2. 10x reaction buffer (Euroclone®): 160 mM (NH$_4$)$_2$SO$_4$, 670 mM Tris-HCl, pH 8.8, 0.1% (v/v) Tween-20, 2 mM MgCl$_2$.

3. EuroTaq DNA polymerase (Euroclone® Life Science Division).

4. Primers for Xnr-3 and ODC

    ODC:5'GTCAATGATGGAGTGTATGGATC 3',

    5'TCCATTCCGCTCTCCTGAGCAC3'

    Xnr-3:5'TGAATCCACTTGTGCAGTTCC3';

    5'GACAGTCTGTGTTACATGTATGGATC 3'

    Primers for other Wnt targets can be obtained from the web site of Eddy De Robertis:

    www.hhmi.ucla.edu/derobertis/protocol_page/oligos 2004.pdf.

2.4.3. Agarose Gel Electrophoresis

1. Agarose (Biozym).
2. Tris-Borate-EDTA (TBE)-buffer: 89 mM Tris-borate, 0.02 mM EDTA.
3. Sample buffer (6 x): 0.25 % (w/v) bromophenol blue (Serva), 30% (v/v) glycerol in H_2O.
4. 100 bp DNA ladder (Fermentas).

2.5. Co-immunopre-cipitation

1. Embryo lysis buffer *(5)*: 10 mM Tris-HCl, pH 7.5, 100 mM NaCl, 2 mM EDTA,1 mM EGTA, 0.5% (v/v) NP-40, 10% (v/v) glycerol supplemented with proteinase inhibitor cocktail (500X solution from Calbiochem, Cat. No. 539134).
2. Protein G/A beads (Pierce).
3. Anti Myc antibody (mouse monoclonal, OP10L; Merck).
4. Anti-myc antibody (rabbit polyclonal, Upstate/Millipore).
5. Anti Flag antibody (mouse monoclonal M2; Sigma).
6. Anti-HA antibody (rat monoclonal, 3F10; Roche).

2.6. Sodium Dodecyl Sulfate-Polyacry-lamide Gel Electro-phoresis (SDS-PAGE) and Western Blotting

1. Minigels (disposable plastic cassettes, Invitrogen).
2. Water-saturated isobutanol.
3. 30% Acrylamide/bis acrylamide (1:37.5) solution such as Rotiphorese® Gel 30 (Roth).
4. Stacking buffer (4X): 0.5 M Tris-HCl, pH 6.8, 0.4% (w/v) SDS.
5. Separating buffer (4X): 1.5 M Tris-HCl, pH 8.8, 0.4% (w/v) SDS
6. Ammonium persulfate (APS): 10% (w/v) solution in water prepared freshly.
7. N,N,N',N'-Tetramethylethylenediamine (TEMED).
8. SDS-PAGE loading buffer (3x): 45 mM Tris-HCl, pH 6.8, 30% (v/v) glycerol, 300 mM DTT, 6% (w/v) SDS, 0.9% (w/v) bromophenol blue (can be obtained from Fermentas as 5X loading buffer).
9. SDS running buffer: 25 mM Tris-base, 0.2 M glycine, 0.02% (w/v) SDS.
10. Transfer buffer: 48 mM Tris-base, 39 mM glycine, 20% (v/v) methanol.
11. Ponceau S dye (Sigma cat. no. P7170)
12. Blocking solution (BS): 5% (w/v) nonfat milk powder (available in most supermarkets) dissolved in PBS with 0.1% (w/v) Tween-20.
13. Phosphate-buffered saline (PBS): 136 mM NaCl, 2.7 mM KCl, 1.5 mM KH_2PO4, 6.5 mM Na_2HPO4, pH 7.4 (can be prepared as 10X solution, autoclaved, and stored at 4°C).

14. Tris-buffered saline (TBS): 136 mM NaCl, 2.7 mM KCl, 50 mM Tris-HCl, pH 8.0.

15. TBS-T: TBS containing 0.1% (w/v) Tween-20.

16. Lumi-Light Western Blotting Substrate (Roche).

17. X-ray film (Cronex, AGFA).

18. Prestained molecular weight markers (PageRuler™ prestained protein ladder from Fermentas).

3. Methods

3.1. In Vitro Fertilization of Xenopus laevis Eggs

1. Remove testes from sacrificed male *Xenopus laevis* and keep them at 4°C in MBS.

2. Stimulate females of *Xenopus laevis* to lay eggs by injecting 400 units of HCG.

3. 15 hours after injection, gently squeeze the frog and collect the eggs in a dry Petri dish (*see* **Note** 2).

4. Macerate a little piece of a testis in 1x MBS and spread over the eggs (a suspension of macerated testes can be stored in 1X MBS for an entire day in the fridge without losing any ability to fertilize the eggs).

5. After 5 minutes of incubation, cover the eggs with 0.1x MBS, which makes the sperm motile.

6. Remove the jelly-coat of the eggs by incubation in 2% cysteine solution 30–60 minutes after fertilization. Stop cysteine treatment after 7 minutes by rinsing several times with 0.1x MBS.

7. Culture embryos in 1% agarose coated plastic culture dishes at 14 to 22°C in 1x MBS prior to injection (low-cost plastic culture dishes used for bacterial culture).

3.2. Microinjection of In Vitro Capped RNA

3.2.1. In Vitro Synthesis of Capped mRNA for Microinjection

1. For microinjection of embryos and oocytes, transcribe capped mRNA from linearized DNA-template containing a RNA polymerase promoter. The DNA plasmid is digested with an appropriate restriction enzyme and phenol/chloroform purified.

2. Synthesize capped mRNA using the relevant (SP6, T3, or T7) RNA polymerase mMessage mMachine kit (Ambion) as per manufacturer's instructions. Plasmid DNA is removed by adding 2 units per reaction of DNase I for 20 minutes at 37°C.

3. Stop the reaction by adding 20 μL of stop solution and diluting with RNAse free water to 200 μL. Enzymes are removed by

phenol/chloroform extraction. To remove unincorporated nucleotides, precipitate the mRNA by adding 0.7 volume of isopropanol, incubating for 1 hour to overnight at −20°C and centrifugation at 12,000×g for 30 minutes at 4°C. Then wash the RNA with 75% ethanol and re-suspend in 16 μL of H_2O.

4. Determine the yield of mRNA photometrically (concentration [μg/mL] = OD_{260nm} × dilution × 40). Integrity of mRNA is assessed by agarose gel electrophoresis. Store aliquots of 1 μL at −80°C and thaw them no more than twice.

3.2.2. Microinjection Into Embryos

`1. At the four-cell stage the blastomeres are injected at the animal pole with RNA solution (**Fig. 28.1A**). The injected volume per blastomere should not be higher than 5 nL. The RNA concentrations injected depends on the protein that is encoded (routinely inject 5–25 μg/mL; *see* **Note 3**).

2. Keep injected embryos in 1x MBS until the animal caps are excised (*see* **Note 4**).

3.2.3. Explantation of Animal Cap Tissue

1. Grow embryos for about 6 hours to NF stage 8.5–9 (blastula).

2. Remove the vitelline membrane with Dumont No.5 forceps (*see* **Note 5**).

3. Excise the animal cap region with sharp forceps and transfer the explants to agar-coated Petri dishes in 0.5x MBS (*see* **Note 6**).

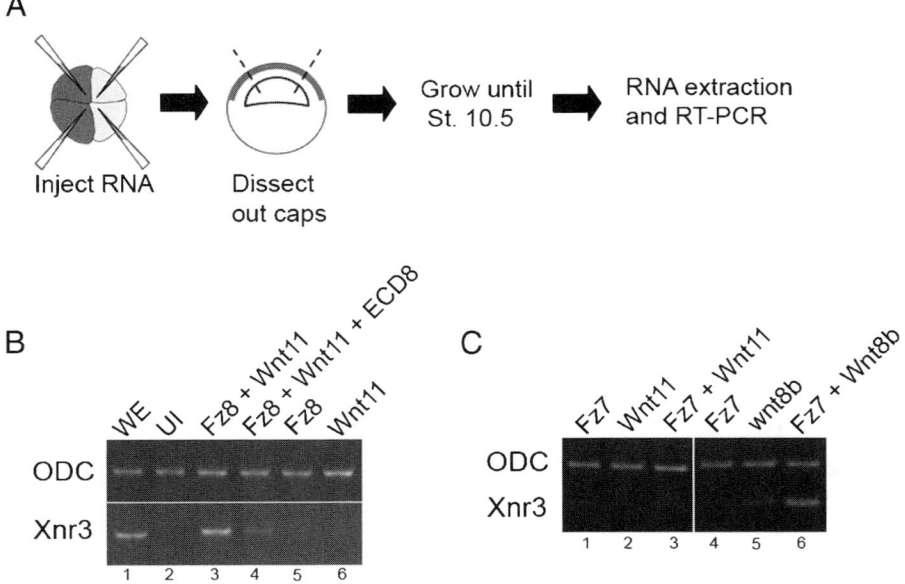

Fig. 28.1. Induction of the Wnt/β-catenin target geneX nr-3 in animal cap explants. (**A**) Animal cap assay scheme. (**B**) Induction of Xnr3 by coinjection of Xfz8 (200 pg/embryo) and Wnt11 (25 pg/embryo), and inhibition by extracellular domain of Xfz8 (ECD8, 500 pg/embryo). (**C**) Coinjection of Xfz7 and Wnt11 does not induce Xnr3 expression. Wnt8b (25 pg/embryo) acts synergistically with Xfz7 (300 pg/embryo) to induce Xnr3.

4. Harvest the explants when the embryos reach the gastrula stage (10.5) and isolate the RNA (**Fig.** 28.1A).

3.3. Isolation of Total RNA

1. Thoroughly homogenize ectodermal explants (15–20) or embryos (3 gastrulae or 2 tailbuds) with a micropestle (Eppendorf Cat. No.022365622) in 0.5 mL or 1 mL TRIZOL reagent. The caps lysed in Trizol can be stored at –20°C or can be processed for extracting RNA (**step 2**).

2. Incubate samples for 10 minutes at room temperature and remove the insoluble material from the homogenate by centrifugation at 12,000×g for 5 minutes at 4°C.

3. Add 0.2 mL chloroform per mL of TRIZOL used for the homogenization to the cleared homogenate solution.

4. Vortex the samples twice for 5 seconds and incubate them for 5 minutes at room temperature.

5. Separate the phases by centrifugation at 12,000×g for 15 minutes at 4° C. Transfer the colorless upper aqueous phase to a fresh tube without disturbing the interface, which contains DNA (the aqueous phase is ~0.5 mL per 1 mL of TRIZOL).

6. Extract once more with chloroform:isoamylalchohol (24:1) and collect aqueous phase.

7. Precipitate total RNA by adding 0.25 mL of isopropanol followed by 0.25 mL of a high salt precipitation solution (0.8 M sodium citrate and 1.2 M NaCl) per 1 mL of TRIZOL and vortexing. Incubate the samples for at least 15 minutes at room temperature and then pellet the precipitated RNA by centrifugation at 12,000×g for 10 minutes at 4°C.

8. After removing the supernatant, the RNA pellet is washed with 1.5 mL 75% ethanol, vortexed, and centrifuged at 10,000×g for 10 minutes at 4°C. Remove the supernatant and allow the RNA pellet to briefly air-dry. Re-dissolve the RNA in 20 μL RNAse free water and incubate for not more than 10 minutes at 55°C.

9. Total RNA is stored at –80°C and 9 μL is used for subsequent reverse transcription reaction.

3.4. Reverse Transcription, PCR, Agarose Gel Electrophoresis

3.4.1 Reverse Transcription

1. Add 1 μL of random 6mer nucleotides to 9 μL (1–5 μg) total RNA. Heat the mixture at 70°C for 5 minutes to destroy secondary structure and place immediately on ice.

2. Add 9 μL of master mix (2 μL of dNTPs, 4 μL 5x RT buffer, 0.5 μL RNAse Inhibitor and 2.5 μL RNAse free H_2O) to the RNA, mix and incubate for 10 minutes at 25°C. Finally,

add 1 μL of M-MLV Reverse Transcriptase, gently mix and incubate the reaction for 120 minutes at 42°C.

3. Stop the reaction by heat inactivation for 15 minutes at 70°C. The cDNA produced is then purified using the JETquick PCR purification kit.

4. The cDNA is eluted from the columns using 100 μL of 10 mM Tris-HCl, pH 8.5 and heated to 70°C. Store the cDNA at –20°C until required.

5. Use 1–3 μL of cDNA for subsequent PCR.

3.4.2. Polymerase Chain Reaction (PCR)

1. For all reactions, a master mix is prepared and all reactions are mixed on ice. A single PCR reaction mix (10 μL) contains: 1–3 μL cDNA, 5 pmol of each oligonucleotide primer, 2 mM of each dNTP, 2 mM magnesium chloride, 1 μL 10 x reaction buffer and 1 unit of EuroTaq DNA polymerase.

2. To avoid evaporation, the reaction mixture is overlaid by an equal drop of mineral oil. DNA is first denatured for 2 minutes at 96°C followed by 25 cycles for ODC and 29 cycles for Xnr3: 30 seconds at 96°C for denaturation, 30 seconds at 60°C for primer annealing, and 1 minute at 72°C for elongation (sufficient for up to 1 kb fragments). The cycles are completed by a final 5 minute elongation at 72°C.

3. The products are then cooled down to 4° C and analyzed by agarose gel electrophoresis.

3.4.3. Agarose Gel Electrophoresis

1. Separate the PCR products in horizontal agarose gels in the presence of 0.5 μg/mL ethidium bromide. The agarose concentration in TBE buffer (0.8–2.5%) depends on the size of the expected DNA fragments. We use 2% agarose gel for separating Xnr3 and ODC (**Fig. 28.1B**).

2. Mix the DNA samples with sample buffer, load the samples into agarose wells and separate at a constant voltage of 100V.

3. Visualize DNA bands on a trans-illuminator screen (366 nm). Pictures were taken with a digital camera and sent to a thermo printer. A 100 bp ladder (0.5–1 μg/lane, GIBCO BRL Lifetechnologies) was used as a size marker.

3.5. Co-immunopre-cipitation

For co-immunoprecipitation (IP) experiments with Wnts and Fzs we co-express Wnt proteins and the extracellular domains of Fz. This gives better results than the full-length receptors (**Fig. 28.2**).

1. Collect embryos in a 1.5 mL Eppendroff tube. Completely remove the culture medium and wash once with cell lysis buffer (without proteinase inhibitors).

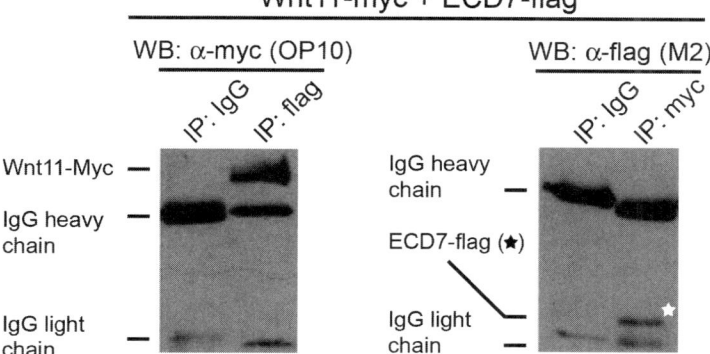

Fig. 28.2. Wnt11 and Fz7 interaction demonstrated by co-immuniprecipitation. Fz7 extracellular domain-flag (ECD7-flag, 500 pg/embryo) and Wnt11-myc (200 pg/embryo) was injected into *Xenopus* embryos at two- to four-cell stage. At stage 10.5–11, the embryos were lysed and processed for immunoprecipitation following the protocol given in this chapter. Immunoprecipitated protein from five embryos was loaded on each lane. Wnt11-myc is co-precipitated with ECD7-flag (left panel) and ECD7-flag is co-precipitated with Wnt11-myc (right panel).

2. Completely remove buffer after wash and add lysis buffer with proteinase inhibitors to the embryos. Use 10 μL lysis buffer/embryo or 5 μL lysis buffer/ animal cap explant cap.

3. Lyse embryos by pipetting up and down through a 200-μL tip several times (*see* **Note** 7).

4. Centrifuge the lysate for 5 minutes at 10,000xg at 4°C. This should separate the lysate into three layers (a pellet containing pigments at the bottom, proteins in the middle, and a thin layer of yolk on the top).

5. Carefully remove the protein without disturbing the pellet or the yolk. Transfer this supernatant into another tube and centrifuge at 10,000xg for 5 minutes. Carefully remove the protein extract without disturbing the yolk. Transfer it to another tube.

6. The embryo extract is then divided into three parts. Incubate two parts with 1 μg of desired primary antibodies (anti-Myc, anti-Flag) and one part with 1 μg nonspecific IgG as control (*see* **Note** 8). Then rotate protein extract in a head-over-head rotator for 1 hour at 4°C.

7. Centrifuge at 10,000xg for 15 minutes. A white/transparent pellet should be visible.

8. Carefully transfer the supernatant to another tube without disturbing the pellet. Add protein G/A beads to the tube and rotate the samples on a head-over-head rotator for another 2 to 3 hours (*see* **Note** 9).

9. Centrifuge the tubes at 4000xg for 2 to 3 minutes. Take out the supernatant without disturbing the beads (*see* **Note** 10).

The beads are resuspended in lysis buffer and left on head-over-head rotator for 5 minutes. Repeat centrifugation and washing three to four times (four to five washes in total).

10. Centrifuge, take out the buffer completely and resuspend the beads 1X SDS-PAGE loading buffer (*see* **Note** 11).

11. Incubate samples at 95°C for 4 to 5 minutes. Centrifuge briefly and load only the supernatant (no beads) on the SDS-PAGE (*see* **Note** 12).

3.6. SDS-Page and Western Blot

3.6.1. SDS-Page

1. To separate proteins, discontinuous SDS-PAGE is performed on minigels (disposable plastic cassettes) according to Laemmli *(7, 8)*. The acrylamide concentration chosen depends on the size of the proteins. Affinity tagged Fz extracellular domains and Wnts can be separated on a 12% SDS-PAGE gel. Pour 6 mL of separation gel solution (8–15% of acrylamide/bisacrylamide [37.5:1], 0.1% ammonium persulfate [APS], 0.01% TEMED and 0.1% SDS) into the cassettes and overlay with water saturated butanol.

2. After polymerization of the separation gel, remove the water-saturated butanol and add 2 mL stacking gel solution (4% of acrylamide/bisacrylamide, 0.1% APS, 0.01% TEMED, 0.1%SDS) and the combs to create wells for protein loading. Allow the gels to polymerize for 15–20 minutes.

3. Mix protein samples with sample buffer, boil them for 5 minutes at 95°C and load them into the gel wells. The gel runs in SDS running buffer. Proteins are stacked with a constant voltage of 100 V and then separated with a constant voltage of 150 V for approximately 1.5 hours. Molecular weight of the proteins is estimated using 5 μL of PageRuler™ prestained or other protein moleculer weight markers.

3.6.2. Western Blot Analysis

For subsequent immuno detection, separated proteins from a SDS-PAGE gel are transferred to and immobilized on a nitrocellulose membrane (wet transfer technique).

1. The membrane and filter papers are carefully soaked in ice-cold transfer buffer and "sandwich" assembled. Transfer the proteins at a constant current of 400 mA for 100 minutes on ice.

2. The transfer is verified by reversible staining with Ponceau S dye (1:10, Sigma) for 5 minutes at room temperature. De-staining is done by several washes with PBS.

3. All subsequent blocking, antibody incubation and washing steps are performed under gentle rocking. To visualize immuno-reactive protein bands, first block nonspecific antibody binding

sites by incubating the transfer membrane in blocking solution (BS) for 2 hours at room temperature.

4. The membrane is then incubated in BS containing the primary antibody overnight at 4°C (we use the recommended antibodies at the following dilutions for western blotting. Mouse anti-myc antibody and rabbit anti-myc antibody at 1:1000 dilution, mouse anti-flag antibody at 1:2000 dilution and rat anti-HA antibody at 1:3000 dilution).

5. After 5 washing steps (10 minutes each) with BS or PBS with 0.1% Tween-20 at room temperature, incubate the membrane in BS containing the secondary antibody for 2 hours at room temperature (we routinely use HRP conjugated goat-anti mouse, goat-anti rabbit, and goat-anti rat antibodies from Dianova at 1:10,000 dilution).

6. Wash the blot membrane again six times (10 minutes each) at room temperature with PBS-T (or TBS-T, and once with PBS [TBS]).

7. Incubate the membrane with Lumi-Light Western Blotting Substrate (Roche) for 1 minute at room temperature. Finally the membrane is fitted between two plastic foils and an X-ray film (Cronex, AGFA), and is exposed for 10 seconds to 30 minutes at room temperature inside an autoradiography cassette. The film is then automatically processed using an X-ray film-processing unit (**Fig.** 28.2).

4. Notes

1. Adjusting the pH is essential because the cysteine solution is very acidic and deadly for the fertilized eggs.

2. In case the unfertilized eggs come in contact with water, the jelly coat swells and fertilization becomes impossible.

3. Each RNA has to be titrated carefully, especially when ligand/receptor/inhibitor combinations are co expressed. One to 10 pg RNA/embryo should be injected for canonical Wnts, such as Wnt3a and Wnt8. Twenty-five to 50 pg/embryo for Wnt5a and wnt11. For Fz, 200–400 pg/embryo. Higher doses of Fz-8 (300–500 pg/embryo) mRNA injection can induce Xnr3 expression without coinjection of Wnts. RNA is injected close to the cleavage furrow otherwise the injected tissue will move away from the animal pole during epiboly.

4. When embryos are kept in 1xMBS during gastrulation they will form so called exogastrulae with severely impaired further development.

5. The embryo can easily get damaged when the vitelline membrane is removed. To keep the animal cap region unharmed, start removal of vitelline membrane from the vegetal pole.

6. Make sure that the explants never come in contact with the buffer–air interphase because the surface tension destroys them.

7. Be quick and do not make too many air bubbles/froth. Leave the lysate on ice for 5–10 minutes (or on head-over-head rotator) in cold room.

8. For example, if one of your proteins is Myc-tagged and the other one is Flag-tagged, then incubate one part with anti-Myc and the other with anti-Flag antibody. The third part should be incubated with Mouse IgG, when both Myc and Flag antibodies are mouse monoclonals. Standardize the amount of antibody needed to pull down the corresponding protein before the actual experiment. To start with, one can use 1 µg antibody to pull-down proteins from extract prepared from 15–20 embryos.

9. At this stage the protein+antibody+Protein G/A beads can also be left overnight at 4°C, but one should be sure that this does not result in protein degradation. Protein G/A beads should be washed twice with lysis buffer before use. Usually 10 µL protein G is sufficient to pull down 1 µg antibody. Refer to the product data sheet for instructions on affinity of beads.

10. Centrifugation involving beads should be done at low speed. Otherwise it will be difficult to resuspend the beads and wash them properly.

11. Resuspend carefully by tapping at the bottom of the tubes. Vortexing or pipetting up and down is not recommended because the beads stick to the tip.

12. While incubating the tubes at 95°C/boiling, tap the bottom of the tubes from time to time to resuspend the beads.

Acknowledgments

We thank Anne Schohl and François Fagotto for sharing their immunoprecipitation protocol and Ana Cristina Silva for providing the scheme of animal cap assay.

References

1. Clevers, H. (2006) Wnt/β-catenin signalling in development and disease. *Cell* 127, 469–480.

2. Widelitz, R. (2005) Wnt signalling through canonical and non-canonical pathways: recent progress. *Growth Factors* 23, 111–116.

3. Kikushi, A., Yamamoto, H., Kishida, S. (2007) Multiplicity of interactions of Wnt proteins and their receptors. *Cellular Signalling* 19, 659–671.

4. Jones, C., Chen, P. (2007) Planar cell polarity signalling in vertebrates. *Bioessays* 29, 120–132.

5. Djiane, A., Riou, J., Umbhauer, M., et al. (2000) Role of frizzled 7 in the regulation of convergent extension movements during gastrulation in *Xenopus laevis. Development* 127, 3091–3100.

6. Swain, R. K., Katoh, M., Medina, A., et al. (2005) Xenopus frizzled-4S, asplicing variant of Xfz4 is a context-dependent activator and inhibitor of Wnt/β-catenin signalling. *Cell Communication and Signaling* 3, 12.

7. Laemmli, U. K. (1970) Ceavage of structural proteins during the assembly of the head ofbacteriophage T4. *Nature* 227, 680–685.

8. Sambrook J et al. (2000) *Molecular Cloning: A Laboratory Manual* (third edition), CHSL.

Section B

Wnt Signaling Function in *Xenopus* Development

Chapter 29

Dorsal Axis Duplication as a Functional Readout for Wnt Activity

Michael Kühl and Petra Pandur

Abstract

The easy accessibility, distinctive features of early cleavage stage embryos and simple manipulation methods make *Xenopus* embryos an ideal model organism to study gene function and deciphering signaling pathways. For many years, investigators have analyzed putative dorsalizing factors by their ability to induce secondary dorsal structures when misexpressed in early *Xenopus* embryos. This assay, among others, has contributed substantially to our knowledge about Wnt signaling pathways and is still the assay of choice to quickly determine whether a factor acts positively or negatively in the Wnt signaling pathway. This chapter describes two experimental approaches to determine canonical Wnt signaling: induction of a secondary axis and analyses of target gene expression.

Key words: Xenopus, Wnts, canonical signaling, -catenin, dorsal axis.

1. Introduction

Historically, Wnt proteins were subdivided into two classes based on their different biological activity in different assays. One of these assays allows the distinction between Wnt ligands that can induce a secondary dorsal body axis when misexpressed on the ventral side of a *Xenopus* embryo from the ones that cannot. The latter comprises, for example Wnt-5a, Wnt-4, and Wnt-11, whereas Wnt-1, Wnt-3a, and Wnt-8 are very potent axis inducing molecules (reviewed in ref. *1*). A combination of biochemical and genetic experimental approaches in various model systems helped to identify the components of the Wnt signaling pathway that elicits the formation of a secondary axis in *Xenopus* embryos.

Elizabeth Vincan (ed.), *Wnt Signaling, Volume II: Pathway Models, vol. 469*
© 2008 Humana Press, a part of Springer Science + Business Media, New York, NY
Book doi: 10.1007/978-1-60327-469-2

One of these components is β-catenin, which becomes stabilized and therefore enriched on the dorsal side of the embryo during the first cell cycle (**Fig.29.1A**; refs. *2* and *3*). Depletion of maternally provided β-catenin resulted in ventralized embryos and demonstrated the requirement of β-catenin for the formation of the endogenous dorsal axis *(4)*. Two target genes have been identified so far, which are activated directly by Wnt/β-catenin signaling. These are the homeobox transcription factor *siamois* and *Xenopus nodal-related-3* (*Xnr-3*), a member of the transforming growth factor-β family *(5–7)*. These two molecules are probably the most suitable markers to analyze activation of a Wnt/β-catenin signaling pathway, preferably by RT-polymerase chain reaction (PCR) as described later. *Xenopus* embryos are ideal to quickly test a putative dorsalizing activity of Wnt ligands

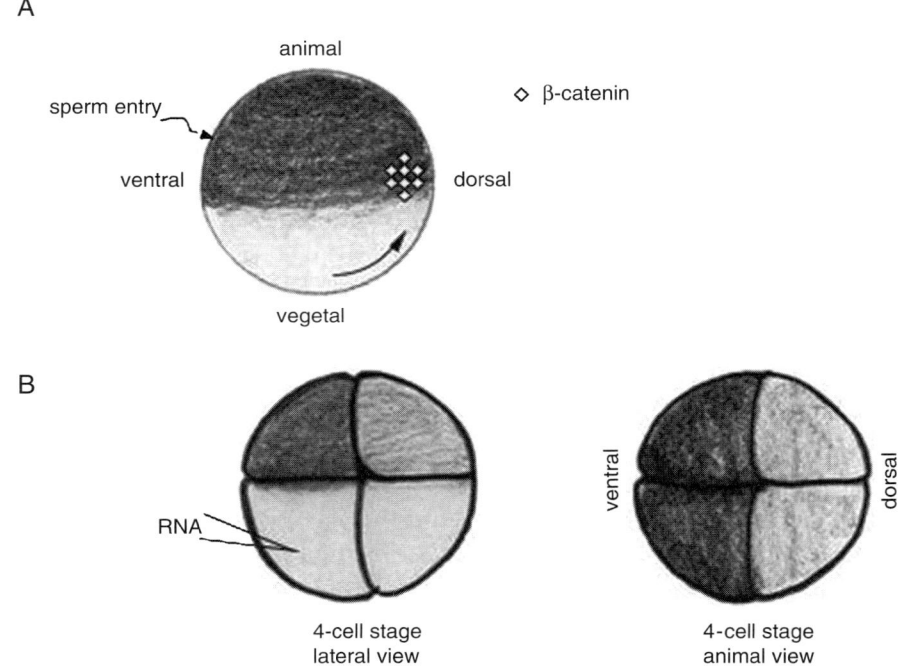

Fig. 29.1.   Establishment of the dorsal axis in *Xenopus* and features of early cleavage stage embryos. (**A**) The initially radial symmetric *Xenopus* egg possesses an animal-vegetal axis with the animal pole being the heavily pigmented upper part of the egg. The radial symmetry is abolished at fertilization when sperm entry determines the dorsal–ventral axis. Sperm entry causes a rearrangement of the cortical cytoplasm (cortical rotation), which results in the translocation of so-called dorsal determinants from the vegetal pole to the future dorsal side of the embryo. The translocation of these factors results in the stabilization of β-catenin on the dorsal side of the embryo. β-catenin is a component of the canonical Wnt signaling pathway and absolutely required for the formation of the endogenous dorsal axis. (**B**) A four-cell stage embryo is shown from a lateral and an animal view. The ventral and dorsal sides can be easily distinguished at this stage by the differences in pigmentation. The animal blastomeres of the ventral side are usually more pigmented than the animal blastomeres of the dorsal side. To test the dorsal axis inducing activity of a particular factor, its RNA is injected into the ventral vegetal region of the embryo.

or any component of the Wnt signaling pathway. It only requires the preparation and injection of RNA of the factor of interest and, due to the fast development of *Xenopus* embryos, a result can be obtained within 2 days. In principle, there are two major ways to evaluate Wnt/β-catenin signaling: expression of components on the ventral side of the embryo and analysis for secondary axis induction and expression of components in the animal cap and analysis for marker gene induction by RT-PCR. The protocols for both methods are described in this chapter.

2. Materials

2.1. In Vitro Transcription for Capped RNA Synthesis

1. The template DNA from which the RNA will be transcribed is a linearized plasmid carrying the cDNA of the molecule whose axis inducing property will be analyzed. There are several vectors available that are suitable for overexpression studies. The most common are pCS2+ and pSP64T. If the cDNA has not been cloned into an expression vector it is advisable to do so. RNAs synthesized from a cDNA that is in, for example, the pBSK vector will not be as efficient.

2. Message Machine Kit from Ambion (containing either SP6, T3 or T7 polymerase) to synthesize capped RNAs for injection.

2.2. Obtaining Xenopus Embryos and Microinjection

1. Human chorionic gonadotropin (HCG; Sigma; *see* **Note** 1) to induce egg laying.

2. 1X MBSH (modified Barth's saline): 10 mM HEPES, 88 mM NaCl, 1 mM KCl, 0.33 mM $Ca(NO_3)_2$, 0.41 mM $CaCl_2$, 0.82 mM $MgSO_4$, 2.4 mM $NaHCO_3$. Rather than mixing the salts before adding them to H_2O, dissolve the salts sequentially in H_2O. It is very important to add the $NaHCO_3$ at the end, when all other salts are dissolved, to avoid precipitation of $CaCO_3$ (*see* **Note** 2). Adjust the pH to 7.4 using 10 M NaOH.
   Prepare 0.1X MBS to culture the embryos.

3. 2% (w/v) L-Cysteine hydrochloride monohydrate (Fluka) solution, pH 8.0-8.2 to remove the jelly coat from the embryos. The jelly coat needs to be removed to make the embryos susceptible for manipulation.

4. For microinjection, the embryos are placed in a 3% Ficoll/1X MBSH solution (Ficoll PM400 from Amersham Biosciences). This will collapse the space between the embryo and the vitelline membrane.

5. Embryos are fixed in MEMFA. MEM solution (5X): 0.1 M MOPS, pH 7.4, 2 mM EGTA, 1 mM MgSO$_4$. Autoclave the solution and store it at 4°C. Of note, the color of the MEM solution will turn yellow after autoclaving. Add formaldehyde freshly to a final concentration of 4% (v/v; *see* **Note** 3).

2.3. Animal Cap Assay for Determination of Canonical Wnt Targets by RT-PCR

1. For dissecting and culturing animal caps: fine forceps and Petri dishes that have been prepared by coating with a layer of 1% (w/v) agarose in 1X MBSH. This prevents the tissue from sticking to the plastic. Use Petri dishes with a diameter of 60 mm for dissecting and smaller Petri dishes for culturing the caps. Use glass Pasteur pipettes to transfer the dissected caps from one dish to another.

2. RNA extraction: there are many RNA isolation kits available which are good for extracting total RNA from small tissue samples. We use the Purescript RNA purification system Cell & Tissue Kit from Gentra Systems (cat. no. R-5500A).

3. Glycogen (Invitrogen). Adding glycogen to the precipitation mixture will yield more RNA.

4. To eliminate contamination of the RNA sample with genomic DNA, the extracted RNA samples have to be treated with DNAse. We use DNAseI (RNAse-free) from Roche.

5. For first strand cDNA synthesis we use all reagents from Invitrogen: random primers, 100 mM desoxynucleotide (dNTP) set (PCR grade), which is diluted 1:10 prior to use, and the SuperScript II Reverse Transcriptase.

6. The PCR is set up using the MasterAmp Taq PCR core kit from Epicentre Biotechnologies.

7. Two direct targets of canonical Wnt signaling are the homeobox transcription factor *siamois (sia)* and *Xenopus Nodal-related-3 (XNr-3)*, a member of the transforming growth factor β family. The product for *histone 4 (H4)* can be used as a loading control. The primer sequences are:

   sia forward: 5′ CTC CAG CCA CCA GTA CCA GAT C 3′
   sia reverse: 5′ GGG GAG AGT GGA AAG TGG TTG 3′
   Xnr-3 forward: 5′ CGA GTG CAA GAA GGT GGA CA 3′
   Xnr-3 reverse: 5′ ATC TTC ATG GGG ACA CAG GA 3′
   H4 forward: 5′ CGG GAT AAC ATT CAG GGT ATC ACT 3′
   H4 reverse: 5′ ATC CAT GGC GGT AAC TGT CTT CCT 3′

8. TBE stock solution (10X) for gel electrophoresis: 0.89 M Tris, 0.89 M boric acid, 0.02 M Na$_4$EDTA. 1X TBE is used for preparing the agarose gels and as running buffer.

9. Loading buffer for the samples: 5 mM Tris-HCl, pH 7.4, 25 mM EDTA, 25% (v/v) glycerol and a few granules of bromophenol blue.

10. Equipment for injecting: we use the PV820 Pneumatic Pico Pump microinjector from World Precision Instruments, a micromanipulator that holds the glass capillary needles with which the embryos are injected can be purchased from, for example, Narishige or Brinkman, a needle puller is available from Sutter Instruments or Narishige (PN-30), and borosilicate glass capillaries for injection (O.D.: 1.0 mm, I.D.: 0.5 mm, 10 cm length) can be purchased from, for example, Fisher or World Precision Instruments.

3. Methods

To investigate whether a component of the Wnt signaling pathway (or any other factor) has dorsal axis inducing activity, the RNA has to be ectopically expressed. Therefore the RNA is injected into the ventral vegetal region of the embryo (**Fig. 29.1B**) opposite of the side that gives rise to the endogenous dorsal axis. The four-cell stage is usually the earliest stage when the ventral and the dorsal side can be unambiguously distinguished by the different intensities of the pigment. The ventral side is characterized by a darker pigmentation of the animal blastomeres compared to the blastomeres on the dorsal side. For an axis duplication assay, four- to eight-cell stage embryos are typically injected with the RNA of interest (**Fig. 29.2**). For the animal cap assay (**Fig. 29.3**), the RNA is injected bilaterally into the animal pole of two-cell stage embryos. The animal cap assay is a very common and valuable assay to investigate the

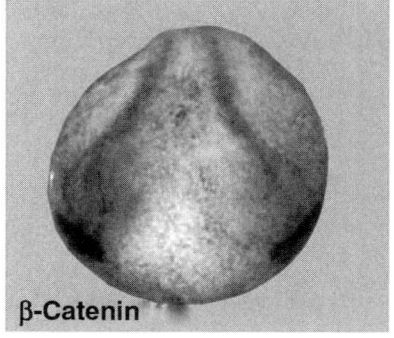

Fig. 29.2. Duplication of the dorsal axis after activation of the canonical Wnt signaling pathway. β-catenin is a potent inducer of a secondary body axis when misexpressed on the ventral vegetal side of an embryo. Here, a secondary axis complete with head structures (cement gland) is seen in a late neurula stage embryo (courtesy of Susanne Gessert).

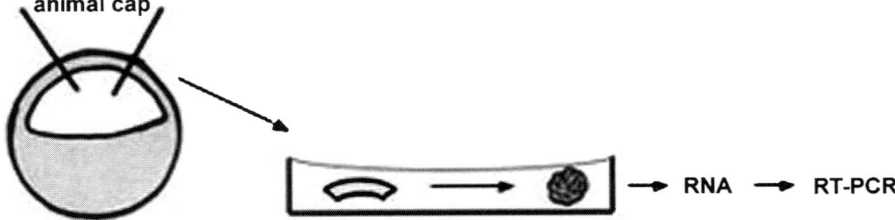

Fig. 29.3. Animal cap assay for detection of canonical Wnt targets by RT-PCR. The animal cap is dissected from embryos at the blastula stage (st. 8) and cultured in 1X MBSH in an agarose-coated Petri dish. When sibling embryos have reached the desired developmental stage, RNA is extracted from the caps and subjected to RT-PCR analyses.

biological activity of any factor. Cells of the animal cap have an ectodermal fate and untreated explanted caps develop into a poorly organized tissue called "atypical epidermis." However, these cells are sensitive to inducing signals and hence will express marker genes that are downstream of the factor whose RNA was injected into the animal pole.

3.1. In Vitro Transcription for Synthesis of Capped RNAs

Capped RNAs are synthesized using the mMessage Machine Kit from Ambion. Prepare the RNA synthesis reaction mixture according to the company's protocol. All reagents should be added to an RNAse-free 1.5 mL microcentrifuge tube at room temperature because Spermidine in the transcription buffer can lead to precipitation of the template DNA if the reaction is assembled on ice. The template DNA is a linearized and purified plasmid containing the cDNA of the molecule to be analyzed. For good laboratory practice when working with RNA (*see* **Note 4**).

3.2. Obtaining Xenopus Embryos and Microinjection

1. Two adult females are injected with 500-800 IU of HCG into the dorsal lymph sac the previous evening. Females should begin laying eggs 12 hours after the HCG injection. One adult male is sacrificed the next day to prepare its testes. The testes are macerated in 1X MBSH buffer and the suspension can be kept in a small Petri dish on ice for one day. For *in vitro* fertilization, dilute 100 µL testes in 900 µL of H_2O and distribute the suspension over the eggs, which were massaged from the female into a Petri dish. Fill the dish with 0.1X MBSH after 3–5 minutes. Leave the eggs undisturbed for approximately 60 minutes before dejellying. When kept at room temperature, the eggs start cleaving 90 minutes after fertilization. Cooling the eggs slows down their development, which is helpful when a lot of eggs need to be injected at a specific stage. However, do not keep the eggs below a temperature of 14°C.

2. To remove the protective jelly from the fertilized eggs, pour off the 0.1X MBSH and incubate the eggs in the 2% L-Cysteine

hydrochloride monohydrate solution. The incubation time should not exceed 4 minutes. This is sufficient time for the jelly coat to be dissolved and any further incubation will actually harm the embryos. To ensure that the jelly coat has been removed properly, swirl the eggs in a small beaker and tilt it so that the eggs accumulate on one side. If there is no space between the eggs, they have been dejellied properly. Wash the eggs at least 5 times with 0.1X MBSH to make sure that they are completely void of any residual L-Cysteine hydrochloride monohydrate solution.

3. Prepare microinjection pipettes by pulling sterile borosilicate glass in a micropipette puller to obtain appropriate needle tip size (~10 μm). Transfer the embryos to a dish filled with 3% Ficoll/1X MBSH and clay or a mesh to hold them in the right position. When injections are done at the four- to eight-cell stage, the injection volume can be up to 10 nL. The amount of RNA to be injected depends on the factor whose activity is being tested and on the quality of the synthesized RNA. For example, with β-catenin, a concentration of 20–50 pg is sufficient to induce a secondary body axis. The concentration for XWnt-8 RNA may vary between 50 and 150 pg. These amounts should get you started to determine the optimal concentration for your assay individually.

4. Leave the injected embryos in the 3% Ficoll/1X MBSH solution for another 1 or 2 hours. This helps the healing of the wound generated by the glass needle and it makes it less likely that the embryo develops blebs. However, the embryos must be transferred to 0.1X MBSH before gastrulation otherwise they will exogastrulate due to the high salt content of the 1X MBSH.

5. Culture the embryos in 0.1X MBSH until they reach the desired developmental stage (*see* **Note** 5), that is, when the expected phenotype becomes clearly visible. Induction of a secondary axis can be seen as early as neurula stages (**Fig. 29.2**). Remember to also culture some uninjected or H_2O-injected embryos as sibling controls. The latter is used in some laboratories as a control to show that the observed effect is not due to the injection process *per se*.

6. Once the embryos have reached the desired stage, they can be processed for further analyses. For whole mount *in situ* hybridization (*see* **Chapter 22** in **Volume 2**), the embryos are fixed for at least 2 hours at room temperature in MEMFA. After fixing, the embryos are dehydrated through a methanol series (25% MetOH, 50% MetOH, 75% MetOH) and stored in 100% MetOH at –20°C until use.

3.3. Animal Cap Assay for Determination of Canonical Wnt Targets by RT-PCR

1. Inject about 30 two-cell stage embryos to ensure obtaining animal cap tissue from 15 to 20 embryos. Some embryos may die and some may become destroyed during the procedure. After RNA injection, the embryos are cultured until the blastula stage when the animal caps are dissected. The exact stage when the caps are dissected may vary from early stage 8 to stage 9 in different laboratories. Care has to be taken when transferring the dissected caps to the smaller Petri dish as to not include an air bubble at the tip of the Pasteur pipette. This air bubble will propel the isolated tissue to the solution's surface where it will disintegrate immediately due to the surface tension.

2. The dissected animal cap tissue will heal and thereby round up to form a little ball. The explanted tissue is then cultured in 1X MBSH until the desired developmental stage, here early gastrulation stages (stages 10 to 11). This is especially important to detect *siamois*. Therefore make sure to keep some embryos from the same batch as stage controls.

3. For RNA extraction, transfer the 15 to 20 animal caps with as little liquid as possible into a 1.5 mL microfuge tube containing the lysis buffer and extract total RNA according to the manufacturer's protocol. Of note, when using the Purescript RNA isolation kit, add an additional centrifugation step after precipitating the DNA/protein components of the extract since there is quite some cell debris present. After precipitating the RNA, resuspend the pellet in only a small amount of H_2O (for example 6–10 µL) to keep it as concentrated as possible.

4. If it is inconvenient to continue directly with the cDNA synthesis the isolated RNA can be stored at −20°C. To check if the RNA isolation was successful run 1 µL of RNA combined with 1 µL of loading buffer briefly on a 1.2% agarose gel (100 V for 5–10 minutes).

5. Perform a DNase digest for 15–20 minutes at 37°C to eliminate any contamination of your RNA sample with genomic DNA.

6. Prepare the cDNA synthesis reaction according to the manufacturer's protocol.

7. Prepare the PCR reaction mixture according to the manufacturer's protocol. Here is an example for a standard PCR program that can be used to detect amplified transcripts of sia and XNr-3. Denature the DNA at 94°C for 4 minutes and perform 30 cycles of PCR amplification as follows: denature at 94°C for 30 seconds, anneal at 54°C for 1 minute, and extend at 72°C for 1 minute. Incubate for

an additional 7 minutes at 72°C and maintain the reaction at 4°C. The protocol may need to be optimized with respect to cycle number to ensure linearity of amplification. To this end, the amplification of H4 is done with less than 30 cycles, for example, 24 cycles.

8. Check for amplified transcripts by mixing 3–5 μL of the PCR reaction with loading buffer and run it on a 1.2% agarose gel.

4. Notes

1. Although tested, it is noteworthy that HCG is derived from human and is classified as biohazard material. Therefore, care has to be taken when injecting the female frogs with this hormone as to not poke yourself with the injection needle.

2. The amounts of precipitated $CaCO_3$ are low; however, reduction of calcium ions in the solution affects the integrity of the tissue. The effect is seen strongest when culturing explanted tissues of *Xenopus* embryos. Of note, MBSH is just one of the described salt solutions to culture *Xenopus* embryos. Other laboratories may use different solutions, such as Normal Amphibian Medium (NAM) or Marc's Modified Ringers (MMR), which basically differ in the composition of the salts.

3. Formaldehyde is highly toxic to you and to the *Xenopus* embryos. Therefore, wear gloves and work in a chemical fume hood. Regarding the embryos, it is important not to transfer drops of formaldehyde solution to a dish with live embryos. Make sure to use different pipettes for transferring live embryos and for pipetting the MEMFA fixative.

4. When working with RNA wear gloves at all times because hands are a major source of contaminating RNAse and make sure pipette tips and microcentrifuge tubes are RNAse-free. Make sure to use reagents and H_2O that are RNAse-free.

5. Staging *Xenopus* embryos is according to Nieuwkoop and Faber *(8)*. If you do not have access to a printed copy of the original table you can look up the developmental stages at, for example, the *Xenopus* home page at www.xenbase.org and click on "Nieuwkoop and Faber stages" or you can go to www.bio.davidson.edu/people/balom/StagingTable/xenopushome.html. The latter also provides photographs of the various stages of development.

Acknowledgments

M.K. and P.P. are funded by the DFG.

References

1. Kühl, M. (2002) Non-canonical Wnt signaling in *Xenopus*: regulation of axis formation and gastrulation. *Semin Cell Dev Biol* 13, 243–249.

2. Larabell, C. A., Torres, M., Rowning, B. A., et al. (1997) Establishment of the dorsal-ventral axis in *Xenopus* embryos is presaged by early asymmetries in beta-catenin that are modulated by the Wnt signaling pathway. *J Cell Biol* 136, 1123–1136.

3. Schneider, S., Steinbeisser, H., Warga, R. M., et al. (1996) Beta-catenin translocation into nuclei demarcates the dorsalizing centers in frog and fish embryos. *Mech Dev* 57, 191–198.

4. Heasman, J., Crawford, A., Goldstone, K., et al. (1994) Overexpression of cadherins and underexpression of beta-catenin inhibit dorsal mesoderm induction in early *Xenopus* embryos. *Cell* 79, 791–803.

5. Lemaire, P., Garrett, N., Gurdon, J. B. (1995) Expression cloning of Siamois, a *Xenopus* homeobox gene expressed in dorsal-vegetal cells of blastulae and able to induce a complete secondary axis. *Cell* 81, 85–94.

6. Brannon, M., Gomperts, M., Sumoy, L., et al. (1997) A β-catenin/XTcf-3 complex binds to the *siamois* promoter to regulate dorsal axis specification in *Xenopus*. *Genes Dev* 18, 2359–2370.

7. McKendry, R., Hsu, S. C., Harland, R. M., et al. (1997) LEF-1/TCF proteins mediate wnt-inducible transcription from the *Xenopus* nodal-related 3 promoter. *Dev Biol* 192, 420–431.

8. Nieuwkoop, P. D., Faber, J. (1967) Normaltabelle von *Xenopus laevis*. Elsevier North-Holland Biomedical Press, Amsterdam.

Chapter 30

Regulation of Convergent Extension by Non-canonical Wnt Signaling in the *Xenopus* Embryo

Lars F. Petersen, Hiromasa Ninomiya, and Rudolf Winklbauer

Abstract

Non-canonical Wnt signaling is an important regulator of gastrulation in *Xenopus*. In particular, it has been implicated in the control of convergent extension movements. Convergent extension in the gastrula occurs primarily in the dorsal tissue of the marginal zone, and explants of this tissue will continue to undergo these movements in isolation. This observation has led to an assay to examine convergent extension movements that is unique to the *Xenopus* system, and is described herein.

Key words: *Xenopus*, convergent extension, non-canonical Wnt, planar cell polarity, dorsal marginal zone, explant.

1. Introduction

The process of gastrulation is accompanied by significant changes in cell adhesion, cell morphology, and cell movements. In the frog, *Xenopus*, region specific movements govern gastrulation: epiboly of the animal cap, active migration of the head mesoderm, vegetal rotation of endoderm, and convergent extension of the marginal zone *(1)*. During convergent extension of the dorsal posterior mesoderm (i.e., the chordamesoderm and presomitic mesoderm) cells intercalate mediolaterally between each other. As a result of this process, the tissue narrows laterally and extends in the antero-posterior direction *(2–4)*.

Members of the Wnt family have been implicated in the regulation of morphogenesis during gastrulation *(5)*. The signaling

Elizabeth Vincan (ed.), *Wnt Signaling, Volume II: Pathway Models, vol. 469*
© 2008 Humana Press, a part of Springer Science + Business Media, New York, NY
Book doi: 10.1007/978-1-60327-469-2

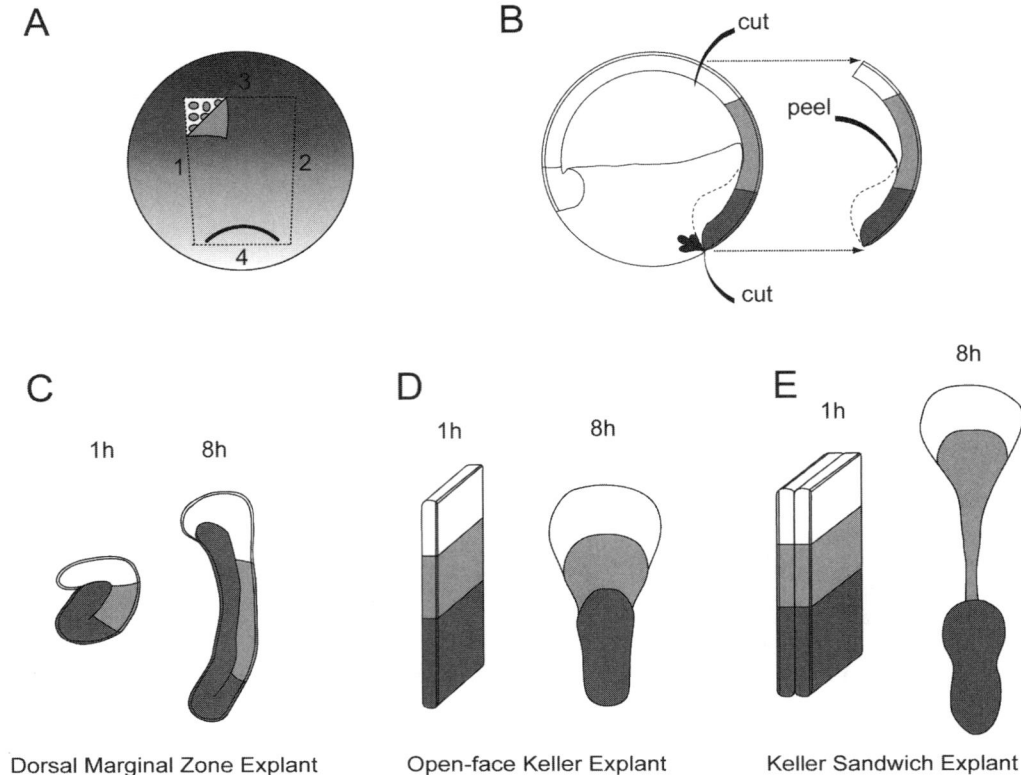

Dorsal Marginal Zone Explant   Open-face Keller Explant   Keller Sandwich Explant

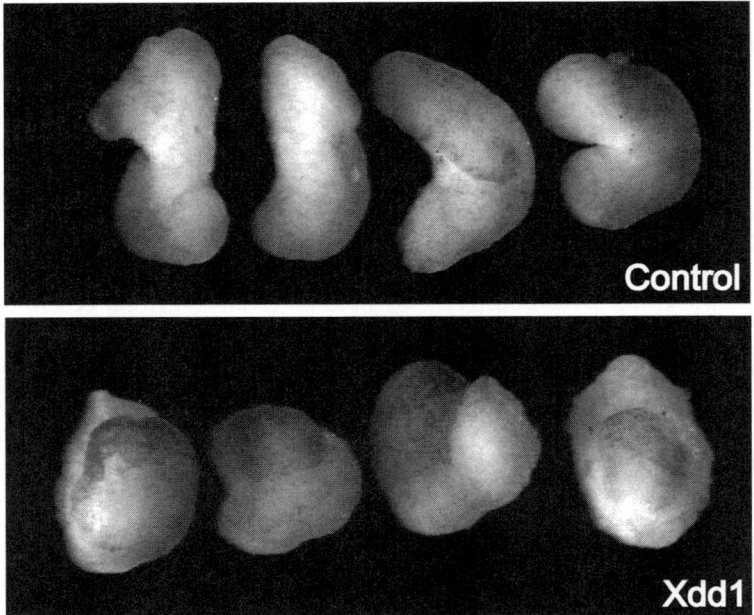

pathways through which these Wnts act are being uncovered, and it appears that the non-canonical Wg/Wnt pathway used for establishing epithelial planar cell polarity (PCP) in *Drosophila* also plays an important role in regulating convergent extension movements in *Xenopus*. This pathway uses canonical Wnt signaling components such as Frizzled and Dishevelled, but deviates downstream and uses different transducers, including Prickle, Strabismus, and JNK *(6–8)*. If these or other PCP transducers are disrupted, convergent extension movements are inhibited *(9)*. Furthermore, specific domains within Dishevelled can be mutated or deleted to inhibit convergent extension (**Fig**. 30.1F), without affecting canonical Wnt signaling *(10–13)*.

Convergent extension in the gastrula occurs primarily in the dorsal marginal zone *(2, 14)*, and if dissected from the embryo, explants of this region are capable of continuing convergent extension in isolation *(15)*. This type of explant is commonly referred to as a Keller explant and is often the first assay used to determine the effect of signaling components or other gene products on convergent extension movements. In addition, explants prepared from embryos injected with fluorescent markers can be used to observe individual or group cell movements and interactions by live imaging *(10–12)*, although this will not be described in this chapter.

Fig. 30.1. Keller explant. (**A**) Preparation of a dorsal marginal zone explant.for culture in isolation or for use in sandwich or open-face Keller explant.assays. Initial cuts 1 to 3 are made so that the "flap" of tissue can be peeled back and the final cut 4 can be made. (**B**) Regions of the early gastrula embryo used to make explants. Stage 10+ embryo in midsagittal cross section, dorsal to the right, animal pole up. The shading represents predicted tissue types based on the *Xenopus* fate map (see refs. *16* and *17*): white = ectoderm that will give rise to epidermis; light grey = prospective neural ectoderm; dark grey = dorsal mesoderm (future notochord and somites). Explants are made by excising a rectangle of tissue spanning from the animal pole to the blastopore (shown by eyebrow knife cuts) and about 60–90° wide around the equator. The head mesoderm, denoted by the dashed line, is removed from the explants (shown by the eyebrow knife "peeling"). (**C**) Approximate representation of a dorsal marginal zone explant.in longitudinal section before and after culturing in isolation. Convergent extension (narrowing and elongating) occurs in the mesoderm and neural ectoderm. (**D**) Representation of an open-face Keller explant. Explants are kept flat and convergent extension only occurs in the dorsal mesoderm. (**E**) Sandwich explant. are made by putting two rectangular explant. together with their inner surfaces contacting each other. These undergo convergent extension in both the posterior neural ectoderm and the mesoderm (modified from ref. *18*). (**F**) Expression of a Dishevelled dominant negative construct (Xdd1) (construct detailed in ref. *19*) inhibits convergent extension in dorsal marginal zone explant. *Xenopus* embryos were injected with 500 pg of control β-galactosidase mRNA or Xdd1 mRNA in the two dorsal blastomeres at the four-cell stage. Dorsal marginal zone explants were excised at stage 10+ and cultured until stage 16.

2. Materials

2.1. Solutions

1. 10x Modified Barth's Saline solution (MBS): 880 mM NaCl, 100 mM Hepes, 10 mM KCl, 25 mM $NaHCO_3$, 10 mM $MgSO_4$, 3.5 mM $Ca(NO_3)_2$, 5 mM $CaCl_2$, 0.1 g/L streptomycin, and 0.1 g/L penicillin in 1 L of distilled water. Adjust pH to 7.4 with NaOH.

2. Sater's modified blastocoel buffer: 50 mM NaCl, 36 mM sodium gluconate, 5 mM Na_2CO_3, 4.5 mM KCl, 1 mM $CaCl_2$, 1 mM $MgSO_4$. Adjust pH to 8.1 with Hepes (~6 mM) and add 50 µg/mL gentamycin sulfate. If desired, 1 mg/mL BSA can be added. Filter sterilize and store in 50 mL aliquots at −20°C.

2.2. Microsurgical Manipulations

1. Fine surgical forceps (eg. Dumont #5 or finer; Fine Science Tools Inc., www.finescience.com).

2. Eyelash cutting knife: mount eyelash in the tip of a fine-gauge needle or fine-pulled glass pipette and fix in place with a small drop of nail polish. Mount the needle on a small syringe.

3. Hair loop: mount a long, fine hair in a fine gauge needle or pulled glass pipette as above, so that a small loop of hair protrudes from the tip.

4. Dissection dish: coat either a 60 × 15 mm or a 95 × 15 mm Petri dish with 1% (w/v) agarose in distilled water.

5. Glass or plastic transfer pipettes with a tip opening diameter of ~2 mm (smaller tips are easier for transferring explants, but are not large enough for whole embryos).

2.3. Open-faced and Sandwich Explants

1. Glass bridges: small glass cover slips or cover slip fragments (pre-treated with 1% (w/v) BSA for 30 minutes; 10 x 10 mm or so).

2. Vacuum grease: Dow Corning High Vacuum Grease, or something similar.

3. Petri dishes (pre-treated with 1% (w/v) BSA for 30 minutes), not coated with agarose, 35 × 15 mm or 60 × 15 mm.

3. Methods

Keller explants were originally used to observe the movements of convergent extension in culture. Left alone, the dorsal marginal zone explants curl up into a ball and the mesoderm involutes beneath the outer ectoderm. While these explants will still undergo elongation movements in later gastrula and neurula stages, the movements associated with early gastrulation cannot be directly observed. To counter this, Keller explants are kept flat by gently pressing them beneath a coverslip resting on vacuum grease (a glass bridge). The explants can be cultured as a single sheet (open-face explant) for the observation of inner mesodermal cell movements, or as two sheets sandwiched together with their inner surfaces apposed (Keller sandwich) to maximize convergent extension movements and elongation.

3.1. Dorsal Marginal Zone Explants

1. Dejelly embryos and remove the vitelline membrane from early gastrula (stage 10-10+) embryos, as outlined in **Chapter 31** of this volume, in an agarose coated dish containing 1x MBS (*see* **Note** 1).

2. Place the devitellinated embryo with the vegetal pole facing up and with the dorsal side (blastopore formation begins here with the appearance of darkly pigmented bottle cells) toward the side in which the eyebrow knife is held.

3. Using a hair loop or forceps, hold the embryo in place, poke the tip of the eyebrow hair downward at one end of the bottle cells and cut outward (toward the animal pole; **Fig**. 30.1A, cut 1; *see* **Note** 2).

4. Make a similar cut at the opposite end of the line of bottle cells (**Fig**. 30.1A, cut 2) and turn the embryo over so that the animal pole is now facing up.

5. Make a horizontal cut at the animal end of the dissection to complete a rectangle with the previous two incisions (**Fig**. 30.1A, cut 3; *see* **Note** 3).

6. Holding the embryo with a hair loop, peel back the dorsal tissues to the dorsal blastopore lip. Use the edge of the eyebrow hair to peel back the involuted head mesoderm (*see* **Note** 4).

7. Turn the embryo over again so that the vegetal pole is up and make a final incision along or just below the line of bottle cells, all the way through to the blastocoel, creating the bottom edge of the rectangle (**Fig**. 30.1A, cut 4; *see* **Note** 5).

8. Remove any pieces of involuted head mesoderm that are left on the inner side of the outer mesoderm (**Fig**. 30.1B). Place the rectangular explant with its inner surface down and trim off any material that protrudes beyond the bottle cells. Then turn the explant over and use the eyebrow hair to gently pick off any loose head mesoderm cells. Trim the explant into a regular rectangle of the desired dimensions (*see* **Note** 6).

9. Transfer the explant into a new Petri dish with 1x MBS and culture at 15–22°C until the explant has clearly extended

(neurula stages). Left alone, the explant will curl up into a rounded ball and extend from the posterior mesoderm and neural ectoderm (**Fig**. 30.1C; *see* **Note** 7).

3.2. Open-face Keller Explants

1. Repeat **steps 1–8**.

2. Transfer the explant to a new dish of Sater's Modified Blastocoel Buffer and lay the tissue flat with the side of interest (usually the inner surface) facing up.

3. Put a small dab of vacuum grease at each end of a sterile rectangle of cover slip or glass bridge (about 5 × 10 mm; *see* **Note** 8).

4. Gently place the cover slip on top of the explant. Exert slight pressure to flatten the explant but do not damage the cells. Avoid contaminating the explant with vacuum grease.

5. Culture explant until the desired developmental stage is reached (**Fig**. 30.1D; *see* **Note** 7).

3.3. Sandwich Keller Explants

1. Dissect two rectangular explants of the same size, as described previously, and immediately press their inner sides together. Make sure that the bottle cells are aligned (*see* **Note** 9).

2. Trim the edges of the sandwich so that the explants are exactly the same size and set aside in a dissecting dish.

3. Transfer the sandwiches to a clean dish and gently press each one under a cover slip, as described above. Be sure not to crush the sandwich (*see* **Note** 10).

4. Culture explant until the desired developmental stage (**Fig**. 30.1E; *see* **Note** 7).

4. Notes

1. Removing the vitelline membrane may often damage the embryo, so to ensure that the region of interest is not damaged, remove the membrane from the ventral side of the embryo.

2. The eyelash tool will pierce the embryo easily and the cuts can be made simply by pressing the eyelash against the dish and flicking outward.

3. If the previous two cuts have reached the animal pole, insert the eyelash under the region where you wish to make the horizontal incision and push the eyelash against the hair loop to complete the cut.

4. When peeling back the explant flap, follow the inner surface with the hair loop or eyelash to peel away the head mesoderm. Continue doing this until you reach the blastopore lip.

5. Do not cut off the bottle cells.

6. If the head mesoderm is not removed, it will migrate inward in the explant, and it may be difficult to observe extension movements.

7. It is recommended to culture a few wild-type embryos with intact vitelline membranes at the same temperature conditions (in 0.1x MBS) to accurately determine the stage of the explants. If embryos are incubated for longer than 24 hours, it is recommended that you replace the medium daily.

8. The vacuum grease will hold the cover slip in place. Use a 5- or 10-cc syringe to dispense the grease in small dabs.

9. The explants will not adhere very well to each other if not pressed together within a few minutes of being exposed to the medium.

10. Ensure that the two halves of the sandwich are aligned correctly and are flat, or the layers will not elongate efficiently.

References

1. Keller, R., Davidson, L. A., Shook, D. R. (2003) How we are shaped: the biomechanics of gastrulation. *Differentiation* 71, 171–205.

2. Keller, R. E., Danilchik, M., Gimlich, R., et al. (1985) The function and mechanism of convergent extension during gastrulation of Xenopus laevis. *J Embryol Exp Morphol* 89 Suppl, 185–209.

3. Shih, J., Keller, R. (1992) Cell motility driving mediolateral intercalation in explants of Xenopus laevis. *Development* 116, 901–914.

4. Keller, R., Davidson, L., Edlund, A., et al. (2000) Mechanisms of convergence and extension by cell intercalation. *Philos Trans R Soc Lond B Biol Sci* 355, 897–922.

5. Tada, M., Concha, M. L., Heisenberg, C. P. (2002) Non-canonical Wnt signalling and regulation of gastrulation movements. *Semin Cell Dev Biol* 13, 251–260.

6. Shulman, J. M., Perrimon, N., Axelrod, J. D. (1998) Frizzled signaling and the developmental control of cell polarity. *Trends Genet* 14, 452–458.

7. Boutros, M., Mlodzik, M. (1999) Dishevelled: at the crossroads of divergent intracellular signaling pathways. *Mech Dev* 83, 27–37.

8. Adler, P. N., Lee, H. (2001) Frizzled signaling and cell-cell interactions in planar polarity. *Curr Opin Cell Biol* 13, 635–640.

9. Wallingford, J. B., Fraser, S. E., Harland, R. M. (2002) Convergent extension: the molecular control of polarized cell movement during embryonic development. *Dev Cell* 2, 695–706.

10. Axelrod, J. D., Miller, J. R., Shulman, J. M., et al. (1998) Differential recruitment of Dishevelled provides signaling specificity in the planar cell polarity and Wingless signaling pathways. *Genes Dev* 12, 2610–2622.

11. Boutros, M., Paricio, N., Strutt, D. I., et al. (1998) Dishevelled activates JNK and discriminates between JNK pathways in planar polarity and wingless signaling. *Cell* 94, 109–118.

12. Wallingford, J. B., Rowning, B. A., Vogeli, K. M., et al. (2000) Dishevelled controls cell polarity during Xenopus gastrulation. *Nature* 405, 81–85.

13. Tada, M., Smith, J. C. (2000) Xwnt11 is a target of Xenopus Brachyury: regulation

of gastrulation movements via Dishevelled, but not through the canonical Wnt pathway. *Development* 127, 2227–2238.

14. Keller, R., Tibbetts, P. (1989) Mediolateral cell intercalation in the dorsal, axial mesoderm of Xenopus laevis. *Dev Biol* 131, 539–549.

15. Keller, R., Danilchik, M. (1988) Regional expression, pattern and timing of convergence and extension during gastrulation of Xenopus laevis. *Development* 103, 193–209.

16. Keller, R. E. (1975) Vital dye mapping of the gastrula and neurula of Xenopus laevis. I. Prospective areas and morphogenetic

movements of the superficial layer. *Dev Biol* 42, 222–241.

17. Keller, R. E. (1976) Vital dye mapping of the gastrula and neurula of Xenopus laevis. II. Prospective areas and morphogenetic movements of the deep layer. *Dev Biol* 51, 118–137.

18. Doniach, T., Phillips, C. R., Gerhart, J. C. (1992) Planar induction of anteroposterior pattern in the developing central nervous system of Xenopus laevis. *Science* 257, 542–545.

19. Sokol, S. Y. (1996) Analysis of Dishevelled signalling pathways during Xenopus development. *Curr Biol* 6, 1456–1467.

Chapter 31

Frizzled-7-dependent Tissue Separation in the *Xenopus* Gastrula

Rudolf Winklbauer and Olivia Luu

Abstract

Formation of tissue boundaries can be studied in a simple, inexpensive system, the *Xenopus* gastrula. Here, the internalized mesoderm and endoderm are separated from the ectodermal blastocoel roof by Brachet's cleft. Non-canonical Wnt signaling mediated by the Wnt receptor, Xfz-7, is essential for this tissue separation event. The function of Wnt pathway components and other factors in tissue separation at Brachet's cleft can be tested in a blastocoel roof assay. Small pieces of mesoderm or endoderm are placed on large blastocoel roof explants, and it is observed whether these test explants remain on the surface of their in vivo substratum, or sink into it.

Key words: *Xenopus* gastrula, tissue separation, Brachet's cleft, Frizzled-7, non-canonical Wnt signaling.

1. Introduction

The mesoderm-ectoderm boundary in the gastrula of the frog, *Xenopus laevis*, can serve as a model for tissue separation processes *(1, 2)*. During gastrulation, the mesoderm moves to the interior of the embryo to migrate as a coherent cell mass across the ectodermal blastocoel roof (**Fig. 31.1**; ref. *3*). While migrating, mesoderm and ectoderm cells are in direct contact, and both cell types express the same cadherins *(4–7)*. To prevent invasion of the substratum layer by migrating cells, a tissue separation mechanism is activated at gastrulation *(2, 8)*. Expression of "separation behavior" by the mesoderm generates a meso-ectodermal interface, Brachet's cleft (**Fig. 31.1**). Non-canonical signaling

Elizabeth Vincan (ed.), *Wnt Signaling, Volume II: Pathway Models, vol. 469*
© 2008 Humana Press, a part of Springer Science + Business Media, New York, NY
Book doi: 10.1007/978-1-60327-469-2

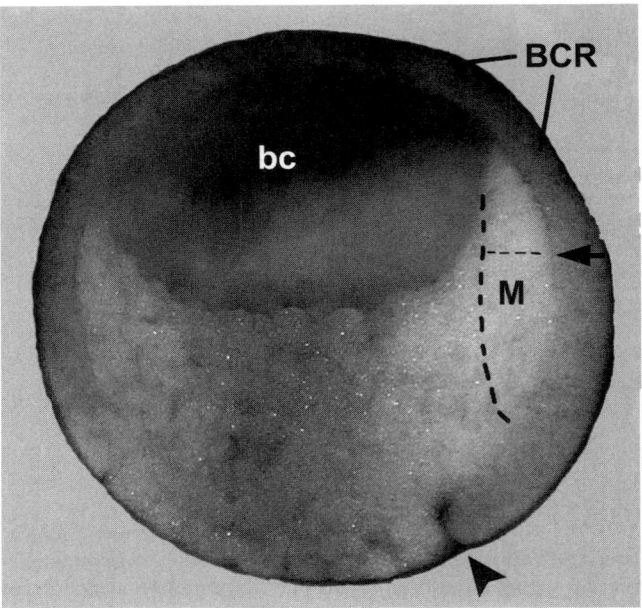

Fig. 31.1. Brachet's cleft. An early (stage 10.5) gastrula was fixed in 4% formaldehyde in MBS for 5 hours, and fractured mid-sagittally through the dorsal blastopore (arrowhead) with the aid of a razor blade. Brachet's cleft (arrow) separates the ectodermal blastocoel roof (BCR) from the mesendoderm (M). bc, blastocoel cavity. Vertical dashed line, mesendodermal region used for assay; dorsal anterior endoderm above, mesoderm below horizontal dashed line.

through the Wnt receptor *Xfz7* and through pertussis toxin-sensitive G protein and PKC downstream of it is essential for the establishment of separation behavior *(9)*. For this function of *Xfz7*, its interaction with the *Xlim*-regulated protocadherin PAPC is necessary *(10, 11)*. In anterior mesoderm, the transcription factors *Mix.1* and *gsc* are also involved in the control of separation behavior *(2)*.

2. Materials

2.1. In Vitro Fertilization

1. 10x Modified Barth's Solution (10x MBS): 25.7 g of NaCl, 11.9 g Hepes, 0.35 g KCl, 1.0 g NaHCO$_3$, 1.0 g MgSO$_4$•7H$_2$O, 0.4 g Ca(NO$_3$)$_2$•4H$_2$0, 0.3 g CaCl$_2$•2H$_2$0, 0.05 g streptomycin and 0.05 g penicillin in 0.5 L of distilled water. NaOH is added to adjust pH to 7.4. Before use, dilute 1:10 (1x MBS) or 1:100 (1/10x MBS).

2. For dejellying, 2% (w/v) cysteine is made by dissolving 1 g of cysteine in 50 mL of 1/10x MBS. pH is adjusted to 8.0 with NaOH.

2.2. Microinjections

1. Ficoll solution is made by dissolving ficoll (Sigma) at 4% (w/v) in 1x MBS.

2. Injection dish. Spread nontoxic modeling clay on the bottom of a 35 × 10 mm or 60 × 15 mm Petri dish and use a mold to make small depressions big enough for embryos to sit in.

3. Nanoject II (Drummond Scientific Company) injector

4. Plastic transfer pipettes (Fisher) for moving embryos between dishes.

2.3. Microsurgical Manipulations

1. 1x MBS, as previously.

2. Operation dish. Spread nontoxic modeling clay on the bottom of a 60 × 15 mm Petri dish.

3. Two fine forceps (Dumont grade 5).

4. Cutting tool. Mount eyelash in tip of injection needle, fix by adding a small drop of nail polish at base of eyelash where it inserts into needle. Mount needle on syringe.

5. Transfer pipettes as previously.

2.4. BCR Assay

1. 1x MBS, eyelash tool, as previously.

2. Tissue culture grade Greiner plastic Petri dishes (35 × 10 mm) or plastic Petri dishes with bottom coated with 1% (w/v) BSA for 30 minutes.

3. Cut ca. 3-mm wide strips from long cover slip with glass cutter. Coat each with BSA by pipetting a drop of 1% (w/v) BSA solution on top of it. After 30 minutes, remove drop, wash in distilled water, and use. Coated strips can also be dried and stored (take care to keep orientation, such that coated surface is known).

4. High vacuum silicone grease (Dow Corning).

---

3. Methods

A simple assay was developed to detect separation behavior in the mesoderm. When small explants of internalized mesoderm are placed on an explanted layer of blastocoel roof (BCR), i.e., on their normal substrate, they remain on its surface while similar explants taken from the BCR sink into the BCR layer (**Fig. 31.2**; ref. *1*).

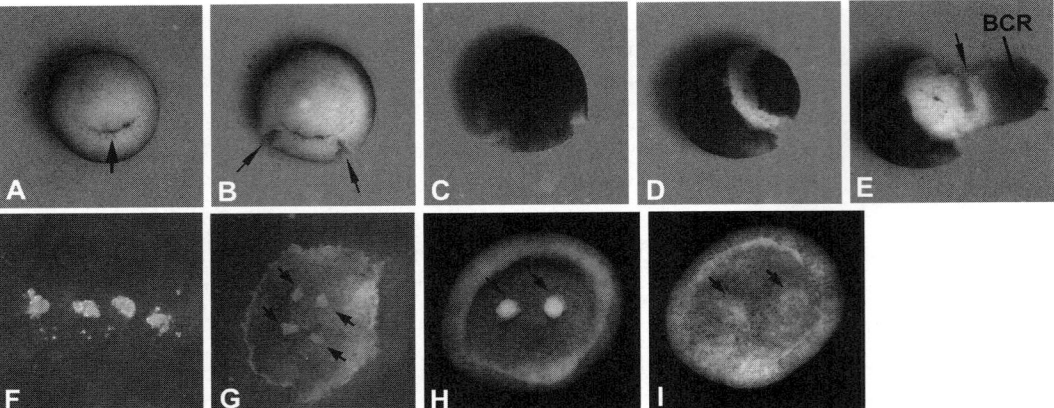

Fig. 31.2. BCR assay. (**A**) Early gastrula, turned upside-down. Dorsal blastopore visible (arrow). (**B**) Cuts lateral to dorsal midline (arrows). (**C**) Same embryo as in (**B**), turned around to be viewed from pigmented animal side. (**D**) Flap of dorsal BCR cut loose. (**E**) Flap of BCR peeled back (BCR), piece of dorsal mesendoderm cut off (arrow), resting on BCR. (**F**) Mesendoderm from (**E**) transferred to BSA-coated dish, cut into four smaller pieces. (**G**) Pieces of mesendoderm (arrows) placed on explanted BCR. (**H, I**) BCR assay after 45 minutes, higher-resolution view, fixed explants viewed under compound microscope under indirect illumination. (**H**) Two mesoderm explants remain on surface of BCR (arrows). (**I**) Two BCR explants have reintegrated into BCR (arrows).

This indicates that there is no physical barrier present that prevents invasion of the BCR cell layer by cells in contact with it, but that cells can nevertheless manage to remain on the surface of the BCR, by expressing what we operationally define as separation behavior *(2)*. By counting the percentage of small cell aggregates remaining on the BCR surface, this assay can be used for semi-quantitative purposes.

To study the effects of non-canonical Wnt signaling on separation behavior, we inhibited the pathway by injecting the blastomeres of four-cell embryos with morpholino antisense oligonucleotides targeting Xfz7, or of various mRNAs (e.g., pertussis toxin as an inhibitor of G protein signaling; dishevelled constructs) *(9)*. Mesoderm from injected embryos was then excised at the early gastrula stage, cut into small pieces, and tested on early gastrula stage BCR explants for separation behavior.

3.1. Injection of Embryos

1. After *in vitro* fertilization (at least 20 minutes later), embryos are dejellied by changing the buffer to 2% cysteine in 1/10x MBS, pH 8, and leaving it for 5 minutes. Embryos are then washed several times with 1/10x MBS and left in 1/10x MBS until they reach the four-cell stage.

2. Embryos are transferred to the injection dish filled with 4% Ficoll solution, placed in the depressions and oriented for injection under the dissecting microscope.

3. A Nanoject II (Drummond) injector mounted on a micro-manipulator is used for injection. It is set up according to manufacturer's instructions. Extend the plunger of the injector by pressing and holding the "EMPTY" button until it is extended 1 cm. Then fill a glass needle with pulled and ground tip with mineral oil (Fisher) and slide the needle through the plunger carefully until resistance is felt. Hold the needle in place while screwing the collet tight. Make sure no bubbles are present in the needle. Extend the plunger again by pressing and holding the "EMPTY" button until it extended for another 1 cm. Take a small piece of parafilm and add a drop of mRNA or morpholino of interest on it. Slowly insert the needle into the drop of sample and press and hold the "Fill" button to fill the needle. To inject, press the "INJECT" button.

4. Under the dissecting microscope, inject four-cell stage embryos into both dorsal blastomeres at the equatorial region, with 4.6 or 6.9 nL of solution each. When viewed from the animal pole, the two dorsal blastomeres appear smaller and less strongly pigmented. Remove injected embryos and put them in another dish of 4% ficoll. Leave them in a 15°C incubator for 2 to 3 hours and then change buffer to 1/10x MBS. Leave overnight in incubator at 15°C until embryos are at about stage 10.5 *(12)*.

3.2. BCR Assay

All operations are performed under a dissecting microscope in 1x MBS.

1. Transfer mesoderm donor embryos (controls or injected) with pipette into operation dish filled with 1x MBS. Remove vitelline membrane by grabbing it simultaneously with both forceps at one point near the animal pole (the center of the darkly pigmented region), and pulling apart. It does not matter if the embryo is damaged at this region in the process, since its BCR will not be used further.

2. Use mounted eyelash to expose anterior dorsal mesoderm. Use forceps to hold, orient, or move embryo. Turn embryo upside down, i.e., with pigmented blastocoel roof facing the bottom of the dish (**Fig. 31.2A**). Make two parallel cuts through the blastopore lip on each side of dorsal midline (the darkly pigmented part of the blastopore), starting at bottle cells of blastopore (**Fig. 31.2B**). Cut by using eyelash, like a knife cutting into butter. Turn embryo around, such that blastocoel roof faces upward (**Fig. 31.2C**). Insert tip of eyelash into one of the two wound edges generated before, and elongate the wound toward the animal pole by ripping the eyelash rapidly outward through the thin blastocoel roof tissue. Do the same for the other wound edge.

Connect the two wounds by cutting across (**Fig. 31.2D**). Peel the flap of the blastocoel roof back toward the vegetal side. This exposes first the blastocoel, then the leading edge of the dorsal mesendoderm. Continue peeling back until the blastocoel roof has been removed from most of the mesendoderm (**Fig. 31.2E**).

3. Excise dorsal mesoderm. Cut through the exposed surface of the mesendoderm (that had been in contact with the blastocoel roof) on each side of the dorsal midline, a few cell diameters deep. Then cut downward, i.e., from animal to vegetal, ca. three to five cell diameters behind the exposed surface (**Fig. 31.1**). At the level of the blastopore lip, cut outward. This isolates a piece of dorsal anterior mesoderm or mesendoderm (**Fig. 31.2E**; for mesoderm vs. mesendoderm regions, *see* **Fig. 31.1** and **Note 1**).

4. Prepare test explants. Transfer excised mesoderm to Greiner Tissue culture or other BSA-coated Petri dish filled with 1x MBS. Use eyelash to cut each of these mesoderm explants into small pieces (**Fig. 31.2F**). Size is not critical, but cutting each original explant into four to eight smaller pieces is convenient. Let small explants heal up for a few minutes. During that time, test explants using differently treated/ injected embryos can be prepared, and BCR explants used as substrate in the assay have to be made. Make sure that small explants are all separated from each other during this time, so that they do not fuse into larger aggregates.

5. BCR substrate explants. Transfer stage 10.5 embryos to MBS-filled operation dish, and remove vitelline membrane as above, except that the vitelline membrane should be grabbed at the non-pigmented vegetal side of the embryo so as to not damage the BCR that will be used in the assay.

6. Push tip of eyelash into the embryo in the lower, more vegetal part of the BCR (use closed forceps to hold embryo against eyelash tool). Make a short cut by ripping out the eyelash through the BCR while holding down the embryo with forceps, insert eyelash again at the end of the wound, and continue cutting around the perimeter of the embryo until the large, central part of the BCR is completely isolated. Trim margin of BCR explant to give it a nice regular shape.

7. Transfer BCR explant into dish where mesodermal test explants have been rounding up. Place BCR explant close to test explants and orient it with the eyelash tool, so that the inner blastocoelic surface faces upward.

8. With eyelash tool, hit test explants in a rapid stroke at its underside, such that it is lifted off the bottom of the dish and

falls onto the BCR explant. Repeat with one to four other test explants, arrange test explants on BCR such that they are nicely separated from each other (**Fig. 31.2G**).

9. Take BSA-coated cover slip strip with forceps, dip both ends in silicone grease, and place over test explant-carrying BCR explant. Press gently down at grease-supported ends with forceps, until it is applied to BCR explant. The explant should be held down at its contracting, elevated margins, while the center with the test explants is not or only lightly touched (this is actually less critical than one would think).

10. Note number, or even spatial pattern of test explants for each BCR explant, and time (for convenient numbers of explants, *see* **Note 2**). After 45 min, observe whether test explants have sunken into the BCR (**Fig. 31.2I**; *see* **Note 3**) or are on its surface (**Fig. 31.2H**; for documentation of results, *see* **Note 4**).

4. Notes

1. If isolated as described, the dorsal anterior mesoderm will most likely contain some endoderm at its leading edge. We found it to behave like the gsc-expressing mesoderm behind it, but this region can be discarded if pure mesoderm is needed (**Fig. 31.1**).

2. For one experimental treatment (one type of injection), we prepare mesoderm from three embryos, cut these into a total of ca. 15–30 test explants in a single dish. We randomly selected 10 good-sized explants from the mixture and placed them on two BCR explants in the same dish. Such an experiment is repeated at least three times, with three different batches of embryos. In our experience, this sufficiently evens out differences due to injection quality and embryo batches.

3. We noted that sinking of test explants into the BCR can occur in different ways. BCR test explants completely fuse with the BCR substratum, as expected, and cells interdigitate and mix. In other instances, test aggregates are sinking in, but remain distinct, and can even be lifted out again with the eyelash tool.

4. For higher-quality photographic pictures, we fix explants after the experiment in 4% formaldehyde in MBS while removing the strip of cover slip. After approximately 30

minutes of fixation, explants are photographed on a compound microscope under indirect illumination using a fiber optic cold light source.

References

1. Winklbauer, R., Keller, R. E. (1996). Fibronectin, mesoderm migration and gastrulation in Xenopus. *Dev Biol* 177, 413–426.

2. Wacker, S., Grimm, K., Joos, T., et al. (2000). Development and control of tissue separation at gastrulation in Xenopus. *Dev Biol* 224, 428–439.

3. Keller, R., Davidson, L.A., Shook, D.R. (2003). How we are shaped: the biomechanics of gastrulation. *Different* 71, 171–205.

4. Choi, Y.S., Seghal, R., McCrea, P., et al. (1990). A cadherin-like protein in eggs and cleaving embryos of Xenopus laevis is expressed in oocytes in response to progesterone. *J Cell Biol* 110, 1575–1582.

5. Angres, B., Müller, A. H. J., Kellermann, J., et al. (1991). Differential expression of two cadherins in Xenopus laevis. *Development* 111, 8229–8244.

6. Ginsberg, D., DeSimone, D., Geiger, B. (1991). Expression of a novel cadherin (EP-cadherin) in unfertilized eggs and early Xenopus embryos. *Development* 111, 315–325.

7. Herzberg, F., Wildermuth, V., Wedlich, D. (1991). Expression of XB-cad, a novel cadherin, during oogenesis and early development of Xenopus. *Mech Dev* 35, 33–42.

8. Ibrahim, H., Winklbauer, R. (2001). Mechanisms of mesendoderm internalization in the Xenopus gastrula: Lessons from the ventral side. *Dev Biol* 240, 108–122.

9. Winklbauer, R., Median, A., Swain, R. K., et al. (2001). Frizzled-7 signaling controls tissue separation during Xenopus gastrulation. *Nature* 413, 856–860.

10. Medina, A., Swain, R. K., Kuerner, K. -M., et al. (2004). Xenopus paraxial protocadherin has signaling functions and is involved in tissue separation. *EMBO J* 23, 3249–3258.

11. Hukriede, N. A., Tsang, T. E., Habas, R., et al. (2003). Conserved requirement of Lim1 function for cell movements during gastrulation. *Dev Cell* 4, 83–94.

12. Nieuwkoop, P. D., Faber, J. (1967). *Normal Table of Xenopus laevis (Daudin)*. North-Holland, Amsterdam.

INDEX

493

Printed in the United States of America